KV-380-529

Electrothermal Atomization for Analytical Atomic Spectrometry

Electrothermal Atomization for Analytical Atomic Spectrometry

Edited by

Kenneth W. Jackson

Wadsworth Center, New York State Department of Health and School of Public Health, State University of New York, USA

JOHN WILEY & SONS LTD

Chichester · New York · Weinheim · Brisbane · Toronto · Singapore

Copyright © 1999 John Wiley & Sons Ltd,
Baffins Lane, Chichester,
West Sussex PO19 1UD, England

National 01243 779777
International (+44) 1243 779777
e-mail (for orders and customer service enquiries):
cs-books@wiley.co.uk
Visit our Home Page on http://www.wiley.co.uk
or http://www.wiley.com

All Rights Reserved. No part of this publication may be reproduced, stored in a retrieval system, or transmitted, in any form or by any means, electronic, mechanical, photocopying, recording, scanning or otherwise, except under the terms of the Copyright Designs and Patents Act 1988 or under the terms of a licence issued by the Copyright Licensing Agency, 90 Tottenham Court Road, London W1P 9HE, UK, without the permission in writing of the Publisher

Other Wiley Editorial Offices

John Wiley & Sons, Inc., 605 Third Avenue,
New York, NY 10158-0012, USA

WILEY-VCH Verlag GmbH, Pappelallee 3,
D-69469 Weinheim, Germany

Jacaranda Wiley Ltd, 33 Park Road, Milton,
Queensland 4064, Australia

John Wiley & Sons (Asia) Pte Ltd, Clementi Loop #02-01,
Jin Xing Distripark, Singapore 129809

John Wiley & Sons (Canada) Ltd, 22 Worcester Road,
Rexdale, Ontario M9W 1L1, Canada

Library of Congress Cataloging-in-Publication Data

Electrothermal atomization for analytical atomic spectrometry / edited by Kenneth W. Jackson.
p. cm.
Includes bibliographical references and index.
ISBN 0-471-97425-0
1. Atomic absorption spectroscopy. 2. Electrothermal atomization.
I. Jackson, Kenneth W. (Kenneth William), 1947– .
QD96.A8E47 1999
543'.0858–dc21 98-53554
CIP

British Library Cataloguing in Publication Data

A catalogue record for this book is available from the British Library

ISBN 0 471 97425 0

Typeset in 10/12 Palatino by Techset Composition Ltd, Salisbury, England
Printed and bound in Great Britain by Biddles Ltd, Guildford
and King's Lynn
This book is printed on acid-free paper responsibly manufactured from sustainable forestry, in which at least two trees are planted for each one used for paper production

Coventry University

Contents

Contributors

Albert Kh. Gilmutdinov, Department of Physics, University of Kazan, 18 Lenin St., Kazan 420 008, Russia

James M. Harnly, U.S. Department of Agriculture, Beltsville Human Nutrition Center, Bldg. 161, BARC-East, Beltsville, MD 20705 USA

James A. Holcombe, Department of Chemistry and Biochemistry, University of Texas at Austin, Austin, TX 78712, USA

Kenneth W. Jackson, Wadsworth Center, New York State Department of Health, and School of Public Health, State University of New York, PO Box 509, Albany, NY 12201-0509, USA

Cornelius J. Rademeyer, Department of Chemistry, University of Pretoria, Pretoria, 0002, South Africa

Ralph E. Sturgeon, Institute for National Measurement Standards, National Research Council of Canada, Ottawa, Ontario, Canada K1A 0R9

David L. Styris, Pacific Northwest National Laboratories, Richland, WA, USA

Preface

This book is intended as a text for the educated student with no prior knowledge of electrothermal atomization, for the researcher who desires more in-depth understanding of some aspect of the subject, and as a reference book. It is the editor's belief that these rather ambitious goals of meeting everyone's needs have been achieved, but ultimately it is for the reader to judge. The aim, in preparing this book, was to present the most comprehensive and authoritative treatise on electrothermal atomic spectrometry since the classic book of Boris L'vov in 1970. As with any advanced level treatment that includes discussion of 'cutting-edge' research, the reader will find much that is undisputed fact, but also material that is hypothetical or occasionally speculative. An obvious pitfall of a single-author book at this advanced level is the perpetuation of that person's possibly biased opinions. Therefore, a team of internationally recognized experts on the subject was assembled, who were (and remain) friends, but readily admit to sometimes disagreeing on those topics that are hypothetical. Each chapter, during its preparation, was peer reviewed within the group of authors in order to present a balanced perspective with the benefit of the authors' collective expertise.

Careful planning and editing of the chapters were undertaken to avoid the fragmented product that may result from multiple authors. Again, the reader may judge if this goal was achieved, but the chapters are presented in a logical sequence and there is extensive cross-referencing between chapters. The student who is new to electrothermal atomization techniques is advised to study the chapters in the order in which they are presented. The seasoned researcher is more likely to selectively study specific sections of the book.

Chapter 1 introduces electrothermal atomization and electrothermal atomic absorption spectrometry (ETAAS). The development of ETAAS from the pioneering work of L'vov to the present day is described. The maturity of modern-day ETAAS, as a highly sensitive and relatively interference-free technique is demonstrated through reference to selected applications. Chapter 2 discusses the theory of electrothermal atomization, starting from a consideration of the chemistry and physics of carbon (the most commonly used atomizer material), and proceeding through discussions of the role of oxygen, the preatomization and atomization processes, spatial distribution of analyte within atomizers, the characteristics of the atomic absorption signal, and implications of atomizer design on its performance characteristics.

Chapter 3 is devoted to the very important subject of heating characteristics and temperature in graphite furnace electrothermal atomizers. Included are the significance of temperature, its spatial and temporal distribution, its effects on interferences and the absorbance signal, ways of delaying atomization until the furnace temperature is higher and more stable, models used to describe temperature and heating characteristics, and temperature control and measurement.

The instrumentation used in ETAAS is the subject of Chapter 4. The influence of instrumental parameters on the analytical signal is discussed, followed by detailed consideration of instrumental components (light sources, atomizers, spectrometers, detectors), the role of electronics and computers, and background correction. Finally, commercial and prototype instruments are described and compared.

An extremely important aspect of modern ETAAS is the use of chemical modifiers, which are discussed in Chapter 5. Described are their role and applicability in ETAAS, and the nature of surface and gas-phase reactions involving modifiers. Detailed mechanisms of the chemical and physical behavior of the more commonly used modifiers are discussed.

Chapter 6 is devoted to the direct analysis of solids and slurries by ETAAS. Differences in atomization mechanisms, compared with the more conventional analysis of solutions, are first discussed. Then practical requirements are considered, including control of absorbance signal characteristics, spectral interferences, dynamic range limitations imposed by the number and size of particles that are introduced to the atomizer, and effects of sample heterogeneity. Selected applications of the direct analysis of solids and slurries are finally presented.

Electrothermal atomization is used in techniques other than atomic absorption spectrometry, and these are the subject of Chapter 7. Discussed are the theory, characteristics, and applicability of thermally excited atomic emission spectrometry, furnace atomic nonthermal emission spectrometry (FANES), furnace atomization plasma emission spectrometry (FAPES), coherent forward scattering (CFS), laser excited atomic fluorescence spectrometry (LEAFS), and laser enhanced ionization (LEI).

Finally, Chapter 8 represents the collective efforts of the authors to look into the future. This was achieved through their answers to the following questions: (1) Can further fundamental research offer improved performance in terms of better sensitivity and reduced interferences? (2) What are the prospects for absolute analysis? (3) Are the current commercial atomizers (end-heated Massmann furnaces with platforms and side-heated tube furnaces with platforms) the best we can get? (4) What improvements are likely to be made in order to make ETAAS more competitive for ultratrace metal determinations? (5) What are the biggest remaining problems in applying ETAAS to the analysis of real samples? How will these problems be minimized or eliminated?

Completion of this book required more years than any of us had envisaged, and it is to the credit of all concerned that friendships survived through this ordeal. The editor wishes to express his personal thanks for the commitment, perseverance, and countless hours of effort provided by the co-authors of this book: Albert Gilmutdinov, Jim Harnly, Jim Holcombe, Cor Rademeyer, Ralph Sturgeon, and Dave Styris. The encouragement and constructive criticism provided by Walter Slavin is also gratefully acknowledged.

Kenneth W. Jackson
Albany, NY
USA

1

Electrothermal atomization and its application in analytical atomic absorption spectrometry

Kenneth W. Jackson

1.1 INTRODUCTION

Electrothermal atomization, particularly when used in analytical atomic absorption spectrometry (AAS), is one of the most widely applied techniques for trace and ultratrace metal determinations. This introductory chapter is designed to prepare the reader, who may not yet be familiar with the technique, for the more complex and complete treatment found in the subsequent chapters. The technique is first described in basic terms, and the way it has evolved into the sophisticated almost interference-free present-day technique is explained. The practising analyst may find the sections on modern applications particularly useful. They are designed to illustrate the present-day capabilities of electrothermal atomic absorption spectrometry (ETAAS), and to guide the reader to the relevant literature where specific methods may be found.

It is hoped that the experienced user and researcher in ETAAS will also find this chapter of interest, as it is the author's belief that a complete understanding of the capabilities of modern ETAAS can only be achieved through studying the developments in instrumentation and technique that have taken place over the years since the inception of analytical ETAAS in 1961.

Electrothermal Atomization for Analytical Atomic Spectrometry. Edited by K. W. Jackson
© 1999 John Wiley & Sons Ltd

1.2 ELECTROTHERMAL ATOMIZERS

These are devices that are heated electrically in order to dissociate metal salts or compounds to free gaseous atoms of the metal for spectrometric measurement. Many configurations of electrothermal atomizer (ETA) have been developed, including graphite rods, metal strips, metal coils, and graphite tubes. Some of these atomizers are described in this chapter. The most common ETA is a graphite tube, approximately 25 mm long and 5 mm internal diameter, heated resistively by mounting it between water-cooled stainless steel electrodes. By far the most widespread use of ETAs is in analytical AAS, and most of this book will be devoted to the technique of ETAAS. Except where otherwise indicated, any reference to ETAAS will imply the use of a graphite tube atomizer. Most commercially available instruments use this type of atomizer, wherein the technique is also commonly called graphite furnace atomic absorption spectrometry (GFAAS). Other analytical techniques that use ETAs are discussed in Chapter 7, and include atomic emission spectrometry (ETAES), furnace atomization nonthermal emission spectrometry (FANES), furnace atomization plasma emission spectrometry (FAPES), coherent forward scattering (ETA-CFS), laser excited atomic fluorescence spectrometry (ETA-LEAFS), and laser enhanced ionization (ETA-LEI). There is increasing use of ETAs as vaporization devices for introducing samples into inductively coupled plasmas and microwave plasmas. However, such uses are not within the scope of this book, as usually they do not involve the ETA as a primary atom reservoir.

1.3 ELECTROTHERMAL ATOMIC ABSORPTION SPECTROMETRY

A typical commercial graphite furnace atomizer is shown in Figure 1.1. The graphite tube is enclosed in a housing and protected from atmospheric oxidation by surrounding it with a protective gas (usually argon). Shown in the figure are two argon gas flows. An internal flow enters from the tube ends and exits through the sample introduction hole halfway along the tube, and an external flow surrounds the outside of the tube. The ends of the housing consist of two water-cooled stainless steel electrodes that are used to resistively heat the tube. The furnace is mounted in the light path of an AA spectrometer, and characteristic radiation from a light source (usually a hollow cathode lamp) passes coaxially through the graphite tube *via* quartz windows. This radiation is isolated by a monochromator, and is subsequently detected and recorded. A sample, typically 20–50 μl of a liquid, is introduced onto the inside wall of the cool tube or onto a graphite platform in the center of the tube through the sample introduction hole. The tube is then heated by means of a programmable power supply. A typical step-wise temperature program

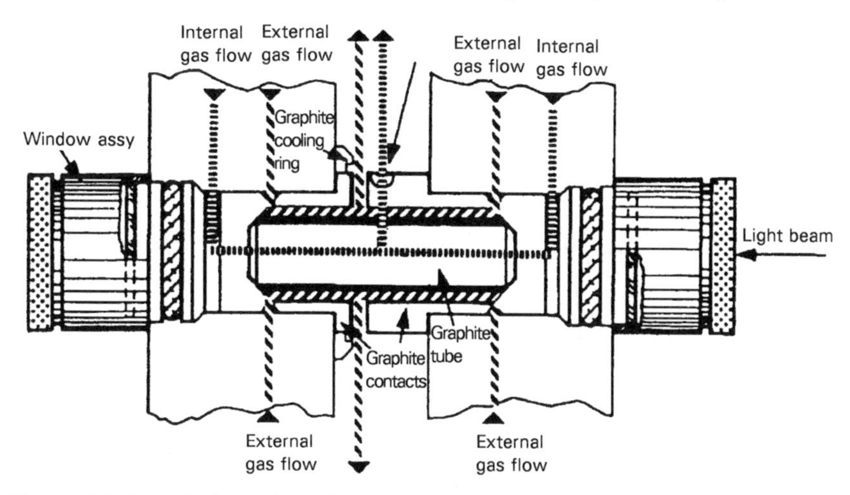

Figure 1.1 A typical graphite furnace atomizer. (Reprinted from R.D. Beaty and J.D. Kerber, *Concepts, Instrumentation and Techniques in Atomic Absorption Spectrometry* , 1993, p. 5–3, with kind permission from The Perkin-Elmer Corporation, Norwalk, CT, USA.)

involves: first a drying stage, in which the sample is heated gently to evaporate the solvent (usually water); second, a pyrolysis stage to remove volatile matrix components without volatilizing the analyte; third, an atomization stage to produce a cloud of free atoms in the tube; and finally a clean-out stage in which the tube is heated to a higher temperature to remove any residual less-volatile matrix components. Frequently, a cool-down step is included immediately prior to the atomization step. The preatomization heating steps are often referred to as thermal pretreatment steps. The argon protective gas flows continuously throughout the heating cycle, except it is usually stopped during the atomization stage, so that gaseous diffusion is the only analyte loss process and optimal sensitivity is realized. More details of temperature programs and the function of each stage, are presented in Chapter 3. Pulse heating of the graphite tube results in a transient cloud of atomic vapor, producing an absorbance signal resembling that shown in Figure 1.2. The integrated absorbance is proportional to the number of atoms, and hence the concentration of the analyte metal in the sample. Standard solutions of the analyte are used to construct a linear calibration graph of integrated absorbance versus concentration. Throughout this book, reference to sensitivity and detection limit will be made in absolute terms, i.e. the mass of analyte rather than its concentration. This convention is followed in ETAAS, because the sensitivity or detection limit expressed as concentration varies according to the volume of sample introduced to the ETA. Sensitivity is usually expressed as characteristic mass, m_0, which is the mass of analyte that gives an

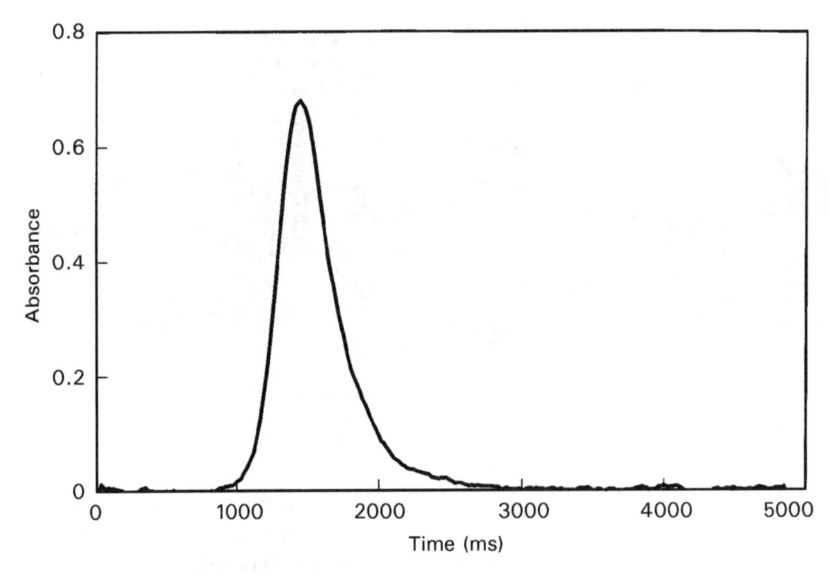

Figure 1.2 Absorbance signal for 1 ng of lead.

integrated absorbance of 0.0044 s [1]. Typical published m_0 values range from 0.1 pg for zinc to 50 pg for titanium.

1.4 EVOLUTION OF ETAAS

A chronological description of the development of modern ETAAS has a place in any book on the subject, but it is of far more importance than just a history lesson. It is a fascinating story that shows a technique with an initial period of development, followed by a stagnation period, and then a tremendous surge of renewed interest and advancement in the design of ETAs and in understanding the underlying chemical and physical processes that lead to the production of free atoms. At the time of writing this book, this rate of progress shows no signs of abating. It is necessary to study these advances in atomizer design in order to understand adequately the advantages and limitations of modern ETAs and their application in chemical analysis.

The ETA as an atomization device for analytical AAS was proposed and developed by L'vov, the first paper in the English language appearing in 1961 [2]. He was inspired by the pioneering work of Walsh in flame AAS, but believed that a 'graphite crucible' (cuvette) would provide better sensitivity and be less subject to matrix effects compared with a flame as an atom cell. L'vov's first graphite cuvette, shown in Figure 1.3, consisted of a graphite tube, 100 mm long, and lined with tantalum foil to eliminate

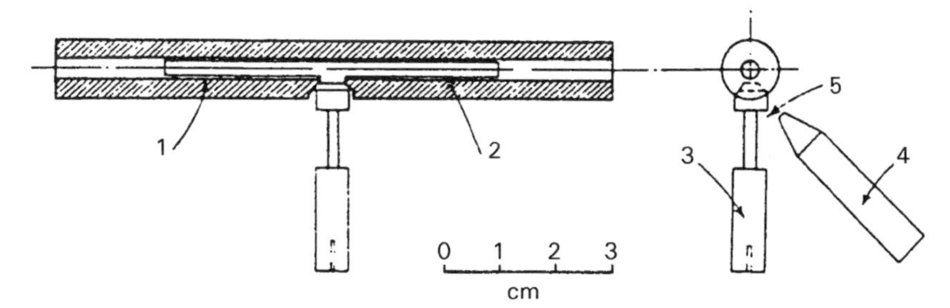

Figure 1.3 The original L'vov furnace: (1) graphite tube; (2) tantalum foil lining; (3) electrode holding the sample; (4) supplementary electrode; (5) arc gap. (Reprinted from *Spectrochim. Acta*, Vol. 17, B.V. L'vov, The Analytical Use of Atomic Absorption Spectra, pp. 761–770, 1961, with kind permission of Elsevier Science–NL, Sara Burgerhartstraat 25, 1055 KV Amsterdam, The Netherlands.)

diffusional loss of atomic vapor through the graphite walls of the tube. It was mounted in a sealed chamber, which was purged with argon at 1 atm pressure. The furnace tube was heated electrically until it had reached the required atomization temperature (up to 2500 K), and then about 100 μg of a liquid sample was introduced on a graphite electrode inserted through the side of the furnace. A supplementary graphite electrode was used to rapidly vaporize and atomize the sample with a d.c. arc. A modified version of L'vov's cuvette [3] dispensed with the supplementary electrode. Instead, the electrode containing the sample was rapidly heated by passing an a.c. current through the electrode and the graphite tube. The graphite tube was also shortened to 30–50 mm, and a pyrolytic graphite coating of the inner surface of the tube replaced the tantalum lining. Solid samples were also analyzed directly in this device. Compared with a flame, in which the sample is continuously nebulized, a much higher concentration of atoms was seen during the pulse atomization in the cuvette. Hence, an absolute sensitivity improvement of 4–5 orders of magnitude, and an improvement in detection limit of 2–3 orders of magnitude was realized. The possibility of absolute or standardless analysis was also suggested. It is important to consider the features of this atomizer, and their influence on its performance. First, this was a constant-temperature cuvette, and, second, diffusional loss of atomic vapor was minimized by rapid atomization and enclosing the vapor in a tube with a tantalum or pyrolytic graphite lining. As the temperature of this tube remained constant throughout the atomization period, and as diffusion was the only loss process, the mean length of time that atoms spent in the analysis (residence time) was the same for all atoms. These are important requirements for accurate analysis with minimal matrix interferences (see Section 6.2). Woodriff and co-workers designed a similar atomizer [4,5] in the late 1960s. The sample was nebulized or introduced *via* a cup through a side-arm into a 150 mm long graphite tube of 7 mm internal diameter. The use of the cup also permitted solid samples to be introduced to the furnace. As with L'vov's cuvette, the tube was preheated to the required atomization temperature before the sample was introduced. Detection limits were in the range from 10^{-10} to 10^{-11} g. Although there was a small-scale attempt to market the Woodriff furnace, it was not popular because, in common with L'vov's furnace, it was inconvenient and slow for routine analysis.

At about the same time, in 1968, Massmann developed a considerably simpler furnace [6], shown in Figure 1.4. A graphite tube, 55 mm long and 6.5 mm internal diameter, was heated electrically from its ends, but unlike the L'vov cuvette and Woodriff furnace, it was not preheated prior to sample introduction. Instead, a liquid sample was deposited on the inside wall of the cold tube by pipetting it through a small hole in the side of the tube. The tube was then heated resistively in stages to atomize the sample. An inert protective gas was flushed continuously through the tube in order to prevent oxidation of the graphite. Because of its simplicity, the Massmann furnace was the first to be produced commercially, by Perkin-Elmer. This furnace, described in 1970 by Manning and Fernandez [7], incorporated three controlled heating steps to dry, pyrolyze and atomize the sample, and was shown to be more sensitive than FAAS for a large number of elements.

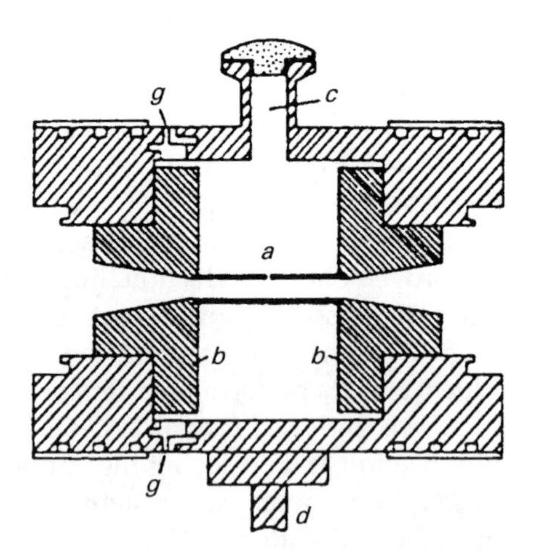

Figure 1.4 The Massmann furnace: (a) graphite tube; (b) electrical contacts; (c) sample introduction port. (Reprinted from *Spectrochim. Acta*, Vol. 23B, H. Massmann, Vergleich von Atomabsorption und Atomfluoresz in der Graphitkuevette, pp. 215–226, 1968, with kind permission of Elsevier Science–NL, Sara Burgerhartstraat 25, 1055 KV Amsterdam, The Netherlands.)

This furnace was destined to become extremely popular and to be the forbear of modern furnaces. However, a price had to be paid for its simplicity compared with L'vov's cuvette. L'vov anticipated this in his critique of the Massmann furnace [8]. The fundamental difference is that L'vov's cuvette was at a constant temperature when the analyte was atomized. However, as the Massmann furnace had to heat from ambient temperature to the atomization temperature while the sample was present, the temperature was changing during atomization. This led to variations in the residence time of atoms, and the tube wall temperature at which the analyte started to vaporize may not have been sufficiently high to adequately dissociate all chemical species containing the analyte. Hence, chemical and spectral interferences could be a problem, and these interferences may not even be consistent between firings of the furnace if the rate of heating was at all variable. A further problem with Massmann furnaces is cooling of the tube ends, caused by the water-cooled electrodes. This causes a temperature gradient along the length of the tube, and this gradient changes with time as the tube is heated. This often leads to condensation of analyte and matrix species at the cooler ends of the tube, and hence to increased spectral and nonspectral (chemical and physical) interferences.

In 1975, Ohta and Suzuki [9] described the use of a molybdenum tube furnace, and metal furnaces (usually molybdenum or tungsten) have continued to be used by these and other researchers. They offer the advantage of improved sensitivity for those analytes that would otherwise form refractory carbides in a graphite furnace, but their popularity has never approached that of graphite furnaces.

For a few years, in the early 1970s, open-filament atomizers were quite popular. West and Williams [10] developed the 'carbon filament atom reservoir', in which a graphite rod or filament (40 mm long and 1–2 mm diameter) was clamped at its ends between stainless steel electrodes, so that the effective length of the filament was 22 mm. The assembly was enclosed in a glass dome and purged with argon. A small volume (5 μl) of sample was pipetted onto the cold filament, which was then pulse heated to atomize the sample. West's group published many papers on this technique, using the improved atomizer design shown in Figure 1.5 [11]. The major advantage of this ETA was its low cost and simplicity. By introducing a laminar flow of argon around the filament, it was possible to dispense with the glass dome, and thus expedite the dispensing of sample aliquots.

However, the open nature of the device was also its major disadvantage. Following its vaporization, the sample immediately entered a lower temperature gaseous environment compared with the Massmann furnace where the gaseous sample was contained within the confines of a heated tube. Hence, the West atomizer was more prone to interferences,

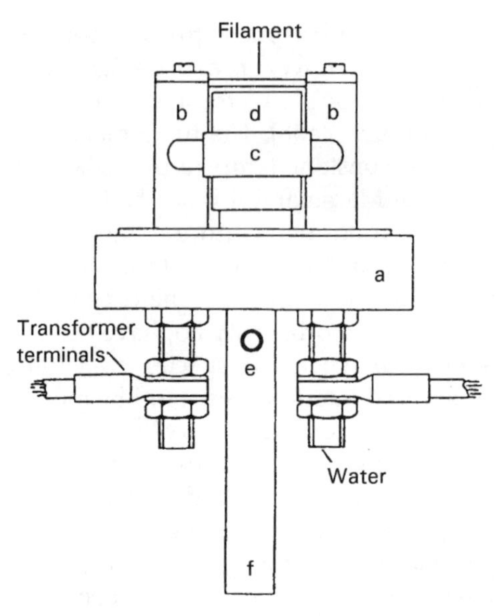

Figure 1.5 The West filament atomizer: (a) base; (b) water-cooled electrodes; (c) water link between electrodes; (d) laminar flow box for protective gas; (e) inlet for protective gas; (f) support stem for reservoir. (Reprinted from *Anal. Chim. Acta*, Vol. 51, J.F. Alder and T.S. West, Atomic Absorption and fluorescence spectrometry with a carbon-filament atom reservoir. Part III. A study of the determination of cadmium by atomic fluorescence spectrometry with an unenclosed atom reservoir, pp. 365–372, 1970, with kind permission of Elsevier Science–NL, Sara Burgerhartstraat 25, 1055 KV Amsterdam, The Netherlands.)

which were reduced by measuring the analyte absorbance at a point very close to the filament surface. Also, the gaseous temperature gradient made it difficult to determine less volatile elements, because recombination to molecular species occurred quickly as the analyte cooled after leaving the graphite surface; for example, vanadium could only be determined by selectively viewing the radiation within 1 mm of the surface [12]. Atomizers based on West's design were marketed in 1970 by Varian, and in 1971 by Shandon Southern.

A further development of the Varian 'carbon rod atomizer' involved drilling a hole of 1.5 mm diameter transversely, and passing the hollow cathode light beam through the hole to create the so-called 'Mini-Massmann' furnace [13] shown in Figure 1.6. This design combined the Massmann furnace's advantage of containing the atom vapor within a heated tube environment with the low power consumption of the rod or filament, but it was inconvenient to use. Sample volumes were limited to 2 μl and sample introduction was difficult.

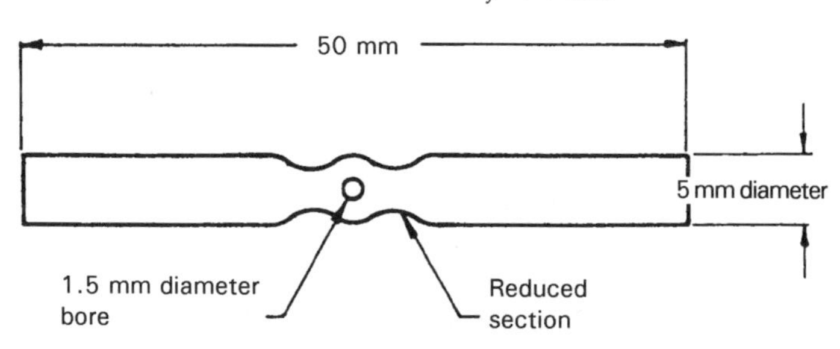

Figure 1.6 The 'Mini-Massmann' furnace. (Reprinted from B.R. Culver, *Analytical Methods for Carbon Rod Atomizers*, 1975, Fig. 1b, with kind permission from Varian Associates, Palo Alto, CA, USA.)

In 1973, Matousek and Brodie [14] modified the carbon rod atomizer by clamping a miniature tube or cup between two carbon rods, as shown in Figure 1.7. The small (9 mm long) tube design then resembled a miniature Massmann furnace, but with side heating rather than conventional end heating. In 1970, Donega and Burgess built the first metal filament atomizer [15] for AAS. It consisted of an electrically heated tantalum boat inside a sealed chamber purged with an inert gas. Based on this design, Hwang *et al.* [16] developed the tantalum strip atomizer shown in Figure 1.8. After slight modification, this design was marketed in 1971 by Instrumentation Laboratory. Perhaps the simplest and least expensive electrothermal atomizer developed was the tungsten filament of Williams and Piepmeier, in 1972 [17]. Made simply by breaking the glass from a commercial light bulb, this device required a current of only 4 A from a 6 V power supply. A graphite braid atomizer was described by Montaser

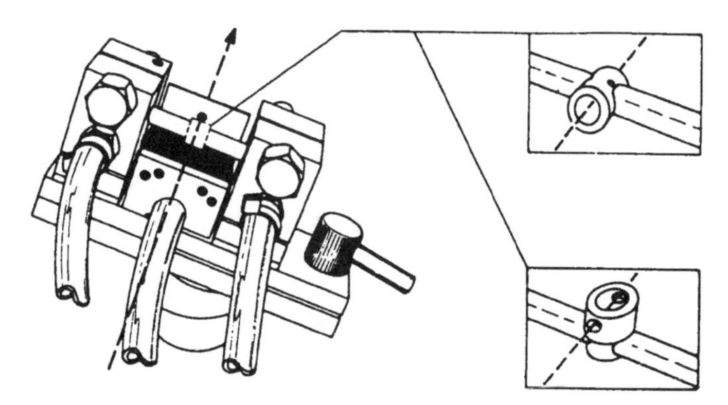

Figure 1.7 The modified Varian carbon rod atomizer, incorporating a tube or cup. (Reprinted from B.R. Culver, *Analytical Methods for Carbon Rod Atomizers*, 1975, Fig. 7, with kind permission from Varian Associates, Palo Alto, CA, USA.)

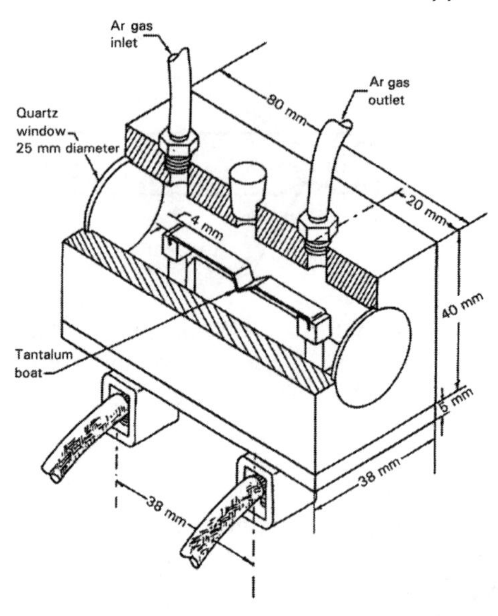

Figure 1.8 The tantalum strip atomizer. (Reprinted with permission from J.Y. Hwang, P. A. Ulluci, S.B. Smith, and A.L. Malenfant, Microdetermination of lead in blood by flameless atomic absorption spectrometry, *Anal. Chem.* **43**, 1319–1321. Copyright 1971 American Chemical Society.)

et al. [18]. Its major advantage over other graphite filament devices was a lower power requirement.

During the 1970s, there were many commercially available electrothermal atomizers, including tube furnaces based on Massmann's design and various filament furnaces as described above. However; during this decade the technique of ETAAS was characterized as interference prone, and its development declined. L'vov pointed out that these problems arose through the use of furnaces that did not provide isothermal atomization conditions. The constant-temperature devices of L'vov and Woodriff, described above, were too complicated for rapid routine analysis. L'vov proposed three modifications that could be made to the Massmann furnace in order to allow it to operate under more nearly isothermal conditions, but without unduly compromising the ease of operation of this furnace.

The first modification was the use of capacitive discharge heating to rapidly pulse heat the furnace to a constant temperature [19]. This was developed further by Chakrabarti *et al.* [20]. Second, L'vov suggested introduction of the sample to a preheated furnace on a graphite probe, inserted through a slot in the side of the tube [21]. Several research groups worked on probe atomization, most notably Ottaway *et al.* [22]. The third idea was also the simplest, and it was destined to revolutionize the

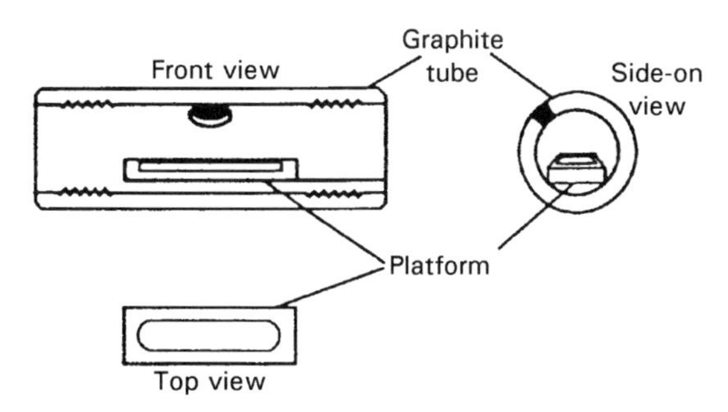

Figure 1.9 The L'vov platform in a Massmann furnace. (Reprinted from R.D. Beaty and J.D. Kerber, *Concepts, Instrumentation and Techniques in Atomic Absorption Spectrometry*, 1993, p. 6–10, with kind permission from The Perkin-Elmer Corporation, Norwalk, CT, USA.)

technique and lead to a resurgence of interest. It consisted of vaporizing the sample from the surface of a graphite platform place loosely inside the Massmann furnace [23], as shown in Figure 1.9. As the platform had only slight contact with the inner wall of the tube, it was heated mainly by radiation from the tube wall. Ideally, this would result in a delay of analyte vaporization until the tube had reached a high and stable temperature. Under these conditions, the analyte absorbance would be much less affected by the heating rate of the graphite tube and by the sample matrix. This should result in fewer matrix interferences compared with wall atomization. Slavin and co-workers [24,25] subsequently developed the technique of platform atomization, and it has now become the standard approach to modern ETAAS. This is discussed in more detail in Section 1.5.1.

Frech and Jonsson [26], in 1982, designed a two-step furnace, which provided nearly isothermal operating conditions and allowed separate control of the vaporization and atomization processes. A more recent version of this furnace is shown in Figure 1.10 [27]. A graphite cup was positioned below a hole drilled in the bottom of a graphite furnace tube, and samples were pipetted into the cup *via* a sample introduction hole in the tube. Separate power supplies, permitting independent control of sample vaporization (cup) and atomization (tube) heated the cup and tube. For normal operation, the tube was first heated to a predetermined constant atomization temperature, and then the cup was heated to the appropriate temperature for sample vaporization, when the sample vapors entered the nearly isothermal tube environment for AA measurement. Integrated graphite contacts on the sides of the tube allowed it to be heated transversely, compared with the conventional end heating of Massmann furnaces. This produced a constant temperature along the length of the tube, unlike the temperature gradient in the Massmann

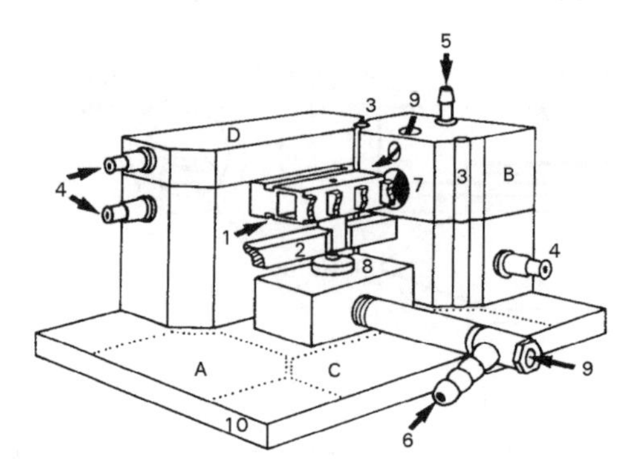

Figure 1.10 The Frech two-step furnace: (1) transversely heated graphite tube; (2) graphite cup; (A,B) cup electrodes; (C,D) tube electrodes. (Reprinted by permission of the Royal Society of Chemistry from W. Frech, D.C. Baxter and E. Lundberg, Spatial and temporal non-isothermality as limiting factors for absolute analysis by graphite furnace atomic absorption spectrometry, *J. Anal. At. Spectrom.*, **3**, 365–372 (1988).)

furnace caused by the water-cooled electrodes at its ends. This furnace more closely approached the isothermal operating characteristics of the L'vov and Woodriff cuvettes than any other design.

In 1986, Frech *et al.* simplified the furnace by dispensing with the separately heated cup [28]. Their transversely heated graphite tube was used with a L'vov platform and operated in a similar way to the Massmann furnace. However, unlike the end-heated Massmann furnace, Frech's furnace was more nearly spatially isothermal along its length. It was shown to be less prone to spectral and nonspectral interferences compared with platform-equipped Massmann furnaces. Perkin-Elmer manufactured an integrated contact furnace based on Frech's design, shown in Figure 1.11.

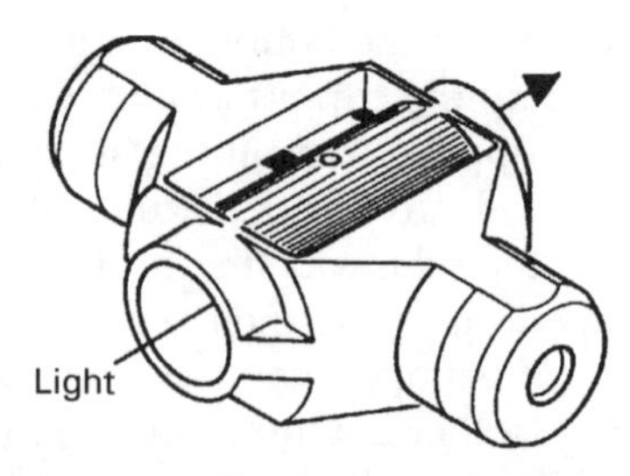

Figure 1.11 Perkin-Elmer integrated contact furnace. (Reprinted from R.D. Beaty and J.D. Kerber, *Concepts, Instrumentation and Techniques in Atomic Absorption Spectrometry,* 1993, p. 5–5, with kind permission from The Perkin-Elmer Corporation, Norwalk, CT, USA.)

1.5 MODERN ELECTROTHERMAL ATOMIC ABSORPTION SPECTROMETRY

1.5.1 Instrumentation

The most popular commercial ETA is an end-heated Massmann furnace with a L'vov platform (described above). Platform atomization is used for all but a few less volatile elements, which are instead atomized from the inner wall of the furnace. The advantages of platform atomization over wall atomization for most elements may be realized by considering Figure 1.12. In this example, the furnace wall has reached its programmed temperature of 1700°C after 0.5 s, and the temperature has remained constant thereafter. The gas temperature in the tube is only a little lower than the wall temperature depicted in the figure, and follows the wall temperature as the tube is heated.

A hypothetical analyte with an appearance temperature of 1000°C if deposited on the tube wall, would start to vaporize after 0.2 s and might be completely vaporized after 0.4 s, to produce the absorbance signal shown in the upper part of the figure. Hence, the analyte would be vaporized completely during the time the furnace was continuing to heat to its optimal temperature. The same analyte deposited on the platform

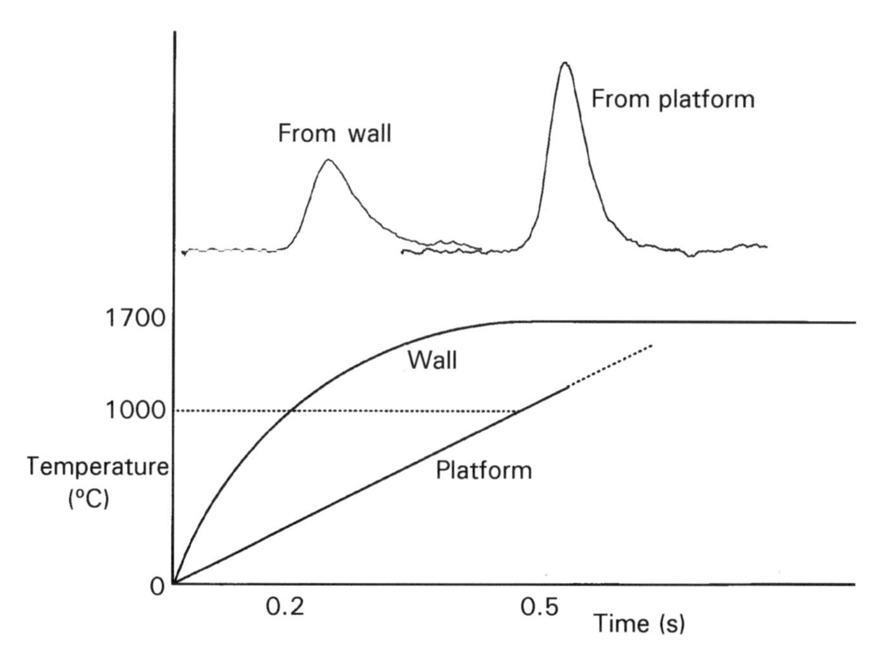

Figure 1.12 Comparison of absorbance signals obtained under conditions of wall and platform atomization. The analyte begins to vaporize when the surface on which it is deposited reaches 1000°C.

would not begin to vaporize until the platform had reached 1000°C (i.e. after 0.5 s), but by this time the tube would have reached its optimal temperature of 1700°C, and would maintain this stable temperature while the analyte vaporized. Hence, the absorbance signal resulting from platform atomization is expected to be much less affected by the heating rate of the tube and by the sample matrix, as atomization occurs at a higher temperature. Platform atomization is often referred to as 'isothermal', but this is not quite true. The tube temperature depicted in Figure 1.12 has become constant after 0.5 s, but this is the temperature at the tube center where the sample introduction hole is located. The tube is always cooler towards the ends, resulting in a temperature gradient along the length of the tube. Also, this temperature gradient changes with time during the atomization stage of the heating cycle. Hence, in a strict sense, this furnace is neither spatially nor temporally isothermal.

Slavin *et al.* [25] realized that the use of a L'vov platform alone may not be sufficient to provide the most accurate analysis with the fewest interferences, and accordingly they developed the 'stabilized temperature platform furnace' (STPF) technique which embodied several additional characteristics: rapid heating and fast digital electronics to accurately capture the transient absorbance signal; measurement of integrated absorbance rather than peak height absorbance; solid pyrolytic graphite platforms and a pyrolytic graphite coated tube; Zeeman-effect background correction (see Chapter 4); and the use of chemical modifiers (see Chapter 5).

Another furnace design that is increasing in popularity is the transversely heated integrated contact furnace, described above in Section 1.4, and shown in Figure 1.11. This is commonly used with an integrated platform. Its improved spatial temperature characteristics, resulting from transverse rather than end-on heating, provide certain advantages. These are briefly discussed below in Section 1.5.2. Recently, there has been

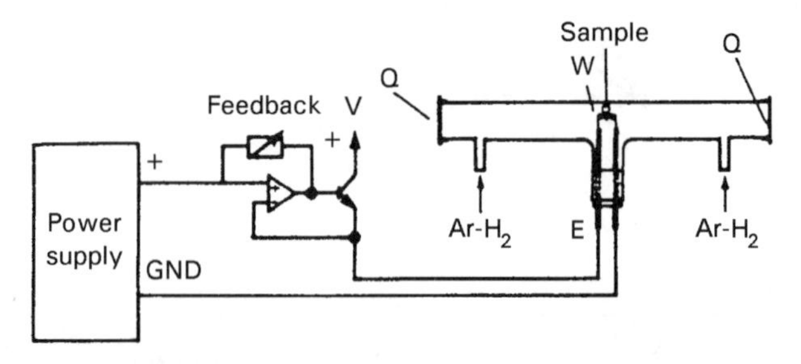

Figure 1.13 The tungsten filament atomizer. (Reprinted by permission of the Royal Society of Chemistry from M.F. Gine, F.J. Krug, V.A. Sass, B.F. Reis, J.A. Nobrega and H. Berndt, Determination of cadmium in biological materials by tungsten coil atomic absorption spectrometry, *J. Anal. At. Spectrom.*, **8**, 243–245 (1993).)

renewed interest in metal filament atomizers, based on the design of Williams and Piepmeier [17], and discussed above in Section 1.4. The modern version [29] uses a coiled tungsten filament, of the type used in projection lamps, enclosed in a quartz tube purged with a mixture of argon and hydrogen, as illustrated in Figure 1.13. The filament is resistively heated by a low-voltage power supply. Open-filament atomizers have disadvantages compared with tube atomizers, as discussed in Section 1.4. Nevertheless, it has been demonstrated that this device can produce accurate data if prescriptive methodology is followed. Of particular note is its successful application to the determination of lead in blood [30–32]. Its low cost and simplicity have made it a possible atomizer for a dedicated portable blood-lead analyzer [31,32].

1.5.2 Applications

The discussion in this section is limited to commercially available instruments, using endheated Massmann or transversely heated integrated contact furnaces. Instruments incorporating these atomizers are almost invariably highly automated. Automatic sample pipetting permits the unattended running of calibration standards, quality control samples, and test samples. Among techniques with commercially available instrumentation, the low detection limits of ETAAS are matched only by inductively coupled plasma mass spectrometry (ICPMS). Disadvantages of ETAAS are twofold. First, sample throughput times are relatively slow, with typical furnace heating cycles taking 2 min or more per sample aliquot. However, fast furnace techniques (discussed in Section 1.5.2.3 below) can significantly reduce the heating cycle time. Second, it is inherently a single-element technique, although there are at least two commercial instruments that permit the simultaneous determination of four to six elements.

As ICPMS is a multielement technique, its use can be justified when many samples have to be analyzed for several metals. However, ETAAS is much less costly, and is often the preferred technique. The end-heated Massmann furnace, operated under STPF conditions, has been shown repeatedly to permit the interference-free determination of metals in a large number of matrices. If used correctly, the technique often has fewer interferences than other techniques for trace metal determinations, and in general it is certainly no more interference prone than its competitors. Applications of the transversely heated furnace demonstrate several advantages over end-heated Massmann furnaces. First, lower atomization temperatures are possible, which may extend the applicability of platform atomization to less volatile elements that usually require wall atomization in a Massmann furnace. Second, improved heating characteristics lead to reduced interferences, such as those caused by sulfate and chloride [33]. Third, detection limits may be improved [33].

Reference to recent applications reviews, such as those appearing regularly in the *Journal of Analytical Atomic Spectrometry,* and *Analytical Chemistry,* show the highly competitive nature of ETAAS in the analysis of a wide range of sample types, including air particulates, water, plants, foods, biological fluids and tissues, soils, sediments, sludges, geological materials, fertilizers, metal alloys, petroleum products, semiconductor materials, ceramics, glasses, etc. A major advantage of ETAAS is the ability to analyze samples of only a few microliters; which makes it particularly suitable for the analysis of clinical and biological samples, where only very small sample volumes may be available. For example, it is the preferred technique for the determination of lead in blood, where children are routinely screened by collecting just a few drops of blood by a fingerstick method. Also, in the analysis of biological samples the multi-element analysis capabilities of techniques such as ICPMS may be less important. Tsalev's book on *Atomic Absorption Spectrometry in Occupational and Environmental Health Practice* [34] includes a comprehensive discussion of the application of ETAAS in the analysis of biological and clinical samples. The author concludes that ETAAS is the preferred technique in many cases. This book also provides extremely useful information on the optimum analysis conditions for a wide range of metals, and much of this information is of general applicability so it can be useful for those working in other fields of application. Although it is now over a decade old, *Graphite Furnace AAS. A Source Book,* by Slavin [35] should also be indispensable to the user of ETAAS. The manufacturers' 'cookbooks' that come with the purchase of an instrument provide optimal conditions for many applications, as does a book by Varma [36]. A compendium of methods for the analysis of a variety of sample types, including natural water, seawater, marine organisms, foods, biological fluids, biological tissues, and bone, is available in a book by Minoia and Caroli [37].

1.5.2.1 Direct analysis using STPF conditions

For most applications, the features of the STPF technique discussed in Section 1.5.1 are important. This applies both to the end-heated Massmann furnace and the transversely heated integrated contact furnace. Of particular note are the addition of chemical modifiers to the sample in the ETA, and the use of background correction. Chemical modifiers are discussed in detail in Chapter 5 and, for the applications scientist, they perform three important functions. First, they may increase the volatility of the sample matrix, permitting its removal during thermal treatment without volatilization loss of analyte. In this class of modifier is ammonium nitrate, which is often added to samples high in chloride to form volatile ammonium chloride. Second, they may thermally stabilize the analyte so that higher treatment temperatures can be used. Palladium salts are the most common modifiers in this class, generally permitting treatment temperatures of 1100–1200°C, when more volatile matrix components are removed. Third, they may stabilize the analyte to a higher atomization temperature (palladium also fulfils this role). This can be advantageous in delaying the analyte absorbance signals from the calibration standards and the samples to similar times. As commercial ETAs are not truly isothermal, this has the effect of ensuring that the analyte is atomized at a similar temperature in the case of both standards and samples. This may be particularly important in the analysis of solids and slurries (see Chapter 6).

Background correction (discussed in Chapter 4) is usually required to remove two types of spectral interference First, the sample matrix may vaporize to produce molecular species with broad-band absorption spectra that overlap with the AA signal from the analyte. Second, large amounts of matrix may also produce an aerosol in the ETA that leads to scatter of the incident radiation, and hence nonatomic absorption at the analyte wavelength.. Of the techniques of background correction described in Chapter 4, Zeeman-effect and continuum source background correction have been used mostly, and both have been shown to be highly effective in eliminating spectral interferences. Zeeman-effect background correction is particularly useful if the background absorbance has a relatively narrow band structure in the region of the analyte wavelength.

The greatest challenges to interference-free analysis using STPF conditions are presented by samples high in inorganic salt content, particularly chloride. As many real samples contain high concentrations of chloride, its effects on the determination of trace metals have been studied extensively. High concentrations of chloride can be responsible for both spectral and nonspectral interferences. Shown in Figure 1.14 is the absorbance signal for thallium in a sodium chloride matrix, together with the nonatomic absorption (background) signal from the salt matrix. These signals were obtained in a Massmann furnace operated under STPF conditions. Characteristic multiple peaks are produced by the salt matrix. The first peak occurs through initial vaporization of the matrix, starting when the platform temperature reaches about 1000°C. The vaporized salt then migrates towards the cooler ends of the furnace, condenses, and then vaporizes again as the ends heat up. The second and subsequent peaks are due to this revaporization process. Under the conditions shown in Figure 1.14, where the thallium and sodium chloride signals coincide, larger amounts of sodium chloride (e.g. 100 μg) would result in a spectral interference due to the nonatomic absorption signal being too large for effective background correction. In practice, however, this interference does not occur if a modifier is used to separate the signals. The addition of a palladium modifier results in the thallium signal being delayed in time. This avoids the temporal overlap of the two signals, and background

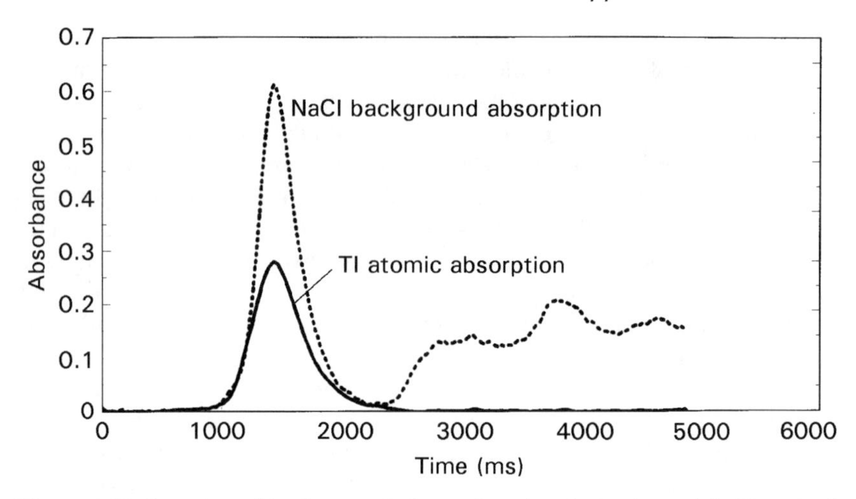

Figure 1.14 Atomic and background absorption signals for 1 ng of thallium in the presence of 10 μg of sodium chloride.

correction is effective. Also, in practice, most of the chloride matrix can be removed in the presence of palladium by use of a pyrolysis temperature high enough to volatilize the salt, but not high enough to result in volatilization loss of thallium. Alternatively, ammonium nitrate may be used as a modifier, to permit the low-temperature volatilization of the matrix as ammonium chloride.

Although spectral interferences are removed in this way, a nonspectral interference may still occur, resulting in low recovery of the analyte. Several possible mechanisms for this interference have been suggested. First, the analyte metal chloride may be formed either in the gas phase or through a condensed phase or heterogeneous reaction on the graphite surface. The volatile chloride may then be lost by diffusion from the furnace before its complete dissociation. Second, if a sufficient amount of the salt matrix remains in the furnace after the pretreatment stages, the gaseous products of its decomposition may rapidly expand during the atomization stage, resulting in coexpulsion of the analyte from the furnace. Third, analyte may be occluded in microcrystals of the salt matrix, which are expelled from the furnace without decomposing.

Much of the evidence for the gas-phase formation of analyte metal chlorides has resulted from gaseous equilibrium considerations. L'vov [19] predicted the magnitude of chloride interference by calculating the degree of dissociation of some gaseous monohalides with respect to the temperature of the furnace, and reasonable agreement with experimental results was obtained. Also, the diminished interference by chloride in the presence of lithium salts and hydrogen was said to result from lithium chloride and hydrogen chloride having higher dissociation energies than the analyte chlorides. Slavin and Manning [24] attributed increased interference by magnesium chloride compared with sodium chloride to the differences in the dissociation energies of the two salts.

Much of the evidence for condensed phase and expulsion interference mechanisms has resulted from experiments where the analyte and sodium chloride were physically separated in the furnace. Welz was the first to do this by redesigning the L'vov platform (Figure 1.9), to provide two cavities on its surface. When the analyte was in one cavity and sodium chloride was separated in the other cavity, there was decreased interference on lead [38], thallium [39], and zinc and cobalt [40] compared with the situation where the analyte and sodium chloride were premixed. These experiments provided strong evidence of a condensed phase mechanism, although some interference occurred when analyte and sodium chloride were separated in the two cavities. This was attributed to expulsion loss [38], but it may have been a condensed phase reaction resulting from some of the analyte migrating from one cavity to the other during the early stages of furnace heating.

Mahmood and Jackson [41] studied the interference of chloride on thallium using 'wall-to-platform migration', whereby the sample was pipetted onto the graphite wall, and then the analyte migrated to the cooler platform on heating, leaving the less-volatile sodium chloride on the tube wall. This permitted the analyte atomic absorption and background non-AA signals to be separated in time. Interference by chloride was seen, even though the analyte and sodium chloride were not present in the gas phase at the same time. Hence, it was established that a condensed-phase interference occurred. They also studied the temperature dependence of the interference, and determined that it occurred in the temperature range 500–700°C during the pyrolysis stage of the furnace heating cycle.

Subsequent work by these authors [42] involved the use of a Frech-type two-step furnace (described in Section 1.4). The analyte was pipetted into the cup and sodium chloride was separately pipetted onto the inside wall of the tube. Separate control of the cup and tube temperatures allowed conditions to be created where the analyte and sodium chloride were present together in the gas phase in the tube, and where they were separated in time. This experiment showed no evidence of a gas-phase interference or expulsion loss of analyte when analyte and sodium chloride were present together in the gas phase. Hence, most of the recent research on this interference indicates the condensed phase formation of the analyte chloride at pretreatment temperatures, and the volatilization loss of this chloride without its complete dissociation to analyte atoms.

Although most of the research on chloride interference has focused on sodium chloride, other chlorides have been found to interfere. Cabon and Le Bihan [43] showed that the magnitude of interference by magnesium and calcium chlorides might be different from sodium chloride, due to their partial hydrolysis during the pretreatment stages of the furnace heating cycle. The literature indicates that chloride interference can be reduced to acceptable levels in many cases, with the use of modern furnace technology and appropriate modifiers. Some modifiers used in the accurate direct determination of trace metals in seawater include: phosphate–nitric acid in the determination of cadmium [44], sodium hydroxide in the determination of cadmium [45] and manganese [46], ammonium nitrate in the determination of copper [47], nickel–ammonium nitrate in the determination of selenium [48], oxalic acid in the determination of zinc [49], ammonium oxalate in the determination of cadmium [50], tetramminepalladium(II) chloride–ammonium oxalate in the determination of lead and manganese [50], palladium in the determination of arsenic, cadmium and lead [51], and palladium–strontium nitrate in the determination of lead [52].

In the analysis of seawater, sulfate may also interfere. Sulfate interference has been attributed to the formation of the analyte sulfide, and its volatilization loss prior to complete atomization. Cabon and Le Bihan [43] found that calcium and magnesium sulfates interfered with the determination of lead, suggesting that they decompose to oxides of sulfur which react with lead to form lead sulfide. Welz *et al.* [53] considered that lead was embedded in the sulfate matrix and then coexpelled during the atomization stage of the furnace heating cycle. Generally, sulfate interference has been removed or reduced to acceptable levels. Lanthanum [54] and a mixture of palladium and strontium nitrate [52] have been shown to be suitable modifiers for the interference-free determination of lead in the presence of sulfate.

With the use of background correction, spectral interferences are few. However, Kurfuerst and Pauwels [55] noted that high concentrations of sulfur caused spectral interference on the lead line at 283.3 nm, when solid samples were analyzed. This interference was said to be due to molecular absorption by S_2. Two well-known spectral interferences are aluminum on the determination of arsenic at 193.7 nm, and iron on selenium when determined at 196.0 nm. The interference of an aluminum matrix on arsenic was eliminated by using a palladium modifier to separate temporally the analyte from the matrix during the atomization stage [56]. Zong *et al.* [57] reported a background overcorrection interference when determining lead in the presence of phosphate by Zeeman-effect ETAAS. They suggested that the problem was caused by molecular absorption of PO, as PO bands appeared to undergo splitting in the magnetic field. Interestingly, this problem occurred when using a longitudinal Zeeman-effect instrument, but not a transverse Zeeman-effect instrument. Epstein *et al.* [58] reported a spectral interference by lead nonresonance AA on both of the primary arsenic lines at 193.7 and 197.2 nm, when analyzing lead-based alloys. They were able to overcome the interference at 197.2 nm by arranging conditions so that temporal overlap of the arsenic and lead absorption lines did not occur.

1.5.2.2 *Analysis requiring separation and/or preconcentration*

For some applications, it may be necessary to separate the analyte species from the sample matrix in order to remove interferences. More often, separation is used as part of a preconcentration procedure. The direct analysis of seawater is discussed in Section 1.5.2.1 above, but most of these direct methods are applicable only to highly polluted seawater samples. Typical concentrations of trace metals in unpolluted seawater are so low that preconcentration is required in order to increase the sensitivity of the analysis. Porta *et al.* [59] described an on-line preconcentration system that involved chelating the analytes, trapping the chelates on an adsorbent in a miniature column in the furnace autosampler arm, and then eluting with an organic solvent into the furnace. Azeredo *et al.* [60] developed a discontinuous micro scale preconcentration method involving chelation of several metals on silica-immobilized 8-hydroxyquinoline in a microcolumn. The elements were then eluted into the graphite furnace, and quantitative recoveries were reported for iron, cadmium, zinc, copper, nickel, manganese, and lead. Silver was complexed and extracted from seawater by its adsorption onto activated carbon [61]. The analyte was then desorbed with nitric acid for determination by ETAAS. Beryllium was similarly complexed and extracted, and the carbon slurry was injected directly into the graphite furnace for analysis [62].

The coupling of flow injection (FI) with ETAAS provides increased automation of analysis, including sample pretreatment and preconcentration steps that otherwise would have to be done manually. A difficulty that had to be overcome was interfacing the continuous flow nature of FI with the discrete sample input requirement of ETAAS. Fang *et al.* [63] solved this problem with an accurately timed multistage FI system that would sequentially preconcentrate the analyte and then transfer it as an aliquot of fixed volume into the ETA. The sample was first mixed online with a chelating agent and loaded onto a microcolumn containing a suitable sorbent. In the next stage the column was washed, and then the chelate was eluted into a collector coil. Finally, the chelate was pumped from the collector coil into the ETA to coincide with the initiation of its heating cycle. This automated preconcentration sequence provided a 26-fold signal enhancement compared with direct injection of the untreated sample.

This system was developed further [64,65], to permit the determination of several metals with adequate sensitivity in seawater reference materials and other water samples. Adams and co-workers used a similar approach for the on-line separation and preconcentration of several elements, including lead [66]. An FI system for the analysis of blood digests [67] involved the on-line coprecipitation of cadmium and nickel, followed by dissolution of the precipitate in an organic solvent and its injection into the furnace. In a similar way, lead and copper were coprecipitated from seawater and the dissolved precipitate was analyzed by ETAAS [68]. Slurried biological and botanical reference materials were microwave-digested on-line in an FI system [69] for the determination of lead. Further applications of the use of FI with ETAAS can be found in the biennial reviews in *Analytical Chemistry* [70,71].

An elegant separation technique for elements that form stable gaseous hydrides involves a combination of hydride-generation (HG) AAS and ETAAS. In HGAAS, the hydride is typically formed by reduction with tetrahydroborate and is then swept by an inert carrier gas into a heated quartz tube atomizer. For the determination of arsenic, selenium, and antimony, HGAAS is usually preferred over ETAAS, and it is used to a lesser extent for the determination of lead, tin, bismuth, germanium, and tellurium. Drasch *et al.* [72] combined the techniques of HGAAS and ETAAS by sweeping the metal hydride into a graphite furnace, where it was trapped. The furnace was then heated to atomize the analyte element and permit its determination by ETAAS. Doidge *et al.* [73] discovered that the hydride could be trapped more efficiently if a palladium salt was first pipetted into the furnace, which was then heated to 600–1000°C as the hydride was swept into it. The mechanism of this trapping was investigated by Sturgeon *et al.* [74], who suggested catalytic decomposition of the hydride and trapping on the reduced palladium deposit by chemisorption.

This technique has become quite popular, providing a preconcentration step and eliminating matrix interferences that might otherwise occur in conventional ETAAS with solution deposition. Applications include the determination of tin in various samples [75], arsenic and selenium in mineral waters [76], germanium in several environmental samples [77], and tellurium in seawater and sediments [78]. Several metals other than palladium have been used for trapping, but often they have been less efficient than palladium, and share the disadvantage of having to be reintroduced after every firing of the furnace. However, iridium [79,80] and iridium–magnesium [80] were effective as 'permanent modifiers', being retained in the furnace for at least 300 firings. Cadmium was determined in seawater by reducing it to a gaseous species using sodium tetrahydroborate and trapping it in an iridium-coated tube [81]. The technique has also been applied to other gaseous species, including alkyllead compounds, and to mercury metal [82]. Matusiewicz and Sturgeon [83] reviewed the advantages and limitations of the *in situ* trapping of gaseous species in graphite furnaces.

In samples where two or more species of an analyte need to be determined individually, their separation is almost invariably required. In common with all atomic spectrometric techniques, the analytical signal in ETAAS is generally independent of the matrix and of the chemical species of the analyte in the original sample. Thus, a sample may contain the analyte as several chemical species and in several oxidation states, but only the total concentration of the analyte will be obtained. Although this is an advantage for interference-free analysis, it means that individual analyte species cannot be determined directly.

Several methods have been used for sample pretreatment in order to separate species prior to analysis by ETAAS. Arsenic(III) and arsenic(V) were removed from environmental samples by coprecipitation with zirconium hydroxide, and converted to their chlorides with hydrochloric acid [84]. The mixture was injected into an ETA, and selective heating allowed the species to be determined differentially. By pretreating at 1100°C for 50 s, arsenic(III) was removed, leaving arsenic(V) in the furnace. The *in situ* trapping technique, described above, was applied to the speciation of arsenic in plant material, urine, and water samples [85]. Chromium(III) was selectively chelated in the presence of chromium(VI) with oxine [86] and the chelate was extracted into methylisobutylketone for analysis by ETAAS. Selenium(IV) was selectively adsorbed onto a resin, and then the resin was injected as slurry into the ETA [87]. Chromium was electrodeposited onto a L'vov platform coated with a film of mercury [88]. Control of the reduction potential allowed the selective reduction of chromium(VI) to chromium(III) and its deposition on the platform.

For speciation, atomic spectrometric techniques such as ICPAES and FAAS are conveniently coupled with a separation technique, such as high performance liquid chromatography (HPLC), when the HPLC effluent is introduced continuously into the nebulizer of the spectrometer. Unfortunately, the discontinuous (pulse) sample introduction requirement of ETAAS makes this difficult, and many workers have avoided direct interfacing of the instruments. Selenium species were separated by HPLC and the eluant was passed into a fraction collector [89]. The vials from the collector were then transferred to the ETA autosampler. Copper and lead species were adsorbed on columns containing silica gel and C_{18} stationary phases prior to selective sequential elution and determination by ETAAS [90]. A manifold system automatically delivered eluant to the ETA.

In spite of the interfacing difficulties, there have been a few reports of successfully coupling chromatographic techniques with ETAAS. Jiang *et al.* [91] selectively determined alkyl selenides extracted from soil by coupling gas chromatography (GC) with ETAAS. The selenide mixture was injected onto the GC column, and as each fraction eluted, it was transferred *via* a quartz tube to the ETA, which was preheated to 500°C and coated with palladium. The alkylselenide was trapped, the quartz tube was removed, the GC carrier gas was stopped, and the complete furnace heating cycle was completed before resuming the flow of GC carrier gas and allowing the next species to elute from the GC column. Sperling *et al.* [92] used the FI-ETAAS system developed by Fang *et al.* [63], and described above, to differentially determine chromium(III) and chromium(VI) in natural water. Chromium(VI) was preconcentrated selectively onto the column, and then eluted to the ETA with ethanol.

1.5.2.3 Fast-furnace Technology

Sample injection, thermal pretreatment, pulse atomization, and the need to cool the ETA prior to injection of the next sample aliquot, make ETAAS a relatively slow technique. Table 1.1 shows a typical heating program for the determination of lead, and the total cycle takes 161 s. Although, the complete cycle shown in the table is best for ensuring maximum removal of matrix during pretreatment, and hence the most accurate analysis, fast-temperature programs have, in many cases, been successful in reducing the analysis time without seriously compromising the accuracy. Over several years, there have been many published methods involving fast-furnace technology and this was reviewed by Halls [93]. Fast-furnace programs are not always applicable and, where used, they require great care to avoid systematic errors. The program should be customized for each analysis. Strategies that may be used include sample injection into a preheated furnace to reduce the drying time, the use of higher drying temperatures, reduction or even total elimination of the pyrolysis stage (often with elimination of modifiers), and reduction of the clean-out time. Typical examples of the application of fast-furnace technology include cycle times of only 10 s for the analysis of water samples [94], and only 45 s for biological reference materials [95].

When samples have to be digested prior to analysis, the atomizer cycle time of 161 s shown in Table 1.1 is very small compared with the sample preparation time. Discussed in Chapter 6 is the direct analysis of solids and slurries by ETAAS, where it is seen that the overall analysis time may

Table 1.1 Typical furnace heating program for the determination of lead using STPF conditions.

Heating stage	*Dry*	*Pyrolyze*	*Cool*	*Atomize*	*Clean-out*
Ramp time (s)	10	20	10	0	1
Hold time (s)	60	30	20	5	5
Temperature (°C)	130	900	200	1900	2700

be shortened considerably by avoiding sample digestion. The direct analysis of slurries has been combined with fast furnace technology to enable extremely rapid analysis [96,97].

1.5.2.4 Absolute analysis

An absolute method of analysis involves a theoretical calculation of the amount or concentration of an analyte, based on a measurement that does not require external calibration.

Metal determinations have long been carried out by absolute methods, principally gravimetric analysis. A classical example is the determination of nickel, which is precipitated from solution by forming a complex with dimethylglyoxime. The complex is then removed by filtration, dried and weighed. Knowledge of the chemistry permits a calculation of the concentration of nickel in the original sample, based on the mass of precipitate and the mass of the sample. Measurement of the mass is made on an instrument (balance) that uses standard weights, and hence does not require calibration prior to use. The much higher sensitivity requirements of trace or ultratrace metal determinations involve instrumental measurements where the output signal is proportional to the mass or concentration of the analyte. Such techniques normally do not lend themselves to absolute analysis, because the magnitude of the signal for a given mass or concentration of analyte is dependent on instrumental conditions that cannot be reproduced, and cannot be predicted theoretically. Hence, calibration is required. In AAS, this is done by measuring the analyte absorbance from several calibration standards, computing the linear relationship between absorbance and concentration, and then determining the concentration of unknown samples from these calibration data. L'vov, in some of his earliest work on ETAAS [3], suggested that absolute analysis should be feasible with his enclosed cuvette, as the analyte was completely vaporized and confined within a graphite tube of known volume. Modern commercial furnaces, even with platform atomization, do not attain the temporal and spatial isothermality of L'vov's cuvette. Nevertheless, Slavin and Carnrick [1] showed that reproducibility of experimental characteristic mass (m_0) values (see Section 1.3) between instruments in different laboratories was within about 15%. They then analyzed several reference materials, using published m_0 values instead of calibration standards, and obtained results that were generally accurate within 10–20%. the theoretical calculation of m_0 values is necessary if they are to be used for absolute analysis. L'vov *et al.* [98] derived a theoretical equation, later refined [99] to

$$m_0 = 5.08 \times 10^{-13} \frac{MD\Delta\tilde{\nu}_D}{H(a,\omega)\gamma'\delta f} \quad \frac{Z(T)}{g_1 \exp(-E_1/kT)} \frac{r^2}{l^2} \tag{1.1}$$

where: M the molar mass of the analyte; D is the diffusion coefficient of atoms in argon at temperature T; $\Delta\tilde{\nu}_D$ is the Doppler line width; $H(a, \omega)$ is the Voigt integral for the point of the absorption line contour that is distant from the line center by $\omega = 0.72a$; γ' is a coefficient accounting for hyperfine splitting in the analytical line and the Doppler line width in the light source; δ is a correction factor to take into account the contribution of radiation from adjacent lines in the light source; f is the oscillator strength; $Z(T)$ is the state sum at temperature T; g_1 and E_1 are the statistical weight and lower energy level of the analytical line; k is the Boltzmann constant; and r and l are the inner radius and length of graphite tube.

In order to apply this equation to atomization in a Massmann furnace operated under STPF conditions, the following simplifying assumptions were made: (1) during atomization, the furnace is temporally and spatially isothermal; (2) sample vaporization occurs from a point source at the tube center; (3) analyte vapor is distributed uniformly over the cross-sectional area of the tube; (4) the atom residence time is determined by diffusion from the tube center towards the ends where the atom concentration is zero; (5) the analyte is totally vaporized and atomized; (6) there is no ionization of analyte atoms; (7) the spectral line width or individual components of the hyperfine structure in the light source are negligible compared with the absorption line; and (8) the absorption line maximum undergoes a specified Lorentzian shift towards longer wavelengths relative to the emission line. The first four assumptions are probably the least accurate. This furnace is not temporally or spatially isothermal (as discussed in Section 1.5.1), liquid samples spread during drying, so point source atomization does not occur, a concentration gradient frequently occurs across the tube (see Section 2.7), and a significant fraction of the analyte atoms is lost by diffusion through the sample injection hole. Nevertheless, L'vov showed [99] that $m_{0\text{calc}}/m_{0\text{exp}}$ (the ratio of calculated to experimental m_0 values) averaged 0.85 ± 0.10 for 35 elements. Low $m_{0\text{calc}}/m_{0\text{exp}}$ ratios for lithium, rubidium and cesium may have occurred through ionization, which would increase the experimental m_0 values, and some of the less volatile elements may have had high experimental m_0 values through the formation of gaseous carbides and monocyanides.

Frech and Baxter [100] used the two-step furnace, discussed in Section 1.4, to measure experimental m_0 values and to compare them with theoretical values. As this furnace operates under isothermal atomization conditions, it allowed the simplifying assumption of isothermality made by L'vov to be tested. Generally, they found good agreement between this furnace and the Massmann furnace with STPF conditions, indicating that L'vov's assumption was valid. Su *et al.* [101] introduced a further correction factor into the calculation of m_0 values by taking into account Zeeman sensitivity ratio and effective stray light. This made the values more precise, as they were less affected by lamp current and slit width. Also, more reliable fundamental parameters, particularly oscillator strength (f), in Eqn (1.1), are making the calculations more accurate [102]. It must also be realized that the ratio of m_{0calc}/m_{0exp} is temperature dependent for some elements [100,103]. The use of more clearly defined instrumental operating conditions and refinement in the calculation of m_0 values have led to improved accuracy and precision of these calculated values. Consequently, the literature is showing increasing numbers of publications that demonstrate accurate absolute analysis.

1.6 CONCLUSIONS

The technique of ETAAS has undoubtedly reached maturity, with instrumentation that is sufficiently reproducible to permit absolute analysis in some cases, although most analyses continue to be done through calibration with standard solutions. Interference-free methods have been developed for most applications of the technique, including the direct analysis of solids and slurries. This area of application is considered sufficiently important to merit its own chapter, and Chapter 6 provides a discussion of the analysis of solids and slurries and associated applications. In spite of its maturity, research on ETAAS continues unabated. Improved knowledge of the physical and chemical processes occurring in ETAs and further improvements in instrumentation can be expected to lead to improved ease of use and even more accurate applications.

REFERENCES

[1] W. Slavin and G.R. Carnrick, *Spectrochim. Acta, Part B* **39B**, 271 (1984).
[2] B.V. L'vov, *Spectrochim. Acta* **17**, 761 (1961).
[3] B.V. L'vov, *Spectrochim. Acta, Part B* **24B**, 53 (1969).
[4] R. Woodriff, R.W. Stone and A.M. Held, *Appl. Spectrosc.* **22**, 408 (1968).
[5] R. Woodriff and G. Ramelow, *Spectrochim. Acta, Part B,* **23B**, 665 (1968).
[6] H. Massmann, *Spectrochim. Acta, Part B* **23B**, 215 (1968).
[7] D.C. Manning and F. Fernandez, *At. Absorpt. Newsl.* **9**, 65 (1970).
[8] B.V. L'vov, *Atomic Absorption Spectrochemical Analysis*, London: Adam Hilger, 1970.
[9] K. Ohta and M. Suzuki, *Talanta* **22**, 465 (1975).
[10] T.S. West and X.K. Williams, *Anal. Chim. Acta* **45**, 27 (1969).
[11] J.F. Alder and T.S. West, *Anal. Chim. Acta* **51**, 365 (1970).
[12] K.W. Jackson, T.S. West and L. Balchin, *Anal. Chem.* **45**, 249 (1973).
[13] J.P. Matousek, *Am. Lab.* 45 (1971).
[14] J.P. Matousek and K.G. Brodie, *Anal, Chem.* **45**, 1606 (1973).
[15] H.M. Donega and T.B. Burgess, *Anal. Chem.* **42**, 1521 (1970).
[16] J.Y. Hwang, P.A. Ullucci, S.B. Smith and A.L. Malenfant, *Anal. Chem.* **43**, 1319 (1971).
[17] M. Williams and E.H. Piepmeier, *Anal. Chem.* **44**, 1342 (1972).
[18] A. Montaser, S.R. Goode and S.R. Crouch, *Anal. Chem.* **46**, 599 (1974).

[19] B.V. L'vov, *Spectrochim. Acta, Part B* **33B**, 153 (1978).
[20] C.L. Chakrabarti, H.A. Hamed, C.C. Wan, W.C. Li, P.C. Bertels, D.C. Gregoire and S. Lee, *Anal. Chem.* **52**, 167 (1980).
[21] B.V. L'vov and L.A. Pelieva, *Zh. Anal. Khim.* **33**, 1572 (1978).
[22] J.M. Ottaway, J. Carroll, S. Cook, S.P. Corr, D. Littlejohn and J. Marshall, *Fresenius' Z. Anal. Chem.* **323** 742 (1986).
[23] B.V. L'vov, L.A. Pelieva and A.I. Sharnopolsky, *Zh. Prikl. Spektrosk.* **27**, 395 (1977).
[24] W. Slavin and D.C. Manning, *Spectrochim. Acta, Part B* **35B**, 701 (1980).
[25] W. Slavin, D.C. Manning and G.R. Carnrick, *At. Spectrosc.* **2**, 137 (1981).
[26] W. Frech and S. Jonsson, *Spectrochim. Acta, Part B* **37B**, 1021 (1982).
[27] W. Frech, D.C. Baxter and E. Lundberg, *J. Anal. At. Spectrom* **3**, 21 (1988).
[28] W. Frech, D.C. Baxter and B. Hutsch, *Anal. Chem.* **58**, 1973 (1986).
[29] M.F. Gine, F.J. Krug, V.A. Sass, B.F. Reis, J.A. Nobrega and H. Berndt, *J. Anal. At. Spectrom.* **8**, 243 (1993).
[30] F.J. Krug, M.M. Silva, P.V. Oliveira and J.A. Nobrega, *Spectrochim. Acta, Part B* **50B**, 1469 (1995).
[31] P.J. Parsons, H. Qiao, K.M. Aldous, E. Mills and W. Slavin, *Spectrochim. Acta, Part B* **50B**, 1475 (1995).
[32] C.L. Sanford, S.E. Thomas and B.T. Jones, *Appl. Spectrosc.* **50**, 174 (1996).
[33] G. Bozsai and M. Melegh, *Microchem. J.* **51**, 39 (1995).
[34] D.L. Tsalev, *Atomic Absorption Spectrometry in Occupational and Environmental Health Practice*, Boca Raton, FL: CRC Press, 1995.
[35] W. Slavin, *Graphite Furnace AAS. A Source Book*, Norwalk, CT: Perkin-Elmer, 1984.
[36] A. Varma, *CRC Handbook of Furnace Atomic Absorption Spectroscopy*, Boca Raton, FL: CRC Press, 1990.
[37] C. Minoia and S. Caroli (Eds), *Applications of Zeeman Graphite Furnace Atomic Absorption Spectrometry in the Chemical Laboratory and in Toxicology*, New York: Pergamon, 1992.
[38] B. Welz, S. Akman and G. Schlemmer, *J. Anal. At. Spectrom.* **2**, 793 (1987).
[39] B. Welz, G. Schlemmer and J.R. Mudakavi, *Anal. Chem.* **60**, 2567 (1988).
[40] S. Akman and G. Doner, *Spectrochim. Acta, Part B* **49B**, 665 (1994).
[41] T.M. Mahmood and K.W. Jackson, *Spectrochim. Acta, Part B* **51B**, 1155 (1996).
[42] K.W. Jackson and T.M. Mahmood, *24th Federation of Analytical Chemistry and Spectroscopy Societies meeting*, Providence, RI, October 1997, abstract no. 331.
[43] J.Y. Cabon and A. Le Bihan, *Spectrochim. Acta, Part B* **51B**, 619 (1996).
[44] H. Chuang and S.-D. Huang, *Spectrochim. Acta, Part B* **49B**, 283 (1994).
[45] C.R. Lan, *Analyst* **118**, 189 (1993).
[46] C.R. Lan and Z.B. Alfassi, *Analyst* **119**, 1033 (1994).
[47] S.D. Huang and K.Y. Shih, *Spectrochim. Acta, Part B* **48B**, 1451 (1993).
[48] Y.Z. Liang, M. Li and Z. Rao, *Fresenius' J. Anal. Chem.* **357**, 112 (1997).
[49] J.Y. Cabon and A. Le Bihan, *J. Anal. At. Spectrom.* **9**, 477 (1994).
[50] S. Sachsenberg, T. Klenke, W.E. Krumbein, H.J. Schellnhuber and E. Zeeck, *Anal. Chim. Acta* **279**, 241 (1993).
[51] E.A.-C. Cimadevilla, K. Wrobel and A. Sanz-Medel, *J. Anal. At. Spectrom.* **10**, 149 (1995).
[52] B. He and Z. Ni, *J. Anal. At. Spectrom*, **11**, 165 (1996).
[53] B. Welz, G. Schlemmer and J.R. Mudakavi, *J. Anal. At. Spectrom.* **7**, 1257 (1992).
[54] M.P. Bertenshaw, D. Gelsthorpe and K.C. Wheatstone, *Analyst* **107**, 163 (1982).
[55] U. Kurfuerst and J. Pauwels, *J. Anal. At. Spectrom.* **9**, 531 (1994).

[56] Z. Ni, R. Zhu and L. Mei, *Can. J. Appl. Spectrosc.* **41**, 129 (1996).
[57] Y.Y. Zong, P.J. Parsons and W. Slavin, *Spectrochim. Acta, Part B* **49B**, 1667 (1994).
[58] M.S. Epstein, G.C. Turk and L.J. Yu, *Spectrochim. Acta, Part B* **49B**, 1681 (1994).
[59] V. Porta, O. Abollino, B. Mentasti and C. Sarzanini, *J. Anal. At. Spectrom.* **6**, 119 (1991).
[60] L.C. Azeredo, R.E. Sturgeon and A.J. Curtius, *Spectrochim. Acta, Part B* **48B**, 91 (1993).
[61] A.K. Avila and A.J. Curtius, *J. Anal. At. Spectrom.* **9**, 543 (1994).
[62] T. Okutani, Y. Tsuruta and A. Sakuragawa, *Anal. Chem.* **65**, 1273 (1993).
[63] Z. Fang, M. Sperling and B. Welz, *J. Anal. At. Spectrom.* **5**, 639 (1990).
[64] M. Sperling, X. Yin and B. Welz, *J. Anal. At. Spectrom.* **6**, 295 (1991).
[65] B. Welz, X. Yin and M. Sperling, *Anal. Chim. Acta* **261**, 477 (1992).
[66] X. Yan and F. Adams, *J. Anal. At. Spectrom.* **12**, 459 (1997).
[67] Z. Fang and L. Dong, *J. Anal. At. Spectrom.* **7**, 439 (1992).
[68] Z. Zhuang, X. Wang, P. Yang, C. Yang and B. Huang, *Can. J. Appl. Spectrosc.* **39**, 101 (1994).
[69] J.L. Burguera and M. Burguera, *J. Anal. At. Spectrom.* **8**, 235 (1993).
[70] K.W. Jackson and G. Chen, *Anal. Chem.* **68**, 231R (1996).
[71] K.W. Jackson and S. Lu, *Anal. Chem.* **70**, 363R (1998).
[72] G. Drasch, L.V. Meyer and G. Kauert, *Fresenius' Z. Anal. Chem.* **304**, 141 (1980).
[73] P.S. Doidge, B.T. Sturman and T.M. Rettberg, *J. Anal. At. Spectrom.* **4**, 251 (1989).
[74] R.E. Sturgeon, S.N. Willie, G.I. Sproule, P.T. Robinson and S.S. Berman, *Spectrochim. Acta, Part B* **44B**, 667 (1989).
[75] G. Tao and Z. Fang, *Talanta* **42**, 375 (1995).
[76] M. Veber, K. Cujes and S. Gomiscek, *J. Anal. At. Spectrom.* **9**, 285 (1994).
[77] Z.-M. Ni and B. He, *J. Anal. At. Spectrom.* **10**, 747 (1995).
[78] M. Grotti and A. Mazzucotelli, *J. Anal. At. Spectrom.* **10**, 325 (1995).
[79] C.P. Hanna, G.R. Carnrick, S.A. McIntosh, L.C. Guyette and D.E. Bergemann, *At. Spectrosc.* **16**, 82 (1995).
[80] Y. Liao and H.O. Hoag, *Microchem. J.* **56**, 247 (1997).
[81] P. Bermejo-Barrera, J. Moreda-Pineiro, A. Moreda-Pineiro and A. Bermejo-Barrera, *J. Anal. At. Spectrom.* **11**, 1081 (1996).
[82] H.W. Sinemus, H.H. Stabel, B. Radziuk and J. Kleiner, *Spectrochim. Acta, Part B* **48B**, 1719 (1993).
[83] H. Matusiewicz and R.E. Sturgeon, *Spectrochim. Acta, Part B* **51B**, 377 (1996).
[84] Y. Chen, W. Qi, J. Cao and M. Chang, *J. Anal. At. Spectrom.* **8**, 379 (1993).
[85] S.N. Willie, *Spectrochim. Acta, Part B* **51B**, 1781 (1996).
[86] E. Beceiro-Gonzalez, P. Bermejo-Barrera, A. Bermejo-Barrera, J. Barciela-Garcia and C. Barciela-Alonso, *J. Anal. At. Spectrom.* **8**, 649 (1993).
[87] M. Chikuma, T. Tanaka and H. Tanaka, *Biomed. Res. Trace. Elem.* **3**, 205 (1992).
[88] J.C. Vidal, J.M. Sanz and J.R. Castillo, *Fresenius' Z. Anal. Chem.* **344**, 234 (1992).
[89] F. Laborda, D. Chakraborti, J.M. Mir and J.R. Castillo, *J. Anal. At. Spectrom.* **8**, 643 (1993).
[90] R.H. Atallah, G.D. Christian and A.E. Nevissi, *Anal. Lett.* **24**, 1483 (1991).
[91] G. Jiang, Z. Ni, L. Zhang, A. Li, H. Han and X. Shan, *J. Anal. At. Spectrom.* **7**, 447 (1992).
[92] M. Sperling, X. Yin and B. Welz, *Analyst* **117**, 629 (1992).
[93] D.J. Halls, *J. Anal. At. Spectrom.* **10**, 169 (1995).

[94] U.K. Kunwar, D. Littlejohn and D.J. Halls, *J. Anal. At. Spectrom.* **4**, 153 (1989).
[95] U.K. Kunwar, D. Littlejohn and D.J. Halls, *Talanta* **37**, 555 (1990).
[96] M.W. Hinds, K.E. Latimer and K.W. Jackson, *J. Anal. At. Spectrom.* **6**, 473 (1991).
[97] I. Lopez-Garcia, M. Sanchez-Merlos and M. Hemandez-Cordoba, *Spectrochim. Acta, Part B* **52B**, 437 (1997).
[98] B.V. L'vov, V.G. Nikolaev, E.A. Norman, L.K. Polzik and M. Mojica, *Spectrochim. Acta, Part B* **41B**, 1043 (1986).
[99] B.V. L'vov, *Spectrochim. Acta, Part B* **45B**, 633 (1990).
[100] W. Frech and D.C. Baxter, *Spectrochim. Acta, Part B* **45B**, 867 (1990).
[101] E.G. Su, A.I. Yuzefovsky, R.G. Michel, J.T. McCaffrey and W. Slavin, *Microchem. J.* **48**, 278 (1993).
[102] P. Hannaford, *Spectrochim. Acta, Part B* **49B**, 1581 (1994).
[103] Y. Zheng, X. Su and Z. Quan, *Appl. Spectrosc.* **47**, 1222 (1993).

2

Fundamental chemical and physical processes in electrothermal atomizers

James A. Holcombe and Albert Kh. Gilmutdinov

2.1 INTRODUCTION

The production of free gaseous atoms, for purposes of spectroscopic investigations within the confines of a heated graphite tube, promulgated the report of the first graphite furnace in the form of the King furnace [1]. It consisted of a long cylindrical graphite tube into which relatively large amounts of material could be heated to relatively high temperatures for studying the basic absorption characteristics of materials. L'vov [2,3] refined this basic concept and clearly demonstrated its applicability for the determination of trace constituents in aqueous solutions. The original constant-temperature cuvette, designed by L'vov, is described in Section 1.4, and illustrated in Fig. 1.3 L'vov's work is appropriately cited as the first demonstration of what became known as graphite furnace atomic absorption spectrometry (GFAAS), and is now generally referred to as electrothermal atomic absorption spectrometry (ETAAS). L'vov conducted a number of investigations of the fundamental processes associated both with atom formation and spectroscopic features of the free metals formed within this environment. These were performed long before ETAAS became an accepted analytical tool. A summary of these basic studies is archived in his classic 1970 book [4].

Since this early work with electrothermal atomization, graphite continues to dominate as the primary construction material of these devices, even though various other high-temperature materials have been used

Electrothermal Atomization for Analytical Atomic Spectrometry. Edited by K. W. Jackson
© 1999 John Wiley & Sons Ltd

successfully in many instances. Similarly, a hollow cylindrical design (e.g. a tube) remains the dominant geometry, although graphite rods, filaments, flats, foils and braids have also shown utility for particular applications or fundamental studies. In an attempt to keep focused on the more widely accepted material and design, graphite tubes will be the primary emphasis of this chapter whenever material or geometry is a consideration.

It is not the intent of this chapter to present a historical perspective of the development of modern electrothermal atomization or to portray a comprehensive tutorial on the entire scope of carbon chemistry. Instead the intention is to highlight the salient features of carbon-based materials commonly used for electrothermal atomizers (ETAs) employed in atomic absorption (AA) spectrochemical analysis, and to consider the basic processes occurring prior to and during the formation of the free metal for the purpose of performing spectrochemical analysis. This chapter discusses some of the basic chemistry and physics associated with graphite, as they relate to possible interactions within the graphite furnace. A detailed look at oxygen and its interactions with graphite will be highlighted, as well as the preatomization chemistry taking place prior to the onset of analyte entrance into the gas phase. Some spatial views will be presented, which will hopefully provide the reader with an appreciation of what is taking place within the furnace at the molecular level. Finally, the atomization process itself and the transport of material within the furnace confines, will be presented as they relate to the formation of the absorbance signal. It is not the intent to cover the vast amount of research focused on atomization mechanisms of particular elements, but rather to provide a broad overview of possibilities and some approaches that have been employed in attempts to elucidate the processes.

Much of the fundamental and applied research that was initially focused on improvements in ETAAS now finds applicability in rapidly generating a transient vapor plume for microanalysis by other techniques. Thus, the basic precepts and designs originally developed for ETAAS now serve as a foundation for use of the device as a general means of sample introduction.

2.1.1 Reason for understanding

As in any scientific discipline, there exists a curiosity, if not a responsibility, to understand the way in which a commonly used tool functions. For many users, it is sufficient to treat the tool as a simple transduction device into which a sample is deposited and out of which an answer emerges. However, in order to stretch an analytical tool to its limits of detectability and applicability for complex samples, a basic understanding of the technique is useful to minimize time-consuming empiricism. Likewise, knowledge of the fundamentals can often provide large benefits in solving problems when the system does not perform as desired, and indicates the direction for improvements in the next generation of instrumentation. An awareness of the fundamental limitations can avoid directing effort towards a futile objective. For example, it is now well recognized that atomization efficiency approaches unity for a large number of elements, so there is little purpose in attempting to alter furnace design or operation in hopes of significantly lowering limits of detection by enhancing the atomization efficiency. Likewise, by recognizing the relationship between signal magnitude and furnace geometry, judicious decisions can be made as to whether time is well invested in pursuing alternative designs.

2.1.2 Common misconceptions

There are a number of common misunderstandings regarding the chemistry and physics associated with the production of free atoms in ETAAS. Many of these are not totally unfounded and are often valid for selected cases. However, the generalities are often carried too far, partially in attempts to create a useful paradigm from which to take the next prognosticative step. At the risk of making a complex issue appear even more complex, some common misconceptions of what takes place during atomization are given below:

- Carbon is a good reducing agent and converts all metal salts to the free metal.
- With the exception of its reducing capability noted above, a pyrolytic graphite surface is chemically inert and free of chemical reactions.
- The free metal, once created, exists on the surface as a solid or liquid and exhibits properties predictable by physicochemical characteristics associated with bulk amounts of the same material.
- The analyte in the system rests on the surface of the graphite if a nonporous graphite such as pyrolytic graphite is employed on the furnace surface.
- The furnace environment into which the analyte is ultimately vaporized is oxygen free, because the hot graphite reduces any residual oxygen and the system is shrouded in an inert gas sheath.
- Once the metal vaporizes, it acts like an ideal gas within an open-ended cylindrical container.
- Analyte loss is from diffusion out of the ends of the furnace.

Some of the items listed are valid for some situations, but there is certainly a complex array of chemistry taking place that invalidates most of the assumptions made. Figure 2.1 portrays some of the possible interactions that can occur within the furnace to a greater or lesser extent, depending on the analyte considered. This figure considers only processes involved

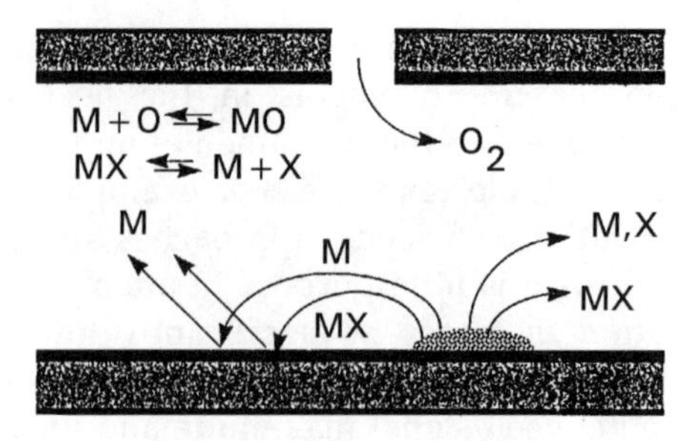

Figure 2.1 Some of the many interactions that can take place within an electrothermal atomizer. Depicted here are: vaporization of metal from a 'bulk' amount of material (e.g. a microdroplet); possible gas-phase interactions to form compounds with species originating from the sample matrix (e.g. X) and with oxygen either from entrained air, sheath gas impurities or sample thermal decomposition; the readsorption onto the walls and desorption from these sites; and compound vaporization (i.e. MX). All gas-phase species can be considered as undergoing reactions in a drive towards equilibrium.

with gaseous species. An equally varied chemistry takes place in the condensed state preceding the production of any vapors.

2.2 CARBON AS A FURNACE MATERIAL

Several attributes of a material that should be considered for the ideal furnace are:

- high melting/sublimation point
- available in high purity/easily purified
- not easily contaminated by samples
- nonporous
- good reducing agent
- chemically inert
- high electrical resistivity
- high thermal conductivity.

Carbon, and more specifically graphite, can be fabricated to exhibit a wide range of the properties noted in this table, and therefore a brief discussion of its desirable features for an atomizer is appropriate.

The need for a high-purity refractory material is obvious, as atomization is generally accomplished through resistive heating of the furnace, and ultratrace analysis is the primary objective. The ability to easily purify the material can greatly affect the cost of the furnace (a disposable item). In the case of graphite, this is generally done through the introduction of a fluorine-containing compound during the fabrication process. This converts many of the metals to metal fluorides, which are generally volatile and easily purged from the furnace through high-temperature treatments.

Oxygen, a common source of potential contamination of the surface of any atomizer, is introduced either through diffusion of air into the furnace or via decomposition of oxyanion salts of the analyte or the matrix:

$$2M(NO_3)_2 \rightarrow 2MO + 4NO_2 + O_2 \tag{2.1}$$

$$2M(SO_4) \rightarrow 2MO + 2SO_2 + O_2 \tag{2.2}$$

Unlike the situation in atomizers made from metals, oxides formed on the graphite surface are readily removed by heating to eliminate this contamination by evolution of carbon monoxide and carbon dioxide. Other species, such as metal carbides, may not be so easily removed from the graphite, although metal contamination can also present problems in metal-based atomizers through the formation of intermetallic compounds or solid solutions with the base metal. Metal liners within a graphite tube have been used to minimize analyte or matrix retention, by elimination of these refractory carbides. An example is the use of a tantalum liner in the determination of molybdenum.

The carbon-based material used in making the furnace should retain totally both the deposited liquid sample and the generated gaseous vapors. Thus, the permeability must be low. The material could be highly porous, but this should present no problems if the pore diameters are sufficiently small to prevent analyte loss through the furnace walls; for example, glassy carbon is a low-density highly porous material with low permeability. In this material the pore sizes can average about 1–3 nm in diameter [5], which is often sufficiently small to minimize significant diffusion of analyte and migration of solution into the furnace material. In contrast, graphite is very nonporous but the spaces between the graphitic domains in electrographite (the base ETA material) may be large enough to present macroscopic pores and access of liquids and vapors to areas below the surface. The main objective is to minimize both the capillary absorption of the sample solution and the extent of analyte vapor diffusion through the furnace walls. The former often causes broadening of the absorbance signal, whereas the latter represents another dissipation route for the analyte and can lead to attenuation of the analytical signal. However, Katskov *et al.* [6] have reported on a novel filter furnace in which such diffusion though a porous graphite barrier within the furnace was used advantageously to improve atomization.

The next two requirements are somewhat contradictory. A material that is a good reducing agent cannot also be considered chemically inert. The former is desirable for the conversion of analyte-containing compounds (such as metal oxides) to the free metal. The reduction capabilities also serve to minimize the partial pressure of any entrained oxygen, and hence to minimize the formation of analyte oxides in the gas phase. It would be ideal if there were no additional interactions between the analyte and the carbon beyond these desirable reactions. However, the reactive character of the graphite can continue to participate in the heterogeneous chemistry within the furnace and result in chemisorption of material to the surface, formation of refractory metal carbides, and formation of intercalation compounds in the interlamellar regions between the graphite planes. There remains some debate as to which type of surface would be preferable, but present indications are toward a less reactive surface, with the hopes that much of the dissociation needed to produce the free gaseous metal vapor will be simple thermal dissociation.

An elevated electrical resistivity (without being insulating) permits delivery of large power requirements (P) without the need for large currents. Impedance matching is generally employed to optimize the power delivery. The rms voltage (V) and current supplied (I) are given by

$$V = \sqrt{PR_T} \tag{2.3}$$

$$I = \frac{V}{R_T} = \frac{P}{V} \tag{2.4}$$

The temperature-dependent resistance of the furnace, R_T, is governed by the type of graphite considered and the manufacturing processes employed. The temperature-dependent resistivity of graphite, shown in Figure 2.2, means that optimal power delivery cannot be ensured over the entire operating temperature range of the furnace. As an example of

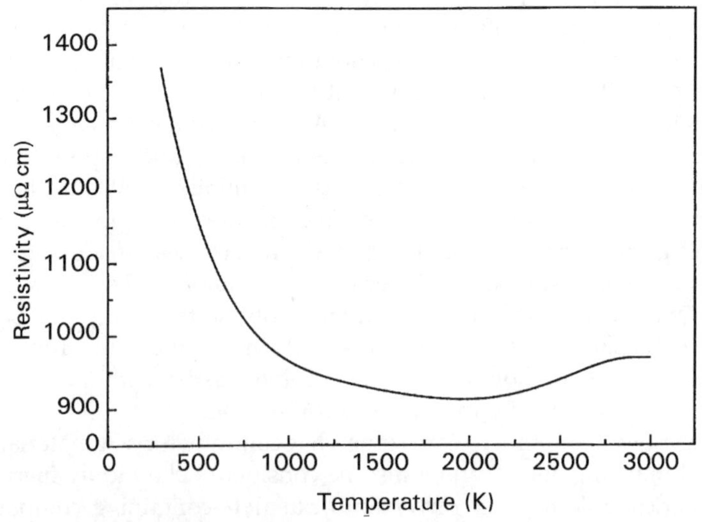

Figure 2.2 Temperature-dependent resistivity of electrographite (reproduced courtesy of Ringsdorff Werke).

typical current and voltages, a furnace whose nominal resistance is 0.010 Ω would require 7 V and 715 A to deliver 5 kW. In contrast, only 71 A at 70 V would be needed for a resistance of 1 Ω. Typical furnaces are presently better characterized by the lower of the two example resistances.

Generally, lower electrical conductivity is accompanied by lower thermal conductivity for graphite. High thermal conductivity minimizes hot and cold spots in the furnace, but introduces problems. For example, where electrical contact is made at the ends of the furnace through water-cooled connections, it assists in transporting thermal energy out of the ends of the furnace to these contacts, and creates a thermal gradient from the center to the ends of the furnace [7,8]. Some furnace designs are discussed below to illustrate how these parameters affect the furnace heating and, to some degree, the analytical signal.

2.2.1 Types of carbon and terminology

There are three allotropic forms of carbon, each having its own structure and physicochemical properties – carbon, graphite, and diamond. The insulating characteristics (and cost!) of diamond discourages its use in ETA designs, and the reactive character and mechanical stability of carbon likewise minimizes its utility. Thus, it would appear that only one material needs to be characterized to adequately understand the role played by the atomizer in the formation of atoms, i.e. graphite. In reality, the microscopic character of graphitic materials comprises a host of subsets. The 24 volume book series *Chemistry and Physics of Carbon* [9], together with the journal *Carbon*, demonstrate the complexity of the material, and illustrate both the wealth of information known about this material and the magnitude of our ignorance.

Graphite is nominally a two-dimensional crystal lattice, composed of hexagonal rings of carbon with the bonding electrons forming a nominal bond order of 1.5 between carbons. Figure 2.3 shows a stylized view of the typical structure often used to illustrate graphite. Also shown are relative

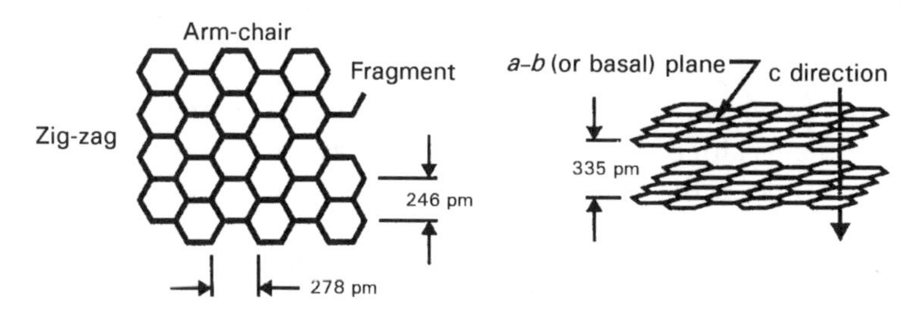

Figure 2.3 Diagram of a basal plane of graphite showing typical interatomic distances and various types of carbon centers.

bond distances as well as the interplanar spacing between graphitic sheets, which are held together by relatively weak van der Waals forces. Figure 2.4 shows a more accurate depiction of graphite, with the staggered orientation of the carbons as viewed along the *c* axis. The material is highly anisotropic as a result of its structure. Reference is commonly given to the *a–b* or basal plane of the graphite, which is considered the surface onto which the sample is deposited. Many materials, including many metals, can enter the interlammelar regions between these planes, which results in intercalation compound formation. [10] These compounds can be relatively stable and often possess well defined crystallographic structure in the interplanar spacing. It is unlikely that atom and molecule migration occurs through the basal plane. More probably, entry occurs at the interplanar region through grain boundaries, cracks and imperfections in the surface. In many cases, the compounds formed in these regions produce an increase in the interplanar spacing and can cause a swelling or delamination of the graphite.

Figure 2.3 also illustrates the edge carbons (arm-chair and zig-zag conformation) as well as fragments, all of which are often generically referred to as the active sites of the graphite. These regions are more chemically reactive than those carbons located in the basal plane of the graphite with sp^2 bonding. They are electron rich and are the first to become involved in chemical interactions between the sample or concomitants and the graphite surface. Such interactions range from chemisorption (dissociative or associative) to chemical reduction.

At least five different types of active sites have been suggested to exist on the graphite surface [11]. As with any other crystalline surface, imperfections exist (kinks, dislocations, etc.), which give rise to many of the active sites. In many fundamental studies of graphite, active sites are often created by conducting a slow, controlled burn-off. A graphitic material is heated in an oxygen environment at a temperature capable of producing slow oxidation or combustion of the carbon to form carbon monoxide and carbon dioxide, and leaving behind a reactive carbon (e.g., edges or fragments, possibly with surface bonded oxygen). Many functional groups (lactones, carboxylic acids, etc.) have been identified on the graphite surface after reactions with oxygen or oxygen-containing compounds such as water [12,13]. A discussion of elementary graphite surface chemistry is well beyond the scope of this chapter. It is only important to point out the wealth of chemistry that can occur on the surface, and that may play a role in the interactions between the surface and the sample introduced for analysis.

It is difficult to accurately characterize graphite, as the various forms and the manner in which it is fabricated can greatly alter its physical and chemical character. Most modern furnaces are composed of a base material composed of a composite graphite, called electrographite. The

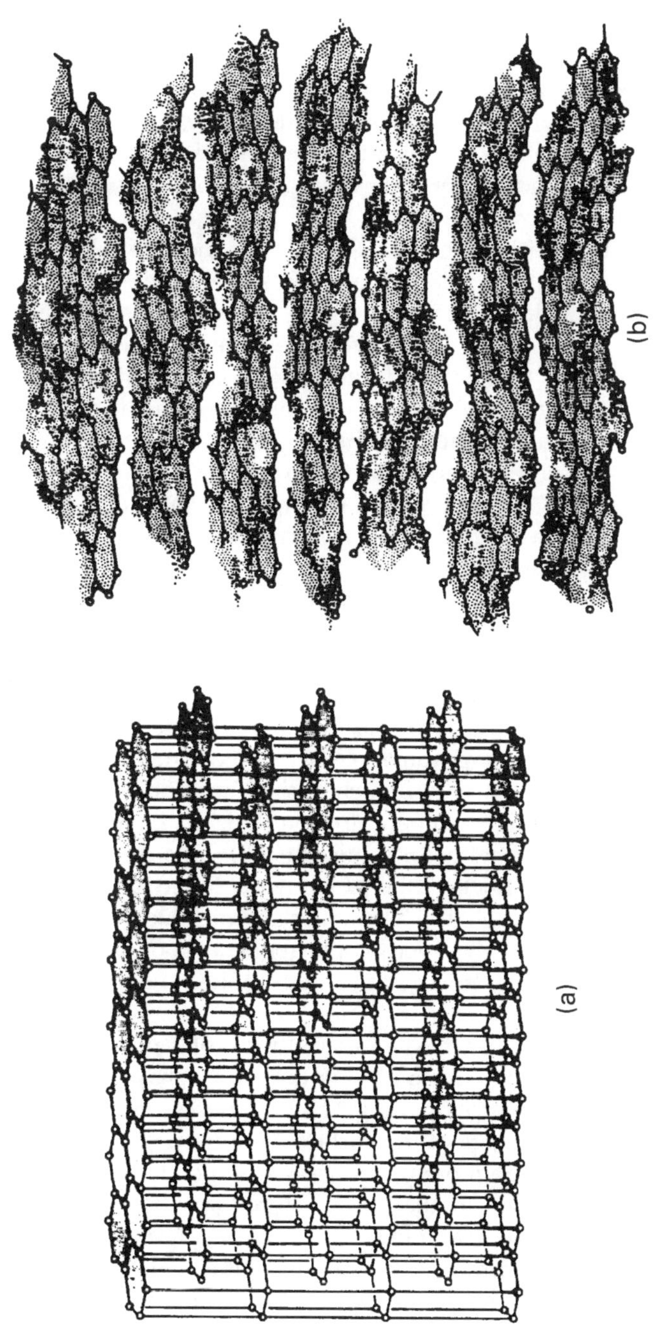

Figure 2.4 Structural diagrams of (a) three-dimensional graphite lattice; (b) turbostratic structure. Notice the imperfections and defects in (b). (Reprinted from Reference 189, p. 191, by courtesy of Marcel Dekker Inc.)

manufacturing process for the formation of this base material varies and is often proprietary. It generally begins with a hydrocarbon base material, such as coke, which is graphitizable, i.e. capable of loosing hydrogen upon heating and taking on the crystalline structure of graphite. As not all stock material can be converted to graphite when heated to high temperatures, the stock plays an important role in the final product. It generally undergoes a number of heating and aging steps leading to the point of forming the graphite stock.

The graphitized stock is then machined into the desired shapes and purified, usually by treatment with a fluorine-containing compound. During this period, additional heating cycles of the resulting furnaces may be performed to improve the stability of the material. This furnace blank will have electrical, thermal and mechanical characteristics that will dominate the final furnace. Although this electrographite furnace consists of relatively dense anisotropic graphite, it is still macroscopically porous and generally not the preferred form for use with modern graphite furnace applications. However, for those studies that require 'uncoated furnaces', this is the furnace material that is often used.

To minimize the porous character and reactivity of the surface, a graphite coating is added. This is accomplished by chemical vapor deposition (CVD) using a hydrocarbon gas and elevated temperatures to lay down a pyrolytic graphite surface. The particular hydrocarbon gas or gas mixture, their partial pressures and the treatment temperatures are once again unique to the manufacturers' process, and are often treated as proprietary information. The pyrolytic graphite surface is more dense than the electrographite base and often approaches the theoretical maximum for graphite. During the high temperature pyrolytic CVD step, the carbon deposits as prismatic growth cones (Figure 2.5). This deposited layer is generally 0.05–0.10 mm thick and, if properly deposited, adheres firmly to the underlying electrographite. The manner in which the pyrolytic coating is deposited gives rise to the microscopic appearance of nodules that are typically seen in scanning micrographs of graphite tube furnace surfaces (Figure 2.6).

Some platforms and furnaces are made from total pyrolytic graphite, which consists of CVD onto a mandrill of the proper shape, such as a tube or plane. Although these pyrolytic graphite surfaces are relatively inert and have a large percentage of the surface comprised of the basal plane, they do not approach the more perfect crystal structure found in natural single-crystal graphite or in the manufactured material referred to as highly oriented pyrolytic graphite (HOPG). Over a limited area, the last two surfaces can be atomically smooth, as witnessed by the scanning tunneling microscopic trace of an HOPG surface (Figure 2.7(a)). In contrast, the typical surface of a pyrocoated tube is much rougher (Figure 2.7(b)). The presence of the imperfections and grain boundaries inherent

Figure 2.5 Prismatic growth cones typical of the macroscopic structure of a graphitic surface produced by chemical vapor deposition. Shown is an electron micrograph of a cross-section through a pyrolytic coating (reproduced courtesy of Ringsdorff Werke).

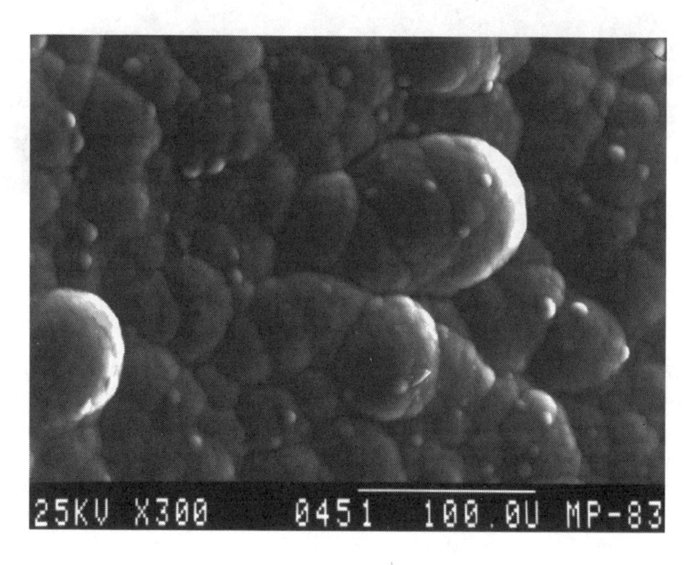

Figure 2.6 Scanning electron micrograph of a pyrocoated graphite surface of a graphite furnace atomizer [190]. (Reprinted from *Spectrochim. Acta, Part B*, Vol. **40B**, H. M. Ortner, G. Schlemmer, B. Welz and W. Wegsheider, Scanning Electron Microscopy Studies on Surfaces from Electrothermal Atomic Absorption Spectrometry – I. Polycrystaline Electrographite Tubes With and Without Pyrographite Coating, pp. 959–977, 1985, with kind permission of Elsevier Science – NL, Sara Burgerhartstraat 25, 1055 KV Amsterdam, The Netherlands.)

Figure 2.7 Scanning tunneling micrographs: (a) 400 × 400 nm area of highly oriented pyrolytic graphite (HOPG) with arrows denoting single layer steps; (b) a topographical display of a 1.1 × 1.1 μm area of a pyrocoated graphite tube surface [191]. (Reprinted by permission of the Society for Applied Spectroscopy from, K.G. Vandervoort, D.J. Butcher, C.T. Brittain and B.B. Lewis, *Scanning Tunneling Microscope Images of Graphite Substrates used in Graphite Furnace Atomic Absorption Spectrometry, Appl. Spectrosc.*, **50**, 928–938 (1996).)

in pyrocoated graphite may result in many of its chemical and physical properties, as they relate to this material's role in ETA, as will be seen later.

For the sake of completion, glassy carbon (or vitreous carbon) should also be mentioned. This is an amorphous material, generally produced by the pyrolysis of polyvinyl alcohols that have been cast into the shape of the final product. This material has been considered in several studies as a possible furnace material because of its microporosity and high electrical resistivity. Although it lacks crystallinity, the covalent bonding gives it a unique hardness and mechanical strength. However, it was found to be relatively reactive at elevated temperatures, and was not pursued further as a serious contender to graphitic materials for the manufacturing of furnaces.

2.2.2 Physical properties and structure

Table 2.1 lists some of the physical properties of graphite. As noted above, many of the characteristics can vary significantly, depending on the manufacturing process used in their production, and some of the values listed should be considered as representative only.

2.2.3 Reactivity

The most commonly studied reaction with graphite is oxidation, usually by oxygen, and is sometimes referred to as gasification. Generally, the oxidation of graphite occurs via the dissociative chemisorption of oxygen, followed by desorption of carbon monoxide or carbon dioxide from the surface at elevated temperatures. Desorption of the oxidation products is relatively rapid at around 1200K. The burning or combustion of graphite

Table 2.1 Selected properties of graphite. Note that some properties vary widely, depending on the manufacturing technique used to fabricate the graphite material. Many properties are temperature dependent and, unless noted, values are given for 298 K.

	Property		
Material	*Density* ($g\ cm^{-3}$)	*Electrical resistivity* ($\mu\Omega\ cm$)	*Thermal conductivity* ($W cm\ K^{-1}$)
Isotropic graphite or electrographite [14]	1.7	880	0.55
Pyrolytic graphite (*a*–*b* plane)	2.18–2.22 [16]	500[17]	3–6 [20,21]
Pyrolytic graphite (*c* direction)	2.18–2.22 [16]	3×10^5–1×10^6 [17]	0.02–0.03 [20]
Single-crystal graphite (*a*–*b* plane)	2.267	40 [18]	2 + 5 [22]
Single crystal graphite (*c* plane)	2.267	1×10^4–1×10^6 [19]	0.4–0.8 [22]
Glassy carbon [15]	1.3–1.55	1000–5000	0.04–0.17

takes place when the reaction sequence becomes rate limited by the mass transport of oxygen to the graphite surface. Thus, even though the oxidation of carbon by oxygen is thermodynamically favored even at ambient temperatures ($C_{(s)} + O_{2_{(g)}} \rightarrow CO_{2(g)}$), the reaction is often rate limited and provides a good example of when thermodynamics are insufficient for a complete understanding of what is occurring within the furnace confines. The gasification of graphite can be catalyzed by a number of chemicals including several metals [23–27]. Likewise, the oxidation can occur from oxidizing solutions, including oxidizing acids. As with oxygen, the oxidation of the graphite by these species often results in surface oxides that are later desorbed as the surface temperature increases. In general, the acid–base character of the surface depends on how the surface is oxidized (e.g. high temperature versus low-temperature oxidation with oxygen), although little use of this fact has been employed in explanations of the fundamental processes leading to free atom formation in electrothermal atomization studies. An oxygenated surface also tends to be more hydrophillic, and solutions often noticeably wet the surface.

Hydrogen is another gas that tenaciously chemisorbs to graphite [28]. Interestingly, the desorption of hydrogen often occurs at a higher temperature than the desorption of oxygen. It has been shown that the residual water from a sample deposited on graphite produces hydrogen when graphite is heated [29,30]. Not surprisingly, hydrocarbons that decompose before they vaporize similarly produce hydrogen. This has been shown to be the case when using modifiers such as citrate salts [31]. A graphite surface treated with hydrogen with the intent to chemisorb the gas to the surface is relatively hydrophobic and causes deposited sample solutions to bead up on the surface, unlike chemisorbed oxygen which causes the sample solution to spread.

In general, the wettability of the surface (i.e. the interfacial tension between sample and surface) varies considerably depending on the surface and the sample. For example, many organic solvents (including alcohols and ketones) and many strongly acidic solutions wet the surface, although some only do so significantly above ambient temperatures. This simple physical effect causes different sample deposition patterns than might be expected from simple aqueous solutions on a clean graphite surface.

Oxygen, as well as other oxidative compounds, can dramatically alter the surface of graphite by creating highly reactive surface sites after desorption of gaseous carbon oxides. Interestingly, the graphite surface also anneals itself at elevated temperatures. This process occurs in an effort to minimize the surface free energy through bond rearrangement. The process becomes significant at temperatures above 2500K [28]. This is another particularly attractive feature of graphite as a furnace material, and probably contributes to the long-term reproducibility associated with electrothermal atomization using graphite-based furnaces.

2.3 OXYGEN IN ELECTROTHERMAL ATOMIZERS

After carbon, oxygen is probably the second element of general importance for ETAAS. The importance originates first because oxygen is present in a variety of forms during the electrothermal atomization of practically all elements. Second, at elevated temperatures oxygen is extremely reactive, and it has been shown to affect almost all aspects of the atomization process [32–42]. Indeed, two major mechanisms of analyte atom formation include analyte oxide thermal dissociation and reduction of analyte oxides by carbon. The former atomization path is directly influenced by gas phase oxygen, as a result of suppression of the rate of condensed phase thermal dissociation of analyte oxides in response to an increase in oxygen partial pressure [32]:

$$M_xO_{y(s,l)} \rightarrow xM_{(g)} + y/2O_2 \tag{2.5}$$

The latter atomization mechanism is affected by oxygen indirectly via alteration of the graphite surface [34–36]. In this case the analyte reduction

$$M_xO_{y(s,l)} + yC \rightarrow xM_{(s,l)} + yCO \tag{2.6}$$

occurs preferentially at the surface active sites and the rate of this reaction determines the appearance temperature of the metal. Extensive shifts in appearance temperature for highly volatile elements, such as cadmium, lead or zinc, that vaporize between 750K and 1200K, may occur through partial blockage of the surface active sites as a result of chemisorption of oxygen onto active sites over this temperature range. Reduction must now occur at less active sites that are kinetically slower or thermodynamically less favorable. Therefore a higher temperature is required before the reaction becomes favorable. This 'oxygen' may manifest itself as a shift of the absorbance pulse to higher temperatures.

In addition to affecting an analyte in the condensed state, oxygen may also noticeably change the free-atom population of some elements by forming stable gas phase oxides (aluminum, silicon, tin), resulting in decreased sensitivity of determination for the analyte:

$$2M_{(g)} + O_2 \rightarrow 2MO_{(g)} \tag{2.7}$$

Finally, oxygen has been suggested for use as a modifier to promote ashing of biological samples [35]. In the presence of an unreactive sheath gas (normally argon), thermal pretreatment may pyrolyze many organic materials, resulting in the probable production of a carbon residue within the atomizer. In the presence of oxygen, many of these same compounds will undergo combustion with the production of carbon monoxide, carbon dioxide, and water.

Previous studies have shown four potential sources of gaseous oxygen in ETAs: (1) impurities in the sheath gas [32]; (2) ingress of oxygen through the sample introduction hole [37,38]; (3) decomposition of the sample matrix [32,39,40]; and (4) other possible sources, such as diffusion through the atomizer walls and ingress through cracks or pores at the contact point between the tube and the graphite support ring [41].

Extensive experimental investigations by Sturgeon and Falk have shown that, for Massmann pyrocoated furnaces (see Section 1.5.1), the major source of free oxygen is the ingress of oxygen through the sample introduction hole in a dry and unloaded furnace [38], and from sample matrix decomposition in a loaded furnace [33,39,40]. The evolution of oxygen during thermal decomposition of oxyanion salts is well documented [43] (Eqns (2.1) and (2.2)). the nitrogen dioxide decomposition product is capable of oxidizing most metals to the metal oxide at elevated temperatures. It has also been proved that oxidants generated during sample matrix decomposition evolve into the gas phase without complete reduction by the graphite [39,40]. The reduction of large amounts of oxidants by carbon is kinetically controlled, and appears to be incomplete in a pyrolytically coated graphite tube even at temperatures as high as 2000K [41].

Although the importance of oxygen in electrothermal atomization was generally recognized, for a long time there was no consensus on the partial pressure of oxygen, P_{O_2}, within the atomizer. In fact, estimates of P_{O_2} in graphite atomizers differed by as much as 12 orders of magnitude at a temperature of 2500K [42].

All early attempts to estimate P_{O_2} in ETAs have been limited by two factors. The first limitation arises from reliance on thermodynamically constructed models which must assume the establishment of thermodynamic equilibrium. However, as has been clearly shown by Sturgeon and co-workers [38,40,41], the graphite furnace is not a closed system and thermodynamic equilibrium is not established until high temperatures are reached. Therefore, a more appropriate description of oxygen in ETAs can be provided by a kinetically constructed model. The second limitation for all previous considerations is that the results obtained are spatially integrated. These results provide no information about the distribution of oxygen within ETAs. As noted in Section 2.6, the distributions of species in tube atomizers may be strongly nonuniform, i.e. dependent on the particular spatial location within the furnace. Therefore, a more complete description of oxygen in ETAs would include information on its distribution within the atomizer.

The concentration of free oxygen molecules and their distribution in the gas phase of a sample-free ETA are determined primarily by three factors [44]: (1) three-dimensional diffusion of the oxygen molecules in the atomizer volume, which is characterized by the temperature-dependent diffusion coefficient, $D(T)$; (2) kinetics of heterogeneous reaction of oxygen at the atomizer walls, which is characterized by the temperature dependent rate constant, $k(T)$; and (3) ingress of oxygen, at a rate of J (molecules per second), through the sample introduction hole combined with oxygen present in the sheath gas as an impurity. Obviously, these parameters should be well defined before developing a model of oxygen distribution in an ETA.

2.3.1 Gas-phase diffusion coefficients

The temperature dependence of the diffusion coefficient for oxygen in argon can be expressed by the power function

$$D(T) = D_0 T^{\alpha} \tag{2.8}$$

where D_0 is the diffusion coefficient at 298K and α is the power dependence that generally falls within the range 1.5–2.0. The relationship generally produces an error not exceeding 3 per cent over a temperature range of 243–10 000K [45].

2.3.2 Rate constants for oxidation of pyrolytic graphite

The $O_{2(g)}$–$C_{(s)}$ interaction is one of the most researched systems in heterogeneous kinetics. In its simplest form, it involves the reaction of gaseous oxygen with solid graphite to form two gaseous products, carbon monoxide and carbon dioxide:

$$2C_a + O_2 = 2CO \tag{2.9}$$

$$C_a + O_2 = CO_2 \tag{2.10}$$

where C_a is an active site on the graphite surface where reaction is possible.

The simplicity of the above stoichiometric relationships are misleading and obscure the complexity of the oxygen–graphite interaction; for example, Harry [46] stated that Eqn (2.9) requires at least 10 intermediate steps to give a complete kinetic description of the processes that are occurring. Qualitatively, the graphite oxidation can be described as follows. The reaction originates with the dissociative chemisorption of oxygen onto graphite active sites. These active carbon atoms have unpaired σ electrons, and therefore their reactivity is several orders of magnitude greater than those of atoms in the basal plane. As a result, a majority of surface chemistry is initiated at the edge carbons and at the lattice imperfections. Also, lattice imperfections tend to concentrate impurities by surface diffusion at elevated temperatures, and small amounts of metal impurities such as copper, chromium, lead and some others are known to catalyze the oxidation of graphite [23]. Once chemisorbed, the oxygen is present as a stable surface oxide that renders the site inactive, and only the sites not covered by oxygen remain available for reaction. Therefore, surface reactivity decreases as the stable surface oxides accumulate on the graphite surface. The optimum temperature for the formation of surface oxides is about 850K. Direct measurements of the amount of oxygen on the graphite substrate have shown that the C:O atom ratio at this temperature is about 60, more than two times smaller than at room temperature [47]. The basal plane can also physically adsorb a significant amount of oxygen, thereby acting as a collection area for these species. The species subsequently migrate to the edge carbon atoms, dissociatively chemisorb to the carbon, and are then desorbed, primarily as carbon monoxide, at a higher temperature. Desorption of surface oxides is essentially complete only at temperatures above 1200K. It is also well recognized that the reactivity of carbon may vary by as much as three orders of magnitude, depending on the type of graphite that is used [48].

The most important type of graphite for studies in electrothermal atomization is pyrolytically coated graphite. A comprehensive description of the interaction of pyrolytic graphite with oxygen at elevated temperatures (1300–2700 K), and at partial pressures of oxygen up to 0.2 atm is available [49,50]. The experimental data can be presented as the rate of oxygen loss (molecules per square centimetre per second) that occurs on a unique pyrolytic graphite [44]:

$$\text{rate of loss of oxygen} = k(T)P_{O_2} \tag{2.11}$$

where $k(T)$ (cm s^{-1}) is the first-order rate constant if P_{O_2} is expressed in molecules cm^{-3}.

Figure 2.8 shows a plot of the rate constant k versus temperature T versus $\log(P_{O_2})$, constructed from the data [49,50]. The figure shows that oxidation of graphite is not first order with respect to oxygen over the entire range of temperatures and partial pressures of oxygen, as the first-order approximation is valid only where $k(T)$ is constant with respect to pressure. There are two regions where the first-order approximation is valid – one at high oxygen partial pressures and relatively low temperatures (region I in Figure 2.8) and the other at low oxygen partial pressures and high temperatures (region III). In the intermediate region II, the reactivity of pyrolytic graphite appears to be anomalous i.e. there is a decrease in reactivity with an increase in temperature. According to Nagle and Strickland-Constable [49], these results can be attributed to the oxidation reaction occurring on two types of sites, reactive A sites and less-reactive B sites on the graphite surface. The B sites are produced from the A sites by a temperature-dependent annealing process that results in the decrease in reactivity at higher temperatures.

Operating conditions for electrothermal atomization are in the low oxygen partial pressure and high temperature region (region III), where first-order kinetics are valid. Also, under these conditions, the major product formed is carbon monoxide and Eqn (2.9) dominates. The first-

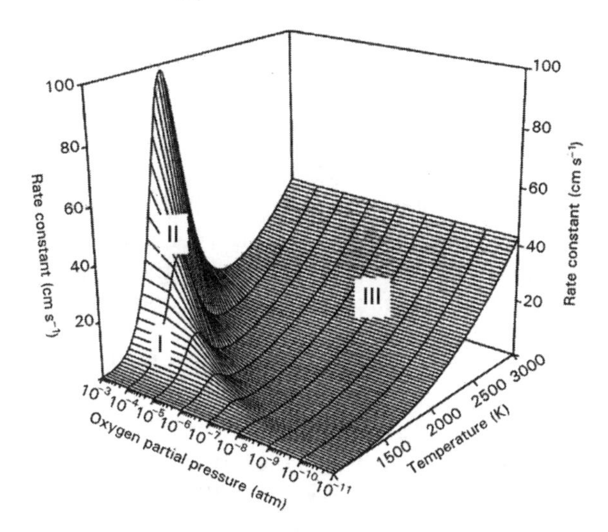

Figure 2.8 The rate constant for the oxygen–pyrolytic graphite reaction versus temperature and partial pressure of oxygen.

order rate constant $k(T)$ can be described adequately by the Arrhenius-type equation

$$k(T) = bT\exp\left(-\frac{c}{T}\right) \tag{2.12}$$

The rate of oxidation of the other high-temperature materials, tantalum and tungsten, used in ETAs can also be expressed with reasonable accuracy as first-order rate equations with a similar Arrhenius-type rate constant [44]. Numerical values obtained from Eqn 2.12 showed surprisingly low values for the rate constants of oxidation of pyrolytic graphite of $4.7\,\mathrm{cm\,s^{-1}}$ and $15.1\,\mathrm{cm\,s^{-1}}$ for the temperatures 1800K and 2300K, respectively. The rate constants for tantalum and tungsten are about three orders of magnitude greater [44] The consequences of these values for both the concentration of free oxygen and its distribution within tube-type atomizers is analyzed below.

2.3.3 Flux of oxygen through the sample introduction hole

The major resistance to oxygen ingress in the Massmann-type furnaces is presented by the external purge gas flow, that has an effective linear velocity of v (cm s^{-1}) within the sample introduction hole. It has been shown [44] that the flux J of ingressing oxygen molecules through the sample introduction hole, that has an effective length of l (cm) and a radius of a (cm), can be presented by

$$J_{O_2}(\text{molecules per second}) = \frac{\pi a^2 D n_0}{l} \times Pe^{-P} = J_{\text{dif}} \times Pe^{-P} \tag{2.13}$$

where $P = (v/D)l$ is a protection parameter, which shows how many times the diffusional velocity for the ingress of oxygen into the atomizer is exceeded by the velocity of convective removal of the oxygen atoms [44]. The equation shows that the flux is directly proportional to the concentration of oxygen in the ambient atmosphere, n_0, (about 0.5×10^{19} cm^{-3} if the ambient atmosphere is air), the sample introduction hole area πa^2, and the value of the diffusion coefficient D, in the region of the sample introduction hole. The first multiplier in this equation, J_{dif} represents the rate of diffusional ingress of oxygen into the atomizer in the absence of the external gas flow. An estimation yields $J_{\text{dif}} \approx 10^{17}$ molecules per second. Under typical atomizer operating conditions, the value of the protection parameter, $P \approx 6$–10. Thus, the use of an external gas flow rate of about 1 l min^{-1} reduces the ingress of oxygen into the furnace by a factor of $(0.5–30) \times 10^2$. This means that, under normal operating conditions, the flux of oxygen into the graphite furnace is approximately 10^{14}–10^{16} molecules per second. This estimation correlates well with that obtained from experimental data presented by Sturgeon and Falk [40].

2.3.4 Three-dimensional distribution of oxygen within electrothermal atomizers

Knowledge of the above processes, determining the behavior of oxygen molecules in ETAS, enabled Gilmutdinov *et al.* [44] to develop a numerical model predicting the three-dimensional distribution of free oxygen molecules within different types of ETA. The following basic assumptions were made for the construction of the model: (1) stationary conditions exist (i.e. the furnace has been heated to its final temperature); (2) the furnace is used under the gas-stop mode of operation; (3) the only oxygen reaction within the atomizer is the heterogeneous reaction of molecular oxygen with the tube walls (i.e. no gas-phase reactions occur that will lead to a change in the oxygen concentration within the furnace); (4) there are only longitudinal temperature gradients within the furnace (i.e. a given furnace cross-section is isothermal); and (5) a dry, unloaded tube is used. Therefore, the only two significant sources of oxygen ingress into the furnace volume are entrance through the sample introduction hole and from impurities in the argon sheath gas. It is shown later that, under certain conditions, the oxygen resulting from analyte oxide decomposition and nonstationary temperature conditions can also be taken into account.

The results obtained from the solution of this model will provide a general background for any atomization situation in ETAs.

The approach [44] is based on a direct solution of the three-dimensional mass-transfer equations, taking into account the boundary conditions that exist at the tube walls, tube ends, and the sample introduction hole. The numerical model was used to obtain three-dimensional distributions of oxygen molecules within both pyrolytically coated and metal-lined (tungsten and tantalum) graphite atomizers. The results obtained for Massmann atomizers (inside radius $R = 0.3$ cm, half-length $L = 1.4$ cm, and $a = 0.1$ cm) are presented in Figure 2.9, which shows distributions for isothermal graphite furnace conditions at 2300K. The calculated distributions are visualized as gray-scale images, where black represents the maximum concentration of oxygen for any given distribution and white represents an absence of oxygen. Again, two sources of free oxygen are taken into account – ingress of oxygen through the sample introduction hole and, as impurities in the sheath gas, through the furnace ends. The presence of oxygen impurities in the sheath gas is simulated here by arbitrarily assuming that the oxygen number density in the sheath gas is 10 per cent of that close to the sample introduction hole. Figure 2.9(a) shows the distribution of oxygen molecules in the longitudinal section, and Figure 2.9(b) shows the distribution integrated along the tube length. Figure 2.9(c)–(e) show cross-sections in (c) the central part of the tube, (d) halfway between the tube center and the tube end, and (e) near the end of the tube. As all measured absorbance values in ETAAS are integrated along the tube length, Figure 2.9(b) can provide information about the integrated effect of oxygen. These results show that the cross-sectional distribution of oxygen is practically uniform near the ends of the graphite furnace (Figure 2.9(d) and (e)). The greatest degree of nonuniformity is in the central cross-section, Figure 2.9(c). However, the degree of nonuniformity, even for a temperature as high as 2300K, is much less than is generally accepted – the concentration of oxygen at the tube bottom of the central cross-section is only about four times lower than the concentration of oxygen near the sample introduction hole. The integrated distribution is also only slightly nonuniform, and the integrated concentration decreases monotonically as the distance from the sample introduction hole is increased (Figure 2.9(b)).

The above results are for isothermal graphite furnace conditions. Taking into account the actual furnace nonisothermality will result in even more uniform distributions of oxygen, as the introduction of temperature gradients leads to a decrease in the effective furnace temperature. In turn, this leads to a decrease in consumption of oxygen by the graphite walls because of a significant decrease in the rate constant (see Eqn (2.12) and Figure 2.8). Thus, all these results suggest that the distributions of oxygen within a pyrolytically coated graphite tube are relatively uniform, even at

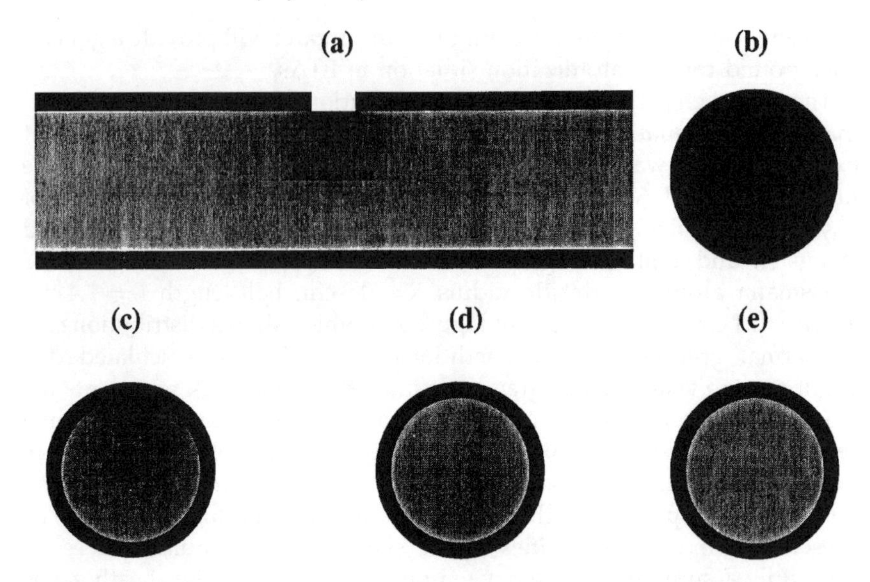

Figure 2.9 Imaged representation of the distributions of oxygen molecules within an isothermal graphite furnace at 2300 K: (a) longitudinal section; (b) cross-sectional distribution of the concentration integrated along the tube length; (c) furnace cross-section at the tube center; (d) furnace cross-section at a location intermediate between the tube end and the tube center; and (e) furnace cross-section near the tube end.

temperatures as high as 2300 K. Calculations performed for lower temperatures yielded even more uniform distributions [44].

As stated above, the rate constant for the interaction of oxygen with metals is significantly greater and should yield substantially different results. For comparisons Figure 2.10 shows the imaged distribution of the oxygen molecular cloud within the tantalum-lined Massmann furnace, obtained under the same conditions as those above for the graphite furnace. Distributions in tantalum-lined furnaces are seen to show a sharp decrease in concentration in both the longitudinal section (a) and the cross-sections (b)–(e). It is interesting that, in this case, the molecules are located almost exclusively near their sources, i.e. there are three local maxima in oxygen concentration – two at the furnace ends and one in the furnace middle near the sample introduction hole, with a low concentration of oxygen between these two extremes.

Note that the imaged distributions presented above are relative, as the concentration of oxygen molecules corresponding to the same density in the gray scale in the case of the tantalum-lined furnace is more than three orders of magnitude lower than that in the graphite furnace. This correlates well with experimental results [51].

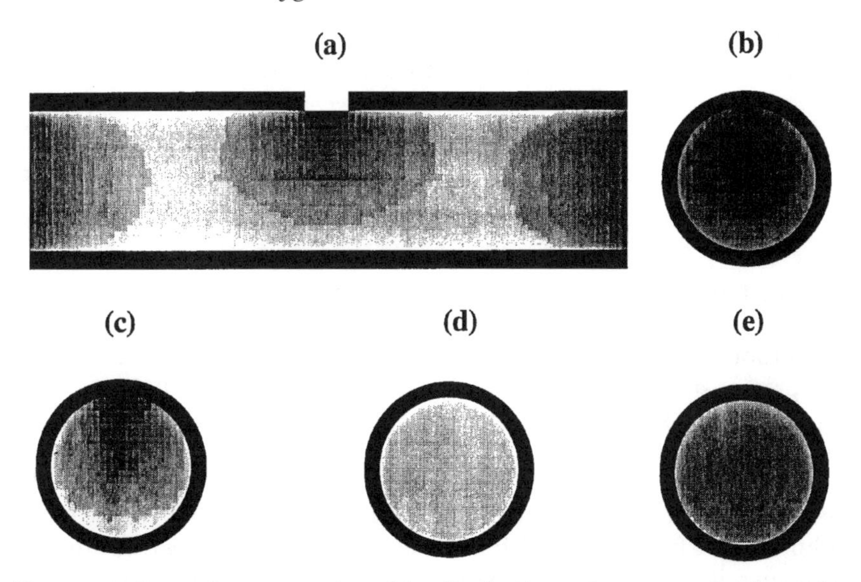

Figure 2.10 Imaged representation of the distributions of oxygen molecules within an isothermal tantalum-lined furnace at 2300K: (a)–(e) are the same sections as in Figure 2.9.

Thus, the oxygen molecular distributions within metal tubes are extremely nonuniform. These are a direct consequence of the relatively high values of their rate constants for heterogeneous oxidation.

2.3.5 Estimation of the total number of free oxygen molecules

The mathematical model described above produces three-dimensional results for a wide range of conditions. These calculations are only possible with the assistance of a computer, but for simplified cases some useful estimations can be made directly. Estimation of the total number of free oxygen molecules, resulting from their ingress through the sample introduction hole, assumes (1) the oxygen is distributed uniformly along a length L^* within a tube-type atomizer and (2) all entering oxygen disappears by heterogeneous reaction at the atomizer walls. Using these assumptions, the following balanced equation can be written:

$$J_{O_2} = n_{O_2} 2\pi R L^* k(T) \tag{2.14}$$

where the right-hand side of the equation gives the total removal of oxygen molecules because of their heterogeneous reaction with the entire inner surface area of the furnace, $2\pi RL^*$. Realizing that, with a uniform distribution of oxygen molecules, the total number of molecules is given by $N_t = \pi R^2 L^* n_{O_2}$, the above equation can be presented as

$$N_t = J_{O_2} \frac{R}{2k(T)} \tag{2.15}$$

Thus, the total number of oxygen molecules is proportional to both the flux of the molecules and the radius of the tube, and is inversely proportional to the rate constant for the heterogeneous reaction. Using the previously calculated value for the flux of oxygen, based on the experimental results of Sturgeon and Falk [38] ($J_{O_2} \approx 10^{16}$ molecules per second), in addition to their experimental conditions (T = 2100K, giving $k(T) = 10.06\ \text{cm s}^{-1}$), Eqn (2.15) gives $N_t \approx 1.5 \times 10^{14}$ molecules. With the assumption that these molecules are distributed uniformly over the entire volume of the furnace ($V = 0.19\ \text{cm}^3$), an estimation of the partial pressure of oxygen in a Massmann furnace under these conditions is obtained as $P_{O_2}(2100\ \text{K}) \approx 5 \times 10^{-5}$ atm. This value can serve as a rough approximation of the oxygen partial pressure in pyrolytic graphite coated graphite furnaces under typical operating conditions. Although the value obtained is much higher than was previously thought, it is in good agreement with experimental results [38].

The model presented above has two limitations. First, all considerations have dealt exclusively with stationary conditions where the furnace has already reached its final operating temperature. However, the results can also be expanded for nonstationary conditions if the heating rate of the atomizer does not exceed a certain value (quasistationary condition). In the quasistationary condition the current value of the total number of oxygen molecules, $N_t(t)$, within the furnace can be estimated using Eqn (2.15) with the current values of $k(T)$ and J. According to proposed criteria for establishing these quasistationary conditions [44], the maximum permissible heating rate for having a quasistationary condition is temperature dependent. If, for example, any species is atomized into the gas phase in a graphite furnace at 1800K, and if the heating rate is less than 1000K s^{-1} at this moment, the species will be under quasistationary conditions in terms of oxygen distributions.

The second limitation is that the model deals with only two possible sources of oxygen ingress, through the sample introduction hole and through the ends of the tube. However, in a furnace that is loaded with a sample, the major source of oxygen is sample decomposition [33,39,40]. This circumstance can also be taken into account in the quasistationary approximation. Indeed, if the change in the rate of decomposition of the oxide of the analyte during the characteristic time of furnace heating is insignificant, then the total number of free oxygen molecules can be estimated according to

$$N_t(t) = \frac{R}{2k(T)}\{J_h(t) + J_a(t)\} \tag{2.16}$$

where $J_h(t)$ is the flux of oxygen molecules entering through the sample introduction hole and $J_a(t)$ is the flux of oxygen molecules from sample decomposition. The quasistationary values of $J_h(t)$ can be estimated using Eqn (2.13) if the time-dependent behavior of the temperature in the sample introduction $T(t)$ is available. The other flux, $J_a(t)$, in Eqn (2.16) is strongly matrix dependent and is determined by peculiarities of matrix volatilization.

2.4 PREATOMIZATION

All atomic spectrometric techniques (emission, absorption and fluorescence) can detect analyte atoms only in the gas phase. However, most samples to be analyzed are introduced into the ETA as liquids (or solids), and are therefore in the condensed phase. Hence, the use of atomic spectrometry for analytical purposes requires prior transformation of the sample into the gas phase. In electrothermal atomization, this conversion usually occurs during the atomization stage that is preceded by the preatomization or thermal pretreatment stage, which includes dry and pyrolysis heating cycles.

The major objective of the preatomization stage is to prepare the analyte for complete atomization, and to minimize the formation of gas-phase species that may chemically or spectrally interfere with the analytical determination. Basic research into the preatomization stage should answer the questions: (1) what are the physical and chemical forms of the analyte immediately preceding the atomization stage and (2) how should the forms be modified in order to provide more complete analyte atomization? Numerous surface techniques, including a variety of electron, atomic and X-ray probes, radiotracers, secondary ion mass spectrometry, Rutherford backscattering spectrometry (RBS), and laser desorption mass spectrometry have been used to provide the answers.

Two factors have to be taken into account when considering data obtained by these techniques. First, the surface techniques are relatively insensitive (except for some based on mass spectrometry) at the nanogram amounts of analyte normally encountered in ETAAS. In order to simulate graphite furnace processes, these less-sensitive techniques require microgram quantities of analyte with subsequent extrapolation of the results to the nanogram or picogram level. The second potential flaw is that reactions are generally not studied during the actual thermal pretreatment, because the sample surfaces are usually cooled prior to analysis. Products stable at the pretreatment temperatures might react to form new products when cooled, or might undergo a reaction upon contact with air during transfer between instruments. Despite the limitations, surface techniques have been extremely useful in many mechanistic studies of preatomization processes. The results obtained were summarized recently by Majidi *et al.* [52]. Two general conclusions can be made from the survey of the results. First, in many cases the processes occurring during the pre-atomization stage ultimately dictate the final efficiency of free gaseous atom formation, the nature of the other gas phase species, and the shape of the analytical signal. Second, the chemistry occurring during the pre-atomization stage is often extremely complex and generally no simpler than the physical and chemical processes associated with the atomization stage.

The superior sensitivity achieved by electrothermal atomization techniques is due primarily to efficient sample use in the atomizer. The sample is normally introduced into the atomizer as a 1–100 μl solution drop, which is distributed over the atomizer surface, and occupies a surface area dependent on the nature of the solvent and the surface. For example, an organic solvent may spread over the entire graphite furnace surface, and the drop of an aqueous solution may take the form of a localized hemisphere. It is normally assumed that, after the drying stage, the residue consists of microcrystals of the salts of the acids that were used in preparing the samples. The formation of the salt crystallites as a result of evaporation of drops of solutions (precipitation) is described in detail in manuals on microanalysis, and consists of the following sequence of processes. The drop evaporation is initially confined to its perimeter and, as a result of oversaturation, the first crystals are formed in this region. As the first crystals are forming, two different flows appear inside the drop: undersaturated less-dense solution goes from the drop periphery to the top of the hemisphere, while heavier saturated solution moves on the surface from the drop center to its periphery. Due to these flows, the crystals being formed take a stretch form and are primarily concentrated on the perimeter of the initial drop. Particular forms of the microcrystals, and their distribution on the graphite atomizer surface, depend strongly on the solution concentration and on the temperature regime in the dry cycle. This simplified scheme of sample drying was confirmed in general for graphite furnace conditions [53–55]. The details of analyte deposition depend on solvent–solvent, solvent–surface, solvent–analyte, analyte–analyte and analyte–surface interactions. These interactions are dynamic in nature and are changed depending on temperature, concentration and chemical environment [52].

Often, sample deposition and desolvation on the graphite surface are accompanied by processes which do not fit the previous picture of microcrystal formation. Holcombe and Droessler [56] described the following simple experiment. A lead solution deposited on the furnace surface, and drawn back into the syringe a few seconds later, resulted in at least 90 per cent of the sample volume being removed from the furnace. However, after a typical atomization cycle, the peak area of the resulting absorbance signal was about 40 per cent of that obtained when the sample solution remained in the furnace throughout the dry, pyrolysis, and atomization cycles. Regardless of the nature of the surface or the composition of the gas phase, an unexpectedly large amount of analyte was retained on the surface. Obviously, the analyte ion was adsorbing to the surface in the absence of the desolvation step. The authors attributed this phenomenon to binding of lead to surface functionalities before the sample is desolvated.

Even more unusual results were presented by Eloi *et al.* [47,57], who used Rutherford Backscattering Spectrometry to measure the depth-dependent concentration profiles of nitrate salts of a number of elements in the pyrolytic graphite-coated graphite substrate at different temperatures. Their results clearly demonstrated that, contrary to popular belief; pyrolytically coated graphite is not impervious; nitrate salts of lead, cadmium, and silver readily diffused through the graphite to a depth of at least 3 μm at room temperature. This low-temperature migration of the samples into the substrate was later attributed to capillary action [58]. As was noted previously (Section 2.2.1), the graphite surface contains a number of macro- and microscopic imperfections. In the unmodified graphite surface, water molecules in the solvent approach the edge carbon atoms and interact due to the hydrophillic nature of oxygen on the graphite active sites. The solvent will then move into the graphite interstitial spaces through these imperfections. When the solvent is removed, the analyte can either remain inside the graphite plane or migrate back to the surface. If the analyte remains in the bulk of the graphite substrate, then at elevated temperatures the analyte vapor pressure inside the graphitic planes becomes very large, possibly resulting in the production of more defects.

The wetting characteristics of the pyrolytically coated graphite surface can be altered significantly by some modifiers, such as the hydrogen pretreatment of graphite (heating the graphite substrate for 60 s at 900 K in a hydrogen–argon atmosphere), which prevents any room-temperature migration of the analyte into the bulk of the graphite [47]. In this case, blockage of the surface active sites by hydrogen significantly reduces the number of lone pair electrons available to interact with the water molecules in the solvent. At higher temperatures, however, non-capillary induced migration is still observed for some elements. As the pretreatment temperature is increased, analyte behavior differs significantly for different elements. Figure 2.11 shows the distributions of lead and silver in the graphite after different thermal treatments, using RBS as the probe

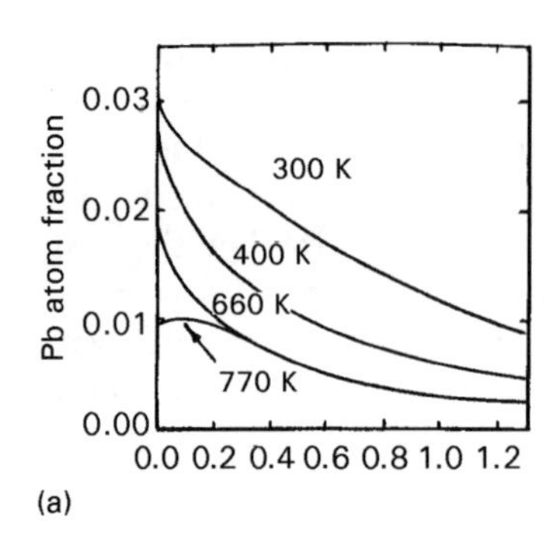

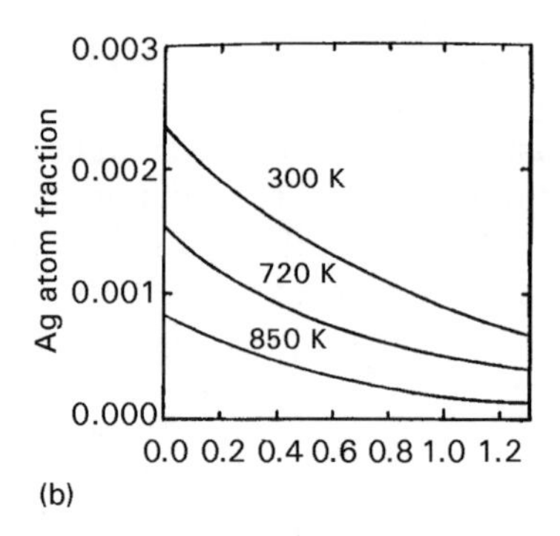

Figure 2.11 Temperature dependent migration of: (a) lead; and (b) silver into the graphite substrate [52]. (Reprinted from *Spectrochim. Acta, Part B*, Vol. **51B**, V. Majidi, R.G. Smith, R.E. Bossio, R.T. Pogue, M.W. McMahon, Observation of preatomization events on electrothermal atomizer surfaces, pp. 941–959, 1996, with kind permission of Elsevier Science – NL, Sara Burgerhartstraat 25, 1055 KV Amsterdam, The Netherlands.)

technique [52]. Interestingly, the material moves several micrometers below the surface. In the case of a cadmium nitrate sample, these researchers found that the cadmium species initially penetrated the graphite substrate, and then migrated back to the surface of the graphite as the graphite was heated to higher temperatures. At still higher temperatures, cadmium desorption occurred only from the top surface layer of graphite. The lead and silver species of Figure 2.11, however, did not appear to return to the surface and were disappearing uniformly from the bulk of the graphite substrate as the temperature was raised and the species were vaporized.

These observations emphasize the complexity of the solution–graphite interaction, even at room temperatures, and discourage the idea that at the ultratrace level, material simply precipitates onto the surface, forming small crystallites. The results presented above suggest that only a part of a sample is present in the furnace in the form of microcrystals. A significant portion of the analyte may be bound to the atomizer surface and/or migrate into the atomizer wall.

Thus, at the preatomization stage the analyte may exist in one or more of the following forms [59]: (1) adsorbed atoms and molecules on the graphite; (2) microcrystallites or molten droplets; (3) intercalation compounds; or (4) occluded analyte in crystallites of the matrix salts. A short description of the forms shown diagramatically in Figure 2.12 is given below.

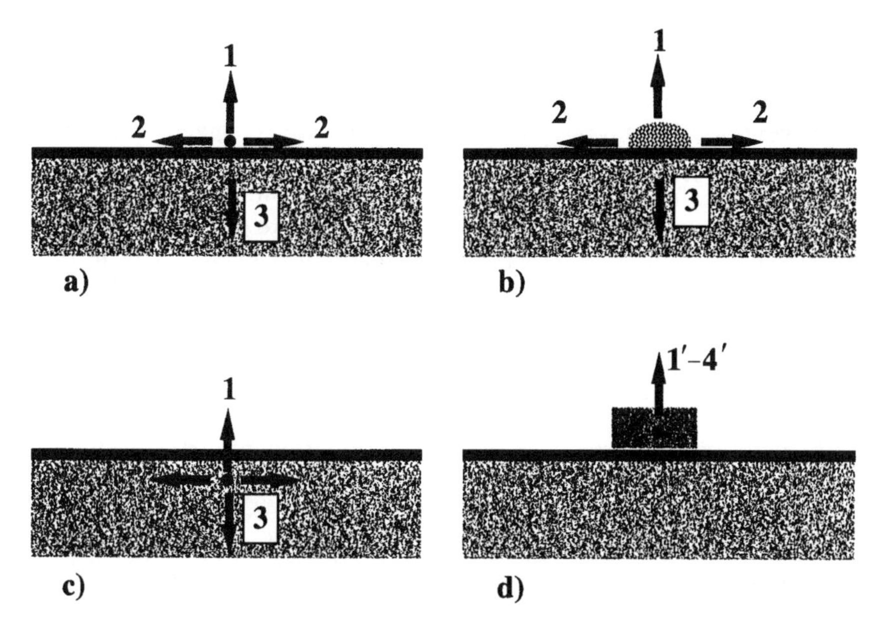

Figure 2.12 Forms in which an analyte may be present in the atomizer during the pretreatment stage: (a) an adsorbed species; (b) a microdrop/crystallite; (c) an intercalation compound; and (d) occlusion into a matrix crystal. Possible physical processes indicated: (1) desorption; (2) surface–diffusion; (3) bulk diffusion; and (4) evaporation. 1′4′ are the same as 1–4 but occurring in the matrix crystal. An analyte atom is shown schematically by a solid dot (•).

2.4.1 Monolayer/microdrops

Intermolecular forces of attraction will, to a large extent, govern the form present in the sample. With strong attractive forces existing between the analyte and the graphite surface, monolayer-type dispersion across the graphite surface is more likely. As the graphite surface is fragmented with existing edge carbons (see Section 2.2.1), it is likely that very strong attractive or bonding forces exist, at least, in localized segments of the graphite surface. The formation of metal carbides on the surface merely represents one form of the metal binding to the surface, albeit a very strong bond in many cases, which may produce a significantly higher appearance temperature than would be expected based on simple vaporization of the metal. However, the existence of a strong metal–carbon bond does not necessarily implicate the formation of a metal carbide with all of its attendant bulk physical chemical properties. Conversely, if intermolecular forces of attraction between the sample species dominate, the possibility of forming two-dimensional surface islands, or even three-dimensional microcrystallites (or microdrops), becomes more likely. In these instances the release energies needed to produce a vapor will be more characteristic of those values needed to remove the molecules or atoms from the microdroplet and not the graphite surface.

Concentration studies, and time and spatially resolved absorbance profiles obtained by McNally and Holcombe [60] suggest that copper desorbs from the graphite surface as individual atoms, resulting in an apparent first-order release. Copper adsorption to the surface occurs during desolvation at the drying step. Later investigations using secondary ion mass spectrometry (SIMS) [61] confirmed this postulate, but also showed that this is true only for relatively low masses of the metal. At copper masses greater than 40 ng, the analyte appears to exist on the graphite surface as microdroplets or some other forms of surface aggregates. Similar transitions from 'dispersed atoms' to 'surface aggregates' with increasing analyte mass were also reported for silver, with the transition occurring at about 0.1 ng of silver [62]. For gold, the formation of microdroplets or hemispherical droplets with desorption occurring from the droplet surface or at the metal–graphite interface is more likely. With an increase in concentration, the average number of droplets remains constant, but the average volume of the aggregate increases [60]. These observations were later confirmed by Lynch *et al.* [63].

2.4.2. Intercalation

As shown in Section 2.2.1, the interplanar distance between two graphite sheets is 335 pm, and the sheets are loosely bound by weak Van der Waals forces. As a consequence, the space between the graphite planes can accommodate some metals and metal-containing molecular species. This property is called intercalation. The exact geometric arrangement within the graphite is dependent on the species incorporated and on the concentrations involved. The alkali metals are particularly subject to intercalation and are known to be able to alter significantly the properties of graphite. Thus, if potassium atoms are allowed to intercalate into the interplanar space of graphite sheets, the distance between the sheets will increase up to 540 pm in the *c* direction. A more complete review of intercalation compounds formed with graphite can be found elsewhere [64]. If an analyte is intercalated, its release to the gas phase may require additional time for diffusional transport of the metal through the graphite to the surface (Figure 2.12(c)).

It is also known [65] that strong acids (nitric, dichromic, etc.) do not directly attack the graphite surface. Rather, they cause oxygen to be bound within the graphite lattice, causing expansion of the lattice, swelling and eventually exfoliation. This probably explains the destructive effect of strong oxidizing agents. The presence of other foreign species within the interlammelar spaces as intercalation compounds, and similar alteration of the interplanar spacing, may likewise account for the destructive nature of some concomitants present either in the sample or when added as a chemical modifier.

2.4.3 Occlusion

In a more complex sample with significant amounts of matrix, precipitation of the matrix salt during the dry cycle may occlude the analyte within the matrix crystallites [55,66]. Similarly, during thermal pretreatment, the analyte may become soluble in the molten salt. The incorporation of the analyte within these matrix crystals, or as a solubilized species within the molten salt, could produce release rates for the analyte which are not necessarily indicative of the metal under study. For example, in considering the same metal in two different matrices, two entirely different release rates could be observed. With a volatile matrix it is possible that significant amounts of analyte would be swept into the gas phase with the vaporization of the matrix. Conversely, a refractory matrix material may delay the appearance of the analyte until a much higher temperature in comparison to that observed for simple standards.

During thermal pretreatment there also may be transformations between different forms of the analyte. A significant increase in the surface area occupied by the sample during the pretreatment stage was attributed to surface diffusion of the analyte when applying pretreatment temperatures [67]. Also, Sabbatini and Tessari [68] presented X-ray photoelectron spectra (ESCA) for lead deposited on graphite, which clearly showed dispersion of lead upon heating the surface. These results suggests that the transition from 'microcrystal' to 'adsorbed species' occurs under these conditions.

The above discussion relates more to the physical aspects of the preatomization stage. Chemistry occurring during this stage is primarily related to the thermal decomposition of analyte salts, a subject that is relatively well documented in high-temperature inorganic chemistry [69,70]. Generally, the decomposition process is presented as

$$\text{analyte salt(s)} \rightarrow \text{analyte oxide(s)} + \text{gaseous products}$$

where the gaseous products include nitric oxide, nitrogen dioxide, dinitrogen tetroxide and oxygen (for nitrates), sulfur dioxide, sulfur trioxide and oxygen (or sulfates) and carbon dioxide (or carbonates). Decomposition of nitrates under the conditions of ETAAS is, however, complicated by the appearance in the gas phase of oxides and hydroxides of some metals in addition to the listed products. The effect was first reported by Sturgeon *et al.* [71] who detected the species PbO during thermal decomposition of lead nitrate. Later, gaseous oxides of copper, nickel and cobalt, and magnesium hydroxide, evolving from thermally decomposing nitrates of the metals, were also observed using mass spectrometry [72]. The effect was attributed either to crystal disintegration, or to desorption of the oxide or hydroxide molecules adsorbed to the graphite surface after the solution evaporation. An alternative explanation, based on the assumption of congruent (gaseous) thermal dissociation of the salts, was proposed by L'vov [73]. A discussion of the appearance of these gaseous oxides well below their expected appearance temperature was presented in a set of papers attaching different interpretation to the same data [74–77].

The sequence of chemical reactions occurring during thermal pretreatment was investigated in a series of papers by Wendl and Mueller-Vogt. In addition to conventional AA techniques, these authors employed X-ray diffraction, electron microscopy, electron microprobe analysis and molecular absorption spectrometry. An example of the reactions that have been found is summarized in the reaction diagram of Figure 2.13 [78]. This shows the metal compound MX hydrolyzing during the drying cycle to MO_x, as discussed above. Loss of MX vapor may occur during this process. The formation of such volatile compounds may be suppressed by changing the oxidation state of the metal, which leads to a reduction of

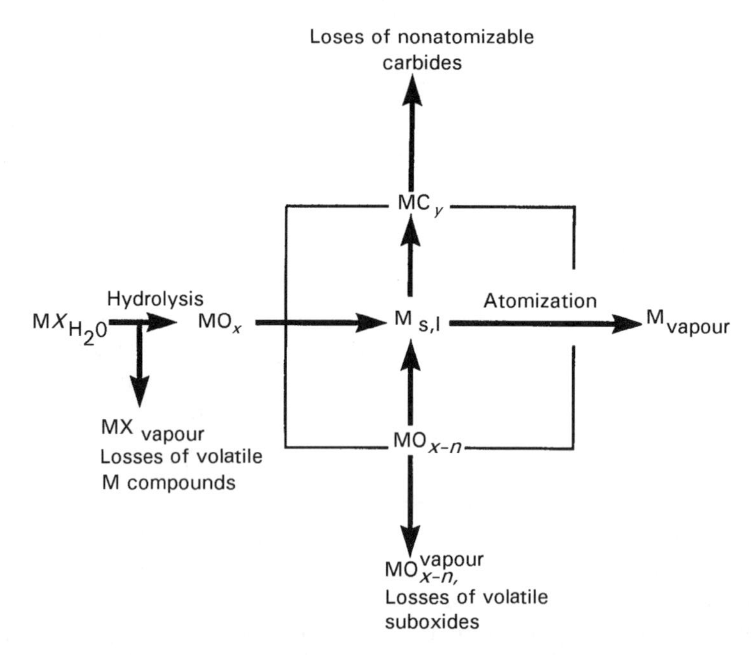

Figure 2.13 Schematic representation of possible reactions preceding the atomization event [78]. (Reprinted with permission from W. Wendl, Investigations on chemical reactions in graphite furnace AAS, *Fresenius' Z. Anal. Chem.* **323**, 726–729 Copyright 1986 American Chemical Society.)

metal losses and therefore to a higher absorbance signal during the atomization cycle. The metal oxide may react with the carbon of the furnace or it may dissociate into the elements. The reduction process of some oxides (such as the oxides of titanium, vanadium, chromium, molybdenum, and tungsten) leads to formation of stable carbides (upper path in the reaction diagram), or strongly chemisorbed species when small concentrations of metal are present. These carbides may dissociate during the atomization step into solid or liquid metal and carbon. Incomplete dissociation of the carbide leads to losses of the atomizable amount of the metal, and therefore reduces the absorbance. The carbides may also be thermally stabilized by the formation of solid solutions with other carbides. This leads to a reduced amount of atomized metal, and subsequently to a lower absorbance signal. An additional reaction path of the metal oxide occurs by reduction to a suboxide, MO_{x-n}, which is further reduced to the metal (e.g. tin and germanium). These suboxides are commonly volatile and their formation during thermal pretreatment can lead to losses of analyte. A higher reduction rate of the suboxides to the metals decreases these losses, and a bigger absorbance signal can also be expected. Thus, it can be concluded from this reaction diagram that every element shows specific reactions in the graphite furnace during the pretreatment stage.

For the formation of compounds during the thermal pretreatment stage, not only are the pretreatment temperatures important but also the dura-

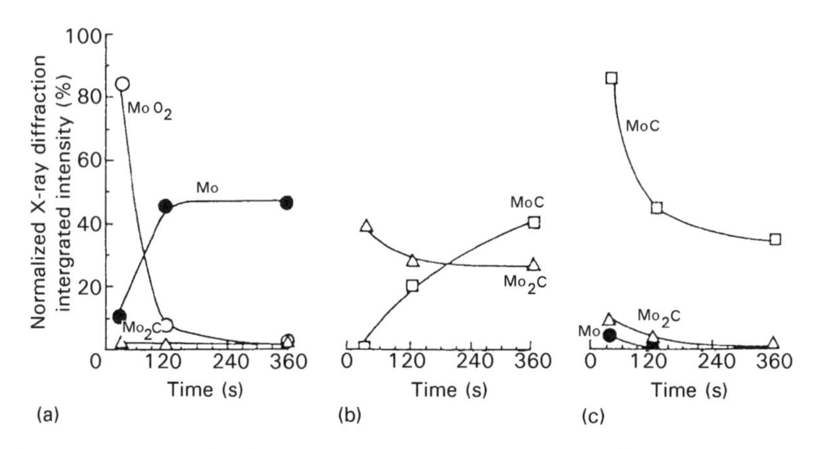

Figure 2.14 Normalized X-ray diffraction intensities of molybdenum compounds identified on the surface of pyrolytic graphite at preset temperatures as a function of heating time. Preset temperatures: (a) 1170 K; (b) 1870 K; (c) 2270 K [79]. (Reprinted with permission from C.L. Chakrabarti, S. Wu, F. Mancantonio and K.L. Headrick, Chemical reactions in the atomization of molybdenum in graphite furnace atomic absorption spectrometry, *Fresenius' Z. Anal. Chem.* **323**, 730–736. Copyright 1986 American Chemical Society.)

tion of the process. This is clearly illustrated in Figure 2.14, which shows normalized X-ray diffraction intensities of molybdenum species identified on the surface of a pyrolytic graphite platform as a function of heating time at a given temperature. Two points are worth noting in this figure. First, different types of solid carbides and their transformation with increasing pretreatment temperature can be seen. Also, at temperatures lower than 1500 K, the authors detected three different molybdenum oxides, $MoO_{3(s)}$, $MoO_{11(s)}$, and $MoO_{2(s)}$ [79]. This indicates that the chemical composition of a sample at the stages preceding volatilization may be rather complicated. Second, it can be seen that the chemical composition is strongly influenced, not only by the preset temperatures, but also the duration of the heating. When using relatively short pretreatment times, some of the compounds may not be formed even though the temperature is high enough for their formation. For example, a vanadium solution undergoes the following transformation during the pretreatment stage in the graphite furnace [80,81]:

$$\text{Solution } VOSO_4 \xrightarrow{\text{dry}} V_2O_3 \xrightarrow[1300\,K]{\text{pyrolyze}} V_2C \xrightarrow[1500\,K]{} VC \xrightarrow[2000\,K]{\text{atomize}} V_{(g)} + C$$

Interestingly, the authors [80] did not detect any measurable amount of vanadium carbides, even though they used pyrolysis temperatures as high as 1900 K. They speculated that the pretreatment times were not long enough for carbide formation.

The sequence and nature of the processes occurring during the pre-atomization stage can be changed dramatically by use of chemical modifiers (discussed in Chapter 5).

2.5 ATOMIZATION

L'vov's original concept for the semienclosed atomizer suggested that, if the vaporization process could be rapid enough and the analytical volume sufficiently confining to minimize gaseous loss, then almost all of the original sample could be isolated within the observation volume before a significant loss occurred [4]. As a result, the peak absorbance signal would approach that expected if the entire mass of analyte were located within the optical path. Using the simple representation of the absorbance signal shown in Figure 2.15, it could be said that the criterion would involve a short atomization time τ_1, compared with the dissipation or loss time τ_2, which is the time for the peak of the signal to decay to $1/e$ of its value. The figure shows that, as the ratio $\tau_1{:}\tau_2$ gets smaller, the fraction of the analyte in the gas phase (N/N_0) approaches unity. Although the initial concept is valid, knowledge of the complexities of both the generation and loss functions has evolved in considerable detail since the early work of L'vov, and is discussed below. Similarly, it was initially assumed that the reducing character of graphite, combined with the chemically inert atmosphere

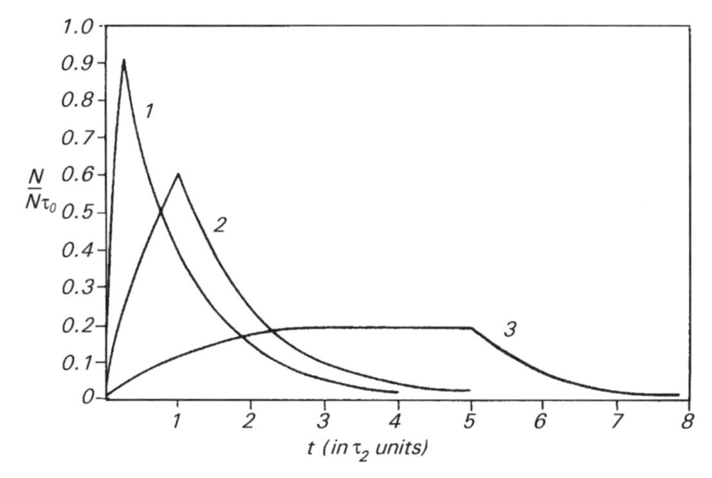

Figure 2.15 Simple representation of the atomization process for different ratios of the atomization time τ_1 and the dissipation time τ_2 [4]: (1) $\tau_1 : \tau_2 = 0.2$; (2) $\tau_1 : \tau_2 = 1$; and (3) $\tau_1 : \tau_2 = 5$.

of the sheath gas and the small analyte mass employed, would provide for a similar, simple vaporization process for nearly all metals, i.e. reduction of any metal salt to the metal with subsequent vaporization from molten metal. The previous section (2.4) dealt with this aspect in detail, and pointed out the varied chemistry and interactions occurring on the surface prior to the vaporization process.

Section 2.4 also indicated the variety of forms of the sample that may exist on the surface prior to the onset of the atomization stage, such as dispersed atoms, microdroplets, and intercalation compounds. The exact nature of the analyte immediately preceding the significant production of gaseous species can have a significant impact on the mechanism leading to atomization, which is a general term that simply implies the formation of the free gas-phase atoms. Free atom formation can occur through a number of pathways, some of which are listed in Table 2.2.

Three terms are used, often indiscriminately, for the conversion of condensed-phase species to a vapor: atomization, vaporization, and desorption. Atomization simply refers to the production of free gaseous atoms within the confines of the atomizer and says little about the mechanism leading to their production. Vaporization is also very general and implies the production of a gaseous species which may or may not be free atoms. Desorption suggests the presence of a species on a surface that can be removed from the surface with the input of energy, such as heat, and often implies a compositional difference between the adsorbed species being desorbed and the substrate or surface, such as 'metal desorbing from graphite'.

Table 2.2 Process occurring with an ETA during the various stages. Although many processes could occur during any of the stages, the designations represent likely occurrences with a system that is well controlled, i.e. good atomization efficiencies with no preatomization analyte loss.

Process	*Drying*	*Pyrolysis*	*Atomization*
Bulk solvent evaporation	✓✓		
Solvent adsorption	✓✓		
Solvent desorption	✓	✓✓	✓✓
Sample precipitation/crystallization	✓✓		
Analyte adsorption to surface	✓✓	✓✓	✓✓
Surface migration of atoms/molecules	✓✓	✓✓	✓✓
Surface migration of particles	✓	✓✓	✓✓
Intercalation formation	✓	✓✓	✓
Thermal decomposition of salts		✓✓	✓✓
Reduction of analyte species by carbon		✓✓	✓✓
Effective atomizer scrubbing of oxygen		✓	✓

2.5.1 Atomization efficiency

The atomization efficiency is often used as an indicator of the effectiveness of free atom production. As an example, this was defined by Sturgeon and Berman [82] as

$$(\varepsilon_{\mathrm{a}})_t = \left(\frac{N_{\mathrm{a}} + N_{\mathrm{i}}}{N}\right) = \eta\beta_{\mathrm{v}}\beta_{\mathrm{a}} \tag{2.17}$$

where $(\varepsilon_{\mathrm{a}})_t$ is the atomization efficiency at a given time, N_{a} is the number of free gaseous atoms, N_{i} is the number of ions, N is the total amount deposited initially, η is the vapor transport efficiency, β_{v} is the fraction volatilized, and β_{a} is the fraction atomized. The number of atoms in the gas phase (at the peak) must be determined from the absorbance signal magnitude, and this is generally done using the formulations of L'vov [83,84], which rely on spectroscopic constants and line shape profile assumptions. Sturgeon and Berman found that the atomization efficiencies at the absorbance peak varied considerably from less than 10 per cent (for aluminum, calcium and gallium) to 62 per cent for magnesium. In contrast, Falk and Tuch [85] recorded values for the tube furnace that were considerably higher, with many elements showing efficiencies in excess of 50 per cent. It is obvious that heating rates and furnace geometries significantly influence the magnitude of the peak absorbance and may partially reflect these differences. Likewise, the assumptions made (such as effective tube length, the accuracy of the physicochemical constants used in relating absorbance to atom density, etc.) will also affect the atomization efficiency calculations.

Other workers have considered efficiencies with respect to the integrated area of the peak (i.e. the total atomization process), and have suggested that atomization efficiencies for a large number of elements approach unity [86]. This is substantiated by the reasonable success of L'vov [87] and others [88–90] in demonstrating the feasibility of using ETAAS for absolute analysis, i.e. the ability to determine the concentration of an analyte within an unknown sample without requiring a calibration curve constructed from standard solutions (see Section 1.5.2.4). In their calculations, assuming an atomization efficiency of unity along with various spectroscopic and geometric constants for the system, they have shown remarkable success in obtaining reasonable agreement between calculations and measured values for the characteristic mass for several elements.

2.5.2 Ionization

As with any atomization source, extensive ionization in an ETA reduces the effective neutral atom density and, hence, the magnitude of the measured absorbance signal. The extent of ionization is governed by the electron density, which has two primary sources within the furnace atomizer: (1) thermionic emission from the heated graphite and condensed phase sample; and (2) ionization of gaseous species. The degree of ionization for an analyte metal at a given temperature is described by the classic Saha equation. This was used by Sturgeon and Berman [91], who used a microwave attenuation technique to measure electron number densities at various temperatures in both a graphite and a tantalum tube. They found electron densities in the range of $1-8 \times 10^{12}\,cm^{-3}$ at temperatures of 2700–3000K for graphite, and 2400–2700K , for tantalum because of the lower work function). They also showed that, for an empty tube, the experimental values of electron density correlated well with those expected from thermionic emission from the tube walls. Hence, they suggested that thermionic emission may be the primary source of electrons in the absence of easily ionized elements within a sample.

Both absorption and emission experiments have confirmed significant ionization in a graphite furnace for barium [92,93]. As with ionization interferences, long known in flame emission and absorption spectrometry, it was found that the addition of an easily ionized element (cesium or potassium) could reduce this interference.

The impact of electron density and ionization on easily ionized analyte elements (such as those with ionization potentials less than about 5 eV) is affected also by their appearance temperatures and the tube wall temperature. For example, a volatile easily ionized element may not show significant ionization, simply because the gas-phase temperature is insufficient to produce significant ionization during the analyte residence time within the furnace. Similarly, a less volatile element with a higher ionization potential may encounter a sufficiently high electron density from thermionic electron emission that, once again, the degree of ionization becomes insignificant.

It is not clear whether the use of platform atomization reduces or exacerbates ionization interference, but Nakamura *et al.* [94] showed clearly such an interference for cesium and rubidium analytes using a platform. They were able to minimize the interference with the addition of excess potassium to their samples.

2.5.3 General expression for the analytical signal

The absorbance signal is roughly proportional to the time-dependent partial pressure (or number density) of free metal atoms within the furnace volume, integrated over the length of the furnace they occupy. As the graphite furnace system is open, the number of free atoms within the furnace observation volume ($N(t)$) is the convolution of a supply and loss function [95]:

$$N(t) = \int_0^t S(t')R(t, t')\mathrm{d}t' \tag{2.18}$$

where $S(t')$ is the atom supply rate and $R(t, t')$ is the removal or loss rate. The generation step could be desorption or vaporization of material from the condensed phase, or production of free metal from homogeneous gas-phase dissociation. The loss process could be an irreversible process, such as material diffusing from the furnace where there is little chance of the material returning to the furnace. Alternatively, it could involve adsorption to the furnace or platform surface, where release into the gas phase at a later time or higher temperature is likely. To understand the atomization signal, all these steps must be accounted for.

2.5.4 Energies associated with release

The energy associated with vaporization is often assumed to be the heat of vaporization of the respective species, ΔH_V, or heats of sublimation, ΔH_S, for phase transitions progressing from solid to gas. For these values to be applicable, it is necessary to assume that the species being vaporized exist on the surface as bulk material (i.e. substances that possess all the physical and chemical properties of large quantities of the same substance) and that equilibrium exists at the interface between the condensed phase analyte and its vapor. However, the assumption of bulk properties may not always be appropriate.

To realize when the existence of bulk properties might not be the case, consider an analyte present in the furnace in an extremely small amount, such that it might be possible to disperse the molecules across the surface with less than unit surface coverage of the exposed graphite. If strong attractive forces exist between the analyte (or analyte-containing molecule) and the surface, it may chemisorb to the surface rather than form crystals or liquid droplets of the material. As an example consider depositing 20 μl of a 5 μg l^{-1} solution (0.1 ng of analyte) on the surface. If 100 g mol^{-1} is the molecular weight, then 6×10^{11} molecules will be present. With a typical monolayer surface coverage of about 10^{15} cm^{-2}, this represents a fractional surface coverage of 0.013 per cent if the entire inside surface of a 0.6 cm diameter by 2.5 cm long furnace is considered.

To transfer this chemisorbed species from the surface to the gas phase, energy equivalent to that responsible for the attraction to the surface (e.g. adsorption energy) must be supplied. In the case of the isolated atom or molecule, this energy is often referred to as the desorption energy. This can be quite different from ΔH_v or ΔH_S, because these enthalpies reflect the energy required to overcome the intermolecular forces of attraction between like-molecules; however, the desorption energy is that energy holding the molecule to a dissimilar material, such as the surface. It is reasonable to expect such a difference when considering a single, isolated molecule. However, as the number of like molecules increases or as the free energies associated with molecular dispersion and aggregation approach one another, islands or other microstructures can form on the surface. If these microstructures have a sufficiently large number of molecules, they eventually must take on the properties of the bulk material and desorption energies will approach ΔH_v. It is not clear how many molecules or atoms are needed to form a cluster with bulk properties but, based on measurements of magnetic properties, it has been suggested that several hundred iron, cobalt or nickel atoms per cluster or microstructure may adequately approximate bulk material [96]. The number of atoms constituting bulk material was shown to be 13–147 for gold, using Mössbauer spectroscopy [97], and 100 using convergence of vibrational information [98].

Whether condensed phase molecules exist as adsorbed species, or as small microdroplets or crystals with bulk properties, can also depend on: (1) how the sample was deposited and dried; (2) modes of dispersion available (surface diffusion, vapor redeposition processes, etc.); (3) the presence of other matrix components; and (4) the relative strength of attraction between the species and itself versus the species and the surface. Some of these various aspects were discussed in Section 2.4 and can govern the pathway taken to place analyte atoms or molecules in the vapor phase.

2.5.5 Atom formation through gas-phase dissociation

The atomization process can also proceed through gas-phase dissociation of a molecular species, MX, vaporized from the surface ($MX = M + X$). The equilibrium constant can be expressed as the respective partial pressures:

$$K = \frac{P_M P_X}{P_{MX}} \tag{2.19}$$

Alternatively, the atomization efficiency could be expressed as the fraction of metal present as the free gaseous metal and, for this simple system, in terms of their respective partial pressures as well as the equilibrium constant from Eqn (2.19):

$$\varepsilon = \frac{P_M}{P_M + P_{MX}} = \frac{K}{K + P_X} \tag{2.20}$$

Thus, if $P_X \ll K$, ε approaches unity. Furthermore, as most dissociation processes are endothermic and K increases with temperature, the atomization efficiency will improve unless there is another source of X which also changes during this heating. Frech and co-workers [99] were the first to suggest the rate of decrease in P_{O_2} with temperature in order to correlate their observed absorbance data with calculations of equilibrium appearance of the free metal, using free energy minimization routines. As another example, it has been shown that selenium dioxide vaporizes at a relatively low temperature in the furnace [100] and can diffuse from the furnace before a significant fraction of the species dissociates, thereby exhibiting a very low atomization efficiency. For the successful determination of selenium, chemical modifiers (see Chapter 5) and platforms are used to increase the atomization efficiency, because both can delay vaporization until the gas phase temperature has reached a higher value where dissociation is more favorable and P_{O_2} decreases.

2.5.6 Kinetic versus thermodynamic approach to free-metal production

The vaporization process, in its simplest form, can be expressed as $M_{(c)} \rightarrow M_{(g)}$, where $M_{(c)}$ represents a condensed form of the metal. This could include bulk solid or liquid, either as microcrystals or droplets, or some form of nonbulk species such as dispersed, adsorbed metal ($M_{(ads)}$) present as islands or microstructures not exhibiting bulk thermodynamic characteristics. In all cases, characterization of vaporization can be viewed either as a kinetic process, in which local thermodynamic equilibrium at the interface is *not* established, or an equilibrium release where local thermodynamic equilibrium is assumed.

2.5.6.1 Kinetic approach

If kinetic expressions are used to evaluate the release, activities of surface species can be used in a format similar to that used in desorption experiments conducted in vacuum, where the activity is equated to fractional coverages (Θ) or surface coverage (σ, expressed in molecules per square centimeter). Thus, the loss rate of material from the surface is given by

$$\frac{\mathrm{d}\sigma}{\mathrm{d}t} = -\nu\sigma^n \exp\left(\frac{-E_\mathrm{a}}{kT}\right) \tag{2.21}$$

where ν is the preexponential term, n is the order of release, E_a is the activation energy of release, k is Boltzmann's constant and T is the absolute temperature. The rate of production of gaseous atoms can be established as

$$\frac{\mathrm{d}N}{\mathrm{d}t} = \frac{-\mathrm{d}\sigma}{\mathrm{d}t}A \tag{2.22}$$

where N is the number of atoms produced and A is the area covered by the sample with surface coverage σ.

The order of release can be indicative of the topology of the sample on the surface [101]. If there exist m circular islands of radius r per area A of the surface, and release occurs at the perimeter of these islands, as shown

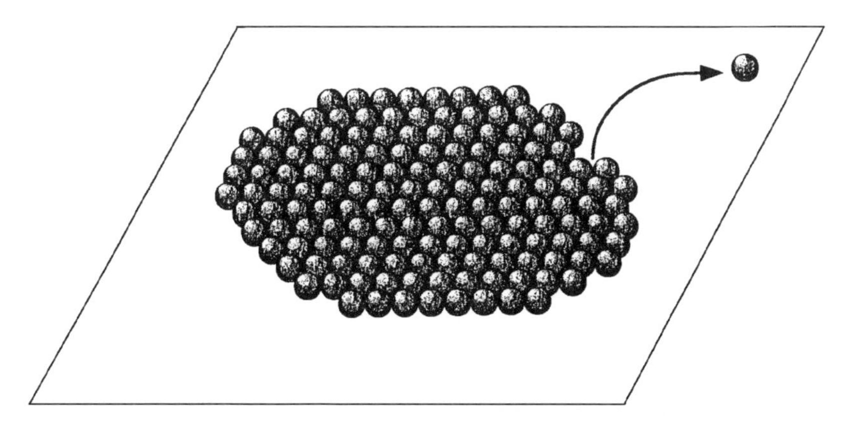

Figure 2.16 Diagram of a surface with adsorbed atoms/molecules forming islands. Adsorption can often occur from the perimeter of these islands.

in Figure 2.16, then in this area of the surface the number of analyte atoms can also be expressed as

$$\sigma A = m\pi r^2 \sigma^{\mathrm{o}} \tag{2.23}$$

where σ^{o} is the adatom density within the islands and is assumed constant and independent of σ and m. Thus, the cumulative circumference C of these islands can be given by

$$C = 2\pi mr = 2\pi m\sqrt{\frac{\sigma A}{m\pi\sigma^{\mathrm{o}}}} = 2\sqrt{\frac{\pi mA}{\sigma^{\mathrm{o}}}}\sigma^{1/2} \tag{2.24}$$

If the rate of release is assumed proportional to the molecules located at the circumference of these islands, then

$$\frac{-\mathrm{d}\sigma}{\mathrm{d}t} \propto C \propto \sigma^{1/2} \tag{2.25}$$

The apparent order of release is 1/2 and the constants from Eqn (2.23) are often simply incorporated into the preexponential factor. Similar relationships can be derived for spherical microstructures ($n = 2/3$), where desorption is assumed to occur from the surface. If large amounts of material exist (as with multilayer coverage) and vaporization occurs from the surface, then it is assumed that loss of material from the surface does not alter the vaporizing surface area and $n = 0$, i.e. a pseudo zero-order release is assumed. This is often the case when large droplets exist on the surface and initial vaporization does not significantly change the surface area of the droplet – the vaporizing surface appears to be of constant area and, to a first approximation, is independent of the material removed.

The relationship between enthalpies and activation energies stems from transition state theory, and the presence or absence of an activation barrier beyond the thermodynamic enthalpy required for gas formation. Only in the absence of such a barrier does $\Delta H = E_{\mathrm{a}}$.

2.5.6.2 *Thermodynamic approach*

If thermodynamic equilibrium is assumed, then

$$K_T = a_{\mathrm{M(g)}}/a_{\mathrm{M(c)}} \tag{2.26}$$

where $a_{\mathrm{M(c)}}$ is the activity of the condensed species and $a_{\mathrm{M(g)}}$ can be approximated by the partial pressure in atm, P_{M} When bulk solids or liquids are assumed $a_{\mathrm{M(c)}}$ can be assumed to be unity. When bulk properties are not assumed, the activity of the condensed species is often taken as the surface coverage of the metal for dispersed or adsorbed atoms. In all cases, for the local environment at any time, Eqn (2.26) can be arranged to reflect the activity of the vapor:

$$a_{\mathrm{M}_{(g)}} = a_{\mathrm{M}_{(c)}} K_T \tag{2.27}$$

Simple replacement of K with its thermodynamic equivalent yields

$$a_{\mathrm{M}_{(g)}} = a_{\mathrm{M}_{(c)}} \exp\left(\frac{-\Delta H_T^{\mathrm{o}}}{RT}\right) \exp\left(\frac{\Delta S_T^{\mathrm{o}}}{R}\right) \tag{2.28}$$

and recognizing that, for ideal gas conditions, the activity of the vapor can be expressed as the partial pressure,

$$P_{\mathrm{M}_{(g)}} = a_{\mathrm{M}_{(c)}} \exp\left(\frac{-\Delta H_T^{\mathrm{o}}}{RT}\right) \exp\left(\frac{\Delta S_T^{\mathrm{o}}}{R}\right) \tag{2.29}$$

where ΔH_T^{o} and ΔS_T^{o} are, respectively, the molar enthalpy and entropy of release at temperature T. To consider the activities as proportional to surface coverage, then equations showing fractional order of release for the kinetic approach would be applicable in the equilibrium model. Thus, if only the molecules at the perimeter of islands could be involved in the heterogeneous equilibrium, the effective concentration on the surface would include only those particular spatial locations, and $a_{\mathrm{M}_{(c)}}$ would be proportional to $\sigma^{1/2}$.

2.5.7 Loss function

The loss or dissipation function can be viewed as any process that reduces the free atom density within the observation volume. It can include irreversible loss, through formation of molecular species, or diffusion from the ends of the furnace or sample introduction hole. Alternatively, it may include adsorption back to the walls of the furnace, where it is released later in time at a higher temperature. The formation of molecular species can be treated using equilibrium expressions similar to Eqn (2.19), assuming that kinetics are rapid enough to establish local thermodynamic equilibrium.

Adsorption to the walls can be a relatively complex process. It requires either the simplifying assumption of local thermodynamic equilibrium or a knowledge of the sticking coefficient of the metal onto graphite in the case where local thermodynamic equilibrium is not likely to exist between the gaseous and condensed state metal due to the small amount of the analyte metal.

L'vov [4] originally proposed a simple relationship for estimating the average time, τ_2, spent by an atom in the graphite furnace. This was based on diffusive loss through the ends of the furnace, because his original design had no sample introduction hole:

$$\tau_2 = \frac{L^2}{8D} \tag{2.30}$$

where L is the furnace length and D is the temperature-dependent diffusion coefficient. Hadgu and Frech [102] showed the use of endcaps, to limit the diameter at the ends of the furnace, was effective in reducing this loss rate, and hence enhanced the magnitude of the absorbance signal.

Monte Carlo simulations [103–107] which take into account the presence of the sample introduction hole, have shown that a significant fraction of sample is lost from the hole, and this fraction increases with increasing furnace length and with the use of a platform [108,109]. Modeling suggests approximately 15–20 per cent of the total analyte mass is present in the furnace at the peak of the absorbance signal. Although this value can be raised with changes in furnace geometry and conditions (longer furnace, absence of a sample introduction hole, more rapid heating, altered release mechanism, etc.), 100 per cent of the analyte generally will not be present in the furnace at the peak. The loss or dissipation of analyte, as well as the time and spatial distribution, is discussed in greater detail in Section 2.6, covering analyte distribution.

2.5.8 Diagnosis of atomization mechanisms

The techniques used to monitor processes occurring during the heating cycle is as varied as the imagination of the researchers attacking this problem. The major dilemma faced for direct determination of other species in the system, besides free atoms, is the inherently high sensitivity of ETAAS. Under typical ETA operating conditions, this requires that any other analytical probe have comparable sensitivity to ETAAS. Otherwise, larger sample masses must be used, and then it is necessary to assume that increases in the mass amount studied by the ancillary probing technique do not alter the chemistry and/or physics of the process taking place under normal analytical conditions, i.e. picogram to nanogram masses under 1 atm pressure of a gas at temperatures between ambient and nearly 3000°C.

Dominating the list of diagnostic probes is the clear uses of the ETAAS signal to speculate on the mechanisms occurring. The shape of the peak, as well as the magnitude of the peak area and thermal shifts in the peak, are routinely combined with innovative mathematical deconvolution techniques to extract information. Such approaches are the least intrusive and, by definition, satisfy the criteria of conducting experiments under exacting analytical conditions. In most cases, assumptions must be made, because direct observations of precursor molecular species or concurrent gas-phase species are not possible by AA. The previous sections discussed approaches used to follow condensed phase processes and many of these results have been extended to speculate on the immediate precursor to the free atom formation.

2.5.8.1 Indirect monitoring

Indirect monitoring of molecular species evolved from the furnace, through detection of the atomic spectroscopic signal by thermally dissociating these molecules, has been used to study preatomization losses and, in general, the release of analyte-containing molecules. As examples, the protective gas from the furnace has been transported into a flame for AA detection [110], and inductively coupled plasmas (ICP) for atomic emission [111] or mass spectrometry [112]. This general approach of secondary atomization assumes reasonable transport efficiency of the evolved molecules to the secondary atomization source. Using either flame or ICP, there is an inherent loss in sensitivity compared with ETAAS, and larger sample amounts are often employed. An analogous approach, with the sensitivity inherent in ETAAS, uses the two-step furnace design (see Section 1.4 and Figure 1.10). In this instance the upper tube is preheated to a temperature that ensures efficient atomization of the suspect molecules. The lower cup is thermally ramped to vaporize the analyte. Any evolved molecular species will hopefully produce an AA signal upon thermal decomposition within the preheated tube. As adequate sensitivity is limited to the atomization of metals, the molecule responsible for the signal must be inferred from the sample composition and physicochemical properties (boiling points, thermal stability, thermodynamic calculations, etc.) of the suspected species. This approach has been used, for example, in the study of aluminum vaporization [113].

2.5.8.2 Direct monitoring

A more direct approach involves molecular absorption, using either atomic lines from a hollow cathode lamp (HCL) which are coincident with molecular bands, or a continuum source (deuterium lamp or xenon arc lamp). Once again, the 100–1000-fold loss in sensitivity compared to AA generally forces the use of elevated analyte concentrations, and any conclusions must be justified or tempered based on this fact. The approach has been used, in many instances, to monitor possible analyte loss by the formation of molecular species either upon vaporization or through gas-phase reactions [114]. The approach is inherently more useful for studies of matrix components or modifiers, which are generally in much higher concentrations in the sample, and hence can be monitored at the levels expected in a typical analysis [115,116]. Ohlsson and Frech [117] used a continuum lamp with spectrographic film detection to study the species evolved when phosphate was used as a modifier. They obtained an absorbance spectrum using this approach, instead of being limited to the traditional monitoring of a single wavelength. This was carried a step further by Majidi *et al.* [118], who used the continuum emitted from a plasma formed from the breakdown of the ambient atmosphere by a pulsed laser, to blacklight the furnace. Laser-induced breakdown provides sufficient intensity that a time-gated absorbance signal was obtained over a broad spectral region, when an optical multichannel analyzer was used in the focal plane of the spectrometer. A three-dimensional plot could be constructed, with a sampling frequency commensurate with the laser pulsing frequency around 0.01–0.1 s. An example of the output of such a

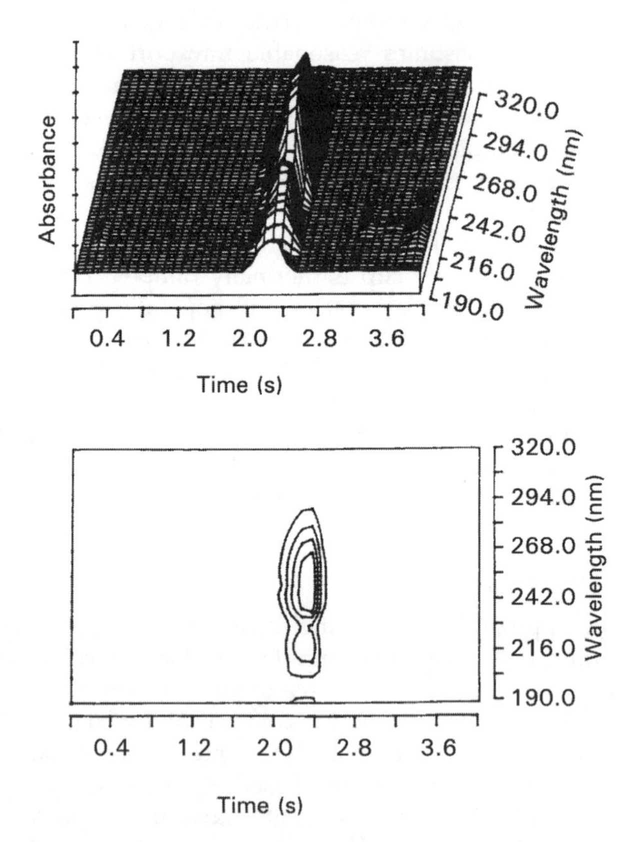

Figure 2.17 Absorbance-time-wavelength plots for a sodium chloride sample showing the clear appearance of NaCl near 254 nm and 2.1 s [118]. (Reprinted by permission of the Society for Applied Spectroscopy from, V. Majidi, J. Ratcliff and M. Owens, Investigation of transient molecular absorption in a graphite furnace by laser-induced plasmas, *Appl. Spectrosc.*, **45**, 473–476 (1991).)

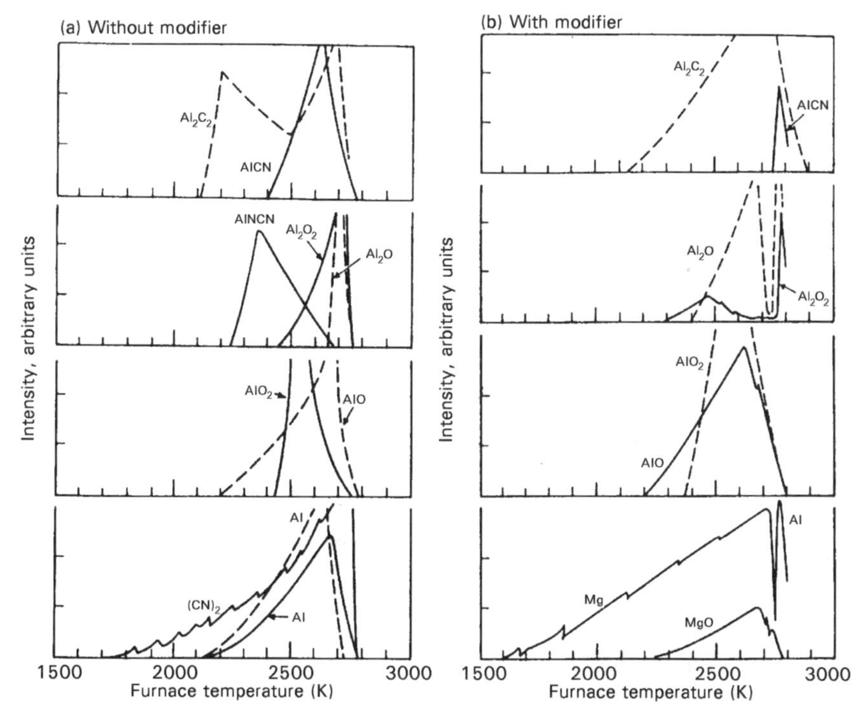

Figure 2.18 Schematic representation from mass spectrometric data for a number of species detected when aluminum is vaporized under 1 atm of argon using the molecular beam system of Styris *et al.* [127]. Sampling is done through a sampling cone inserted into the sample introduction hole. (Reprinted with permission from D.L. Styris and D.A. Redfield, Mechanism of graphite furnace atomization of aluminum by molecular beam sampling mass spectrometry, *Anal. Chem.* **59**, 2891–2897 Copyright 1987 American Chemical Society.)

system is shown in Figure 2.17 for the analysis of a sample containing 20 μg of sodium chloride [118].

Gilmutdinov and co-workers [119–121] applied their technique of spectral shadow filming to obtain time and spatial molecular absorption information. In the most recent version of this technique, the backlighted furnace is imaged onto a CCD detector. This system can provide both time and spatial information on the distribution of atomic and molecular species, depending on the selection of wavelengths monitored. As is noted in more detail in Section 2.6, they have looked at several systems, and have considered both atomic and molecular absorption. Spatial viewing provides an additional perspective that permits deductions regarding possible molecular formation/dissociation pathways. Because of the increased molecular signal near the sample introduction hole and the lower signal near the wall, they concluded that the ingress of oxygen is responsible for AlO_x formation, and that the hot graphite wall promotes dissociation of this species within the furnace [122].

Mass spectrometry provides one of the more powerful diagnostic tools to study molecular species evolved from the graphite surface. It has the inherent sensitivity to work at nanogram levels, and can provide more conclusive molecular identification compared with the techniques discussed above. Most studies have used either a furnace or graphite flat that is heated in vacuum below a quadruple mass filter that employs electron impact ionization [123–126]. Vacuum vaporization can provide isolation of the actual species vaporized from the surface, and can discriminate against species that may have resulted from secondary gas-phase reactions formed in a more conventional furnace operated at atmospheric pressure. This selectivity results from the large mean free path in vacuum that usually precludes gas phase reactions between vapor species after leaving the surface. Secondary collisions between evolved gases and the graphite wall are also minimized with the graphite flat geometry, and are relatively small with the tube geometry. An example of some of the signals captured mass spectrometrically for an aluminum-containing sample, vaporized at 1 atm and sampled through a skimmer cone, is shown in Figure 2.18. Several elements have been studied, including aluminum [127,128], selenium [71,124,129,130], cobalt [131], chromium [132], PO [133] and palladium [130,134]. Styris and Kaye [123] have also interfaced mass spectrometry with a graphite furnace operated at atmospheric pressure, through the use of a sampler cone inserted into the sample introduction hole of the furnace. With this system, the gas-phase products can be monitored, and the impact of any gas-phase reactions within the ETA are also taken into account.

The combination of atmospheric pressure and vacuum vaporization mass spectrometric studies can potentially isolate those products resulting from gas-phase collisions between molecular species directly vaporized from the surface. Although mass spectrometry is applicable to the study of basic vaporization processes and the determination of precursor species, the complex matrix of many real samples can render the interpretation of the resulting mass spectral signal extremely difficult.

2.5.8.3 *Use of the atomic absorption signal*

The most widely used techniques for indirectly discerning the mechanism for analyte release involve deconvolution of the time and temperature dependent absorbance signal. This has the distinct advantage of being able to use experimental conditions (e.g., sample masses and atomization conditions) which are identical to those used for analysis. Tessari and co-workers [135–137] first proposed such an approach and employed the entire absorbance signal for nickel from a heated carbon rod atomizer. A number of other techniques were used, including the use of the back edge of the signal by Fuller [138], and a more comprehensive use of the same region by Katskov [139]. Both these techniques are potentially viable, as the vaporization process continues even during the declining signal later in time. Additionally, these approaches have the advantage of measuring a signal once the furnace temperature has reached a nearly constant value. However, neither approach provides information during the initial release of material, and some concern must exist with end-heated Massmann furnaces, where a substantial temperature gradient along the tube length has been shown to exist once the center of the furnace has reached a steady-state value [7,140]. The more commonly used approach was proposed by Sturgeon *et al.* [141], who employed the leading edge of the absorbance signal and suggested that the time or temperature-dependent absorbance signal, A_t, early in the profile was proportional to the rate of release of material from the surface. Working from Eqn (2.21), the following relationships could be derived:

$$A_t \propto \frac{-\partial\sigma}{\partial t} = v\sigma \exp\left(\frac{-E_\mathrm{a}}{kT}\right) \tag{2.31}$$

Assuming that σ does not change significantly early in time during the vaporization process, they simplified the relationship to

$$\log(A_t) = \frac{-E_\mathrm{a}}{kT} + c \tag{2.32}$$

where c incorporates v, σ, and the various spectroscopic constants needed to relate absorbance to partial pressure or concentration of the metal in the furnace. Thus, a plot of $\log(A_t)$ versus $1/T$ would produce a straight line whose slope is given by $-E_\mathrm{a}/k$. This approach has become known as an *Arrhenius plot*, because of its similarity to the more traditional approach used in other kinetic areas. Another common variation on this theme was suggested by Smets [142], who proposed a means of accounting for changing surface coverage during the vaporization process. He assumed that the surface coverage at any time, t, could be determined by considering the difference between the total absorbance signal and that part of the sample removed up to that time:

$$\sigma \propto \int_0^\infty A_t\,\mathrm{d}t - \int_0^t A_t\,\mathrm{d}t \tag{2.33}$$

Referring back to Eqn (2.32), the substitution can be made to result in what is commonly referred to as a Smets plot:

$$\log\left(\frac{A_t}{\int_0^\infty A_t\,\mathrm{d}t - \int_0^t A_t\,\mathrm{d}t}\right) = \log\left(\frac{A_t}{\int_t^\infty A_t\,\mathrm{d}t}\right) = \frac{-E_\mathrm{a}}{kT} + c' \tag{2.34}$$

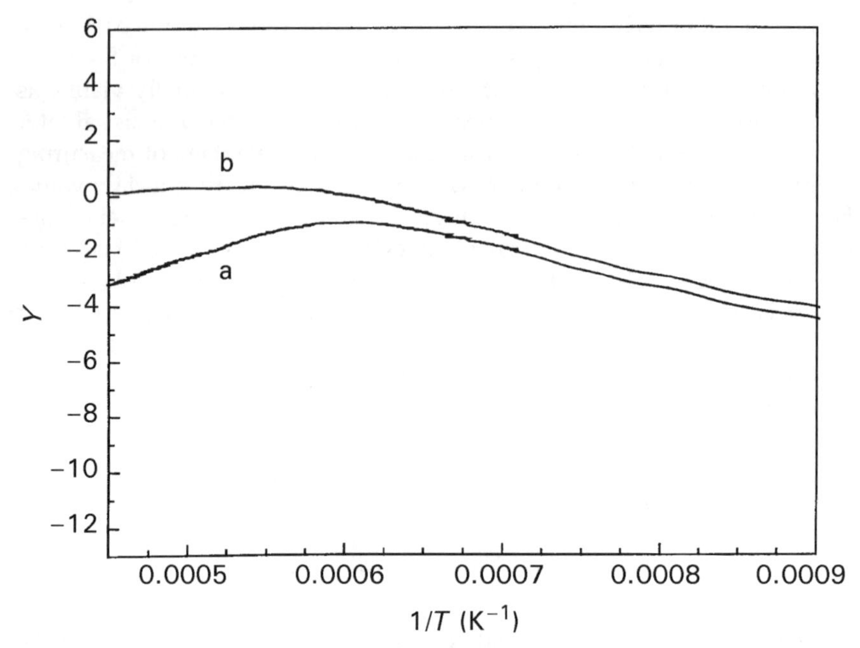

Figure 2.19 Comparison of: (a) an Arrhenius plot, Eqn (2.32); (b) Smets' [142] modified form, using Eqn (2.36), for copper. The ordinates are scaled for the respective methods, but the slopes of the lines should be directly comparable with the slope, indicative of the activation energy of the release process. It should be clear that a small increase in the linear region results from Smets' modification; (a) $Y = \ln(A_t)$; and (b) $Y = \ln(A_t \int A_t dt$. (Reprinted from *Spectrochim. Acta Part B*, Vol. **35B**. B. Smets, Atom formation and dissipation in electrothermal atomization pp. 33–42, 1980, with kind permission of Elsevier Science – NL, Sara Burgerhartstraat 25, 1055 KV Amsterdam, The Netherlands.)

This approach produces an improved linearization of the curve as it approaches the temperature where the absorbance peak maximum appears, but generally does not noticeably alter the slope of the line at the low-temperature end. Figure 2.19 shows plots of the same data set using Eqns (2.32) and (2.34) [143]. On the back end of the peak, the linearity clearly fails because the assumption is no longer valid that the loss rate is insignificant relative to the generation rate. Frech *et al.* [144] considered these two approaches and others in their predictive ability. They found that all were satisfactory, and that linearity was excellent for a given absorbance data set, but there were large sample-to-sample deviations in the calculated activation energy. They suggested that nonreproducible sample deposition and/or changes in the surface from shot-to-shot were the major source of uncertainty.

Experimental order of release

Referring to Eqn (2.21) and assuming desorption occurs, it was suggested that the order of release can provide indications of the topology of the sample prior to atomization, such as droplets, islands, dispersed atoms, etc. McNally and Holcombe [60] suggested that the order of release could be estimated by looking at the alignment of the absorbance signals as the concentration is altered. They concluded that copper was desorbed from dispersed atoms on the surface, but that silver and gold vaporized from microdroplets. Attempts to deduce the order of release were further refined by Rojas and Olivares [145] and by Yan *et al.* [146]. These authors derived a numerical value for the order of release. A comparison of their approaches was provided by Fonseca *et al.* [147]. Two independent studies also suggested that the order of release could be altered as the analyte concentration changed [148,149]. Interestingly, it was found that silver produced an activation energy value approaching ΔH_v (300 kJ mol^{-1}) with larger sample masses deposited on the surface, and approached a constant, low value (113 kJ mol^{-1}) at low concentrations. The suggested explanation was a transition from bulk silver droplets to desorption of silver from dispersed atoms [149].

Comparison of experimental and tabulated values.

Although Arrhenius plots suggest that kinetics are being measured, most researchers. compare their results with thermodynamic values of ΔH to deduce the mechanism. If an activation barrier is absent, this may be valid in many situations. It is important to note that, if production proceeds from adsorbed atoms on the surface, it is unlikely that thermodynamic values exist in the literature for the desorption energy of a metal from a graphite surface, with or without the existence of an activation energy barrier. As a result, many erroneous conclusions have probably been drawn, because only tabulated enthalpy data for bulk reactions are considered for comparison purposes. Additionally, Holcombe [150] pointed out that, if the free atom formation process is governed by a gas-phase process (e.g. dissociation of $MO_{(g)}$ or $M_{2(g)}$), then activation energy values obtained from slopes in Arrhenius-type plots will not reflect the enthalpy change for this homogeneous gaseous reaction responsible for free atom formation. In fact, he goes further and suggests that if such a coincidence in measured values and tabulated values exist then that cannot be the origin of the free metal.

Although the derivation and the use of activation energy implies a kinetic process, it has been pointed out that a thermodynamic assumption controlling release will yield similarly linear plots. Consider, for example, equilibrium control when Eqn (2.29) is active. Assuming the absorbance signal is proportional to the partial pressure of the metal in the vapor state, P_M, Eqn (2.29) can be rearranged as

$$A_t = c' \exp\left(\frac{-\Delta H_T^o}{RT} + \frac{\Delta S_T^o}{R}\right) \tag{2.35}$$

and

$$\log(A_t) = \frac{-\Delta H_T^o}{RT} + \left(\frac{\Delta S_T^o}{R} + \log(c')\right) \tag{2.36}$$

where c' incorporates various spectroscopic constants and it is assumed that the activity of the condensed metal is constant. Thus, reasonably assuming that ΔH_T^o and ΔS_T^o are relatively constant over the temperature range explored, a plot of log A_t versus $1/T$ would again produce a linear plot with slope proportional to ΔH. The intercept again contains proportionality constants as well as an entropy term.

It is interesting ΔS_T^o reflects the change in randomness in proceeding from a surface-bound species to a gas. Thus, if the surface species were an adsorbed atom with some degree of mobility on the surface, the entropy change would not be as large as would be the case for rigidly bound species on the surface. Looking at the kinetics, one sees similar trends where the preexponential term ranges from about 10^{14}–10^{18}, depending on whether the species desorbs from a species weakly bound to the surface (i.e. a two-dimensional gas) or rigidly fixed on the surface [151]. Thus, parallel connotations exist, regardless of looking at the problem from an equilibrium or kinetic vantage point.

Extending this reflection further, and studying its impact on the shape of the absorbance profile, reveals that adsorbed metals have enthalpies of desorption that are less than ΔH_v from the bulk liquid. Using metal vapor deposition onto a single graphite crystal and vaporization in vacuum, Arthur and Cho [152] found that ΔH_v for copper changes from 196 kJ mol^{-1} to 264 kJ mol^{-1}, depending on whether vaporization proceeded from the perimeter of small droplets or from bulk copper when larger amounts of the metal were deposited, respectively. With ETA studies where a less perfect graphite surface is employed, and where a larger number of active sites are likely, the lower energy range with small amounts appears to be from <140 kJ mol^{-1} to ΔH [152]. These lower values probably represent dispersed atoms rather than microstructures, and are supported by the first-order release suggested for this metal [60,107,138,142]. The preexponential (or entropy differences) would also be expected to increase if adsorbed species aggregate to form liquid droplets. Looking at Eqn (2.21) or (2.29) shows that these trends compensate one another in governing the rate of release (or partial pressure), and this is often referred to as the compensation effect. It is for this reason that the appearance temperatures for the metals do not shift as much as

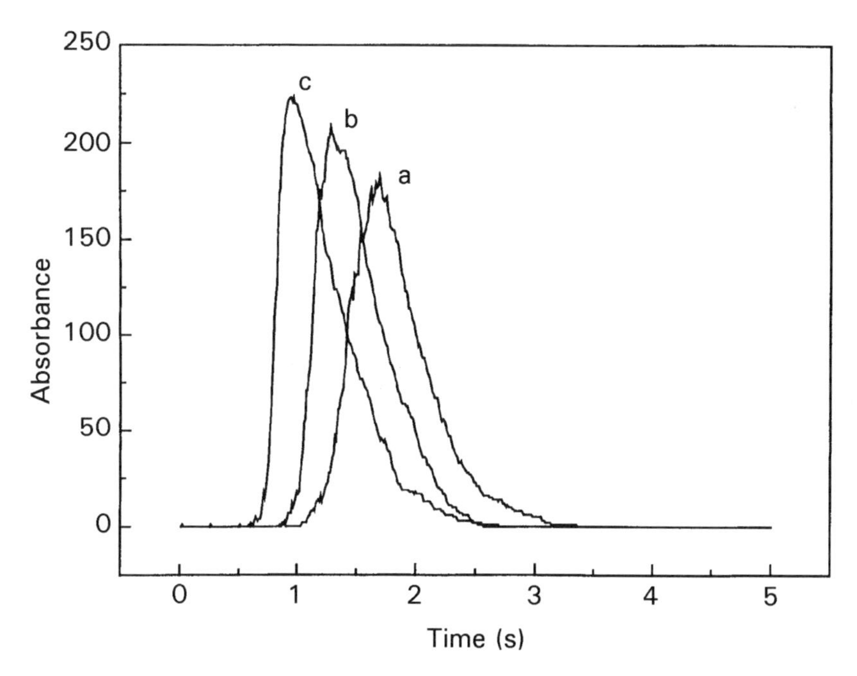

Figure 2.20 The effect of changing activation energy for release ($n = 1$) using Monte Carlo simulations. The preexponential factor has been adjusted to place the appearance temperature in the same general temperature region. Represented are curves for E_a values of: (a) 105; (b) 167; and (c) 230 kJ mol^{-1}.

might initially be expected if only a change in the enthalpy of desorption or vaporization is considered. Figure 2.20 shows some hypothetical peaks from a Monte Carlo simulation where the activation energy values and preexponential terms have been altered. With increasing values of activation energy, the preexponential term is increased to keep the curves within the same general temperature/time region. A change in activation energy from 105 to 167 kJ mol^{-1} without any adjustment in the preexponential factor could cause a shift in the appearance temperature of about 600 K. In general, increasing the activation energy (with adjustments in the preexponential term such that the appearance temperature is unaltered) produces a sharpening of the peak.

2.5.9 Conclusion

Studies to date on a variety of chemical systems suggest one central theme – the behavior of the analyte during the thermal heating process is complex and very element dependent. Preceding the actual vaporization, the analyte may be present on the surface, either as molecules or atoms, and they may exist either as dispersed or adsorbed species, as microcrystallites, or as microdroplets. There is the possibility that many species migrate below the surface, where they are trapped in the interlammelar spaces of the graphite. The energetics for removal of the analyte from these interplanar regions is not clear at this time; however, it is suspected that the attractive forces to the graphite may be weak.

The development of atomization mechanisms has reached a point where the relatively simple processes can be analyzed and modeled reasonably well with varying degrees of certainty. Diagnostics of species on the surface during the preheating stages have been probed by a number of techniques, but all have assumptions that must be considered when evaluating their reliability. Often, a key assumption is that the behavior of analytical amounts are similar to the much larger masses that must be used in the diagnostic study. Other papers make the assumption that the situation on the surface after heating and cooling (to make the measurement) are indicative of the situation at the elevated temperature.

The actual atomization process has relied heavily on deconvolution of the absorbance profile, i.e. Arrhenius plots. Although the assumptions needed for many of these diagnostic deconvolutions are minimal, the assignment of a chemical or physical process to the number is often difficult.

2.6 ANALYTE DISTRIBUTION IN ELECTROTHERMAL ATOMIZERS

In ETAAS, the formation of an analytical signal $U(t)$, recorded at the spectrometer output, can be represented by the sequence

$$m \xrightarrow{\text{atomization}} S(t) \xrightarrow{\textit{transport}} n(t; x, y, z) \xrightarrow{\text{absorption}} A(t) \xrightarrow{\text{recording}} U(t) \qquad (2.37)$$

During thermal atomization, the analyte of mass m is converted into the gaseous phase at a rate of $S(t)$ atoms per second, and is then transported into and out of the atomizer volume. These processes combine to form a transient and nonuniform absorbing layer with time and spatial dependence on the number density of analyte atoms $n(t; x, y, z)$. The exposure of the atomic cloud to line radiation from a primary source provides an optical absorbance $A(t)$, at the atomizer output, which is then transformed into an electrical output signal $U(t)$, the spectrometer read-out system.

The first step of this sequence has been described in the previous sections of this chapter. Here, the interest is on the second step of the sequence, i.e. the dynamics of formation and dissipation of the analyte layer $n(t; x, y, z)$ in the atomizer gas phase. For electrothermal atomization, both features of the dynamics, i.e. temporal evolution of the layer and its spatial structure, are equally important. In general, the factors that affect dynamics of atomic and molecular layers in ETAs can be divided into two

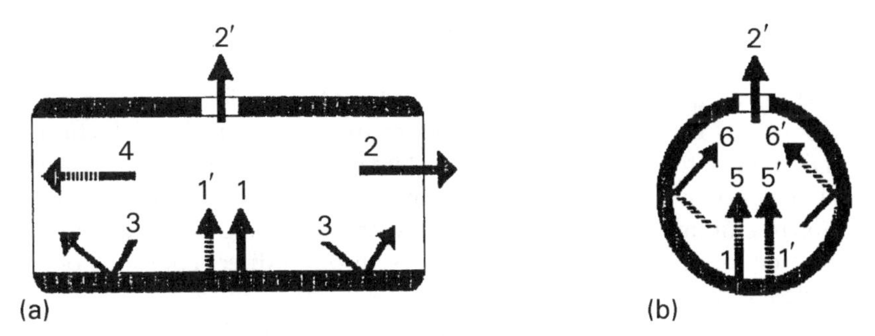

Figure 2.21 Processes occurring within the atomizer volume: (1, 1′) primary generation; (2, 2′) analyte loss through the furnace ends and through the dosing hole, respectively; (3) atom adsorption–desorption at the furnace wall; (4) analyte condensation at cooler parts of the furnace; (5, 5′) homogeneous gas-phase reactions consuming and producing free analyte atoms, respectively; and (6, 6′) heterogeneous reactions of analyte vapor with the furnace wall producing and consuming free analyte atoms. Nonatomic species containing the analyte are depicted by a broken line.

groups. First, there are the physical factors that do not change the nature of species and are characteristic of all elements. Second, there are the chemical factors that change the nature of the species and are specific to each element. Figure 2.21 is a schematic representation of the basic processes that govern the analyte distribution within the tube atomizer. The basic processes and a brief definition are given below, with a more complete discussion in the following sections.

The physical factors are represented by processes 1–4 in Figure 2.21. The primary generation of analyte vapor represents the initial vaporization of the analyte from the site of sample deposition. The analyte may be vaporized as an atomic (1) or a molecular (1′) species. Loss, through the ends (2) and through the sample introduction hole (2′), is the only true, irreversible, analyte loss from the furnace. Description of this loss factor must include different kinds of diffusive loss, as well as forced convection and thermal expulsion of the analyte by the expansion of the sheath gas within the furnace because of its rapid heating. Physical adsorption/desorption at the graphite surface (3) in this discussion is concerned only with the interaction of the free analyte atoms at the surface. Obviously, if there is no interaction between the analyte and the graphite surface, a simpler model of elastic collisions at the wall could be invoked. Condensation (4) ($M_{(g)} \rightarrow M_{(s,l)}$) may occur at the cooler parts of the furnace as most real atomizers have substantial temperature gradients.

Processes 5 and 6 are the chemical factors. Gas phase (homogeneous) reactions (5), which reduce the free atom density below the maximal value, are not true loss mechanisms, but merely represent pseudo-equilibrium situations that reduce the effective residence time of the free atoms within the atomizer. The most typical example for this kind of reaction is oxidation of a metal vapor ($MO_{(g)} + O = MO_{(g)}$). Similarly, gas-phase reactions (5′) can increase the free atom density, as in the reduction of metal oxides by carbon monoxide. Heterogeneous reactions between analyte vapor and the atomizer walls include both production (6) and loss (6′) of free atoms at the furnace wall, such as metal oxide reduction by carbon or formation of gaseous carbides through direct collisions of free analyte atoms with the graphite walls.

The processes described qualitatively above may best be viewed as microscopic steps. The mechanism leading to species present in the observation zone may involve a combination of these processes. Although a molecular species may be vaporized directly from microdroplets on the surface (primary generation, 1′), it is highly likely that the molecule will undergo many collisions with the furnace wall, and a heterogeneous reaction (process 6) may produce the free metal prior to net diffusion of the analyte into the analytical volume. Thus, to the observer, the primary generation incorporates both processes 1′ and 6. In contrast, a secondary generation of free atoms may be produced at the furnace wall at a great distance from the original site of sample deposition. Examples are presented below, as are experimental investigations and theoretical modeling of the evolution of analyte distributions within tube ETAs.

2.6.1 Experimental investigations of analyte distribution

Various approaches have been used to measure the distribution of species during atomization in a graphite furnace. Salmon and Holcombe [153] used a spatial isolation wheel technique, that allowed collection of temporally and spatially resolved absorbance data for nine distinct zones in a West-type carbon filament atomizer (see Section 1.4 and Figure 1.5). When applied to a tube furnace, the technique is limited by providing information only about a narrow region along the vertical diameter of the tube (one-dimensional spatial resolution), and not providing information about the entire cross-section of the atomizer.

Two-dimensional resolution of sodium atom distributions within a graphite tube was obtained by Huie and Curran [154] who used a laser-based silicon Vidicon imaging system. This only allows for a single frame to be taken from each atomization event, so these two-dimensional measurements are subject to the uncertainty caused by run-to-run variations in the atomization process. The shadow spectral filming (SSF) technique developed by Gilmutdinov *et al.* [119] allows the recording of many images of the whole atomizer cross-section during a single atomization event. The image of the atomizer volume is obtained by backlighting with the analytical line of the element of interest and recording, by use of a cine camera, the shadow produced by the absorbing vapor. Major features of analyte distributions obtained using this technique [119–122] are discussed below.

Generally, the above two groups of physical and chemical factors act simultaneously, the influence of the chemical factors appearing against the background of the physical processes. Therefore, it is desirable initially to isolate the effect of the physical processes. For this purpose, an element that is fairly inert to the chemical processes that occur within the graphite furnace should be investigated. The structure of the absorbing layer of a chemically inert element can only be determined by the physical processes and a suitable element could be used to investigate this group of factors. Once the influence of the physical processes has been established, it should be much easier to investigate the influence of the chemical processes.

A suitable test element is silver [121], and the results of shadow spectral imaging of the atomization of 2.5 ng of silver in a Massmann furnace equipped with a platform are shown in Figure 2.22(a). The sharp rise in the absorption pulse (Figure 2.22(b) indicates that platform atomization of silver is fairly fast. Nevertheless, as can be seen from both the images, and also more quantitatively from Figure 2.22(c) and (d), the cross-sectional distributions of silver atoms are practically uniform, except for the earliest moments in the atomization process. This suggests that analyte diffusion is relatively rapid relative to the time-scale of the vapor generation process, resulting in a nearly uniform distribution across the furnace diameter. Silver atomization occurs at temperatures below 1600°C. Thus, it is reasonable to assume that chemically inert elements evolved at high temperatures from a platform surface should also be distributed nearly uniformly over the atomizer radial cross-section. Further, it is reasonable to conclude that cross-sectional nonuniformities in the analyte distribution within the furnace strongly indicate the presence of gas-phase or heterogeneous interactions.

Figure 2.2(a) also illustrates the anisotropy of longitudinal and transverse propagation of silver vapor in the atomizer volume. Frames 58–63 clearly show the absence of atoms under the platform, while the upper part of the furnace is completely filled by atoms. This means that silver atoms propagate faster in the furnace cross-section than in its longitudinal section [121]. The most likely reason for the delay in atom propagation in the longitudinal direction is the remarkable decrease in the temperature of the atomizer towards the ends of the tube (see Section 3.1.4). This can lead to adsorption/condensation of the vapor on the graphite walls at distances shorter than the length of the platform. The anisotropy of vapor propagation is also characteristic of practically all the other elements investigated by the 5SF technique [119–121].

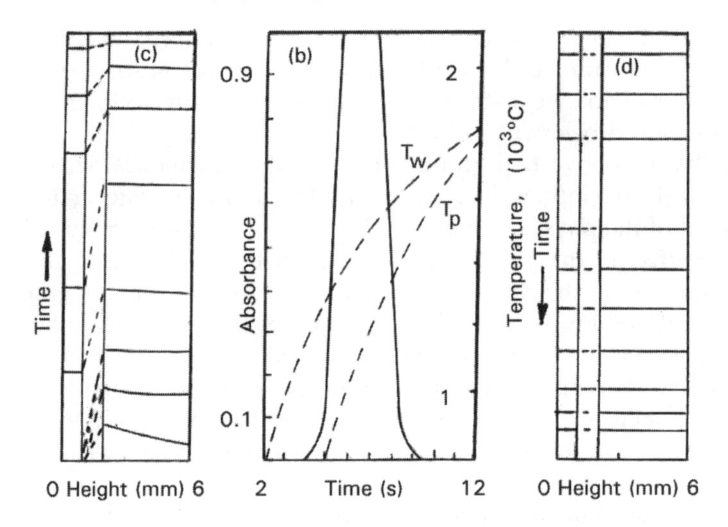

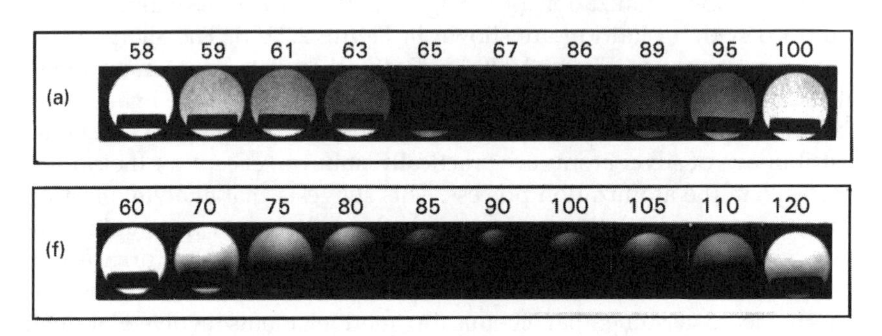

Figure 2.22 Dynamics of the formation and dissipation of the absorption layer of silver atoms using atomization from a platform under (a)–(d) gas-stop and (f) gas-flow conditions. For the gas stop and gas flow conditions, 2.5 and 50 ng of the metal was used respectively: (a) images of the process recorded under gas-stop conditions using the SSF technique; (b) the peak absorbance; (c) and (d) absorbance versus height contours at 0.17 s time intervals prior to, and after, the peak absorbance, respectively; and (f) images of the flow rate of 300 ml min^{-1}. Curves T_w and T_p represent the change of wall and platform temperatures, respectively. The figures over the frames (for this figure and subsequent figures) are their numbers beginning at the onset of atomization. The images were recorded at 12 frames per second [121]. (Reprinted by permission of the Royal Society of Chemistry from, A.K. Gilmutdinov, Y.A. Zakharov and A.V. Voloshin, Shadow Spectral Filming: A Method of Investigating Electrothermal Atomization. Part 3. Dynamics of Longitudinal Propagation of an Analyte within Graphite Furnaces, *J. Anal. At. Spectrom.*, **8**, 387–395 (1993).)

Figure 2.2(f) presents the results of shadow imaging of the dynamics of silver atom formation and dissipation when using an internal gas flow of 300 ml min^{-1} for the atomization 50 ng of the metal. Frames 70–110 of the figure show strong cross-sectional nonuniformities of the absorbing layer for the duration of the atomization process, with a sharp decrease in atom concentration when going from the platform to the top of the atomizer. The results presented here reveal an interesting feature of gas dynamics within the atomizer: the atoms appear in the region under the platform and remain there until the very end of atomization (frames 75–120). Comparison of this observation with Figure 2.22(a) shows that, when an internal gas flow is used, the atoms reach the region under the platform even earlier than in the case of gas-stop conditions (compare frames 59–67 in Figure 2.22(a) and frames 70–75 in Figure 2.22(f). This indicates that the internal gas flow (at a volume velocity of 300 ml min^{-1}) is not laminar in the central part of the atomizer where the two flows of argon gas coming from the tube ends meet. The earlier appearance of the atoms under the platform in this case can be explained only by the presence of a turbulent flow of argon gas in the central region of the furnace, which can transport the atoms under the platform.

An internal gas flow of 300 ml min^{-1} is typical for the pyrolysis stage in ETAAS. The above observations suggest that the sample matrix, which is assumed to be carried away by the internal gas flow, might partially remain under the platform. This may potentially lead to interferences during the subsequent atomization stage.

The results presented above illustrate the influence of the group of physical factors on analyte distribution. The following examples deal with the additional effect of the chemical factors. Figure 2.23(a) presents shadow spectral imaging for the atomization of 50 ng of gallium from a platform, recorded at the 403.3 nm atomic line. The results show an apparent anomaly: the gallium atoms enter the gas phase, not from the platform where the sample was deposited, but from the upper furnace wall (frame 99). This leads to an inverse distribution of gallium atom concentration in the furnace cross-section with a maximum not above the atom source (the platform) but at the opposite furnace wall. The probable reason for this phenomenon is that the sample first vaporizes as suboxides (e.g. Ga_2O, process 1′ in Figure 2.21) which do not absorb at the gallium 403.3 nm atomic line [119,120]. Subsequently, the molecules are reduced (or dissociate) on the higher temperature graphite wall and reenter the furnace as atoms. This is a direct illustration of process 6 in Figure 2.21.

Another pronounced feature, shown in Figure 2.23(a) is transverse oscillation of the analyte vapor during the course of platform atomization. Frames 99–103 clearly show a sharp decrease in the concentration of gallium atoms near the upper surface of the platform. As the temperature of the platform at this time is approximately 400°C lower than that of the

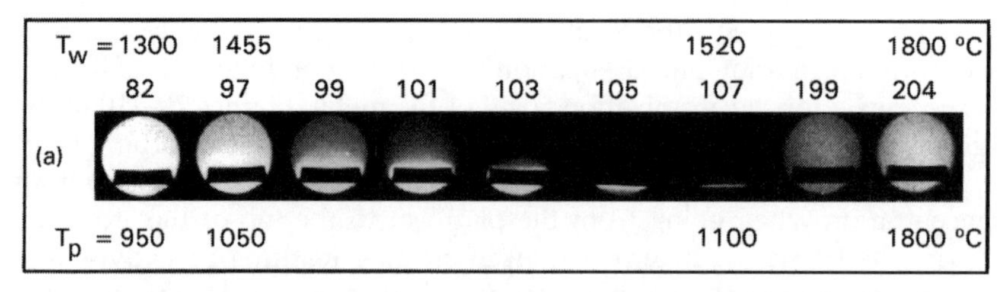

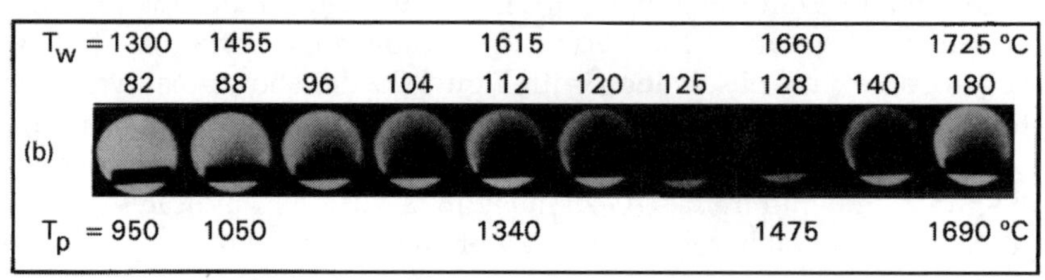

Figure 2.23 Dynamics of formation and dissipation, in a graphite furnace, of the layer of (a) gallium atoms and (b) molecules. In all figures the numbers over the frames are the number of frames taken since the onset of atomization. Filming speed is 16 frames per second. In all figures the numbers labeled T_w and T_p indicate the temperature of wall and platform, respectively, for a given frame [120]. (Reprinted by permission of the Royal Society of Chemistry from, A.K. Gilmutdinov, Y.A. Zakharov, V.P. Ivanov, A.V. Voloshin and K. Dittrich, Shadow Spectral Filming: A Method of Investigating Electrothermal Atomization. Part 2. Dynamics of Formation and Structure of the Absorption Layer of Aluminium, Indium and Gallium Molecules, *J. Anal. At. Spectrom.*, **7**, 675–683 (1992).)

walls, as noted in the figure, it can be assumed that the gallium vapor is being efficiently adsorbed back onto the cooler platform. Interestingly, this condensation process occurs later in the region under the platform (frames 105–107), when the vapor reaches this area. Subsequently, as the platform is further heated, the condensed species reenter the gas phase in the furnace (frame 199).

It follows from the above consideration that gallium molecules play an important role in gallium atomization. To complete the spatially resolved analysis, it is necessary to obtain the images of gallium-containing molecular distributions within the graphite furnace, as shown in Figure 2.23(b) [120]. Analysis of this figure reveals significant transverse and longitudinal nonhomogeneity of the molecular distribution. The molecular concentration is always highest near the platform, continually decreases moving away from it, and is always minimal near the graphite walls. This is characteristic of oxide molecules, and this distribution indicates that the oxide molecules are formed directly by the process of sample evaporation:

$$Ga_2O_{3(s)} \rightarrow Ga_2O_{(g)} + O_2 \tag{2.38}$$

Molecular (Figure 2.3(b)) and atomic (Figure 2.23(a)) distributions are seen to be complementary to one another; i.e. atoms appear near the top of the atomizer where the molecules disappear. This complementary distribution indicates that both processes are related.

The spatial distributions can be explained by a two-step atomization mechanism. First, the sample vaporizes according to Eqn (2.38) (process 1′ in Figure 2.21), to give a high concentration of gallium suboxide near the platform. The suboxide is then reduced to form free gaseous gallium atoms on close encounter or collision with the opposite graphite wall for the following reasons: (1) the temperature of the wall is about 400°C higher than the temperature of the platform, leading to increased thermal dissociation; (2) the equilibrium of the reaction

$$Ga_2O_{(g)} = 2Ga_{(g)} + 1/2O_2 \tag{2.39}$$

is shifted to the right near the graphite walls because of a decrease in the partial pressure of oxygen (reaction 5′ in Figure 2.21); and (3) there may also be a direct reduction of the suboxide (reaction 6 in Figure 2.21), according to

$$Ga_2O_{(g)} + C_{(s)} = 2Ga_{(g)} + CO \tag{2.40}$$

Figure 2.23(b) also indicates only a small number of molecules under the platform throughout the duration of the atomization event, and therefore, most of the molecules must be confined to the central part of the atomizer. This could occur because of the reduction of the gallium suboxide by the graphite surfaces, according to Eqn (2.40), as the molecules move towards the ends of the furnace, and/or the oxidation of the gallium suboxide into a solid-phase oxide as a result of Eqn (2.38) proceeding in the opposite direction at the cooler end of the tube.

Of 13 elements investigated by the SSF technique, effects similar to those presented in Figure 2.23(a) and (b) were recorded [119–121] for aluminum, indium, thallium, and germanium. Thus, the described features are rather typical for electrothermal atomization.

The imaging experiments were further modified by replacing the film-based camera with a charge-coupled device (CCD) camera [122] that enabled digitized images with temporal and spatial resolution of the atomization processes to be obtained. Figure 2.24(a) shows the atomic absorbance profile recorded by the conventional AA spectrometer for the atomization of 15 ng of aluminum from a graphite platform, and Figures 2.24(b) and (c) present the distribution of aluminum atoms along the vertical diameter of the graphite tube, prior to and after the absorbance maximum, respectively. These distributions were obtained by reading out the appropriate portion of the complete image obtained by the CCD imaging system. The dotted lines connect the absorbances that were measured both above and below the platform at the same moment of time. There are two noteworthy features in Figure 2.24(b). First, a dis-

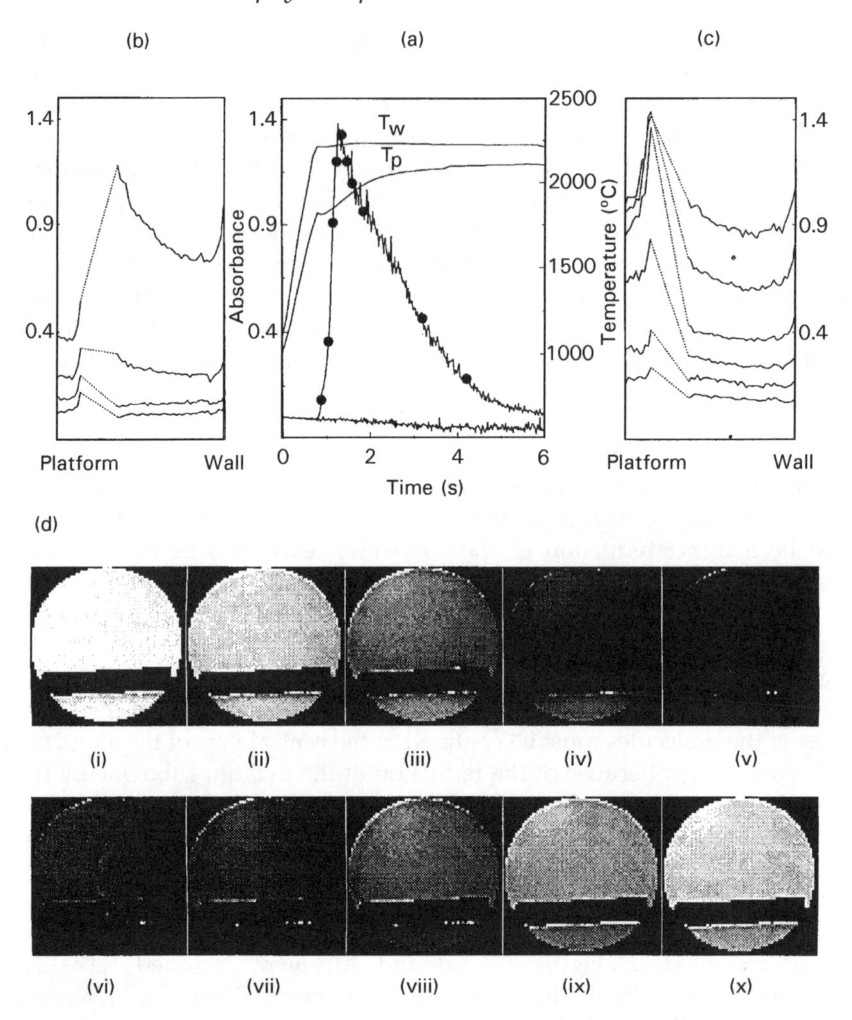

Figure 2.24 (a) Conventional recording of the atomization of aluminum from the platform and the associated temperature profiles; (b and c) distribution of aluminum atoms along the vertical diameter of the graphite tube during atomization from the platform, prior to and after the absorbance maximum respectively; (d) sequence of images for atomization of aluminum from the platform.

tinctly high aluminum atom density is seen immediately under the graphite platform, and the distributions show that initial atomization occurs underneath the graphite platform. Second, in the distribution recorded just before the absorbance maximum, there are far fewer atoms under the graphite platform than above it, even though the graphite platform used in this experiment is the type that only contacts the graphite tube surface

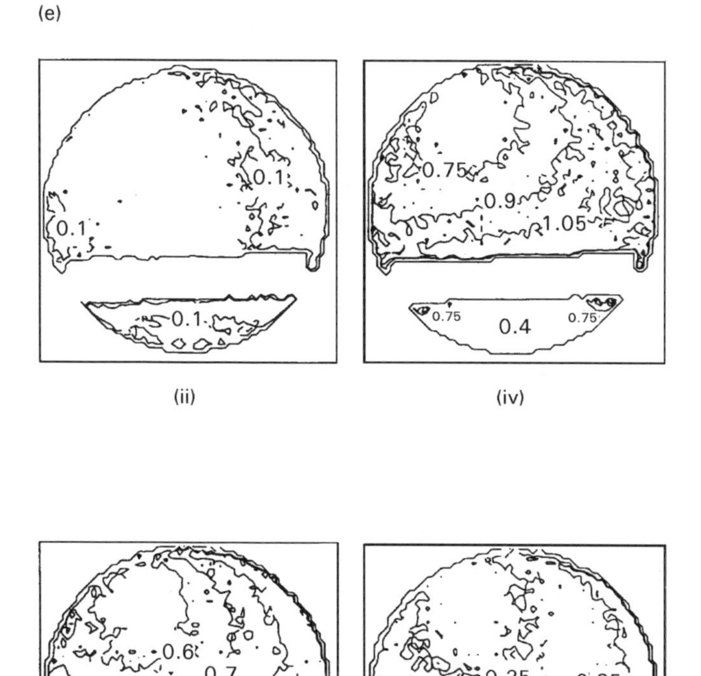

Figure 2.24 (e) absorbance contour maps for atomization of aluminum from the platform [122]. (Reprinted with permission from C.L. Chakrabarti, A.K. Gilmutdinov and J.C. Hutton, Digital imaging of atomization processes in electrothermal atomizers for atomic absorption spectrometry, *Anal. Chem.* **65**, 716–723. Copyright 1993 American Chemical Society.)

at four corners. For the distributions recorded after the absorbance maximum (Figure 2.24(c)) the opposite is true, i.e. there is a greater concentration of atoms below the graphite platform. Thus, it can be assumed that there is a redistribution of the analyte during atomization.

The shadow images of the process recorded by the CCD-based spectrometer are given in Figure 2.24(d). The distributions were recorded at the times indicated by the filled-in circles on the conventionally recorded atomic absorbance profile (Figure 2.24(a)). Atomization is initiated primarily from the graphite tube wall that is furthest from the sample introduction hole (frame (ii)), with a smaller amount originating from the graphite platform surface. However, as the atomization that is occurring from the graphite tube wall does not appear along the vertical diameter of the graphite tube, it is not recorded in Figure 2.24(a). The fact that the sample is initially deposited on the graphite platform, and that atomization starts from the wall, indicates that atomization does not occur directly from the sample. It appears that aluminum is vaporized as both gaseous atoms and molecules. The molecules that are vaporized are reduced to gaseous atoms when they come in contact with the heated graphite walls.

The sample introduction hole is seen to play an important role in determining the distribution of aluminum atoms. There is a pronounced decrease in the concentration of gaseous aluminum atoms near the sample introduction hole (in the upper left-hand comer of the images), which is probably a result of the ingress of oxygen through the sample introduction hole, as discussed previously. The ingress results in oxidation of the gaseous aluminum atoms, and this gives a direct illustration of process 5 in Figure 2.21. Later frames (Figure 2.24(d) (v–viii)) show the above-mentioned redistribution of the aluminum atoms. The maximum concentration of gaseous aluminum atoms changes its location, and occurs underneath the platform. This is probably a result of the more reducing environment underneath the platform, which sponsors reduction of aluminum oxides.

All of the above-mentioned features can be seen quantitatively in the absorbance contour maps in Figure 2.24(e), which correspond to the images in Figure 2.24(d) that have the same Roman numeral identification. The locations for the onset of atomization are easier to identify in absorbance contour maps, and image. Figure 2.24(e) frame (ii) clearly shows the simultaneous onset of atomization taking place, from both the graphite tube walls and the graphite platform. The onset of atomization is occurring both above and below the platform. In image (iv), the supply of aluminum atoms to the underside of the graphite platform can be seen along the edge of the bottom corners of the graphite platform, and images (vi) and (viii) clearly show that the region of highest aluminum atom concentration is located under the graphite platform at later times.

The absorption images of aluminum-containing molecules are shown in Figure 2.25(a) [120]. A sharp decline in molecular concentration near the graphite walls throughout the duration of the atomization event can be seen (transversely inhomogeneous distribution), and no molecules are detected under the platform during the course of atomization, which indicates that they are primarily confined to the central part of the graphite tube (longitudinally inhomogeneous distribution). These two features are similar to the aforementioned observations for gallium and suggest that atomization mechanisms for aluminum and gallium are similar. However, there are some additional features in the molecular distribution of aluminum. The mechanism of formation of aluminum vapor is not the same throughout the atomization event. There are two definite stages, initial vaporization from the platform (frames 72–90) and subsequent formation of a cloud in the gaseous phase (frames 94–114). The cloud has a complex structure, the center not absorbing as strongly as the perimeter (frames 94–114), and the perimeter having the same geometry as the inner surface of the furnace. This torroidal-shaped cloud is henceforth called the 'donut' [120]. Figure 2.25(b) shows the dynamics of formation and dissipation of the aluminum vapors, recorded by emission at 465 nm [120]. The initial formation of aluminum vapor, clearly seen in absorption (frames 72–90 in Fig. 2.25a), cannot be detected in emission. However, emission images that are formed are similar to the donut measured by absorption (frames 94–114 in Figure 2.25(a). This confirms that the process of formation of aluminum vapor incorporates two separate stages. The authors concluded that the cloud consists of finely dispersed condensed particles, most probably solid or liquid aluminum oxide [120]. Thermodynamic analysis of the Al–O–C system [155] predicts the existence of considerable amounts of aluminum oxides, not only in the gas phase, but also in condensed phases. The formation of condensed particles in the furnace gas phase was also reported by L'vov and co-workers [156,157] for a number of other elements. Thus, it can be

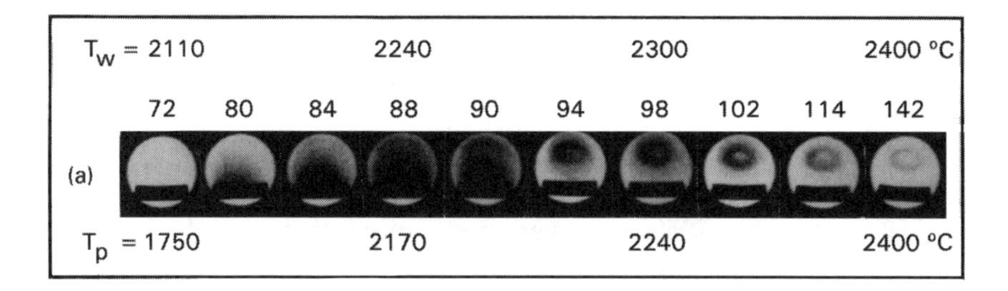

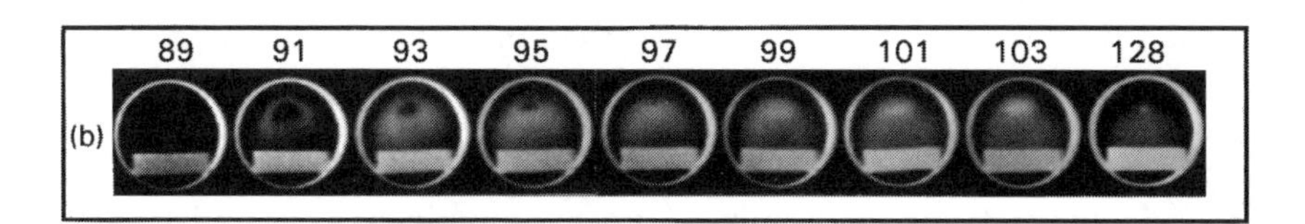

Figure 2.25 Sequence of (a) absorption and (b) emission images for the vaporization of aluminum-containing molecules from the platform [120]. (Reprinted by permission of the Royal Society of Chemistry from, A.K. Gilmutdinov, Y.A. Zakharov, V.P. Ivanov, A.V. Voloshin and K. Dittrich, Shadow Spectral Filming: A Method of Investigating Electrothermal Atomization. Part 2. Dynamics of Formation and Structure of the Absorption Layer of Aluminium, Indium and Gallium Molecules, *J. Anal. At. Spectrom.*, **7**, 675–683 (1992).)

concluded that, within ETAs, there may exist not only atoms and molecules but also condensed-phase particles even at temperatures as high as 2400°C (Figure 2.25(b)).

The results presented above testify to the complexity of atom formation and dissipation processes. The processes occur against a background of rapidly increasing atomizer temperature with pronounced spatial temperature gradients. Using a variety of tools discussed thus far, many of the mechanisms responsible for the formation of the free AA signal can be deduced. In an attempt to make *a priori* predictions of the impacts of changing various parameters in the ETA system, it would be ideal to completely understand all interactions that can occur and, with this information, construct a modeling program. Armed with such a model, rapid assessment could be made of the analytical impact of changing such parameters as heating rate, furnace geometry, sheath gas composition, etc. A complete model will also permit the monitoring of parameters which cannot easily be measured experimentally, such as atom densities, time-dependent spatial distributions of the atomic vapor, analyte adsorption rates on the furnace wall, etc.

As attractive as such prognosticative capabilities would appear from the construction of such a model, the discussions and illustrations thus far clearly indicate that an accurate model of the system, including analyte distribution, is a rather complex task. One step in the development of an accurate model involves the basic steps of analyte formation and dissipation in the atomizer gas phase.

2.6.2 Theoretical modeling of analyte transfer

The first attempt at modeling analyte distribution within an ETA was made by L'vov [2] in 1959, in the very first paper introducing these devices into the analytical arena. A considerable number of papers have been devoted to the solution of the problem since then. Historically, two clear stages in the development of the models can be seen. The first stage was dealing primarily with accounting for the nonstationary character of the atom layer formation. All the models constructed at this stage were one-dimensional models, i.e. they considered only longitudinal analyte distributions. The other important features of the transfer process, such as real tube geometry of an atomizer with the sample aperture in its center and significant cross-sectional nonuniformity of the analyte layer, were completely ignored because of the obvious complexity that they introduced. One of the major results obtained at this stage of development was that, at any given moment in time, the number of atoms, $N(t)$, in the atomizer volume is described by the convolution of the two independent functions given in Eqn (2.18). The supply function, $S(t')$, represents the rate (atoms per second) the primary generation of analyte atoms from the deposited sample (process 1 in Figure 2.21), whereas the removal function, $R(t, t')$, characterizes subsequent transfer of the vaporized atoms in the atomizer volume. As follows from this relationship the general problem of modeling analytical signal formation involves two independent problems, determining the sample supply function and calculating the analyte removal function. The supply function has been discussed extensively in the preceding sections. This section focuses on the features of the removal function.

The removal function describes the probability that an analyte atom, vaporized at time t', will be present in the atomizer volume at a later instant, $t > t'$ [158]. Early in time, the probability is equal to unity and decreases with increasing time. In general, the shape of the removal function is governed by all the physical and chemical processes occurring in the atomizer (processes 2–6 in Figure 2.21). Of these processes describing analyte transfer, process 2, including convection, concentration diffusion and thermal diffusion, has been most intensely investigated. Figure 2.26 presents the change in the removal function when taking into account all the above transfer mechanisms for a typical heating regime shown by curve 1 [158]. The first and simplest approximation (curve 2) is obtained when only concentration diffusion is taken into account, while neglecting the rapid increase in the diffusion coefficient as the temperature increases. Allowance for the temperature dependence of the diffusion coefficients transforms the removal function into curve 3. Introduction of the convective transfer of atoms yields curve 4 and, finally, the total removal function allowing for all the factors including thermal diffusion is presented by curve 5. From these results, it is seen that the predominant transfer mechanism in ETAs is concentration diffusion, and accounting for the temperature dependence of the diffusion coefficient is essential. The other transfer mechanisms appear to play a secondary role.

Knowledge of the removal function allows the mean residence time τ analyte atoms in the atomizer volume to be calculated. This parameter is defined as

$$\tau(t') = \int_0^\infty R(t, t')\mathrm{d}t \tag{2.41}$$

and gives the life-time of vaporized atoms in the atomizer volume. Obviously the analyte residence time will be dependent on t', the time of onset of evaporation. For the established temperature profile, the parameter can be estimated [158] as follows

$$\tau = \frac{1}{8D_{\mathrm{ef}}}\left(L^2 - \frac{a^2}{3}\right) \tag{2.42}$$

where a is the linear dimension of the area occupied by the sample, L is the furnace length, and D_{ef} is the effective diffusion coefficient in the long-

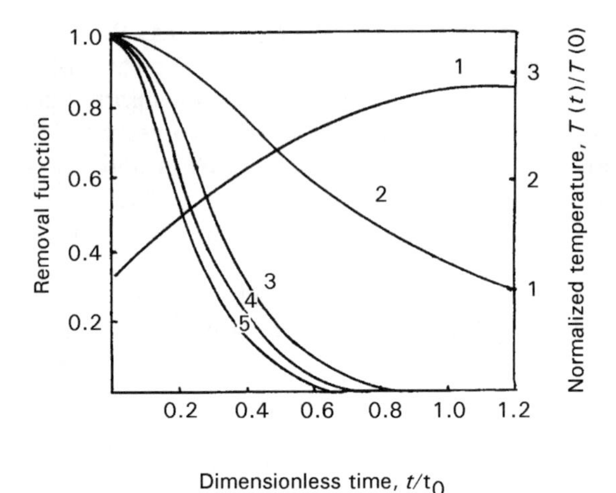

Figure 2.26 The change of the removal function as the determining factors are taken into account: (2) only concentration diffusion at a constant diffusion coefficient at the initial temperature is taken into account; (3) only concentration diffusion, but the increase in diffusion coefficient with increasing temperature is taken into account; (4) as (3) plus convective transfer because of gas expansion; and (5) as (4) plus thermodiffusion mechanism. Curve 1 shows the assumed heating curve [158]. (Reprinted from *Spectrochim. Acta, Part B*, Vol. **39B**, A.K. Gilutdinov and I. Fishman, The Theory of Sample Transfer in Semienclosed Atomizers for Atomic Absorption Spectrometry, pp. 171–192, 1984, with kind permission of Elsevier Science – NL, Sara Burgerhartstraat 25, 1055 KV Amsterdam, The Netherlands.)

itudinally nonisothermal furnace. The effective diffusion coefficient can be estimated [158] as

$$D_{ef} = D_0(T_e/T_c) \tag{2.43}$$

where D_0 is the diffusion coefficient at the temperatures T_c of the furnace center and T_e the temperature of the furnace ends. It is seen from Eqn (2.42) that, when assuming a point source of analyte atoms ($a = 0$) and an isothermal atomizer ($T_e = T_c$), this relationship reduces to the simplified relationship shown in Eqn (2.30). Spreading of the sample over the whole furnace surface ($a = L$) reduces the atom residence time by one third, and if the actual nonisothermality in Massmann atomizers ($T_c \approx 2T_e$) is taken into account, this leads to a corresponding increase in the mean residence time τ.

The complex chemistry and geometry associated with the graphite furnace means that further progress in modeling analyte distributions could be done only by employing computer simulation. This second stage of modeling the analyte layer formation began when Holcombe and

Rayson [59] introduced Monte Carlo simulation techniques to model the atom formation and dissipation processes within a graphite furnace. The approach enabled not only accurate description of temporal features of analyte layer formation, but also the full extent of spatial nonuniformities of analyte distributions. Additionally, computer simulation readily allowed for changes in furnace geometries, including geometrical perturbations such as a sample introduction hole. Specific interactions (wall adsorption, convective gas flows, etc.) could easily be added or deleted from the computer program to evaluate their impact on the analytical signal.

In essence, the Monte Carlo simulation is a stochastic approach, that monitors individual atoms through the time period being viewed. As a consequence of tracking the individual particles, the fraction of the total number that are located on the wall, in the gas phase, or outside the furnace, can easily be monitored as a function of time. Similarly, the approach is capable of recording the specific location of particles within the gas phase, and hence allows the monitoring of both axial and radial distributions. Figure 2.27 presents cross-sectional views of the furnace during simulated atomization of copper [104]. The position of every atom can be monitored, not only in the gas phase (b), but also on the graphite surface (c). It is seen that a significant fraction of initially vaporized copper atoms is retained on the graphite surface because of adsorption (process 3 in Figure 2.21).

In the course of running the simulation program, a number of parameters can be traced independently as a function of time. Figure 2.28

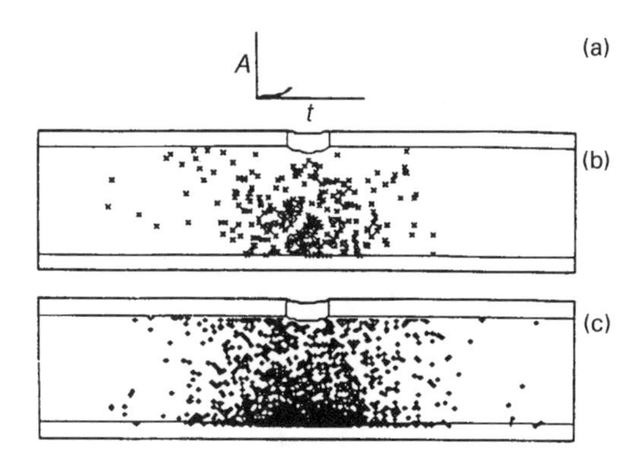

Figure 2.27 Explicit Monte Carlo simulation of the early stage of copper atomization: (a) absorbance–time profile; (b) distribution of particles in the gas phase; (c) distribution of particles adsorbed on the surface [104]. (Reprinted with permission from O.A. Güell and J.A. Holcombe, Analytical applications of Monte Carlo techniques, *Anal. Chem.* **62**, 529A–542A. Copyright 1990 American Chemical Society.)

shows additional information that has been extracted from the simulation [104], i.e. the fractional number of atoms in the gas phase (a) and on the wall (b), and the number of atoms lost through the sample introduction hole (c) and through the furnace ends (d). Curve (c) shows that significant numbers of atoms are lost through the sample introduction hole. For a hole diameter of 0.15 cm and a furnace length of 1 cm, more than 30 per cent of the atoms placed within the furnace are lost through the sample introduction hole, whereas simple area considerations of the hole versus the open ends of the tube predict that only 11 per cent would escape. This is a logical result of the initial placement of the sample under the sample introduction hole.

The impact of furnace geometry and initial sample distribution on the absorbance profiles were simulated by Güell and Holcombe [159,160], who showed that increasing furnace length produces nearly proportional increase in peak area:

$$\int_0^\infty A(t)\mathrm{d}t \propto L^\Omega \tag{2.44}$$

where Ω varies with changes in the sample introduction hole, heating program and readsorption barrier (i.e. analyte–graphite interactions). Therefore the simplified equation, Eqn (2.42), predicting the L^2 dependence, can only be used as an estimate for the atom residence time, as it does not account for many important features of the process. The Ω factor also seems to be altered by using a platform, but not by changes in the furnace diameter. Increasing the furnace diameter increases the fraction of

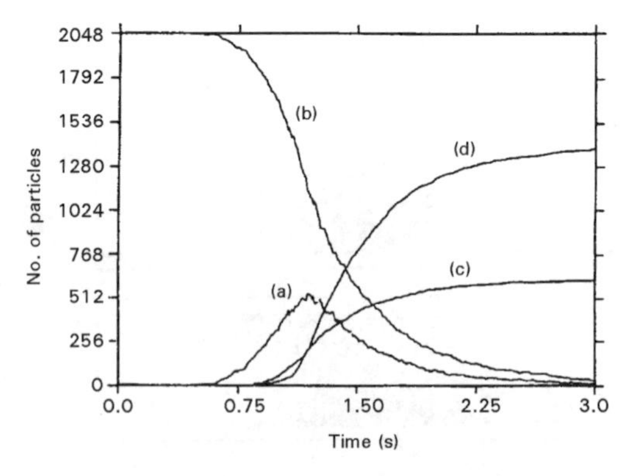

Figure 2.28 Explicit Monte Carlo simulation of copper atomization. Distribution of particles during the transient heating pulse: (a) in the gas phase; (b) adsorbed on the surface; (c) lost through the sample introduction hole; (d) lost through the furnace ends [104]. (Reprinted with permission from O.A. Güell and J.A. Holcombe, Analytical applications of Monte Carlo techniques *Anal. Chem.* **62**, 529A–542A. Copyright 1990 American Chemical Society.)

the sample found in the gas phase, but decreases the peak absorbance signal. An increase in the length and a decrease in the diameter of the furnace would significantly increase the signal but would also magnify the impact of the sample introduction hole. An improvement in precision and/or accuracy may be realized if the precision in initial sample positioning can be improved. This may prove to be particularly noticeable with solvents that wet the graphite. All improvements noted can be realized if other factors are held constant; for example, if an increase in the furnace length is accompanied by a reduction in heating rate or increasing non-uniformity, then a net loss in signal may actually be observed.

Another approach, also providing three-dimensional evolution of an analyte within tube atomizers, has been developed by Gilmutdinov *et al.* [161]. The approach is based on numerical solution of the corresponding equation for analyte transfer, together with a set of boundary conditions. Although the accuracy and flexibility of the program are not as good as the Monte Carlo approach, the computation time is greatly reduced and many of the key parameters contributing to the time and spatially dependent free atom densities can be included. The distribution of analyte concentration in the atomizer gas phase is described by the continuity equation, which is actually a mathematical expression of the law of mass conservation:

$$\frac{dn}{dt} + \operatorname{div} J = \left(\frac{\delta n}{\delta t}\right)_v \qquad (2.45)$$

The atom flux J includes all of the above transfer mechanisms (convection, concentration and thermal diffusion). The right-hand side of this equation expresses the rate of variation of the concentration of analyte atoms due to the volume processes (processes 4 and 5 in Figure 2.21). Processes 3 and 6, on the atomizer walls (Figure 2.21), and also the presence of the sample introduction hole are taken into account as a set of proper boundary conditions.

All the predictions of the model regarding the effect of atomizer geometry, sample introduction hole etc., are in agreement with the previously discussed Monte Carlo simulation [59,104,105,160]. However, the model was also designed to account for condensation of the analyte vapor in the furnace gas phase, which is an additional process affecting the analyte distribution. Significant temperature gradients that may exist in ETAs are favorable for formation of condensed particles near the cooler ends of the furnace. In general, condensation may occur both in the furnace gas phase (homogeneous condensation) and at the furnace walls (heterogeneous condensation), and there is experimental evidence for both types of processes occurring within the graphite furnace. Figure 2.25 clearly shows formation of alumina microparticles in the gas phase of the graphite furnace. L'vov and co-workers [156,157] reported condensation of silver, gold, copper, magnesium, manganese, and palladium when using excessive masses of the metals. Direct scanning electron microscopy and radiotracer techniques [162] have shown that a significant portion of the sample, initially located at the furnace center, can be detected at the furnace ends after the atomization cycle.

The results of computer simulation of the atomization of 10^{13} silver atoms under the conditions of a Massmann end heated furnace (length 2.8 cm, inner tube diameter 0.6 cm, and diameter of the sample introduction hole 0.2 cm), are presented in Figures 2.29 and 2.30. It was assumed that the longitudinal distribution of the atomizer temperature can be described by a Gaussian function, with the temperature at the furnace center two times higher than the temperature of the ends. Figure 2.29 shows the spatially integrated characteristics of the process. Here, curve $N_a(t)$ presents the change in the total number of free silver atoms in the atomizer gas phase, whereas the curve $N_c(t)$ represents the number of silver atoms that have condensed to form microparticles by time t. In this model, only homogeneous condensation is considered and the process is assumed to be irreversible. Therefore, $N_c(t)$ monotonically increases and

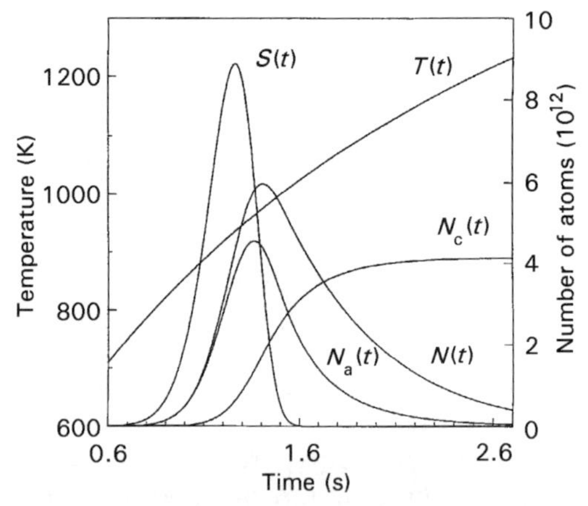

Figure 2.29 Explicit simulation of silver atomization. The first-order release of a silver atom from the surface $S(t)$ for a given heating curve $T(t)$. Curves $N_a(t)$ and $N_c(t)$ represent the number of silver atoms in the furnace gas-phase and in the condensed phase, respectively. Curve $N(t)$ represents the change in the number of silver atoms in the furnace gas-phase when neglecting the condensation [161]; (Reprinted from *Spectrochim. Acta, Part B*, Vol. **50B**, A.K. Gilmutdinov, R.M. Mrasov, A.R. Somov, C.L. Chakrabarti and J.C. Hutton, Three-dimensional Modeling of the Analyte Dynamics in Electrothermal Atomizers for Analytical Spectrometry: Influence of Physical Factors, pp. 1637–1654, 1995, with kind permission of Elsevier Science – NL, Sara Burgerhartstraat 25, 1055 KV Amsterdam, The Netherlands.)

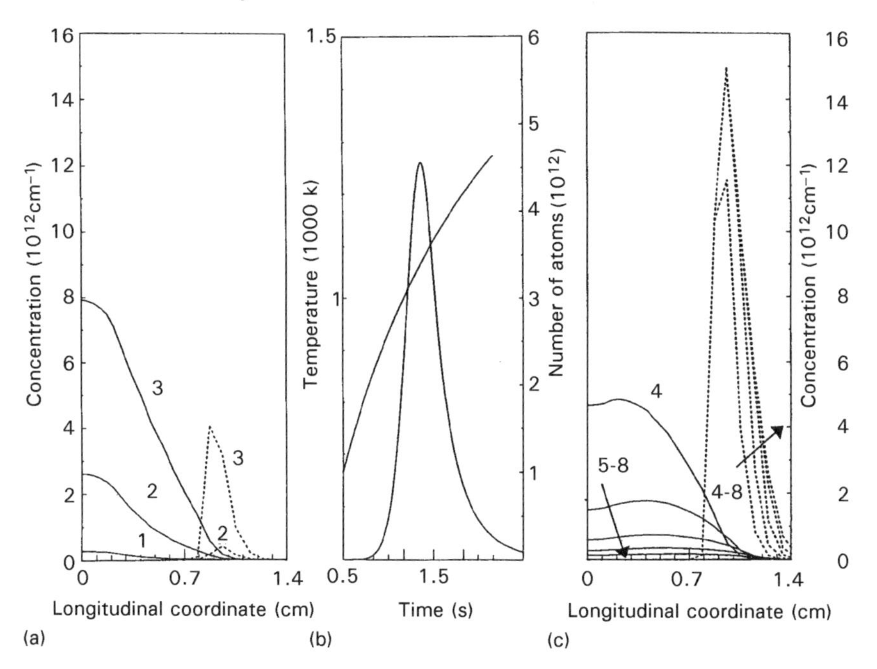

Figure 2.30 Computer simulation of longitudinal distributions of silver atoms presented in the furnace in the form of monomer (solid lines) and in the form of microdrops (dashed lines): (b) The change of total number of free atoms and corresponding heating curve; (a) and (c) the atom density versus distance from the furnace center for time intervals preceding and after the peak atom percentage respectively. Time contours are drawn every 0.2 s [161]. (Reprinted from *Spectrochim. Acta, Part B*, Vol. **50B**, A.K. Gilmutdinov, R.M. Mrasov, A.R. Somov, C.L. Chakrabarti and J.C. Hutton, Three-dimensional Modeling of the Analyte Dynamics in Electrothermal Atomizers for Analytical Spectrometry: Influence of Physical Factors, pp. 1637–1654, 1995, with kind permission of Elsevier Science – NL, Sara Burgerhartstraat 25, 1055 KV Amsterdam, The Netherlands.)

produces an upper limit to the condensation process. Curve $N(t)$ represents the change in the total number of silver atoms without accounting for condensation. These data show that condensation is significant, with more than 40 per cent of the initially atomized silver condensing on the cooler parts of the furnace.

Figure 2.30 gives more detailed insight to the simulated process. The figure consists of three parts, where the curves in the central part (b) coincide with the curves $T(t)$ and $N_a(t)$ in Figure 2.29. Figures 2.30(a) and (c) show silver atom distributions both as free atoms (solid lines) and as aggregates or microdrops (dashed lines) from the center to the end of the furnace. It is seen that extensive condensation occurs about 1 cm from the furnace center, and that the effective length of the atom layer for the present conditions does not exceed 2 cm. As noted above, more than 40 per cent of the analyte is condensed on the cooler furnace ends. Another 60 per cent of the analyte escapes from the furnace through the sample introduction hole. This leads to interesting profiles in the longitudinal atom distribution, as can be seen from solid curves 4–8 in Figure 2.30(c) where the gas-phase atom concentration in the central part of the furnace is somewhat lower than at greater distances from the center.

Although the computer models of analyte distribution are very powerful, they are primarily dealing with the physical factors, in terms of the definitions given above. The chemical factors (processes 5 and 6 in Figure 2.21) have not yet been modeled. The major reason for this is not due to inherent difficulties in the proposed models, but the lack of reliable data on the rate constants for the corresponding gas-phase and heterogeneous reactions. An example of accounting for the wall reaction of oxygen in the framework of this three-dimensional model was given in Section 2.3, dealing with oxygen in ETAs.

2.7 ATOMIC ABSORBANCE SIGNAL

Atomic absorption spectrometry (AAS) is based on interaction of the probing radiation beam from a primary source with analyte atoms in the optical path, as within an atomizer. Generally, a nonuniform radiation beam is used to probe a nonhomogeneous absorbing layer, that develops in a nonisothermal atomizer. As a result, the interaction of the probing beam with the analyte atoms is dependent on both the spectral features and spatial distribution of the probe beam and analyte.

First, the spectral profile, $J(\lambda)$ of the analytical line emitted by the primary source interacts with the profile $k(\lambda, T)$ of the analyte atoms in the atomizer, as illustrated in Figure 2.31(a). It is apparent that absorption of incident radiation is strongly dependent on the relative position of these two spectral profiles. In one extreme, there will be no absorption if the emission profile does not overlap the absorption profile and, in the other extreme, the absorption is at a maximum if the emission profile coincides with the absorption peak. For measurements in argon at atmospheric pressure there is a shift of the absorption profile to longer wavelengths relative to the emission profile (a red shift). This is shown schematically in Figure 2.31(a). This part of the interaction is called the spectral interaction.

Second, there is the interaction originating from the spatial nonhomogeneity of the radiation beam and the analyte distribution within the atomizer volume. The probing radiant flux, having a nonuniform intensity distribution $J(x, y)$ over the beam cross-section, is absorbed by an analyte $n(x, y)$, nonuniformly distributed in the atomizer. Figure 2.31(b) shows schematically the bell-shaped spatial intensity profile (solid line) characteristic for a hollow cathode lamp (HCL), and a nonuniform

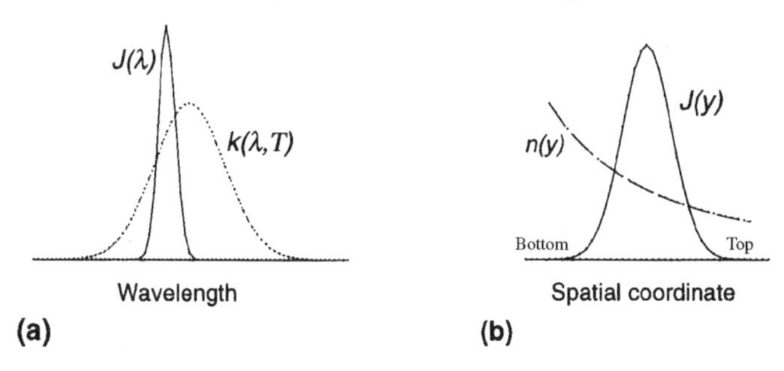

Figure 2.31 Diagrams of (a) spectral and (b) spatial interaction of radiation from the primary source with the analyte in the atomizer. $J(\lambda)$ and $k(\lambda, T)$ are the emission and absorption spectral profiles, respectively; $J(y)$ and $n(y)$ are the spatial distribution of intensity and analyte number density along the atomizer vertical diameter, respectively [161]. (Reprinted from *Spectrochim. Acta, Part B*, Vol. **50B**, A.K. Gilmutdinov, R.M. Mrasov, A.R. Somov, C.L. Chakrabarti and J.C. Hutton, *Three-dimensional Modeling of the Analyte Dynamics in Electrothermal Atomizers for Analytical Spectrometry: Influence of Physical Factors*, pp. 1637–1654, 1995, with kind permission of Elsevier Science – NL, Sara Burgerhartstraat 25, 1055 KV Amsterdam, The Netherlands.)

(broken line) analyte distribution along the furnace vertical diameter characteristic for wall atomization. If the analyte distribution in the atomizer cross-section is uniform, the result of the absorbance measurement is the same, regardless of the position of the sampling beam. If, however, the analyte is distributed nonuniformity over the atomizer radial cross-section, the radiation absorption will depend also on the relative positions of these two spatial profiles. This part of the interaction is called the spatial interaction. Thus, in conventional AAS, the measured absorbance A depends not only on the number of absorbing atoms N, but also on the spectral features of the analytical lines used to probe the absorbing layer, and the distribution of analyte atoms and radiation intensity in the probing beam:

$$A = f(N; \text{spectral, spatial}) \tag{2.46}$$

Detection systems of most AA spectrometers are based on the use of photomultiplier tubes (PMTs) that can only provide spatially integrated radiant intensities falling on their working surfaces. The atomic absorbance signal A provided by such a detection system is the logarithm of the ratio of the incident radiant flux $\hat{O}_0$ to the transmitted radiant flux $\hat{O}$ that is transmitted through the absorbing layer of the analyte atoms. A radiant flux passing through a surface S is expressed in terms of intensities as

$$\Phi = \int_S \int_\lambda K(x, y) J(\lambda) \mathrm{d}S \, \mathrm{d}\lambda \tag{2.47}$$

where $J(\lambda)$ is a function describing the spectral composition of the radiation, and $J(x, y)$ is a function describing the intensity distribution in the plane of the surface S. Taking this into account, the general expression for the absorbance recorded by a conventional AA spectrometer can be expressed [163] as

$$A = \log\frac{\Phi_0}{\Phi} = \log\left\{\frac{\int_0^l \int_{-b/2}^{1/2} \int_{-\lambda^*}^{\lambda^*} J(x, y)J(\lambda)\mathrm{d}\lambda\,\mathrm{d}x\,\mathrm{d}y}{\int_0^{2R} \int_{-b/2}^{b/2} \int_{\lambda^*}^{\lambda^*} J(x, y)J(\lambda)\exp\int_{-L1/2}^{L/2} K(\lambda; x, y, z)\mathrm{d}z)\mathrm{d}\lambda\,\mathrm{d}x\,\mathrm{d}y}\right\} \tag{2.48}$$

Here, b is the width of the part of the illuminating beam that is recorded by the PMT, R and L are the radius and the length of the furnace, respectively, $J(x, y)$ is the intensity distribution in the radial cross-section of the probing beam, $J(\lambda)$ is the spectral profile of the analysis line, and the limits of $-\lambda^*$ to λ^* are the spectral bandwidth isolated by the monochromator (Figure 2.32). The Beer–Lambert law was used for expressing the transmitted radiant flux $\hat{O}$. The absorption coefficient, $K(\lambda; x, y, z)$, depends on the concentration of absorbing atoms in the ground state, n_0 (cm^{-3}), and the spectral profile, $k(\lambda)$, of the absorption line, and can be presented as follows [4]:

$$K(\lambda; x, y, z) = \frac{2\sqrt{\pi \ln 2}e^2}{mc}\frac{f}{\Delta\nu_D}n_0(x, y, z)k(\lambda) = c(\lambda, T)n_0(x, y, z) \tag{2.49}$$

where e and m are the charge and mass of the electron, respectively, c is the velocity of light, f is the oscillator strength, and $\Delta\nu_D$ is the Doppler width of the absorption line. The coefficient $c(\lambda, T)$ denotes the combination of values that depend only on the spectral features of the analysis lines.

From the basic relationship in Eqn (2.48), it follows that the absorbance recorded by the PMT-based detection system is dependent on three

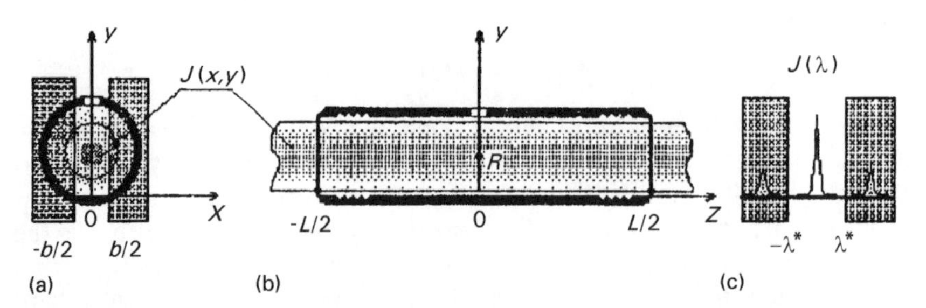

Figure 2.32 (a) Lateral and (b) longitudinal sections of a graphite furnace atomizer and the schematic path of the incident beam of radiation (hatched region); (c) is the analysis line isolated by the monochromator within its spectral bandwidth ($-\lambda^*, \lambda^*$).

groups of factors: (1) the spectral characteristics of the analysis line (dependence of J and K on λ); (2) the cross-sectional distribution of intensity in the incident beam $J(x, y)$; and (3) the spatial distribution of the analyte and temperature within the atomizer (dependence of K on x, y, z). Therefore, before analyzing the absorbance signal, all the above features must be well defined. The spatial distribution of the analyte is discussed in detail in previous sections of this chapter. Considered below will be the first two items, i.e. the spectral line profiles and the spatial distribution of the radiant intensity.

2.7.1 Profiles of atomic lines

Questions related to the shapes of atomic lines are, perhaps, the most studied part of the general theory of AAS. Therefore, this point is discussed only briefly below. Complete descriptions of atomic line broadening mechanisms are also available [4,164–166].

When describing spectral profiles, wavenumbers (cm^{-1}), or wavelengths (nm), can be used. Wavelengths are used in experimental work, because most spectrometers are calibrated in this dimension. Theoretical descriptions of atomic lines are usually expressed in wavenumbers, because the resulting equations are somewhat easier to solve than those using wavelengths. In the following, both units will be used. Conversions between these units and the atomic line widths are given by the well known relationships

$$\nu(\mathrm{cm}^{-1}) = \frac{1}{\lambda(\mathrm{cm})}; \quad |\Delta\nu| = \frac{|\Delta\lambda|}{\lambda_0^2} \tag{2.50}$$

where λ_0 is the wavelength at which the maximum intensity of the spectral line occurs.

Spectral profiles in AAS are basically determined by Doppler broadening, collisional broadening, hyperfine splitting, and isotopic shifts. Other broadening mechanisms (natural broadening, Stark broadening, etc.) can generally be neglected. The origin of Doppler broadening is thermal agitation of emitting and absorbing atoms, that results in a spectral line with a Gaussian line shape. The full width, $\Delta\nu_D$ at one-half of the maximum intensity (FWHM) of an atomic line which is only broadened by the Doppler effect is given [4] by

$$\Delta\nu_D\,(\mathrm{cm}^{-1}) = \frac{7.16}{\lambda\,(\mathrm{nm})}\sqrt{\frac{T(\mathrm{K})}{M\,(\mathrm{g\,mol})}} \tag{2.51}$$

The Doppler line width is inversely proportional to the wavelength of the transition, the square root of the absolute temperature T, and the reciprocal of the square root of the molecular weight of the analyte M. Absorption line widths predicted from Eqn (2.51) range from 1 pm to 10 pm in ETAs. Although the Doppler widths of the lines emitting from conventional primary sources (HCLs and electrodeless discharge lamps (EDLs)) vary from a few picometers to several tens of picometers, they are generally 2–3 times smaller than those for absorption profiles in ETAs atomizers.

The second major broadening mechanism in ETAAS is collisional or pressure broadening, as a result of interatomic collisions. Two types of collision are classified by the identity of the colliding species: (1) collisions between two analyte atoms leads to Holtzmark broadening, which is negligible in analytical AAS because the concentration of analyte is small; (2) Lorentz broadening, which involves collisions between analyte atoms and foreign species (called perturbers, normally argon atoms), and is a significant contributor to the FWHM.

An atomic line that is collisionally broadened displays a distinctive line profile. First, compared to a Doppler spectral profile, the Lorentzian profile is broader and has lower peak height. Second, the intensity maximum undergoes a red shift (in an argon environment). The shift is proportional to the collisional width, and theory predicts a value of 2.76 for the width-to-shift ratio. The shape of a collisionally broadened line is described by a Lorentzian function with the FWHM given [4] by

$$\Delta\nu_{\mathrm{L}} = N_{\mathrm{p}}\sigma\sqrt{\frac{8RT}{\pi}\left(\frac{1}{M_{\mathrm{A}}}+\frac{1}{M_{\mathrm{p}}}\right)} \tag{2.52}$$

where N_{p} is the number density of perturbers (cm^{-3}), σ is the collisional cross-section (cm^2), R is the gas constant, and M_{A} and M_{p} are the molecular weights ($\mathrm{g\ mol}^{-1}$) of the analyte and the perturbers, respectively. The collisional line width, $\Delta\nu_{\mathrm{L}}$ predicted by Eqn (2.52) in ETAs is of the same order of magnitude as the Doppler line width (from 0.001 to 0.01 nm). The values of collisional line widths for the lines emitted from conventional sources are much lower than those for the absorption profiles in ETAs, because the width is proportional to the total pressure in the cell. Therefore, this part of the broadening is normally neglected in the primary sources. However, at high optical densities of an absorbing layer, when the wings of the emission line plays an important role, the Lorentzian component in the broadening of the emission line also becomes important. This is discussed later.

The total spectral profile of an atomic line is determined by the contribution of the two broadening mechanisms, as given by the convolution of the Gaussian profile (due to Doppler broadening) and the Lorentzian profile (due to pressure broadening). The result is the Voigt distribution function $H(\omega, a)$ given by

$$H(\omega, a) = \frac{a}{\pi} \int_{-\infty}^{\infty} \frac{e^{-y^2}\, dy}{a + (\omega - y)^2} \tag{2.53}$$

where

$$a = \sqrt{\ln 2} \frac{\Delta \nu_L}{\Delta \nu_D}$$

is the damping constant determining the shape of the Voigt profile. As the damping constant increases, the profile broadens and its peak height decreases. In the limiting cases of $a = 0$ and $a = \infty$, the Voigt function reverts to the purely Gaussian function and purely Lorentzian function, respectively. To make the consideration more general, the dimensionless frequency ω, normalized by the Doppler width $\Delta \nu_D$ of the absorption profile, is often introduced:

$$\omega = \left(\frac{\nu}{\Delta \nu_D} \right) a \sqrt{\ln 2} \tag{2.54}$$

Finally, the third broadening mechanism is hyperfine splitting, which is caused by nuclear spin splitting and isotope shifts. Generally, a spectral line consists of n hyperfine components with relative intensities b_i

$$\sum_{i=1}^{n} b_i = 1$$

located at a distance $\Delta \nu_i$ (cm^{-1}) from the component with the minimum frequency. Results of calculations of the hyperfine structure (HFS) for 21 analysis lines [163] are given in Table 2.3. The table presents a simplified HFS layout, where the closely spaced components are unified and the weak components with $b_i < 0.01$ are neglected; for example, the copper 324.8 nm line consists of 12 components which form two groups of six components. The separation of the components within each of these groups is less than $0.01\, cm^{-1}$. Therefore, this line may be considered a two-component line with good accuracy. This situation is also characteristic of a number of other lines. The total number of HFS components n is given in the last column of Table 2.3. As can be seen from the table, the hyperfine splitting for half of the presented elements is $\geq 0.2\, cm^{-1}$, that is, greater than Doppler broadening in ETAs. The total FWHM of such lines will be determined by the hyperfine splitting.

Each component of the HFS is both Doppler and collision broadened and can he described by the Voigt function (Eqn (2.53)). Thus, in the most

Table 2.3 Hyperfine structure of the analysis lines.

Element	*λ (nm)*				$\Delta\nu_i$ (cm^{-1}) B_i			*n*
Ag	328.1	0	0.022	0.064	0.077			6
		0.120	0.130	0.361	0.389			
Al	308.2	0	0.013	0.023	0.064	0.074	0.080	6
		0.375	0.104	0.104	0.188	0.104	0.125	
Au	242.8	0	0.204					4
		0.625	0.375					
Bi	306.8	0	0.062	0.140	0.892	0.970	1.064	6
		0.175	0.165	0.110	0.060	0.165	0.325	
Ca	422.7	0	0.013	0.026				5
		0.969	0.007	0.024				
Cd	228.8	0	0.009	0.018				5
		0.289	0.324	0.387				
Cs	852.1	0	0.303	0.310				4
		0.562	0.328	0.110				
Cu	324.8	0	0.390					12
		0.625	0.375					
Hg	253.7	0	0.160	0.340	0.490	0.730		10
		0.124	0.110	0.231	0.342	0.203		
In	303.9	0	0.381					4
		0.556	0.444					
K	766.5	0	0.010	0.014	0.018			8
		0.446	0.034	0.466	0.034			
Li	670.8	0	0.009	0.016	0.349	0.376		8
		0.022	0.031	0.022	0.578	0.347		
Mg	285.2	0	0.024	0.047				3
		0.790	0.100	0.110				
Na	589.0	0	0.059					6
		0.625	0.375					
Pb	283.3	0	0.190	0.264	0.347	0.444		5
		0.073	0.014	0.241	0.524	0.148		
Rb	780.0	0	0.045	0.144	0.228		8	
		0.174	0.421	0.301	0.104			
Sb	231.1	0	0.041	0.213	0.237			4
		0.239	0.187	0.240	0.334			
Sn	286.3	0	0.079	0.240				5
		0.110	0.835	0.055				
Te	225.9	0						2
		1.000						
Tl	276.8	0	0.038	0.704	0.747			6
		0.125	0.625	0.074	0.176			
Zn	213.9	0	0.106	0.022	0.033			7
		0.489	0.278	0.041	0.192			

general case, the emission $J(\omega)$ and absorption $K(\omega)$ profiles can be presented in the scale of the dimensionless frequencies ω [163] as

$$J(\omega) = \sum_{i=1}^{n} b_i H_i\left(\frac{\omega - \Delta\omega_i}{\alpha}; a_e\right), \qquad K(\omega) = \sum_{j=1}^{n} b_j H_j(\omega - \Delta\omega_j + \Delta\omega_s; a) \tag{2.55}$$

where $a = \sqrt{T_e/T}$ is a parameter equal to a ratio of the Doppler width of the emission line to the Doppler width of the absorption line, T_e and T are the translational temperatures of the emitting layer in the primary source and the absorbing layer in the atomizer, respectively. Equation 2.55 takes into account that the emission profile $J(\omega)$ is approximately $1/\alpha$ times narrower than the absorption profile $K(\omega)$ in the scale of dimensionless frequency ω, and that the absorption profile is shifted relative to the emission profile to a value $\Delta\omega_s$.

In ETAAS, the damping constant a varies from 0.17 (beryllium resonance line at 234.9 nm) to 3.27 (cesium at 852.1 nm) [167] for absorption lines, and its value for emission lines a_e in HCLs or EDLs varies from 0.01 to 0.05 [163]. A relatively small value of the Voigt parameter means that the profile of the emission line is almost, but not exactly, coincident with the Doppler profile.

Table 2.3 shows that the analysis lines have very different HFSs. Therefore, for a general consideration it makes sense to combine these elements into three groups: (1) ordinary lines that do not have HFS; (2) lines that are composed of HFS components that are closely spaced, so that $\Delta\nu_i \ll \Delta\nu_D$ (these components are unresolved and the profile does not appear to have a complex structure, i.e. quasi-ordinary lines); and (3) lines that have a developed HFS with widely spaced components, so that $\Delta\nu_i \geq \Delta\nu_D$. The typical representatives of each of these three groups under the conditions of electrothermal atomization are: (1) the 225.9 nm tellurium line which can be considered an ordinary line; (2) the 308.2 nm aluminum line, a quasi-ordinary line; and (3) the 283.3 nm lead line. Figure 2.33 shows the emission and absorption Voigt profiles of these analysis lines [163]. These profiles were calculated using typical conditions for ETAAS, and the HFSs of these analysis lines taken from Table 2.3 are shown by vertical lines.

In a review of experimentally measured atomic line profiles, Larkins [168] showed that calculated profiles are in good agreement with measured ones, provided that correct account is taken of all the above spectral features (HFS, Voigtvs parameters, etc.). This makes it possible to take the calculated profiles for the following analysis of the concentration curves.

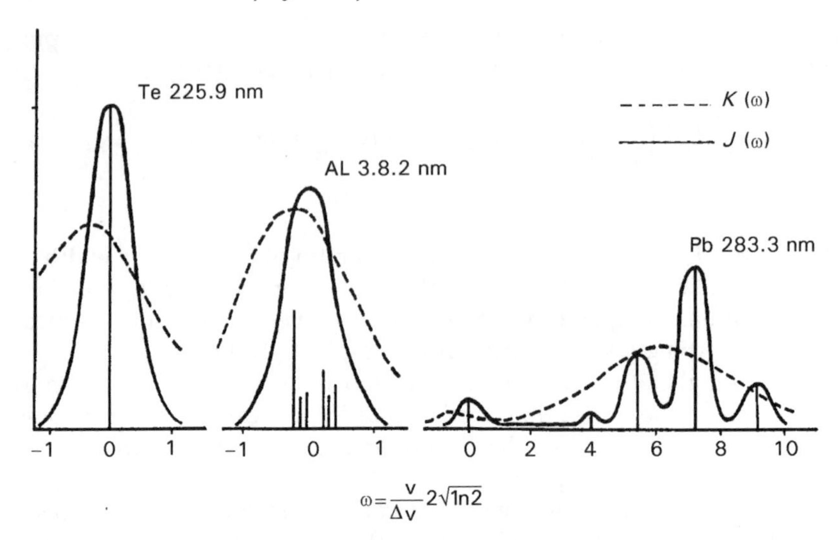

Figure 2.33 Calculated emission (solid curves) and absorption (dashed curves) profiles of different kind of analysis lines for typical conditions in GFAAS: $T_e = 550\,K$; $a_e = 0.01$; $T = 2300\,K$ (Te), 2700 K (Al), and 2100 K (Pb); $a = 0.48$ (Te), 0.34 (Al), and 1.24 (Pb) [163]. (Reprinted from *Spectrochim. Acta, Part B*, Vol. **47B**, A.K. Gilmutdinov, T.M. Abdullina, S.F. Gorbachev and V.L. Makarov, Concentration Curves in Atomic Absorption Spectrometry, pp. 1075–1095, 1992, with kind permission of Elsevier Science – NL, Sara Burgerhartstraat 25, 1055 KV Amsterdam, The Netherlands.)

2.7.2 Intensity distribution of the incident radiation

When discussing the spatial structure of the probing radiation beam, the two basic planes of the beam are of special interest are the radial cross-section at the atomizer center and the longitudinal cross-section along the vertical beam diameter. The first cross-section is important because it corresponds to the image of the primary source, and a major portion of the analyte is also located near the atomizer center. The second cross-section is important because radiation located in the vicinity of this plane is normally transmitted to the detector through the monochromator exit slit.

Hollow cathode lamps and EDLs are normally used as primary sources of line radiation in AAS. Figure 2.34(a) presents the spatial distribution of the radiant intensity recorded in the radial cross-section of the beam produced by a lead HCL at the location of the atomizer center. The illumination system of an AA spectrometer is normally designed so that the image of the primary source is projected on the atomizer center. Hence, the presented distribution is, in fact, a monochromatic image of the lead HCL recorded at the 283.3 nm resonance line [169]. The distribution is bell shaped with a sharp maximum at the beam axis. The intensity goes sharply to zero when moving to the edges of the beam, and the base of the

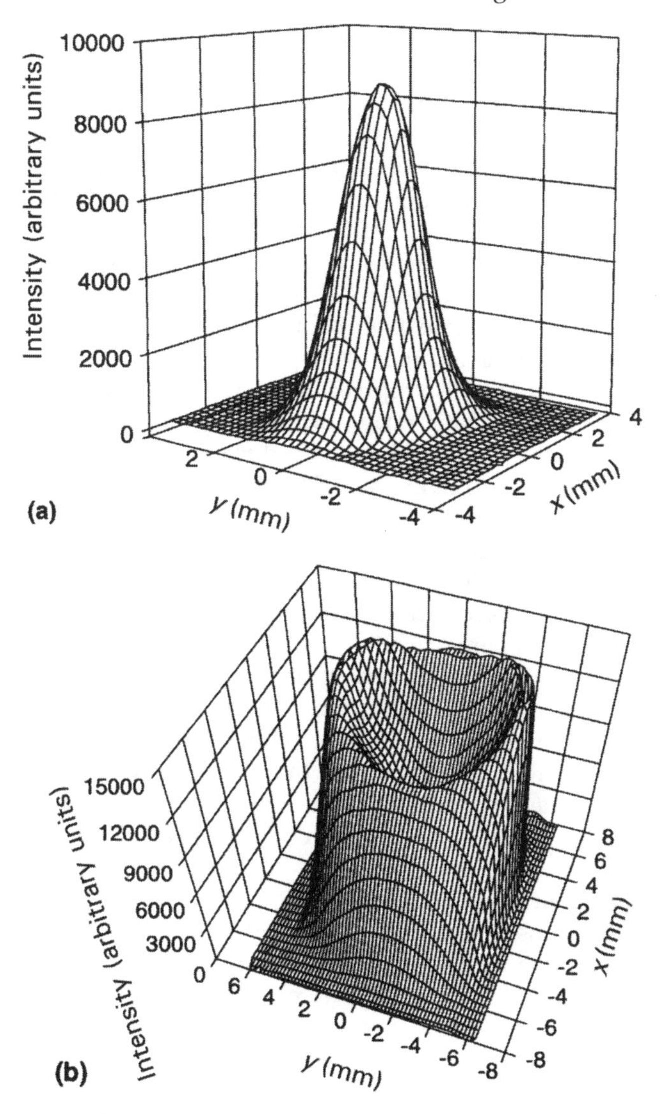

Figure 2.34 Distribution of radiant intensity in the radial cross-section of the free beam produced by: (a) a lead HCL; and (b) a lead EDL. The cross-sections are taken at the beam location corresponding to the atomizer center [169,170]. (Reprinted by permission of the Society for Applied Spectroscopy from, A.K. Gilmutdinov, B. Radziuk, M. Sperling, B. Welz and K.Y. Nagulin, Spatial distribution of radiant intensity from primary sources for atomic absorporption spectrometry. Part I. hollow cathode lamps, *Appl. Spectrosc.*, **49**, 413–424 (1995) and A.K. Gilmutdinov, B. Radziuk, M. Sperling, B. Welz and K.Y. Nagulin, Spatial distribution of radiant intensity from primary sources for atomic absorption spectrometry. Part II: electrodeless discharge lamps, *Appl. Spectrosc.*, **50**, 483–497 (1996).)

distribution is circular with a diameter close to the inner diameter of the lamp's hollow cathode.

Although EDLs are much brighter than the corresponding HCLs, the spectral profiles of the lines emitted by these two radiation sources are similar. However, because the geometry of an EDL, as well as the plasma generated in it, differ significantly from those in an HCL, the beam structures produced by these two different kinds of primary sources also differ significantly [170]. The intensity distribution in the radial cross-section of the free beam (i.e. the beam that is not attenuated by the tube atomizer) produced by the lead EDL is presented in Figure 2.34(b). The cross-section is taken at the location corresponding to the atomizer center where the image of the lamp is projected. Thus, the presented distribution is in essence a monochromatic image of the lamp recorded at the lead 283.3 nm resonance line. As could be expected from the axial symmetry of the bulb and the exciter coil in the EDL, the measured intensity distribution is symmetrical relative to the beam axis. However, the cross-sectional structure of the beam differs significantly from that of the lead HCL. When going from the beam edge towards its axis, the intensity increases sharply and reaches the maximum at about 1.5 mm from the beam edge. When moving further towards the axis, the intensity decreases again, reaching a minimum at the beam axis. The depression in the measured distributions is significant – the axial intensity for the lead line is only half of that near the edge. Such a distribution is caused by the radial cataphoresis effect that results in the metal excited species being located in a narrow near-wall region of the lamp (optical skin effect) [170]. Insertion of the atomizer into the beam produced by the EDL may change significantly the intensity distribution, as the tube presents a physical constraint for the much wider beam.

The change of intensity distribution along the vertical diameter of the beam produced by the lead HCL when moving along the beam axis is presented in Figure 2.35. There is a remarkable change in the beam structure when approaching the atomizer center where the image of the source is located–the intensity increases sharply while the beam decreases in its lateral size. After passing the atomizer center, intensity goes down again while the beam expands. The arrows in the figure indicate the locations of the center, entrance and exit ends of the tube. It can be seen that there is a significant change in intensity by about a factor of two along the 2.8 cm length of Massmann furnaces.

2.7.3 Concentration curves

The primary task of the theory of AAS is to establish a relationship between the measured atomic absorbance signal A and the number N of analyte atoms in the analysis volume. According to the notation of

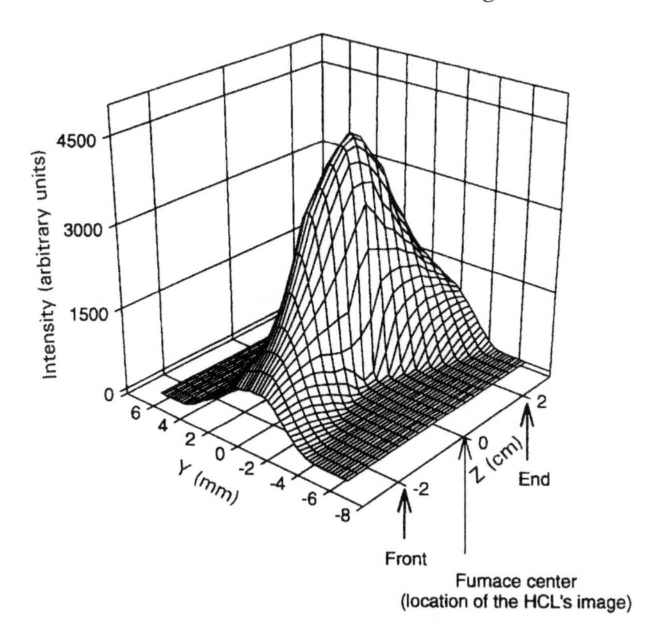

Figure 2.35 The change of intensity distribution along the atomizer diameter in the longitudinal section of the beam produced by a lead HCL.

Gilmutdinov *et al.* [163] the relationship $A = f(N)$ is the concentration curve. The general relationship (Eqn (2.48)) describing the concentration curve is simplified significantly if the following assumptions are made: (1) radial cross-sectional distributions of the radiation beam and the analyte are uniform, so that $J(x, y) = \text{constant}$ and $n(x, y, z) = n(z)$; (2) the atomizer is spatially isothermal, so that $T(x, y, z) = \text{constant}$; and (3) the emission line is a single and infinitely narrow line at wavelength λ_0, so it is described by Dirac's delta function, $J(\lambda) = J_0\delta(\lambda - \lambda_0)$. Substitution of these simplifications into Eqn 2.48 gives

$$A = (\log e)\int_{-L/2}^{L/2} K(z; \lambda_0)\mathrm{d}z = (\log e)c(\lambda_0, T)N \qquad (2.56)$$

where $c(\lambda_0, T)$ is a constant for a given transition and for a given temperature that follows from Eqn (2.49), and

$$N(\mathrm{cm}^{-2}) = \int_{-L/2}^{L/2} n(z)\mathrm{d}z$$

is the number absorbing atoms per unit cross-sectional area along the radiation beam (in the case of uniform analyte distributions, this value is proportional to the total number of analyte atoms in the atomizer). Thus, if the above assumptions are valid, the recorded absorbance at any instant of time is proportional to the total number of absorbing atoms in the atomizer. This circumstance is an actual basis for AAS to be used as an analytical technique, because the recorded analytical signal A is directly proportional to the unknown number N of analyte atoms.

The actual situation, however, differs substantially from that presented by the simple expression in Eqn (2.56). The spatial and spectral non-uniformities lead to measured absorbances depending not only on the number of analyte atoms, but also on their distribution in the atomizer volume and the spectral features of the analysis lines. Thus, an accurate depiction of the concentration curve has both spectral and spatial dependencies, and is described by Eqn (2.46). The effect of these factors on the recorded absorbance are described in greater detail elsewhere [163].

2.7.3.1 *Effect of spectral features of the analysis lines*

At this point, a uniform absorbing layer to analyze the influence of the spectral features, without the additional effect of spatial non-uniformities, will be assumed, i.e. the concentration curve $A = f(N;$ spectral, spatial $= 0)$ will be considered. In the calculation of the concentration curves, approximations with different degrees of completeness can be used. Figure 2.36 shows how the concentration curves of the above three analysis lines are affected by sequentially accounting for more spectral features of the analysis lines [163]. The simplest approximation is that the emission line may be considered as a single and infinitely narrow line, and that the absorption profile is not pressure shifted. In this case, absorbance will increase linearly without any curvature, following Eqn (2.56). This is shown as the set of curves labeled 1. Accounting for broadening of the emission line causes the concentration curves to bend and also decreases their initial slopes (curves 2). Introducing HFS ($n = 1$ for tellurium, $n = 6$ for aluminum, and $n = 5$ for lead) into the analysis causes additional bending of the curves, and the curvature is much more pronounced for greater separation of HFS components (curves 3). The final case (curves 4) accounts for all spectral features, including the shift of the emission line relative to the absorption line, i.e. the atomic line profiles are calculated on the basis of the most general Eqn (2.55). An interesting feature, seen in these results, is that the slope and curvature of the concentration curves are largely dependent on HFS in the spectral transition. It has been found that the separation of the HFS components is of greater importance than the broadening of these components. The 308.2 nm aluminum line consists of six HFS components that are closely spaced, i.e. $\Delta\omega < 0.7$ (see Figure 2.33). In this case, the curvature of the concentration curve is mainly caused by broadening of the emission line and not by HFS (as shown by curves 2 and 3 in Figure 2.36 for the case of aluminum). The opposite situation defines the concentration curve for the 283.3 nm lead analysis

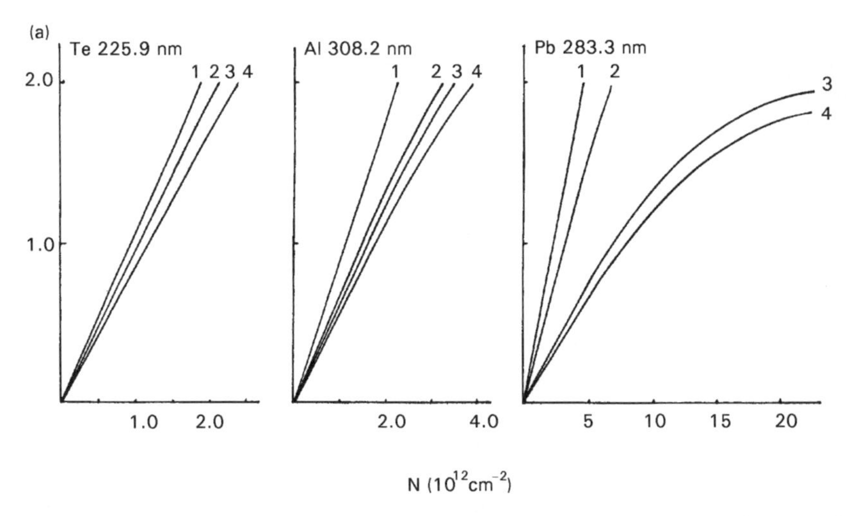

Figure 2.36 The evolution of the concentration curves as the spectral features that determine the concentration curves are successively taken into account (see text for comments) [163]. (Reprinted from *Spectrochim. Acta, Part B*, Vol. **47B**, A.K. Gilmutdinov, T.M. Abdullina, S.F. Gorbachev and V.L. Makarov, Concentration Curves in Atomic Absorption Spectrometry, pp. 1075–1095, 1992, with kind permission of Elsevier Science – NL, Sara Burgerhartstraat 25, 1055 KV Amsterdam, The Netherlands.)

line. In this case, accounting for HFS (curve 3) gives much greater contribution to the final result (curve 4) than accounting for broadening only (curve 2).

The broadening of the emission line, seen in standard AAS radiation sources (HCLS and EDLs), is not purely due to the Doppler effect because of the finite pressure of the inert filler gas. To analyze the effect on the line shape, as it is changed from a purely Doppler-broadened line to a line that has a large contribution from Lorentzian broadening, the dependence of the concentration curves on the Voigt parameter a_e of the emission line from the primary source as it is varied from 0 to 0.05 (this situation reflects real conditions) has been studied [163]. Calculations for all of the above lines have shown that neglecting the Lorentz component of broadening (i.e. the adoption of $a_e = 0$) slightly overpredicts the absorbance A in the linear portion of the concentration curve with a maximum deviation of 2 per cent. However, at greater values of atom concentration, when absorption by the wings of the analysis lines starts to play a significant role, the error of neglecting the Lorentz component increases, reaching, at $A = 1.5$, a value of the order of 10 per cent.

The line profiles in Figure 2.33 and the concentration curves in Figure 2.36 were calculated assuming that the temperature T_e in the radiation source is 550 K. However, it has been shown that the temperature of the emitting layer of the HCL for different elements can undergo significant variations. Calculations [163] show that, by increasing the source temperature from 400 K to 700 K, the value of absorbance for the linear part of the concentration curve decreases by up to 4 per cent. Therefore, the assumption that the temperature in the radiation source is 550 K for all elements is fairly reliable. However, these results are true only under conditions where no self-absorption occurs in the radiation source.

Among the factors that define the profile of the absorption line, the value of absorbance is most influenced by the temperature of the absorbing layer, *T*. Variation of the absorbing layer temperature by ± 10 per cent causes approximately the same relative change in absorbance. Variations of the collisional $\Delta\omega_s$ and damping constant a within limits of 10 per cent, cause the absorbance of all the elements considered to change by up to 2 per cent and 5 per cent respectively.

Therefore, the results of the described simulation for three elements with quite distinct spectral characteristics suggest that the absorbance is only minimally affected by variations in the magnitude of the parameters defining it. This conclusion supports the idea of absolute AA analysis [87,172]. However, it should be emphasized that the high stability of the concentration curve is true only for the linear portion of this curve – the higher the value in the optical density of the absorbing layer, the more sensitive is the absorbance to variations of the parameters defining it [163].

All the above considerations were based on the assumption that there is no self-absorption broadening or self-reversal in the analysis lines emitted by the primary source. At increased currents in the primary source, however, the emitted lines can exhibit self-absorption that can result in significant changes in the concentration curves. Figure 2.37(a) represents measured profiles of the magnesium 285.2 nm resonance line, emitted by a HCL, at three different operating currents [168]. The operating currents reported are the average values for square-wave modulation. This atomic line consists of three closely located hyperfine components (see Table 2.3), which are unresolved in the presented measurements. Thus, in terms of the above classification, the line can be specified as a quasi-ordinary line. Magnesium has high sputtering efficiency and large oscillator strength. As a result, even at the recommended lamp operating current, the magnesium line is significantly self-reversed Figure 2.37(a). The highest current shown for magnesium (5 mA) is only one quarter of the maximum recommended current.

The changes in line profiles with operating current produce significant changes in the calibration graphs, as shown in Figure 2.3(b) for magnesium in the air–acetylene flame. At low lamp current (1 mA), when the emitted line is not self-absorbed, the calibration graph shows only a small curvature up to an absorbance of 2. Such behavior is in accordance with the theoretical prediction (see Figure 2.36 for the aluminum 308.2 nm line

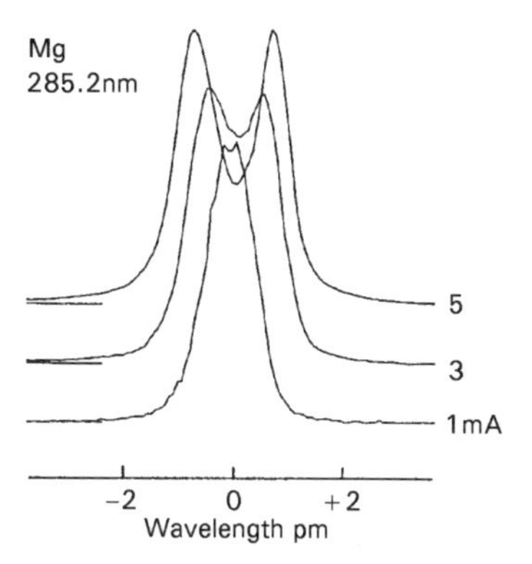

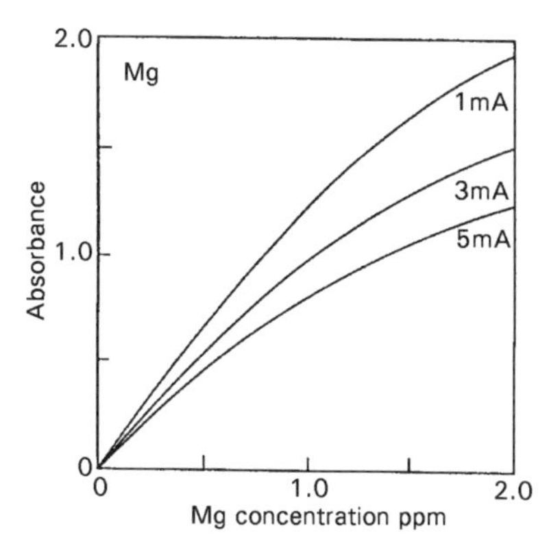

Figure 2.37 (a) Profiles of the magnesium 285.2 nm resonance line from a Varian HCL at different lamp currents. (b) Corresponding calibrations graphs for magnesium in an air–acetylene flame [168]. (Reprinted by permission of the Royal Society of Chemistry from, P.L. Larkins, Atomic Line profiles – Their Measurement and Importance in Analytical Atomic Spectroscopy, *J. Anal. At. Spectrom.*, **7**, 265–272 (1992).)

which represents a quasi-ordinary line). However, as the current is increased, the sensitivity decreases and the curvature increases. This is primarily caused by broadening and self-absorption. At a current of 5 mA the curvature is much greater than that calculated on the assumption of the Doppler profile for the emission line, and the reason for this increased curvature is the following. The Lorentzian contribution to the line shape due to pressure broadening is normally small, as can be seen in the small wings on the line profiles at low current. Self-absorption, however, only produces a significant reduction in the intensity of the central part of the line, and therefore increases the relative contribution from the wings. This increase is particularly obvious in the presented profile of magnesium. As these wings extend outside the main body of the absorption profile, the effect is similar to that produced by the presence within the monochromator's spectral bandwidth of a weak line of very low sensitivity, i.e. the wings result in substantial curvature of the calibration graph at high absorbance values.

The presence of stray light or nonabsorbed radiation, which falls within the spectral bandwidth of the monochromator, is known to produce severe curvature and, if such problems exist, they may well dominate the overall effect observed in the calibration graph [173].

2.7.3.2 *Influence of spatial nonuniformities*

In this section, the emission line is considered initially as a singlet of infinitely narrow width. This will allow an analysis of the effect of spatial nonuniformities within the atomizer without having to consider the additional effects of spectral features. Thus, the concentration curve is analyzed initially for the situation $A = f(N;\text{ spectral} = 0, \text{spatial})$. Then, the combined effect of both spectral and spatial features is considered.

As noted previously, graphite tube atomizers are nonisothermal devices. For the Massmann furnace, the temperature gradients along the tube length can be as high as $1000\,\mathrm{K\,cm^{-1}}$ when the temperature at the furnace center is about 2500 K (see Chapter 3). In addition, as was shown in Section 2.6, the analyte distribution can also be highly nonuniform in both lateral and longitudinal cross-sections of the atomizer. As follows from the general Eqn (2.48) these circumstances directly affect the measured absorbance. The spatial nonuniformities act in two different ways – longitudinal concentration and temperature gradients result in a decrease in the slope of the concentration curves, and cross-sectional nonuniformities lead to curvature of the concentration curves [163].

The effect of the longitudinal nonuniformities can be understood in a qualitative fashion as follows. Although the measured absorbance is proportional to the total number of absorbing atoms along the radiation beam, the atoms located in different locations within the confines of the nonisothermal atomizer absorb the radiation differently because the spectral features of the analysis lines are temperature dependent. As shown in the previous section, a relative increase in the atomizer temperature of 10 per cent leads to an increase in the observed absorbance by 5–13 per cent, depending on the analytical line. Because the furnace temperature significantly decreases towards the furnace ends, the effective temperature of the absorbing layer is also decreasing, which results in a decrease in the absorbance value. Thus, the effect of the longitudinal nonuniformities is actually a spectral effect. The longitudinal gradients of analyte distributions would not affect the absorbance in an isothermal atomizer. The estimations performed previously have shown that the longitudinal nonuniformities result in a decrease in the slope of the concentration curves from 0 to 10 per cent depending on the value of the gradients [163].

The effect of the radial cross-sectional gradients in analyte distribution on the concentration curves is much more pronounced, and it can be understood from the following consideration. Neglecting all the features of concentration curves except the cross-sectional nonuniformities (i.e. assuming uniform irradiance of an isothermal atomizer by monochromatic radiation), the more general Eqn (2.48) yields

$$A = \log\left\{\frac{1}{\int_0^{2R} \exp\left(-c(\lambda_0, T)\int_{-L/2}^{L/2} n(y, z)\mathrm{d}z\right)\mathrm{d}y}\right\}$$
$$= -\log\left\{\int_0^1 \exp[-c(\lambda_0, T)N_0N^*(y')]\mathrm{d}y'\right\} \qquad (2.57)$$

where: N_0 is the total number of analyte atoms in the furnace volume;

$$N^*(y') = N(y)/N_0, \qquad N(y')(\text{cm}^{-2}) = \int_{-L/2}^{L/2} n(z, y')\text{d}y'$$

is the analyte distribution along the vertical furnace diameter, $y' = y/2R$ is the normalized vertical coordinate; and R and L are the furnace radius and length, respectively (see Figure 2.32). For uniform analyte distribution ($N^*(y') = 1$) this relationship reduces to the previous Eqn (2.56). Finally, taking into account Eqn (2.56) and $c(\lambda_0, T)N_0 = (\ln 10)A^{(u)}$, Eqn (2.57) may be rewritten as

$$A^{(n)} = -\log\left\{\int_0^1 10^{-A^{(u)}N^*(y')}\,\text{d}y\right\} \tag{2.58}$$

This relationship represents directly the effect of the cross-sectional nonuniformities. The superscript (n), denotes that the absorbance in this case is the absorbance that would be produced by the actual nonuniform absorbing layer. The above expression relates the real value of absorbance, $A^{(n)}$ produced by a given number of atoms taking into account the transverse gradients, with the value $A^{(u)}$, which would be produced by the same number of atoms with a uniform distribution.

Equation (2.58) is perhaps clearer if a step atom distribution is used:

$$N^*(y', h) = \begin{cases} \frac{1}{h}, & y' < h \\ 0, & h < y' \leq 1 \end{cases} \tag{2.59}$$

Here, the dimensionless parameter h characterizes the localization of atoms in the general direction – a value of $h = 0.5$ means that the atoms fill uniformly the lower half of the atomizer radial cross-section, and are completely absent in the upper half of the atomizer (Figure 2.38(a), curve 2). The particular case where $h = 1$ means uniform atom distribution in the atomizer radial cross-section (curve 1 in Figure 2.38(a)). Note that the total number of atoms is the same for any of these distributions, i.e.

$$\int_0^1 N^*(y', h)\text{d}y = 1$$

for any value of h. As can be seen from Eqn (2.58), in the case of a uniform distribution, $A^{(n)} = A^{(u)}$ and the concentration curve is a straight line (curve 1 in Figure 2.38(b).

For the step distribution, Eqn (2.58) simplifies further:

$$A^{(n)} = -\log\{(1 - h) + h10^{-A^{(u)}/h}\} \tag{2.60}$$

Curve 2 in Figure 2.38(b) shows the concentration curve corresponding to the step analyte distribution. The curve is strongly nonlinear and exhibits a saturation effect. In the case of incomplete filling by the absorbing atoms of the illuminating beam cross-section, the absorbance will not exceed a

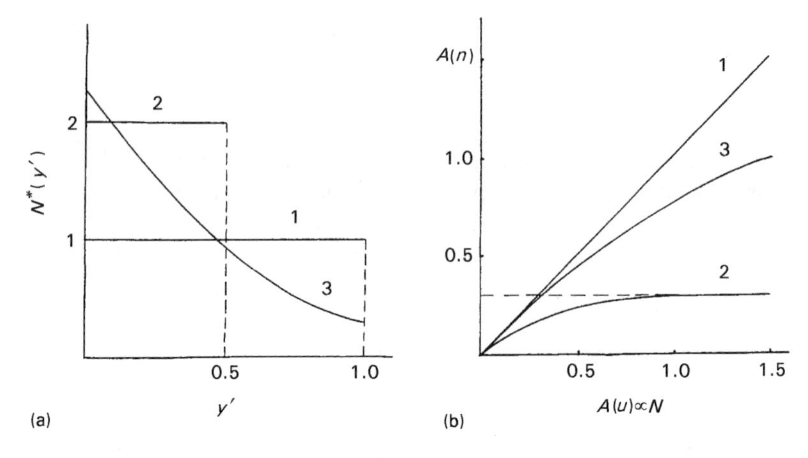

Figure 2.38 (a) Analyte cross-sectional distributions of atom densities: (1) a uniform distribution; (2) a step distribution; and (3) a continuously decreasing concentration. (b) The respective calibration curves.

limiting value A_{lim}, no matter how large the number density of absorbing atoms becomes within the atomizer. This result can be explained using Eqn (2.60), from which it follows that for $A^{(u)} \propto N_0 \to \infty$, $A^{(n)} = -\log(1 - h)$. Therefore, absorbance will always be limited to a maximum value, due to the propagation of the illuminating radiation through a fraction $(1 - h)$ the atomizer radial cross-section that is not filled with the vapor of the element being measured. Curve 2 approaches this value at high values of $A^{(u)}$. It is interesting that the presence of stray light in the spectrometer results in the same distortion of the concentration curve up to its saturation point [173].

The situation described above was characterized by sharp transverse gradients of analyte distribution. However, the actual distributions are smoother than those presented by the step function.

As was shown earlier in this chapter, the most common case occurs when the atoms fill the complete furnace cross-section, but there is considerable localization in the lower part of the furnace. This smoother type of distribution is simulated in Figure 2.3(a) by curve 3. Once again, this distribution is related to the same total number of analyte atoms as for the distributions represented by curves 1 and 2. Curve 3 is intermediate between curves 2 and 1, corresponding to the uniform and the step atom distributions. Although more realistic concentration curves do not exhibit complete saturation, and are less curved than curve 2 (Figure 2.38(b)), their deviation from the straight line is still significant. The figure clearly shows that the observed absorbance depends not only on the number of absorbing atoms but also on how these atoms are distributed in the atomizer cross-section.

2.7.3.3 Combined effects of spectral and spatial features

The results of combining the effects of both spectral and spatial features, i.e. the concentration curve in the most complete form of $A = F(N;$ spectral, spatial) are presented in Figure 2.39. Curve 1 is the concentration curve of the aluminum 308.2 nm resonance line that accounts for all the spectral features of the line (broadening, HFS, etc.) but is constructed for the case of uniform analyte distribution. It coincides with the final curve 4 in Figure 2.36. When cross-sectional nonuniformities of analyte distribution are also present, curves 2 and 3 can result, depending on the particular gradients of the distributions (curves 2 and 3 in Figure 2.39(a)). Once again, the total number of analyte atoms represented by these curves is the same for any of these distributions. It can be seen from the figure that the higher the degree of the cross-sectional nonuniformities, the higher is the curvature of the concentration curves.

The results presented in Figure 2.39(b) show that the observed absorbance depends strongly on the distribution of the atoms in the atomizer volume. Curve 1 shows that 3×10^{12} aluminum atoms produce an absorbance of about 1.5 when the atoms are distributed uniformly in the furnace cross-section. At the same time, the same number of absorbing atoms when distributed nonuniformly, as shown by curve 3, will produce an absorbance of only about 1.0. The higher the degree of cross-sectional nonuniformity (shown by the curves in Figure 2.39(a)), the lower is the corresponding absorbance (shown by the curves in Figure 2.39(b)). The relative decrease in absorbance due to a nonuniform cross-sectional

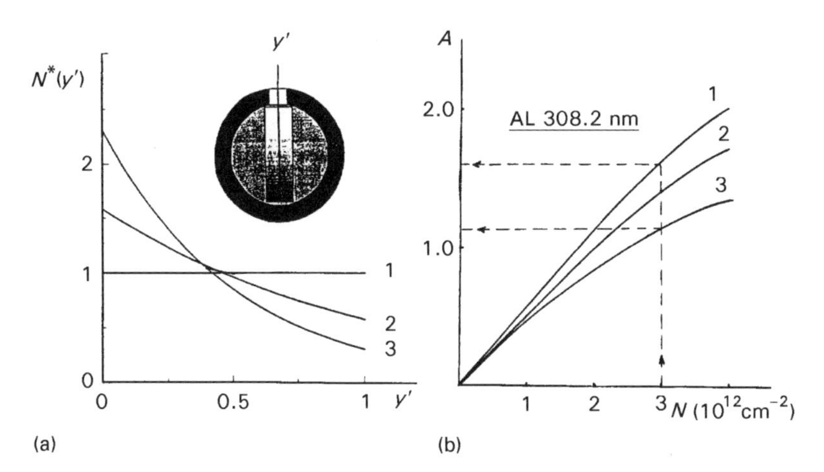

Figure 2.39 Dependence of aluminum concentration curve (b) on cross-sectional non-uniformities of the absorbing layer (a). The insert in (a) is the diagram of the radial cross-section of the tube atomizer and the monochromator exit slit.

redistribution of the same number of atoms can reach a few tens of per cent.

2.7.3.4 *Influence of cross-sectional nonuniformities of radiant intensity*

Taking into additional account cross-sectional nonuniformities of intensity distribution in the probing radiation beam further complicates the situation. Complete quantitative analysis of the effect has shown [174] that the nonuniformity in the primary radiant beam somewhat compensates for the effect of nonuniformity of the analyte distribution. This is because the increase in the degree of beam nonuniformity means in essence a decrease in its effective diameter. This results in a nonuniform beam, having the same spot size as a uniform beam, sampling a smaller region within the atomizer.

The uncertainty in absorbance measurements caused by the nonuniformities in both analyte and probing beam distributions can be avoided when using spatially resolved detection of intensities transmitted through the atomizer [174]. This can be achieved by replacing the PMT with a linear solid-state detector (photodiode allay, linear charge coupled device, charge injected device) located vertically along the exit slit of the monochromator. Such an arrangement allows monitoring of transient intensities, $J_i(t)$ transmitted through different parts along the vertical diameter of the atomizer. This, in turn, permits calculation of local absorbances $A_i(t) = \log J_{i0}/J_i(t)$ that are proportional to the number of absorbing atoms in the ith part of the atomizer. The spatially resolved absorbance, $A^{(SRD)}$, is then defined [174] as

$$A^{(SRD)} = \frac{1}{n}\sum_{i=1}^{n} \log\frac{J_{i0}}{J_i} \tag{2.61}$$

The difference in this definition of absorbance, as compared with the conventional one, is that the spatially resolved absorbance is the sum of the logarithm of intensities, while conventional absorbance (defined by Eqn 2.48) is the logarithm of the sum of intensities. It was shown [174] that absorbance defined in such a way is proportional to the number of analyte atoms within the probing beam, irrespective of their distribution and distribution of radiant intensity in the probing beam. At this time, analytical benefits of spatially resolved detection in AAS are not documented and are under intensive investigation.

2.8 IMPLICATIONS FROM ETA DESIGN AND OPERATION

The geometry, heating rates and material of the ETA can dramatically affect the shape of the AA signal, the spatial distribution of analyte vapor within the furnace and the interferences associated with analysis. In this section, the impact of these parameters and their relationships to the fundamental aspects discussed previously is explored. Many of the options have been explored experimentally with differing degrees of speculation on the global meaning of the empirical observations. This section draws on the experiments and simulations conducted for the various parameters affecting the absorbance signal, and speculates where appropriate.

2.8.1 Geometry

2.8.1.1 Tube versus flat

Although L'vov's initial design was based on a cylindrical geometry, a number of studies persisted in the early days of ETA development using graphite rods, which featured a flat surface onto which the sample was deposited (see Section 1.4). The simplicity of the design made it easy to fabricate in the laboratory. Unfortunately, the lack of confinement also had more negative side-effects, including substantial cooling only short distances above the surface, and extreme signal attenuation as a result of diffusion on moving further from the atomizer surface [153]. As a result, the absorbance signal rarely persisted for more than a few hundred milliseconds. Flat surfaces continued to be used for some mechanistic studies, with mass spectrometry and vacuum vaporization, when secondary collisions with the surface were being minimized [175,176]. However, the tubular design generally optimizes analyte containment within the optical path and provides a more nearly isothermal high-temperature environment in which to promote free atom formation.

L'vov's initial intent was to provide sufficiently rapid sample heating that the analyte would vaporize completely and all of the analyte vapor would be present in the optical path before diffusional loss began. In his perspective, if τ_1 (time to vaporize) was much less than τ_2 (a time characteristic of the loss rate), then this assumption was reasonably valid (see Figure 2.15). To approach this situation, the semiconfining geometry of a tube was the reasonable choice, since it maximized τ_2.

2.8.1.2 Length and diameter

With the tubular geometry as a reasonable approach, the length and diameter of the furnace must be optimized. Certainly, some constraints would appear to exist based on the power requirements that may be needed if too large a furnace were used. However, Woodriff [177] showed clearly that a 25 cm long furnace was a possibility, and had the design closest to the original L'vov concept where the sample was placed on a pedestal and inserted into the preheated furnace (see Section 1.4). Interestingly, even with the Woodriff furnace, where there was no sample introduction hole for analyte loss and the diffusional path was considerably longer than the original design of L'vov, the absorbance signal never reached a constant maximum value. Working on the simplified vaporization model, and assuming ideal gas behavior of an analyte elastically colliding with the furnace walls, it might have been expected that the sample would vaporize and be completely confined within the furnace for a finite time. This would generate a plateau at the absorbance peak, and this constant value would then represent the absorbance from the entire sample. This was never observed, partly because many metal vapors exist in an equilibrium with their adsorbed state on the graphite. The longer tube provided additional sites for adsorption and probably contained an even larger fraction of adsorbed metal than would have been the case with a smaller furnace. In short, very long furnaces may have a counterbalancing impact on the magnitude of the analytical signal (at least the peak height) that is not completely offset by the longer analyte residence time within the furnace.

Figure 2.40 shows a Monte Carlo simulation of the integrated absorbance signal that results from altering the tube length, as well as several other sets of conditions [178]. Copper represents a metal exhibiting moderately strong interactions with the graphite surface, and it is assumed to stick when colliding with the surface. By considering copper in the absence of a sample introduction hole, curve b shows a larger signal and a stronger dependence on the length than is the case for curve a, which includes the presence of a sample introduction hole. The influence of the hole on the signal magnitude also shows the greatest impact for longer furnaces. When a metal whose vapor is assumed to act weakly with graphite (such as silver) is simulated with a sample introduction hole present, the resulting length dependence is represented by curve c. Also shown is the curve expected if there is a squared dependence of the absorbance signal on the length, i.e. $A \propto l^2$, (curve e) or a power dependence of 1.23 on the length (curve d). Interestingly, the absorbance signal for copper with a sample introduction hole present is approximately a power dependence of 1.23, whereas in the absence of a sample introduction hole copper approaches a squared dependence.

The tube diameter also effects the cross-sectional area in which the sample is contained. As the diameter decreases for a given tube length, the gas phase atom density must increase for the same number of atoms in the gas phase. Hence, the absorbance signal should also increase. In simulations, the signals increase with decreasing diameter as expected, but not in proportion to the decrease in cross-sectional area. This is a result of increased loss through the sample introduction hole, as it is much closer to initial sample location.

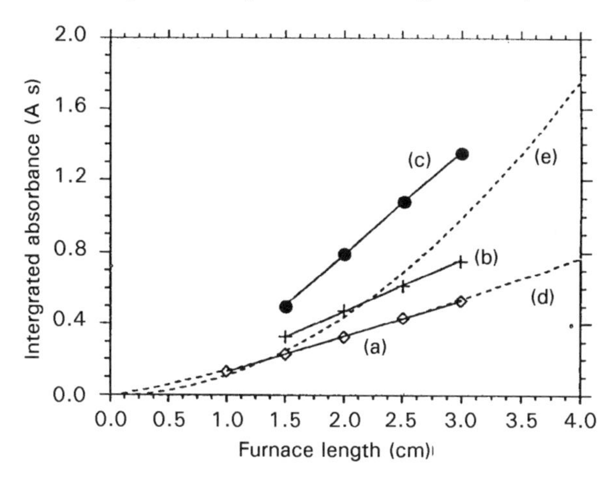

Figure 2.40 Monte Carlo simulations showing integrated absorbance signal changes for different length furnaces. The curves represent different atomization conditions: copper which adsorbs to the graphite (a) with and (b) without a sample introduction hole; (c) silver (which does not strongly adsorb to graphite) without a sample introduction hole; curves which represent a signal dependence on the length as 1^{Ω} for(d) $\Omega = 1.23$ and (e) $\Omega = 2$.

Thus, longer furnaces with smaller diameters appear to provide improved signals, although not in the proportional fashion that the simplified concepts would have predicted. There are other practical considerations that may discourage long furnaces and small diameters. As noted previously, the power required to heat the larger mass furnaces may be prohibitive, and the greater length produces a signal that persists for a longer time period, requiring the furnace to be maintained at elevated temperatures for longer times and shortening the useful number of firings available on each furnace. The smaller diameter furnace can also limit the sample volume that can be deposited.

Finally, the smaller diameter and greater length may result in the furnace serving as a light-limiting aperture for the optical system, thus increasing the noise level as the furnace length increases and the diameter decreases. Güell and Holcombe [178] considered the impact of this parameter in arriving at a signal-to-noise ratio (SNR) for various furnace shapes, and making the assumption of a shot noise limited detection system and the furnace serving as the light limiting aperture. An example of the SNR contours is shown in Figure 2.41.

As noted earlier, Hadgu and co-workers [102,179] attempted to retain the diameter needed to accommodate the sample while limiting diffusive loss from the ends by placing end caps on the furnace. This resulted in

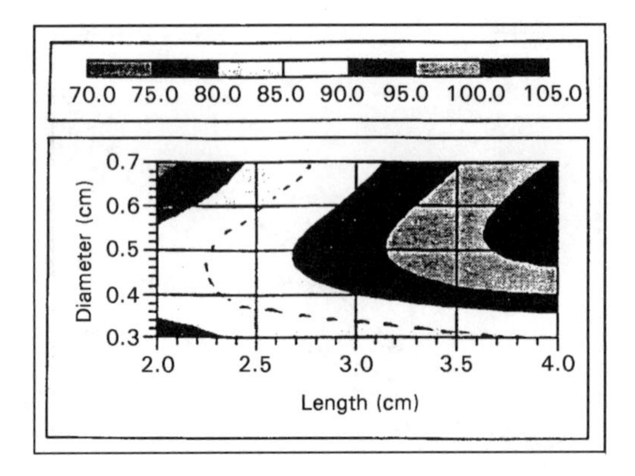

Figure 2.41 Contour plot for the SNR for various furnace lengths and diameters using wall atomization and isothermal furnace heating. The model employs copper as the analyte with 800 K s^{-1} heating rate.

improved limits of detection, in keeping with their theoretical prediction, and suggests that the furnace was not serving as the light-limiting aperture for their system.

2.8.1.3 Effect of the sample introduction hole

Loss out of the sample introduction hole was studied by Sturgeon and Chakrabarti [180], who estimated that 20 per cent sample left via this route for a 2 mm diameter sample introduction hole in a furnace 2.8 cm long and 0.6 cm internal diameter. The area of the sample introduction hole is only 5 per cent of the area of the opening of the ends plus the hole. Thus, these estimates suggest that a disproportionately large sample loss occurs from this opening. Monte Carlo simulations for the same furnace suggest that 34 per cent is lost from this opening during stopped flow conditions. Both these findings can be explained by initial placement of the sample below the hole, and thus increasing loss rate. Interestingly, these same simulations showed that longer furnaces and furnaces using a platform had a larger fraction of the sample lost from the sample introduction hole. Although these are interesting facts, they may serve little utility unless a means of easily introducing sample without the sample introduction hole can be found.

2.8.1.4 Platforms and other inserts

Many inserts have been employed within the furnace to foster the interference-free production of the analyte. L'vov first suggested the use of a platform as a means of attempting to delay the release of the sample from the surface until the gas phase had reached a higher temperature. As explained in Section 1.5.1, delayed heating of the sample with the use of a platform will result in the analyte being present at a higher temperature in the gas phase, which should assist in the production of free atoms. This concept would probably be most successful for those elements or analyte-containing compounds that volatilize at relatively low temperatures. Although temperature measurements are the subject of Chapter 3, it is sufficient to note at this time that a gas-phase temperature gradient exists between the platform and the wall (i.e. it is not an isothermal environment) and rapid heating of the wall produces a greater temperature difference between wall and platform. Likewise, beginning the heating ramp at a low temperature (such as ambient) also promotes a larger temperature gradient.

Another approach involves vaporization from the wall and transfer to a second surface that acts as a trap for the condensable species, as described by Holcombe and Sheehan [181]. This was carried a step further by providing external cooling to this second surface site which, when turned off, caused the rapid secondary release into a high-temperature furnace [182]. These second surface trapping approaches permitted dispersion of the analyte onto the second surface and the loss of much of the gaseous matrix products during the initial transfer.

The two-step furnace, discussed in Section 1.4, provides the ultimate control of vaporization and atomization steps. This system would be particularly useful for metals that form stable gas-phase molecules at the temperature they vaporize. With the two-step design the temperature gradient between vaporization from the cup and atomization in the tube furnace can be much larger than that associated that any of the previous designs and result in efficient free-atom formation.

More recently, Katskov *et al.* [6] built into the furnace a graphite filter containing coils of carbon fiber as well as the sample (Figure 2.42). Upon heating, the sample diffuses through the porous graphite into the optical. path where absorbance measurements are made. Like the other devices, a delay is experienced between the entrance of the analyte into the optical path and vaporization of the material. This approach provides considerable graphite–analyte contact because of the nature of the porous filter, and may prove very beneficial for elements which do not form strong interactions with carbon and as a means of minimizing nonspecific absorption.

2.8.1.5 *Sample placement*

The importance of reproducible positioning or spreading of the sample is a difficult parameter to measure because it is difficult to control. Early work by Sturgeon and Chakrabarti [180] with aluminum discussed the impact of sample positioning. Simulations once again suggest that there is

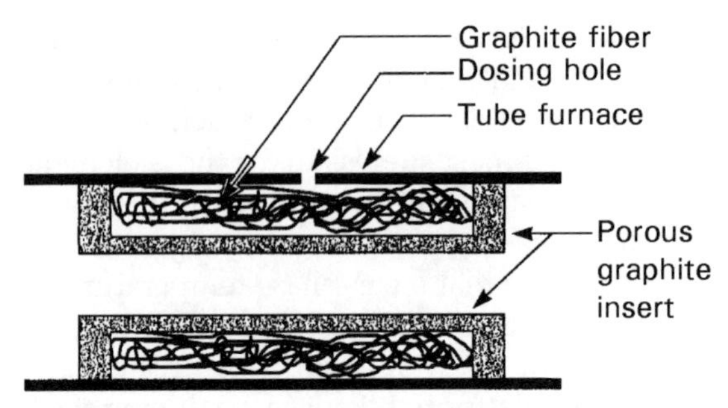

Figure 2.42 Schematic diagram of the carbon fiber filter furnace of Katskov. The analyte is initially located in the space between the porous insert and the graphite furnace. With heating, the analyte vaporizes and diffuses though the porous carbon and enters the analytical observation zone in the center of the tube. (Reprinted from *Spectrochim. Acta, Part B*, Vol. **50B**, D.A. Katskov, R.I. McCrindle, R. Schwarzer and P.J.J.G. Marais, The graphite filter furnace: a new atomization concept for atomic spectroscopy, pp. 1543–1555, 1995, with kind permission of Elsevier Science – NL, Sara Burgerhartstraat 25, 1055 KV Amsterdam, The Netherlands.)

not a significant impact of symmetrical spreading until the sample gets very close to the ends of the furnace [159]. Along similar lines, L'vov *et al.* [183] suggested that the position of the recondensed sample after thermal pretreatment is responsible for the increased stabilization seen in the pyrolysis curves for many elements. This effect was shown quantitatively by Chen and Jackson [184].

2.8.1.6 Heating rate

It is sufficient to note at this time that rapid heating produces a sharper and taller absorbance peak. However, when the time axis is converted to temperature, it is observed that the average temperature experienced by the sample is higher with increased heating rates. This fact alone would recommend the use of higher heating rates, assuming that the processes needed to promote free atom formation are improved at higher temperatures. As noted earlier, higher heating rates also have some distinct advantages when using platform vaporization.

With increased heating rates, there exists an improved chance of temporal coincidence between analyte and potential matrix interferences. This must be weighed against the advantages of a higher gas phase temperature promoting dissociation, and most workers currently prefer elevated rates during the atomization cycle. It has also been suggested that a higher heating rate will increase the rate of gas expansion in a constant pressure (i.e. open or semi-enclosed) atomizer design, as a result of heating of the gas within the furnace volume. However, significant analyte loss is not expected unless coincident vaporization of large amounts of matrix components occur.

2.8.1.7 *Isothermal versus nonisothermal*

This topic is explored in greater depth in Chapter 3. It is sufficient to note at this time that the obvious advantages of a constant high-temperature environment include the promotion of molecular dissociation, the minimization of condensation (for example, possible memory effects) on the cooler parts of the furnace, and more efficient vaporization of refractory materials. The potential disadvantage is increased diffusional loss due to the overall higher temperature. It is not clear at this time whether secondary condensation along the nonisothermal furnace provides any advantage.

2.8.1.7 *Protective gas*

The protective gas is an inert gas used to provide a nonreactive environment into which the sample is vaporized. It also minimizes the diffusion of oxygen to the furnace during the heating steps where oxidation of the graphite can occur. Argon is the preferred gas, although nitrogen is also sometimes used for elements which do not form stable nitrides or cyanides.

Alternative gases have been mixed with the argon or nitrogen. These gases are reactive in character and provide chemical modification of the analyte or the sample concomitants. A commonly used gas modifier is oxygen, which is used during the pretreatment stage to assist in the combustion (i.e. ashing) of organic or biological materials. If used properly, this can eliminate much of the organic matrix as well as stabilize many of the more volatile metals, although the exact cause for the stabilization is not known at this time. Gases used to provide a more reducing environment include hydrogen, carbon monoxide, and various gaseous alkanes. Fluorocarbons have also been used in an attempt to enhance the volatility of many metals, either to assist in the removal of a matrix or to assist in the determination of low volatility analytes. However, this can significantly enhance the volatility of many metals, resulting in low-temperature vaporization. In contrast, the gaseous metal–fluoride bond can be quite strong, thus requiring relatively high temperatures for dissociation.

Many systems have two protective gas flows, one around the furnace and one through the furnace. The former is used primarily to minimize air entrainment and thus oxidation of the furnace. The latter serves this function as well as to sweep unwanted products from the furnace during the drying and thermal pretreatment stages. Although both flow systems serve their function relatively well, diffusion of air into the furnace still occurs, as noted earlier in this chapter. The protective gas generally enters the furnace from the ends, and exits through the sample introduction hole. This directional flow also helps minimize condensation of pyrolysis products on the cooler ends of the furnace.

Protective gas flow during atomization has been used as a means of attenuating the analytical signal by increasing the rate of analyte loss. In this way, higher concentrations can be determined without sample dilution. One must proceed with caution in this approach, since the flowing gas stream may not reach the elevated temperature available with a static (stop-flow) mode and interferences may arise in some situations.

2.8.1.8 Pressure

From the previous discussions, it is clear that the local protective gas pressure can affect both the absorption line profiles and the removal rate of analyte. It has been shown that elevated pressures of several atmospheres cause a reduced sensitivity and increased linearization of the working curve [4,185]. The primary impact of elevated pressures is to broaden and shift the absorbing line profile, thereby reducing the absorption coefficient at the wavelength of the relatively narrow line source, such as an HCL. This effect seems to dominate over the increased residence time resulting from slower diffusion at the higher pressure.

Reducing the pressure tends to narrow the absorbing line widths and enhances diffusive loss. The initial work reported results to a pressure of 1 torr [186]. Other studies confirmed loss in sensitivity by as much as four orders of magnitude, and results were mixed on the effects of interfering matrices [187,188].

With the exception of special applications, it is not clear that the added engineering efforts needed for pressurizing or depressurizing the analytical furnace chamber yield substantial advantages.

2.8.1.9 Modifiers

Chapter 5 covers this topic in detail. It is sufficient to note at this point that the pragmatic purpose of modifiers is quite simply to promote the atomization efficiency and make this efficiency matrix independent. Having 100 per cent atomization efficiency is not a requirement for good analytical results. It is only required that the atomization efficiency is constant. Improved atomization efficiency will naturally lead to improved sensitivity, but will not necessarily make the system more immune to matrix interferences.

In general, modifiers are involved in one of several tasks. A very common function is to promote matrix volatility, thereby providing a matrix-free environment when the analyte enters the gas phase. Alternatively, the modifier may stabilize the analyte, which results in vaporization at a higher temperature where molecular dissociation is more likely and, possibly, after the matrix has volatilized and diffused from the furnace. As would be expected, a universal modifier (i.e. one that thermally stabilizes all elements) would be less effective, and most ideally, one would be wanted that only stabilizes the analyte.

Although the chemical role of some modifiers is relatively well understood, many remain elusive in their physical/chemical involvement. In most cases the modifiers react directly with the analyte or the matrix. However, it would not be surprising if, in some cases, the modifier also alters the behavior of the graphite substrate. The simplest example is the use of modifiers which form strong metal–carbon bonds (e.g. tantalum modifier) to reduce the likelihood of carbon–analyte interactions. The role of oxygen or hydrogen gas, if considered as a modifier, likewise alters many of the graphite surface characteristics.

2.9 CONCLUSIONS

The chemistry and physics of the processes involved in the production of the transient signal in ETAAS are, in many instances, multitudinous and complex. The development of an understanding of these processes has progressed steadily since the conception of the graphite furnace as an analytical atomizer for use with AA spectrochemical analysis. The analytical perspective of ETA can be restated in a more chemical perspective by looking at the time-dependent analytical steps.

The sample solvent affects the wetting of the graphite and, potentially, the ultimate form of the analyte location on the surface prior to the drying step. Acids and other major concomitants or modifiers present in the sample can affect the analyte in a number of different ways. Acids and surfactants can alter the interfacial tension of the sample solution and the graphite, causing spreading and possibly altering the crystal size or nature (i.e. adsorbed versus crystal deposition) of the analyte on the surface. Some of these acids may also migrate into the graphitic planes to form intercalates and persist well beyond temperatures where one would normally expect them to decompose or vaporize. Oxidizing solutions are capable of altering the graphite surface chemistry, as well as contributing significantly to the oxygen burden in the gas phase. The anionic component of the acid and anions of the matrix may determine the chemical form of the analyte deposited during the drying step and may affect the way in which the analyte is ultimately vaporized.

The drying step appears to be less critical, as long as it is not so rapid as to expel material from the furnace. However, very rapid drying may promote the formation of smaller analyte or matrix crystals which may be more dispersed on the graphite surface. This in turn may affect the shape, but not the area, of the analytical profile.

The pyrolysis step is important if there is any time-dependent process that must occur for improved results. This could include kinetically limited reactions or processes, such as diffusion, which require time to complete. Thus, matrix removal by volatilization, solid-state reactions between analyte and modifier, and other slow reactions are impacted by judicious use of this step. For many of the reactions and decompositions, the rate of reaction can be strongly temperature dependent and system settings should be empirically tested rather than relying on reported values in a manual or in the literature, as there is limited accuracy on all temperature settings for commercial instruments. During this thermal pretreatment, movement of the analyte may occur either into the graphite or along the tube's surface. These possibilities also may make the time spent at any given treatment temperature important and a parameter worth optimizing. On nonisothermal furnaces, analyte may move from its initial location to cooler areas of the tube during this stage if the temperature is sufficient to promote. some analyte vaporization. Interestingly, there is a built-in thermal trap because of this gradient that may prevent analyte loss through use of an inadvertently high thermal pretreatment temperature. The isothermal cuvettes do not have this safeguard against loss during the pretreatment step. Thus, it is reasonable to expect that procedures and methods proposed for one type of furnace may not work in the same fashion in a different furnace design.

The atomization step should raise the furnace to a temperature that is sufficient to vaporize the analyte without overheating the system. Excessively high atomization temperatures may reduce the furnace lifetime, but will probably not cause any significant alteration in the peak area. The use of a rapid ramp is beneficial for most applications, but is particularly important when using platform-type atomization in order to optimize the temperature difference between the wall and the platform.

The clean step should be used to remove any of the residual, refractory nonanalyte-containing species. However, if the analyte is also vaporized during this process, the atomization temperature should be raised. The clean step may also serve a secondary useful function. The high temperature employed for this cycle may also partially anneal the graphite surface and serve to promote a more reproducible surface from sample to sample in the event that the sample has caused alteration in the graphite surface.

REFERENCES

[1] A.S. King, *Astophysical Journal* **27**, 353 (1908).
[2] B.V. L'vov, *Inzhenerno Fizichecidi Zhurnal* **2**, 44 (1959).
[3] B.V. L'vov, *Spectrochim. Acta.* **17**, 761 (1961).
[4] B.V. L'vov, *Atomic Absorption Spectrochemical Analysis*, London: Adam Hilger, 1970.
[5] G.A. Beital and D.C. Benson, *J. Vac. Sci. Techn.* **10**, 201 (1973).
[6] D.A. Katskov, R.I. McCrindle, R. Schwarer and P.J.J.G. Marais, *Spectrochim. Acta, Part B* **50B**, 1543 (1995).
[7] H. Falk, A. Glismann, L. Bergann, G. Minkwitz, M. Schubert and J. Skole, *Spectrochim. Acta, Part B* **40B**, 533 (1985).
[8] W. Slavin, D.C. Manning and G.R. Carnrick, *At. Spectrosc.* **2**, 137 (1981).
[9] P.L. Walker Jr. (Ed.), *Chemistry and Physics of Carbon*, Marcel Dekker, New York, 1965, Vol. 1.
[10] K. Kinoshita, *Carbon; Electrochemistry and Physicochemical Properties*, John Wiley and Sons, New York, 1988.
[11] R.O. Lussow, F.J. Vastola and P.L. Walker, *Carbon*, **5**, 591 (1967).
[12] C.A.L. Leon and L.R. Radovic, in *Chemistry and Physics of Carbon*, (Ed.) P.A. Thrower, Marcel Dekker, New York, 1994, Vol. 24, p 231.
[13] B.R. Puri, in *Chemistry and Physics of Carbon*, (Ed.) P.L. Walker Jr., Marcel Dekker, New York, 1970, Vol. 6, p 191.
[14] A.R. Ford, *Engineer* **224**, 444 (1967).
[15] D.E. McKee, *Annu. Rev. Mat. Sci.* **3**, 195 (1973).
[16] W.H. Smith and D.H. Leeds, in *Modern Materials, Advances in Development and Applications*, (Ed.) B.W. Gonser, Academic Press, New York, 1970, Vol. 7, pp 139.
[17] C.A. Klein, *J. Appl. Phys.* **33**, 3338 (1962).
[18] D.E. Soule, *Phys. Rev.* **112**, 698 (1958).
[19] I.L. Spain, in *Physics of Semimetals and Narrow Band Gap Semiconductors*, (Ed.) Carter, Bate, Pergamon Press, New York, 1971, pp 177.
[20] C.N. Hooker, A.R. Ubbelohde and D.A. Young, *Proc. Roy Soc. (London)* **A276**, 83 (1963).
[21] M.G. Holland, C.A. Klein and W.D. Straub, *J. Physics and Chem. of Solids*, **27**, 903 (1996).
[22] B.T. Kelly, in *Chemistry and Physics of Carbon*, Dekker, New York, 1969, Vol. 5, pp 119.
[23] R.T. Baker and J.J. Chludzinski, *Carbon* **19**, 75 (1981).
[24] E.A. Heintz and W.E. Parker, *Carbon* **4**, 473 (1966).
[25] P.J.J. Walker, F.J. Ruotino and L.G. Austin, *Adv. Catal.* **11**, 133 (1959).
[26] D.J. Suh, T. Park and S. Lhm, *Carbon* **31**, 427 (1993).
[27] A. Tomita, N. Sato and Y. Tamai, *Carbon* **12**, 143 (1974).
[28] R.E. Sturgeon and S.S. Berman, *Anal. Chem.* **57**, 1268 (1985).
[29] W. Frech and A. Cedegren, *Anal. Chim. Acta.* **82**, 93 (1976).
[30] W. Frech and A. Cedergren, *Anal. Chim. Acta.* **82**, 83 (1976).
[31] J.P. Byrne, C.L. Chakrabarti, D.C. Gregoire, M. Lamoureux and T. Ly, *J. Anal. At. Spectrom.* **7**, 321 (1992).
[32] B.V. L'vov and G.M. Ryabchuk, *Spectrochim. Acta, Part B* **37B**, 673 (1982).
[33] W. Frech, E. Lundberg and A. Cedegren, *Prog. Anal. At. Spectrosc.* **8**, 279 (1985).
[34] S.G. Salmon and J.A. Holcombe, *Anal. Chem.* **54**, 630 (1982).

[35] S.G. Salmon, R.H. Davis and J.A. Holcombe, *Anal. Chem.* **53**, 324 (1981).
[36] R.E. Sturgeon and S.S. Berman, *Anal. Chem.* **57**, 1268 (1985).
[37] G.D. Rayson and J.A. Holcombe, *Anal. Chim. Acta.* **136**, 249 (1982).
[38] R.E. Sturgeon and H. Falk, *Spectrochim. Acta, Part B* **43B**, 421 (1988).
[39] M.S. Droessler and J.A. Holcombe, *Can. J. Spectrosc.* **31**, 6 (1986).
[40] R.E. Sturgeon and H. Falk, *J. Anal. At. Spectrom.* **3**, 27 (1988).
[41] R.E. Sturgeon, K.W.M. Siu, G.J. Gardner and S.S. Berman, *Anal. Chem.* **58**, 42 (1986).
[42] R.E. Sturgeon, *Fresenius' Z. Anal. Chem.* **324**, 807 (1986).
[43] K.H. Stern, *J. Phys. Chem. Ref. Data* **1**, 747 (1972).
[44] A.K. Gilmutdinov, C.L. Chakrabarti and J.C. Hutton, *J. Anal. At. Spectrom.* **7**, 1047 (1992).
[45] T.R. Marrero, *J. Phys. Chem. Ref. Data* **1**, 3 (1972).
[46] M. Harry, *Chem. Soc. Pub.* **32**, 133 (1978).
[47] C.C. Eloi, J.D. Robertson and V. Majidi, *Anal. Chem.* **67**, 335 (1995).
[48] R.J. Day, P.L. Walker Jr. and C.C. Wright *Industrial Carbon and Graphite* 348 (1958).
[49] J. Nagle and R.F. Strickland-Constable, *Proceedings of the 5th Carbon Conference* **1**, 154 (1962).
[50] J.R. Walls and R.F. Strickland-Constable, *Carbon* **1**, 333 (1964).
[51] R.E. Sturgeon, K.W.M. Siu and S.S. Berman, *Spectrochim. Acta, Part B* **39B**, 213 (1934).
[52] V. Majidi, R.G. Smith, R.E. Bossio, R.T. Pogue and M.W. McMahon, *Spectrochim. Acta, Part B* **51B**, 941 (1996).
[53] J. Sire and I.A. Voinovitch, *Analusis* **7**, 275 (1979).
[54] R.B. Cruz and J.C. van Loon, *Anal. Chim. Acta.* **72**, 231 (1974).
[55] D.R. Churella and T.R. Copeland, *Anal. Chem.* **1978**, 50, 309 (1978).
[56] J.A. Holcombe and M.S. Droessler, *Fresenius' Z. Anal. Chem.* **323**, 689 (1986).
[57] C. Eloi, J.D. Robertson and V. Majidi, *J. Anal. At. Spectrom.* **8**, 217 (1993).
[58] J.G. Jackson, R.W. Fonseca and J.A. Holcombe, *Spectrochim Acta, Part B* **50B**, 1837 (1995).
[59] J.A. Holcombe and G.D. Rayson, *Prog. Anal. Atom. Spectrosc.* **6**, 225 (1983).
[60] J. McNally and J.A. Holcombe, *Anal. Chem.* **59**, 1105 (1987).
[61] J.G. Jackson, R.W. Fonseca and J.A. Holcombe, *J. Anal. At. Spectrom.* **9**, 167 (1994).
[62] R.W. Fonseca, J. McNally and J.A. Holcombe, *Spectrochim. Acta, Part B* **48B**, 79 (1993).
[63] S. Lynch, R.E. Sturgeon, V.T. Luong and D. Littlejohn, *J. Anal. At. Spectrom.* **5**, 311 (1990).
[64] A.W. Moore, *Chem. and Physics of Carbon* **17** (1981).
[65] H.B. Kagan, *Chem. Tech.* **Aug.,** 510 (1976).
[66] W. Slavin, G.R. Carnrick and D.C. Manning, *Anal. Chem.* **54**, 621 (1982).
[67] V.I. Sherbakov, Y.I. Beljaev, B.F. Mjasoedov, I.N. Marov and N.B.Z. Kalinichenko, *Analyt. Khim. (Russia)* **37**, 1717 (1982).
[68] L. Sabbatini and G. Tessari, *Ann. Chim. (Rome)* **74**, 779 (1984).
[69] D.A. Young, *Decomposition of Solids*, Pergamon Press, London, 1966.
[70] K.H. Stem and E.L. Weise, *High Temperature Properties and Decomposition of Inorganic Salts, Part 1; Sulfate.* National Bureau of Standards Bulletin 7, US Government Printing Office, Washington, DC 1966.

[71] R.E. Sturgeon, D.F. Mitchell and S.S. Berman, *Anal. Chem.* **55**, 1059 (1983).
[72] P. Wang, V. Majidi and J.A. Holcombe, *Anal. Chem.* **61**, 2652 (1989).
[73] B.V. L'vov, *Mikrochim. Acta.* **II**, 299 (1991).
[74] J.G. Jackson, A. Novichikhin, R.W. Fonseca and J.A. Holcombe, *Spectrochim. Acta, Part B* **50B**, 1423 (1995).
[75] J.G. Jackson, R.W. Fonseca and J.A. Holcombe, *Spectrochim. Acta, Part B* **50B**, 1449 (1995).
[76] B.V. L'vov and A. Novichikhin, *Spectrochim. Acta, Part B* **50B**, 1427 (1995).
[77] B.V. L'vov and A. Novichikhin, *Spectrochim. Acta, Part B* **50B**, 1459 (1995).
[78] W. Wendl, *Fresenius' Z. Anal. Chem.* **323**, 726 (1986).
[79] C.L. Chakrabarti, S. Wu, F. Marcantonio and K.L. Headrick, *Fresenius' Z. Anal. Chem.* **323**, 730 (1986).
[80] W. Wendl and G. Muller-Vogt, *Spectrochim. Acta, Part B* **40B**, 527 (1985).
[81] J.P. Matousek and H.K.J. Powell, *Spectrochim. Acta, Part B* **43B**, 167 (1988).
[82] R.E. Sturgeon and S.S. Berman, *Anal. Chem.* **55**, 190 (1983).
[83] B.V. L'vov, *J. Quant. Spectrosc. Radiat. Transfer* **12**, 651 (1972).
[84] B.V. L'vov, B.G. Nikolaev, B.G. Norman, L.K. Polzik and M. Mojica, *Spectrochim. Acta, Part B* **41B**, 1043 (1986).
[85] H. Falk and J. Tilch, *J. Anal. Atom. Spectrom.* **2**, 527 (1987).
[86] W. Frech and D.C. Baxter, *Spectrochim. Acta, Part B* **45B**, 867 (1990).
[87] B.V. L'vov, *Spectrochim. Acta, Part B* **33B**, 153 (1978).
[88] D.C. Baxter and W. Frech, *Spectrochim. Acta, Part B* **42B**, 1005 (1987).
[89] W. Frech, D.C. Baxter and E. Lundberg, *J. Anal. At. Spectrom.* **3**, 21 (1988).
[90] W. Slavin and G.R. Carnrick, *Spectrochim. Acta, Part B* **39B**, 271 (1984).
[91] R.E. Sturgeon and S.S. Berman, *Anal. Chem.* **52**, 1049 (1980).
[92] J.M. Ottaway and F. Show, *Analyst* **101**, 582 (1976).
[93] M.S. Epstein, T. Rains and T. O'Haver, *Appl. Spectrosc.* **30**, 324 (1975).
[94] T. Nakamura, H. Oka, II. Morikawa and J. Sato, *Analyst* **117**, 131 (1992).
[95] W.M.G.T. van den Broeck and L. de Galan, *Anal. Chem.* **49**, 2176 (1977).
[96] I.M.L. Billas, A. Chätterlain and W.A. deHeer, *Science* **265**, 1682 (1994).
[97] F.M. Mulder, T.A. Stegink, R.C. Thiel, L.J. de Jongh and U. Schmid, *Nature* **367**, 716 (1994).
[98] M. Moskovits, *Annu. Rev. Phys. Chem.* **42**, 465 (1991).
[99] A. Cedergren, W. Frech and F. Lundberg, *Anal. Chem.* **56**, 1382 (1984).
[100] X. Shan and Z. Ni, *Acta Chimica Sinica* **37**, 261 (1979).
[101] P.A. Redhead, *Vacuum* **12**, 203 (1962).
[102] N. Hadgu and W. Frech, *Spectrochim. Acta, Part B* **49B**, 445 (1994).
[103] O.A. Güell and J.A. Holocmbe, *Spectrochim. Acta, Part B* **43B**, 459 (1988).
[104] O.A. Güell and J.A. Holcombe, *J. Anal. Chem.* **62**, 529A (1990).
[105] T.E. Histen and J.A. Holcombe, *Spectrochim. Acta, Part B* **51B**, 1279 (1996).
[106] C. Rohrer and W. Wegscheider, *Spectrochim. Acta, Part B* **48B**, 315 (1993).
[107] S.S. Black, M.R. Riddle and J.A. Holcombe, *Appl. Spectrosc.* **40**, 925 (1986).
[108] O.A. Güell, Ph.D. Thesis, The University of Texas At. Austin, Austin TX, USA, 1990.
[109] O.A. Güell, J.A. Holcombe and C.J. Rademeyer, *Anal. Chem.* **65**, 748 (1993).

[110] T. Kántor and L. Bezúr, *J. Anal. At. Spectrom.* **1**, 9 (1986).
[111] T. Kántor and G. Záray, *Fresenius' Z. Anal. Chem.* **342**, 927 (1992).
[112] M.M. Lamoureux, C.L. Chakrabarti, D.M. Goltz and D.C. Grégoire, *Anal. Chem.* **66**, 3208 (1994).
[113] D.A. Redfield and W. Frech, *J. Anal. At. Spectrom.* **4**, 685 (1989).
[114] W. Wendl and G. Müller-Vogt, *Spectrochim. Acta, Part B* **39B**, 237 (1984).
[115] J.P Erspamar and T.M. Niemezyk, *Anal. Chem.* **54**, 538 (1982).
[116] J.M. Shekiro, R.K. Skogerboe and H.E. Taylor, *Anal. Chem.* **69**, 2578 (1988).
[117] K.E.A. Ohlsson and W. Frech, *J Anal. At. Spectrom.* **4**, 379 (1989).
[118] V. Majidi, J. Ratcliff and M. Owens, *Appl. Spectrosc.* **45**, 473 (1991).
[119] A.K. Gilmutdinov, Y.A. Zakharov, V.P. Ivanov and A.V. Voloshin, *J. Anal. At. Spectrom.* **6**, 505 (1991).
[120] A.K. Gilmutdinov, Y.A. Zakharov, V.P. Ivanov, A.V. Voloshin and K. Dittrich, *J. Anal. At. Spectrom.* **7**, 675 (1992).
[121] A.K. Gilmutdinov, Y.A. Zallharov and A.V. Voloshin, *J. Anal. At. Spectrom.* **8**, 387 (1993).
[122] C.L. Chakrabarti, A.K. Gilmutdinov and J.C. Hutton, *Anal. Chem.* **65**, 716 (1993).
[123] D.L. Styris and J.H. Kaye, *Spectrochim. Acta, Part B* **36B**, 41 (1981).
[124] D.L. Styris, *Fresenius' Z. Anal. Chem.* **323**, 710 (1986).
[125] D.A. Bass and J.A. Holcombe, *Anal. Chem.* **59**, 974 (1987).
[126] D.L. Styris and J.H. Kaye, *Anal. Chem.* **54**, 864 (1982).
[127] D.L. Styris and D.A. Redfield, *Anal. Chem.* **59**, 2891 (1987).
[128] J.A. Holcombe, D.L. Styris and J.D. Harris, *Spectrochim. Acta, Part B* **46B**, 629 (1991).
[129] M.S. Droessler and J.A. Holcombe, *Spectrochim. Acta, Part B* **42B**, 981 (1987).
[130] D.L. Styris, L.J. Prell, D.A. Redfield, J.A. Holcombe, D.A. Bass and V. Majidi, *Anal. Chem.* **63**, 508 (1991).
[131] N.S. Ham and T. McAllistair, *Spectrochim. Acta, Part B* **43B**, 789 (1988).
[132] R.W. Fonseca, K.I. Wolfe and J.A. Holcombe, *Spectrochim. Acta, Part B* **49B**, 399 (1994).
[133] D.C. Hassell, V. Majidi and J.A. Holcombe, *J Anal. At. Spectrom.* **6**, 105 (1991).
[134] D.L. Styris, L.J. Prell and D.A. Redfield, *Anal. Chem.* **63**, 503 (1991).
[135] S.L. Paveri-Fontana, G. Tessari and C. Torsi, *Anal. Chem.* **46**, 1032 (1974).
[136] G. Torsi and G. Tessari, *Anal. Chem.* **47**, 839 (1975).
[137] G. Tessari and G. Torsi, *Anal. Chem.* **47**, 842 (1975).
[138] C.W. Fuller, *Analyst* **99**, 734 (1974).
[139] D.A. Katskov, *Zh. Priki. Spectrosk.* **30**, 619 (1979).
[140] B. Welz, M. Sperling, G. Schlemmer, N. Wenzel and G. Marowsky, *Spectrochim. Acta, Part B* **43B**, 1187 (1988).
[141] R.E. Sturgeon, C.L. Chakrabarti and C.H. Langford, *Anal. Chem.* **48**, 1792 (1976).
[142] B. Smets, *Spectrochim. Acta, Part B* **35B**, 33 (1980).
[143] D.A. Bass and J.A. Holcombe, *Anal. Chem.* **60**, 2680 (1988).
[144] W. Frech, N.G. Zhow and E. Lundberg, *Spectrochim. Acta, Part B* **37B**, 691 (1982).
[145] D. Rojas and W. Olivares, *Spectrochim. Acta, Part B* **47B**, 387 (1992).
[146] X. Yan, Z. Ni, X. Yang and G. Hong, *Spectrochim. Acta, Part B* **48B**, 605 (1993).

[147] R.W. Fonseca, L.L. Pfefferkorn and J.A. Holcombe, *Spectrochim. Acta, Part B* **49B**, 1595 (1994).
[148] S. Lynch, V.T. Luong and D. Littlejohn, *J. Anal. At. Spectrom.* **5**, 311 (1990).
[149] R.W. Fonseca, O.A. Güell and J.A. Holcombe, *Spectfrochim. Acta, Part B* **47B**, 573 (1992).
[150] J.A. Holcombe, *Spectrochim. Acta Part B* **44B**, 975 (1989).
[151] G.A. Somoijai, *Chemistry in Two Dimensions: Surfaces*, Cornell University Press, Ithaca, NY, USA, 1981.
[152] J.R. Arthur and A.Y. Cho, *Surf Sci.* **36**, 641 (1973).
[153] S.G. Salmon and J.A. Holcombe, *Anal. Chem.* **51**, 648 (1979).
[154] C.W. Huie and C.J. Curran, *Appl Spectrosc.* **42**, 1307 (1988).
[155] D.A. Katskov, A.M. Shtepan, I.L. Grinshtein and A.A. Pupyshev, *Spectrochim. Acta, Part B* **47B**, 1023 (1992).
[156] W. Frech, B.V. L'vov and N.P. Romanova, *Specfrochim. Acta, Part B* **47B**, 1461 (1992).
[157] B.V. L'vov and W. Frech, *Spectrochim. Acta, Part B* **48B**, 425 (1993).
[158] A.K. Gilmutdinov and I.S. Fishman, *Spectrochim. Acta, Part B* **39B**, 171 (1984).
[159] O.A. Guell and J.A. Holcombe, *Spectrochim. Acta, Part B* **43B**, 459 (1988).
[160] O.A. Guell and J.A. Holcombe, *Spectrochim. Acta, Part B* **44B**, 185 (1989).
[161] A.K. Gilmutdinov, R.M. Mrasov, Y.A. Zakharov, C.L. Chakrabarti and J.C. Hutton, *Spectrochim. Acta, Part B* **50B**, 1637 (1995).
[162] V. Krivan, *J Anal. At. Spectrom.* **7**, 155 (1992).
[163] A.K. Gilmutdinov, T.M. Abdullina, S.F. Gorbachev and V.L. Makarov, *Spectrochim. Acta, Part B* **47B**, 1075 (1992).
[164] A.C.G. Mitchell and M.W. Zemansky, *Resonance Radiation and Excited Atoms*, MacMillan, New York, 1934.
[165] C.T.J. Alkemade, T. Hollander, W. Snelleman and P.J.T. Zeegers, *Metal Vapours in Flames*, Pergamon Press, Oxford, 1982.
[166] J.D. Winefordner, S.G. Shulman and T.C. O'Haver, *Luminescence Spectroscopy in Analytical Chemistry*, Wiley Interscience, New York, 1970.
[167] B.V. L'vov, *Spectrochim. Acta, Part B* **45B**, 633 (1990).
[168] P.L. Larkins, *J Anal. At. Spectrom.* **7**, 265 (1992).
[169] A.K. Gilmutdinov, B. Radziuk, M. Sperling, B. Welz and K.Y. Nagulin, *Appl Spectrosc.* **49**, 413 (1995).
[170] A.K. Gilmutdinov, B. Radziuk, M. Sperling, B. Welz and K.Y. Nagulin, *Appl Spectrosc.* **50**, 483 (1996).
[171] H.C. Wagenaar, C.J. Pickford and L. deGalan, *Spectrochim. Acta, Part B* **28B**, 157 (1973).
[172] W. Slavin, D.C. Manning and G.R. Carnrick, *J Anal. At. Spectrom.* **3**, 13 (1988).
[173] M.T.C. DeLoos-Vollebregt and L. deGalan, *Appl Spectrosc.* **34**, 464 (1980).
[174] A.K. Gilmutdinov, K.Y. Nagulin and Y.A. Zakharov, *J Anal. At. Spectrom.* **9**, 643 (1994).
[175] D.L. Styris, L.J. Prell, D.A. Redfield, J.A. Holcombe, D.A. Bass and V. Majidi, *Anal. Chem.* **63**, 508 (1991).
[176] D.A. Bass and J.A. Holcombe, *Anal. Chem.* **59**, 1105 (1987).
[177] R.W. Woodriff, *Appl Spectrosc.* **28**, 413 (1974).
[178] O. Guell and J.A. Holcombe, *Spectrochim. Acta, Part B* **47B**, 1535 (1992).
[179] N. Hadgu, A. Ohlsson and W. Frech, *Spectrochim. Acta, Part B* **50B**, 1077 (1995).
[180] R.E. Sturgeon and C.L. Chakrabarti, *Prog. Anal. At. Spectrosc.* **1**, 5 (1978).

[181] J.A. Holcombe and M.T. Sheehan, *Appl Spectrosc.* **36**, 631 (1982).
[182] T.M. Rettberg and J.A. Holcombe, *Spectrochim. Acta, Part B* **39B**, 249 (1984).
[183] B.V. L'vov, L.K. Polzik and L.F. Yatsenko, *Talanta* **34**, 141 (1987).
[184] G. Chen and K.W. Jackson, *Spectrochim. Acta, Part B* **51B**, 1505 (1996).
[185] T.M. Rettberg and J.A. Holcombe, *Spectrochim. Acta, Part B* **41B**, 377 (1986).
[186] H.M. Donega and T.E. Burgess, *Spectrochim. Acta, Part B* **25B**, 1521 (1970).
[187] P. Wang and J.A. Holcombe, *Spectrochim. Acta, Part B* **47B**, 1277 (1992).
[188] D.C. Hassell, T.M. Rettberg, F.A. Fort and J.A. Holcombe, *Anal. Chem.* **60**, 2680 (1988).
[189] J.C. Bokros, in *Chemistry and Physics of Carbon*, (Ed.) P.L. Walker, Marcel Dekker, New York 1970, Vol. 6, p. 191.
[190] H.M. Ortner, G. Schlemmer, B. Welz and W. Wegsheider, *Spectrochim. Acta, Part B* **40B**, 959 (1985).
[191] K.G. Vandervoort, D.J. Butcher, C.T. Brittain and B.B. Lewis, *Appl Spectrosc.* **50**, 928 (1996).

3

Temperature: its significance, control and measurement

Cornelius J. Rademeyer

3.1 SIGNIFICANCE OF TEMPERATURE IN ELECTROTHERMAL ATOMIZATION

3.1.1 Introduction

In practice, the technique of ETAAS is used by placing the sample at the center of the furnace, onto the inside wall of the graphite tube, or onto a platform that has been inserted into the graphite tube. An electric current is then passed through the tube in increasing steps to eventually atomize the analyte. Analyte that was placed on the tube wall will be directly heated through contact with it. The platform is not intended to be directly heated by the electric current, but rather by the absorption of energy radiated from the furnace wall. As the platform is in nominal electrical contact only at its four corners with the tube, conduction of the electric current through the platform may also contribute to the total current through the platform-in-tube combination, but to a smaller degree as this contact area is very small. The only other possible source of energy delivered to the platform is convection through the argon protective gas. However, heating of the platform by conduction of current and convection by the argon is considered to be insignificant. Therefore, as a result of the nature by which the platform is heated, the temperature of the platform always lags behind the temperature of the wall as indicated in Figure 1.12 (Section 1.5.1).

The tube, or the platform-in-tube combination, is heated through a series of increasing temperature steps. During the first of these steps, the sample undergoes a physical process during which the analyte is

Electrothermal Atomization for Analytical Atomic Spectrometry. Edited by K. W. Jackson
© 1999 John Wiley & Sons Ltd

desolvated. This process is referred to as the drying stage. As the analyte is in physical contact with the tube wall or the platform surface, the temperature of these surfaces may be considered the decisive quantity that governs the drying of the sample. After completion of the drying stage, the dried sample is heated to a temperature at which the bulk of the matrix is removed, called the pyrolysis stage. Again, the important parameter controlling this process is the temperature of the graphite wall or the platform surface. Finally, the analyte is heated through contact with the atomizer surface, until analyte vapor is produced (the atomization stage). This vapor then enters the light path where the atomic absorption process takes place. Intermediate temperature steps are frequently used to enhance the amount of analyte that eventually reaches the light path.

Accompanying these steps are physical and chemical changes to the analyte that are strongly temperature dependent (see for example Chapters 2 and 5). Consequently, the course of all these processes are of utmost analytical importance. The drying and pyrolysis stages determine the reproducibility and the accuracy of the determination, whereas the atomization stage controls the volatilization of the sample into the gas phase and influences the sensitivity of the determination. The temperature and duration of each of these stages should be optimized carefully for a given element in a given matrix. This is done by studying pyrolysis curves and atomization curves, shown in Figure 3.1, and described below in Section 3.1.2. These steps, and the temperature dependence thereof, are discussed in more detail in the following paragraphs.

3.1.2 Drying and pyrolysis temperatures

The value of the selected programmed temperature for the drying stage is chosen so that the solvent evaporates at an acceptable rate leaving the analyte in a dried form. It is important that the analyte droplet is heated carefully and slowly, so as not to cause any violent boiling with a chance for sample expulsion as aerosol droplets. Additionally, with too rapid drying, the analyte may spread over a relatively large surface area and be subjected to different temperatures, as a temperature gradient exists over the length of the furnace tube.

Selection of the pyrolysis temperature depends on numerous factors, of which the composition of the sample (matrix etc.) and the nature of the element in question are most important. As it is necessary for good analytical results to atomize the analyte in a matrix-free environment, a primary purpose of this step is to eliminate as much of the matrix as possible before atomization commences. This is not always possible, because components of the matrix may atomize at the same time as the analyte. In cases where the bulk of the matrix is more volatile than the analyte, the chosen temperature should be high enough to evaporate the

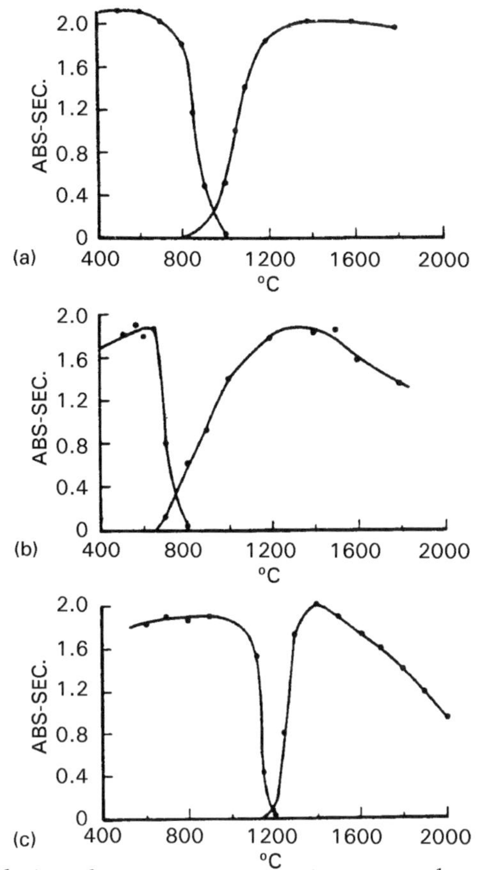

Figure 3.1 Pyrolysis and atomization curves for 0.1 mg l^{-1} lead: (a) no additions, lead in the nitrate form; (b) in 0.04 % NaCl; (c) in 0.01 % NaH_2PO_4. For the pyrolisis cycles, the atomization temperatures were 1400 °C. For the atomization cycle of (a) and (b) the pyrolysis temperature was 600 °C and for (c) it was 1000°C [1]. (Reprinted with permission from W. Slavin and D.C. Manning, Reduction of matrix interferences for lead determination with the L'vov platform and the graphite furnace, Anal. Chem. **51**, 261–265. Copyright 1979 American Chemical Society.)

bulk of the matrix before the atomization stage. Too low a temperature may not remove a sufficient quantity of the matrix. Too high a temperature may lead to loss of the analyte through evaporation, particularly in the case of the more volatile elements having relatively high vapor pressures. The nature of the analyte element in the sample, such as the vapor pressure, the ease with which the compound decomposes, and the length of time that the sample is heated, will determine the state of the analyte after the pyrolysis stage has been completed.

The pyrolysis temperature is therefore chosen for a particular element and matrix so that the analyte may be in its optimum form for the subsequent atomization stage, ensuring that the optimal amount of analyte reaches the light path in atomic form. Temperature is therefore the dominant factor that determines the quantity of interferent still present at the onset of the atomization stage, thus determining the population of atoms in the graphite furnace during this step. To illustrate this, Figure 3.1 shows the influence of the pyrolysis and atomization temperature on the integrated absorbance of lead in different matrixes. The left-hand curves in each set of curves (a), (b) and (c) in Figure 3.1 are often designated as pyrolysis curves, and show the absorbance for different pyrolysis temperatures and a relatively high fixed atomization temperature of 1400°C in each case. The right-hand portion of each set of curves is called an atomization curve. The atomization temperature is varied while the pyrolysis temperature is fixed at the values indicated in the figure caption, close to the maximum signal, i.e. before loss of lead occurs during the pyrolysis stage. For each matrix, the atomization curve shows that the signal begins to appear at about the same temperature that the lead signal has begun to decrease in the pyrolysis curve. It is clear that this temperature is different for the different matrices studied. Obviously, for lead determination with samples atomized from the furnace wall, these variations in appearance temperature will cause the lead to be volatilized into the gas atmosphere at a temperature that will depend upon the matrix, thus producing different signals for different matrices. This is true not only for lead but also for many other elements. It is also true for platform atomization.

3.1.3 Atomization temperature

In the Massmann furnace, as it is used in commercial instruments, the formation and dissipation of analyte atoms proceed in an environment where the temperature increases rapidly. Assuming that the majority of atoms in such a furnace remain in their electronic ground energy state, the relative magnitude of the atomic absorption signal will correspond closely to the change in population of atoms with respect to time. This maximum in signal corresponds to a particular tube temperature and time after initiation of the atomization stage (see Figure 1.12).

The optimum atomization temperature depends on the element to be determined – ca. 800°C for cadmium, ca. 1700°C for chromium, and ca. 2700 °C for vanadium. At these temperatures, atomization is fairly rapid. However, atomization commences at a much lower temperature depending on the relative vapor pressure of the element; for example, cadmium starts to evaporate at a significant rate at 500°C. Also, it is important that the surface from which atomization occurs (tube wall or platform) reaches a sufficiently high temperature before a significant fraction of the element is lost from the furnace – at atomization surface temperatures between 500 and 800°C the signal for cadmium will be very dependent on the temperature and the heating rate. However, the final atomization temperature should not be higher than that needed to completely vaporize the element, because other ions or sample components may interfere with the atomization process. Importantly, elevated temperatures increase the rate of diffusional loss and decrease furnace lifetime.

For atomization from the wall, different sensitivities are observed for different elements at a specific atomization temperature when peak height and integrated absorbance measurements are compared. Elements such as cadmium, lead, cobalt, and nickel are relatively insensitive, whereas molybdenum and vanadium are rather sensitive to temperature. Shown in Figures, 3.2 and 3.3 are the peak and integrated absorbance, respectively, as a function of wall temperature for these elements. Shown in Figure 3.4 is the effect of tube wall temperature on atomization from a platform. Chakrabarti *et al.* [2] offered an explanation for higher peak height measurements at lower heating rates for wall atomization, such as the vanadium curve in Figure 3.2. Loss of vanadium by expulsion, due to rapidly expanding vapor, is greater at the higher heating rate of 7.0 K ms^{-1}. Such an explanation presupposes that vanadium species exist in the vapor phase at a temperature lower than the atomization temperature, and that the rate-determining step in the atomization mechanism is a vapor-phase reaction such as dissociation. In Figure 3.2, molybdenum shows a very high increase in the peak height absorbance with increasing temperature at a heating rate of 7.0 K ms^{-1}, probably because molybdenum remains in the condensed phase at the temperature at which vanadium is in the vapor phase, and hence the loss of molybdenum species by expulsion with the rapidly expanding vapor phase is avoided. In addition, however, there may be other factors, such as interfering reactions or interferents as in the formation of relatively involatile carbides in the condensed phase.

Shown in Figure 3.5 is the effect of atomization temperature on the integrated absorbance of thallium as a function of the percentage of interferent (magnesium chloride) using wall atomization. The absorbance signal is larger at lower atomization temperatures because the residence time is longer in the furnace. However, the chloride interference is more severe at the lower temperature. In the case of atomization from the platform, this effect is even more pronounced due to the delayed heating of the platform.

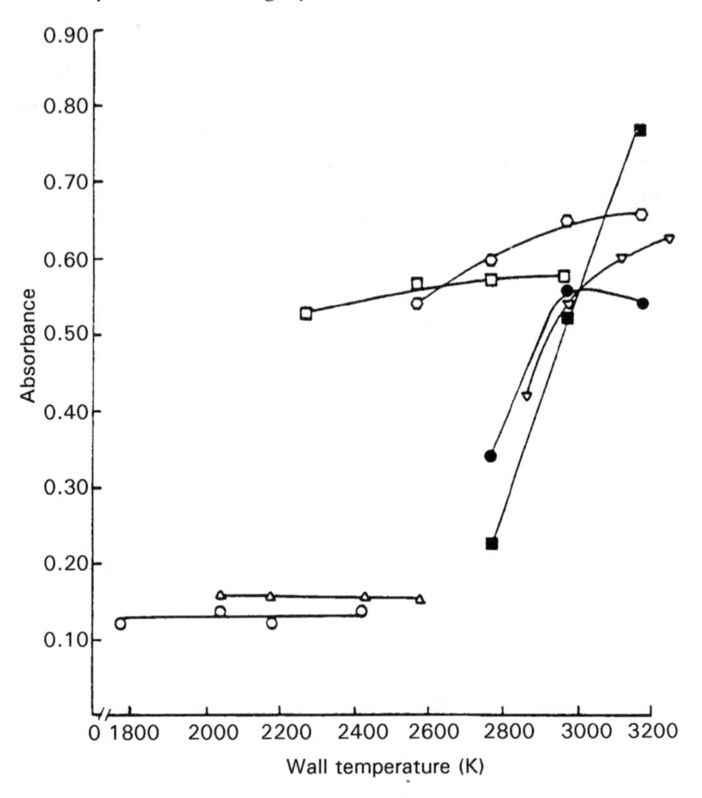

Figure 3.2 Effect of the final wall temperature on the peak absorbance for wall atomization. The wall heating rate was 3.0 K ms^{-1} for cadmium (○), lead (△), cobalt (□), nickel (⬡), vanadium (▽), and was 7.0 for vanadium (●) and molybdenum (■) [2]. (Reprinted from *Spectrochim. Acta*, Vol. **39B**, C.L. Chakrabarti, S. Wu, R. Karwowska, J.T. Rogers, L. Haley, P.C. Bertels and R.H. Dick, *Temperature of platform, furnace wall and vapor in a pulse-heated electrothermal graphite furnace in atomic absorption spectrometry*, pp. 415–448, 1984, with kind permission of Elsevier Science – NL, Sara Burgerhartstraat 25, 1055 KV Amsterdam, The Netherlands.)

3.1.4 Spatial and temporal distribution of temperature in the graphite furnace atomizer

Processes in the graphite tube are affected not only by the final temperature and the ramp rate used to attain that temperature. Temporal and spatial temperature variations on the surface of the furnace and any inserts, and in the gas phase, must be known in order to intelligently study mass transport, gas phase reactions and chemical mechanisms occurring during the programmed heating cycle.

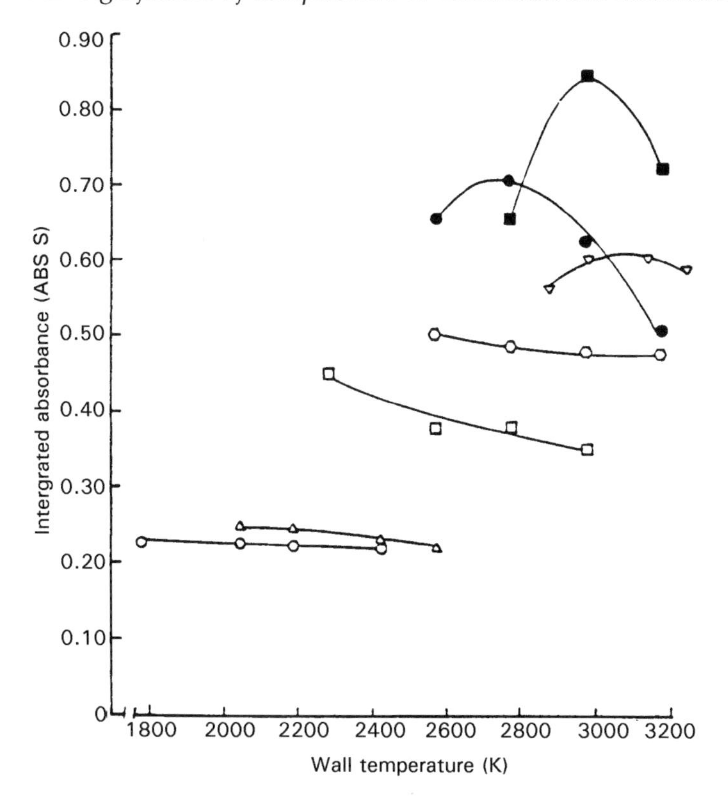

Figure 3.3 Effect of the final wall temperature on the integrated absorbance for wall atomization. The wall heating rate was 3.0 K ms^{-1} for cadmium (○), lead (△), cobalt (□), nickel (⬭), vanadium (▽), and was 7.0 for vanadium (●) and molybdenum (■) [2]. (Reprinted from *Spectrochim. Acta*, Vol. **39B**, C.L. Chakrabarti, S. Wu, R. Karwowska, J.T. Rogers, L. Haley, P.C. Bertels and R.H. Dick, Temperature of platform, furnace wall and vapor in a pulse-heated electrothermal graphite furnace in atomic absorption spectrometry, pp. 415–448, 1984, with kind permission of Elsevier Science – NL, Sara Burgerhartstraat 25, 1055 KV Amsterdam, The Netherlands.)

It is important to consider the origin of the temperature gradient that develops over the tube length. In graphite furnaces that are based on the Massmann design (or end-heated furnace), the graphite tube atomizer ends are in close contact with water-cooled graphite end pieces. The result of this configuration is a temperature gradient over the length of the tube that forms during relatively slow heating. This temperature gradient increases with time as the temperature of the tube increases, and persists while a steady state is maintained. As the gas inside the tube is heated by radiation and conduction from the wall, its temperature will also show a similar gradient. This temperature gradient, which changes with time, is

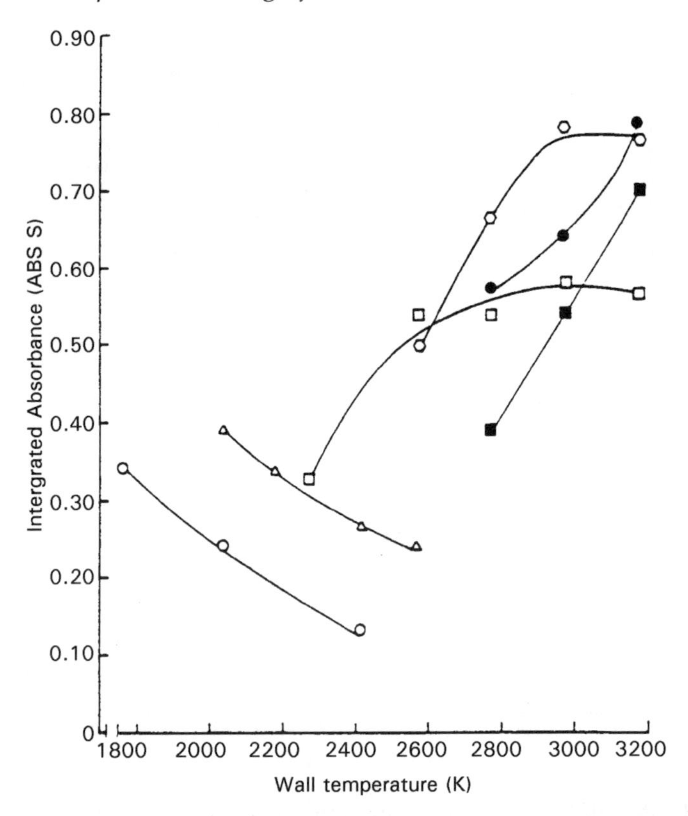

Figure 3.4 Effect of the final wall temperature on the integrated absorbance for platform atomization. The wall heating rate was 3.0 K ms^{-1} for cadmium (○), lead (△), cobalt (□), nickel (⬡), vanadium (▽), and was 7.0 for vanadium (●) and molybdenum (■) [2]. (Reprinted from *Spectrochim. Acta*, Vol. **39B**, C.L. Chakrabarti, S. Wu, R. Karwowska, J.T. Rogers, L. Haley, P.C. Bertels and R.H. Dick, Temperature of platform, furnace wall and vapor in a pulse-heated electrothermal graphite furnace in atomic absorption spectrometry, pp. 415–448, 1984, with kind permission of Elsevier Science – NL, Sara Burgerhartstraat 25, 1055 KV Amsterdam, The Netherlands.)

largely responsible for vapor condensation and molecular recombination towards the cooler ends of the graphite tube. This causes memory effects (because condensed material often vaporizes during subsequent heating cycles), spectral interferences, and recombination of atoms in the cooler regions of the tube. Furthermore, the temperature of the gas phase determines directly the rate of diffusional loss and therefore the residence time of the absorbing atoms in the measurement volume. This in turn affects the absorption signal. Diffusional coefficients, which are element specific, are functions of temperature. Hence, the lack of temporal and spatial isothermality influences the diffusional loss directly. Therefore, it is

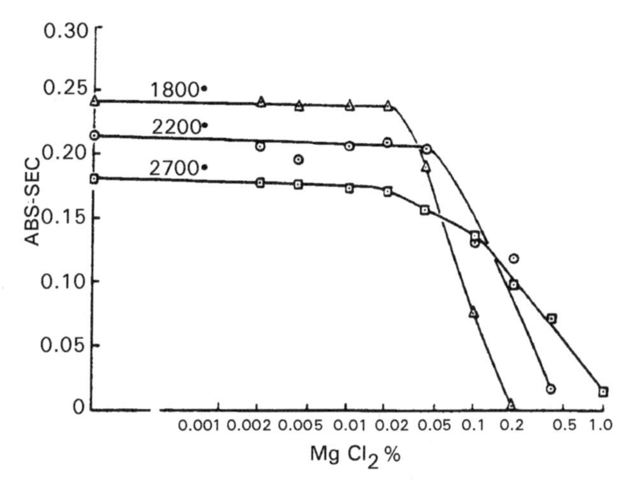

Figure 3.5 Effect of atomization temperature on $MgCl_2$ interference of thallium absorption [3]. (Reprinted with permission from D.C. Manning and W. Slavin, Sampling at constant temperature in graphite Furnace atomic absorption spectrometry, *Anal. Chem.* **51**, 2375–2378. Copyright 1979 American Chemical society.)

not only the final temperature of the furnace wall and the temperature of the gas in the center of the furnace that governs the efficiency of AA measurements. The temporal and spatial distribution determines both the shape and the amplitude of the absorption peak. This distribution of

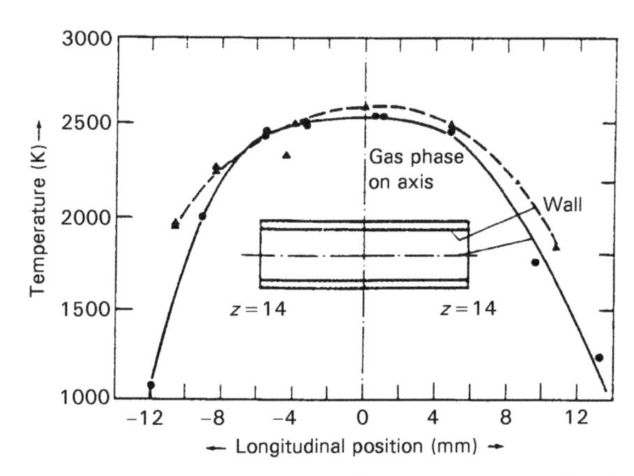

Figure 3.6 Longitudinal distribution of (Coherent Anti-Stokes Raman Scattering) CARS-monitored gas-phase temperature on the tube axis ($r = 0$) and pyrometric wall temperature, measured in a graphite tube under steady state conditions [4]. (Reprinted from *Opt. Commun.*, Vol. **68**, N. Wenzel, B. Trautmann, H. Grosse-Wilde, G. Schlemmer and B. Welz, CARS temperature studies of the gas phase in a Massmann-type graphite tube furnace, pp. 75–79, 1988, with kind permission of Elsevier Science – NL, Sara Burgerhartstraat 25, 1055 KV Amsterdam, The Netherlands.)

temperature is shown in Figure 3.6 for a typical case, as measured by Wenzel *et al.* [4]. An understanding of the influence of temperature on the atomization process of an analyte in a graphite furnace also depends on a knowledge of the longitudinal distribution of temperature with time.

Furthermore, any understanding of other processes, such as loss of analyte vapor by expansion and expulsion of gas, the dynamics of combustion, and interpretation of thermodynamic data, also depends on a knowledge of the spatial and temporal distribution of temperature over the tube length. Many of these aspects were discussed in Chapter 2. This changing temperature with time may be measured or, alternatively, a theoretical computer model may be used to predict the dynamic temperature [5]. This temperature gradient is dependent on numerous factors, of which the heating rate is one of the most important. Falk *et al.* [6] indicated that a near-constant temperature over the tube axis may be obtained during the ramp step if a high heating rate is used. This is discussed in more detail below. A typical temperature-time curve (Figure 3.7) shows that after about 0.5 s, when the wall has reached 2200 K, the gas at the tube center is approximately 550 K cooler. This figure shows the gas temperature lagging the wall temperature by less than 0.2 s.

The design of the tube is also an important factor that determines the temperature and the temperature distribution. Rademeyer and Human [7] indicated that the shape of the temperature gradient might be changed by altering the tube design. It has been shown that the temperature gradient

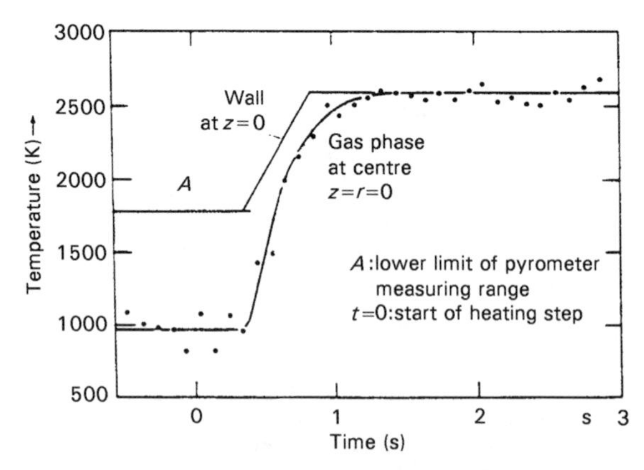

Figure 3.7 Typical time-development of pyrometric wall temperature at tube center and the CARS-monitored gas phase temperature at the same location [4]. (Reprinted from *Opt. Commun.*, Vol. **68**, N. Wenzel, B. Trautmann, H. Grosse-Wilde, G. Schlemmer and B. Welz, CARS temperature studies of the gas phase in a Massmann-type graphite tube furnace, pp. 75–79 , 1988, with kind permission of Elsevier Science – NL, Sara Burgerhartstraat 25, 1055 KV Amsterdam, The Netherlands.)

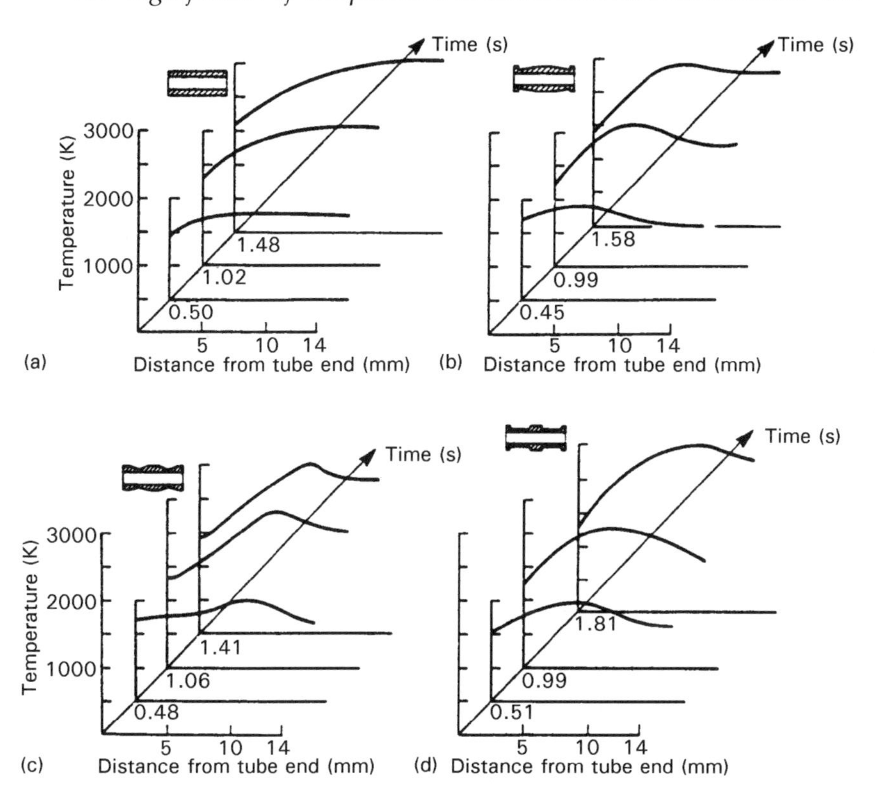

Figure 3.8 Calculated temperature distributions across the length of the furnace tube at three different times after commencement of the atomization cycle. Considered are four different tube designs, as illustrated in the figure [5]. (Reprinted from *Spectrochim. Acta*, Vol. **41B**, C.J. Rademeyer, H.G.C. Human and P.K. Faure. The dynamic wall and gas temperature distributions in a graphite furnace atomizer, pp. 439–452, 1986, with kind permission of Elsevier Science – NL, Sara Burgerhartstraat 25, 1055 KV Amsterdam, The Netherlands.)

may even be reversed so that a region of maximum temperature may be created on either side of the tube center. This may help confine the atomic vapor in the tube center to minimize diffusion out the ends of the tube. To illustrate this, Figure 3.8 shows the temperature gradient as its changes with time for a standard tube and four contoured tube designs [7].

3.1.5 Temperature and interferences

Atomization will always be prone to some degree of interference. Numerous attempts have been made to overcome these interferences employing a variety of approaches. The use of chemical modifiers has had a high degree of success and is discussed in detail in Chapter 5. Although the mechanisms by which these modifiers operate are not yet fully understood, it is known that some modifiers ensure atomization at a time when the thermal environment inside the furnace is more favorable. The advantages of atomizing the sample into a high and more nearly isothermal environment (such as from a platform surface) is well established and discussed elsewhere in detail (see Section 1.5.1). The temperature of a platform lags behind that of the wall, so an analyte that atomizes from the platform will enter an environment of higher and more constant temperature. A combination of platform atomization, chemical modifiers and integrated absorbance is probably the best general set of conditions for accurate analysis.

Various approaches, including manipulation of the prevailing temperature conditions, have been made to create a favorable environment for the introduction of analyte atoms. Such an environment would minimize the loss of free, neutral (absorbing) atoms. Some researchers introduced the analyte into the furnace with a probe after the furnace had reached a steady state temperature [8]. Near constant temperature conditions with respect to time may also be obtained by using rapid heating of the furnace employing a capacitive discharge method, as reported by Chakrabarti *et al.* [9]. However, neither of these approaches provides spatially isothermal conditions. Frech and co-workers [10,11] designed a side-heated two-step furnace in pursuit of constant temperature conditions (this is described in Section 1.4). Instruments that employ this side-heated principle are currently available commercially.

A more favorable temperature environment may also be achieved by shaping the tube contour as indicated by Rademeyer and Human [7]. Using an analytical model, they calculated the contours of tubes that had the desired temperature profiles. This enabled a tube with essentially the same temporal and spatial temperature characteristics as that of a platform-equipped Massmann furnace to be designed. However, reduction of matrix effects was achieved in only 50 per cent of the cases investigated, and no general elimination of matrix effects was found. Ajayi *et al.* [12] confirmed these studies and found that a modified tube with an inverted temperature distribution showed little improvement in its ability to curb interferences.

3.1.6 Influence of temperature on the absorbance signal

The temperature of the tube wall, as well as the temperature of the gas phase in the tube, determines the number of absorbing atoms. The absorbance signal recorded during the atomization cycle is directly proportional to the absorbing analyte atom number density. This assumption is true provided that the temperature of the tube wall and the gas in the furnace is constant, that a continuum source is used, and that the analysis volume is entirely filled with analyte atoms during the measurement period. Furthermore, for analytical purposes, the residence time in the measurement volume should be constant. The absorbance signal recorded under these conditions is also a function of the absorption coefficient, which in turn is a function of temperature. In practice, the situation is more complex, because the absorbance signal is dependent on numerous factors, and because during the atomization stage the temperature of the tube wall and the gas inside increases rapidly. Hence, most elements will atomize while the temperature of the furnace is still increasing, and before it has reached its final temperature.

Lovett [13] studied the effect of temperature on the absorption coefficient, as shown in Figure 3.9. At any point of time during the atomization cycle, the absorbance signal is a direct measure of the absorbing atom number density. Relative absorbances within any single peak are, however, not exact measures of the atom number densities at the respective times. This lack of linear relationship between any two absorbances and their respective number densities is partially due to the temperature dependence of the absorption line profile. As with the excitation spectral line, temperature has an effect on the width and shape of the absorption lines. Absorption profiles are dictated by natural line widths, their Doppler widths and by collisional widths and shifts. In atmospheric pressure atomization the natural linewidth is generally negligible compared to Doppler and collisional broadening effects. Therefore, the actual line profile is presented by the convolution of these contributions and expressed by the Voigt function. Lovett modelled the temperature dependence of the Doppler and collisional broadening, assuming that the natural line broadening and stark broadening are negligible. Hence, he calculated the absorbance using integrated overlap of emission profiles and absorption profiles. The intensity of a spectral line at increasing temperature intervals across the line, using the Gaussian Doppler profile for the hollow cathode lamp emission line and the Voigt profile for the absorption line, was calculated followed by the transmittance at each point and summed over the emission line. Finally, the absorbance was obtained from the total transmittance. The density-normalized absorbance is a function of the absorption coefficient, the damping constant (a parameter characterizing the Voigt function) and the shift, as well as the hyperfine structure for constant pressure and constant volume. Figure 3.9 shows the calculated change in density-normalized absorbance with temperature at constant damping constant for both constant pressure and constant volume conditions. The relationship between the peak absorbance and the absorbances of the peak shoulders is usually assumed to be linear, but was shown to vary over the temperature range that prevails in the furnace during the analysis cycle. Greatest variation of shape-nor-

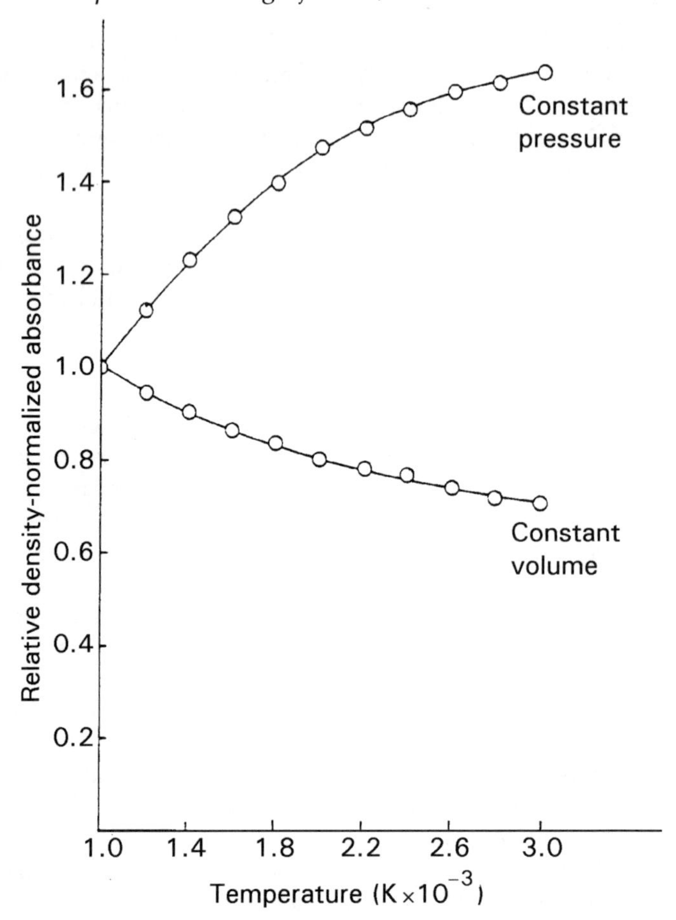

Figure 3.9 Change in density-normalized absorbance as a function of temperature for $a_0 = 2$ at 1000 K for constant pressure and constant volume for the nickel 232.0 nm peak [13]. (Reprinted by permission of the Society for Applied Spectroscopy from, R.J. Lovett, The influence of temperature on absorbance in graphite furnace atomic absorption spectrometry. I. General considerations, *Appl. Spectrosc.*, **39**, 778–786 (1996).)

malized absorbance from a Gaussian release peak is for points between where 1 per cent of the atoms are released and where 1 per cent of the atoms are left in the system (Figure 3.10).

Without considering self-absorption, Lovett's model showed that only above an absorbance of 0.7 did a variation in the linear relationship between density and absorbance occur. He concluded that temperature-induced variations in the absorbance signal have no negative influence for general analytical applications, but have implications for diagnostic measurements when line sources are used in ramped systems.

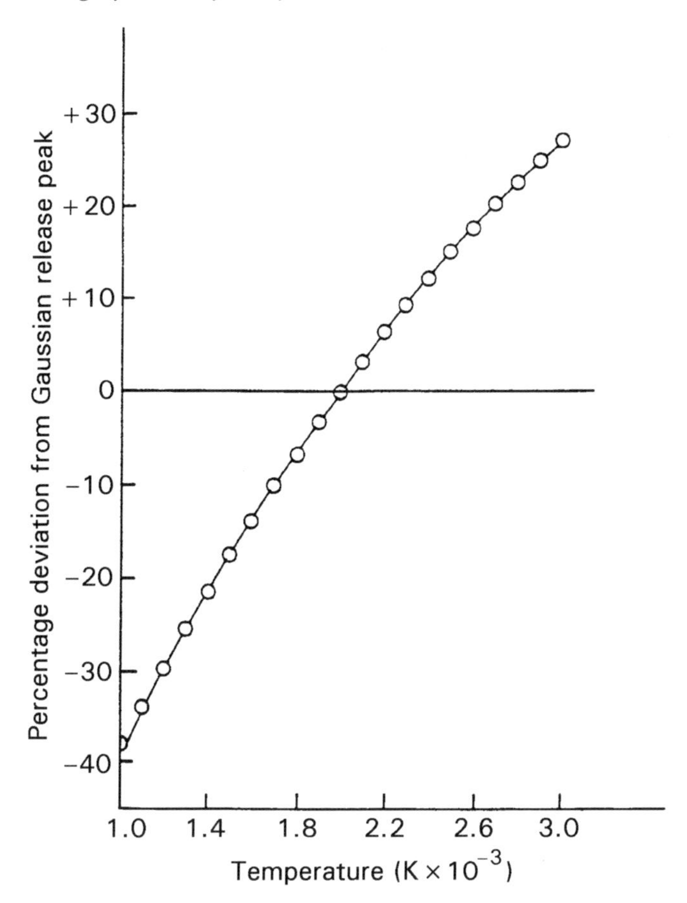

Figure 3.10 Deviation from shape-normalized output peak at constant pressure from the Gaussian atom release peak as a function of temperature for the nickel 232.0 nm peak [13]. (Reprinted by permission of the Society for Applied Spectroscopy from, R.J. Lovett, The influence of temperature on absorbance in graphite furnace atomic absorption spectrometry. I. General considerations, *Appl. Spectrosc.*, **39**, 778–786 (1996).)

3.1.7 Gas temperature

The processes occurring in the gas phase were discussed in detail in Chapter 2. After evaporation from the furnace wall or the platform surface, the analyte enters the gas phase either as atoms or molecules. While in the vapor phase it undergoes further chemical and/or physical processes. Prior to this, the temperature of the atomization surface governs the extent and nature of any temperature-dependent processes. Obviously, once in the gas phase localized vapor temperatures are of interest. Furthermore, these processes, which include the formation and dissipation of analyte atoms, mostly take place in an environment in which the temperature is increasing rapidly. This is truer for atomization from the tube surface than for atomization from the platform, when the temperature of the wall and the gas inside it may have already equalized, as discussed in Section 1.5.1. The temperature of the gas is directly influenced by the temperature and heating rate of the tube wall and the platform. As the gas temperature and heating rate of the gas influence the free atom density and residence time of the vapor cloud, the magnitude of the atomic absorption signal is directly affected. However, it is not only the gas temperature at the center of the tube that is important but also the gas temperature distribution over the length of the tube, as indicated earlier.

3.1.8 Heating rate

Perhaps more important than the temperature is the duration of the different heating stages and the rate at which these temperatures are attained. The duration and temperatures at which drying and pyrolysis are performed contribute largely to the reproducibility and accuracy of the determination. During the atomization stage, atoms are formed principally by vaporization of molecular species from the surface of the graphite tube followed by atom formation through decompostion. These atoms are lost from the analysis volume by diffusion in all directions, even through the graphite tube walls. The tube may be coated with pyrolytic graphite to reduce this effect. In furnaces at elevated temperatures, whose gas temperatures change slowly with time, diffusion is the dominant loss mechanism. Thermal expansion of the gas inside the tube represents a further significant mechanism for expelling analyte from the furnace and is increasingly important at high heating rates and especially for the volatile elements. Furthermore, apart from diffusion and expulsion, the atoms are being swept from the analysis volume by a forced flow of argon. During practical analysis, this loss of analyte atoms is counteracted by stopping the flow of gas during the atomization stage.

The rate at which any given molecular species leaves the surface of the graphite tube depends firstly on the temperature at which the molecular species exhibits a significant vapor pressure. The degree of atomization that follows this evaporation process depends on the temperature and on the rate of temperature increase of the atomization surface, as well as on the temperature of the gas inside the tube. Not only does the degree of atomization depend on these quantities, but also the rate of formation and dissipation of analyte atoms. These processes take place in an environment in which the temperature of the surface, and perhaps more important, the temperature of the gas is increasing rapidly. Therefore, because reaction kinetics are critically dependent on temperature, the temperature and the rate of temperature change are critical parameters affecting the performance of the atomizer. Numerous experimental studies have confirmed that the heating rate of the tube furnace has indeed this effect on the relative magnitude of the observed absorbance.

Gregoire *et al.* [14] established the relationship between peak height measurements and heating rate for a modified version of a Massmann atomizer. On the basis of previous models, Zsako [15] proposed a relationship by considering the supply function only. Fuller [16] and Chakrabani *et al.* [17] derived useful equations expressing the heating rate. Later, Zhou *et al.* [18] assumed that first-order reactions dominate, and that all analyte is convened to free atoms. They proposed relationships expressing the peak absorbance A_p as a function of the heating rate (among other factors) under nonisothermal conditions:

$$A_{\mathrm{p}} = \beta N_0 \frac{k_{\mathrm{p}}}{k_{\mathrm{R}}} \exp(-[(t - t_0)k_{\mathrm{m}} + (t_{\mathrm{p}} - t)k_{\mathrm{p}}]) \tag{3.1}$$

where β is the heating rate, N_0 the number of analyte atoms, K_{p} is the rate constant for atom formation, k_{R} is the rate constant for atom dissipation at time t, t_0 is the time at onset, t_{p} is the time at peak, k_{m} is the rate constant for atom formation at temperature T_{m}, T_0 is the temperature at onset and T_p the temperature at peak where $T_0 < T_{\mathrm{m}} < T_{\mathrm{p}}$. It should be remembered that all rate constants are temperature dependent and, therefore, are changing with time as the system heats. For isothermal conditions,

$$A_{\mathrm{p}} = \beta N_0 \frac{k_{\mathrm{s}}}{k_{\mathrm{R}}} \exp(-k_{\mathrm{s}}(t_{\mathrm{p}} - t_0) \tag{3.2}$$

Integrated absorbance is given by

$$\int_{t_0}^{t_x} A_t \, \mathrm{d}t = \frac{\beta N_0}{k_{\mathrm{R}}} \tag{3.3}$$

It is clear from the equations that peak height measurements are dependent on several factors. Experimental verification is rather complicated, because isothermal conditions can only be approximated by using high heating rates, or by introducing the sample into a preheated isothermal furnace. Should such isothermal conditions be obtained, peak height measurements will depend only on the rate at which atoms are dissipated and on the rate at which atoms are formed. In practice, such high temperatures cannot be attained, and the conditions where the rate of supply of atoms is greater than the rate of removal of atoms for obtaining $A_{\mathrm{p}} = \beta N_0$ cannot be met. The result is that peak absorbance remains a function of the rate at which atoms are formed. The presence of a matrix in the sample may influence the vaporization pathway and, thus, the values of k_{p} and k_{m}.

Zhou *et al.* [18] studied absorbance values obtained for silver, with and without a sodium nitrate matrix, for several isothermal tube temperatures (sample introduction on a probe), and for nonisothermal atomization from the wall. A decrease in the peak area with increasing tube temperature was attributed to an increase of the rate constant for atom removal. A steady increase of peak height with tube temperature was caused, not only by the increase of the rate constant of atom formation with increasing temperature, but also by a faster heating rate of the probe at higher tube temperatures. The same trend was found for lead under isothermal conditions, i.e. the peak area decreased and the peak height increased with increasing tube temperatures.

Zhou *et al.* [18] studied atomization of silver under different heating rates. Figure 3.11 shows that the peak height kept increasing over the range of heating rates, whereas the peak area and appearance temperatures were largely independent of heating rate. They concluded that, for a fixed amount of sample, the peak height would only depend on the atomizer heating rate and the peak temperature for a given sample. If these are kept constant, the final atomizer temperature will not alter the peak height absorbance. In order to obtain maximum peak absorbance,

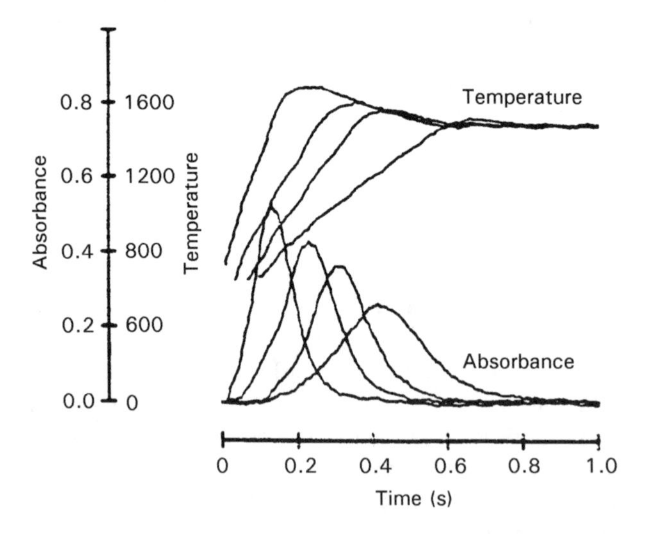

Figure 3.11 Peak absorbance for 0.05 ng of silver at different heating rates ($1 = 1620\,K\ s^{-1}$; $2 = 4250\,K\ s^{-1}$; $3 = 3600\,K\ s^{-1}$ and $4 = 2880\,K\ s^{-1}$). The corresponding integrated absorbance values were: 1 = 0.072; 2 = 0.069; 3 = 0.078 and 4 = 0.070. [18]. (Reprinted from *Spectrochim. Acta*, Vol. **39B**, N.G. Zhou, W. Frech and L. de Galan, On the relationship between heating rate and peak height in electrothermal atomic absorption spectroscopy, pp. 225–235, 1984, with kind permission of Elsevier Science – NL, Sara Burgerhartstraat 25, 1055 KY Amsterdam, The Netherlands.)

they estimated that the rate at which atoms are formed should be about $35\,s^{-1}$. This corresponds to a temperature of 2400 K and to a heating rate of $30\,kK\,s^{-1}$. Peak absorbance can also be increased by decreasing the rate at which atoms are removed, for example, by increasing the tube length. Not only is the temperature of the tube center dependent on the heating rate but the temperature profile over the length of the tube is strongly dependent on the heating rate. Falk *et al.* [6] found that, at high heating rates, a pronounced overshoot of the temperature at the center occurs. In the case of very fast heating rates, starting from room temperature, the temperature along the length is nearly equal to that of the center during the ramp (Figure 3.12). However, when ramping starts at higher temperatures, this effect is not observed – for example the temperature of the tube center is not reached, as can be seen from Figure 3.13.

When a platform is placed in a graphite tube, its performance can also be expected to be affected by the heating rate of the tube. Usually, the temperature of the platform is lower than that of the furnace tube wall,

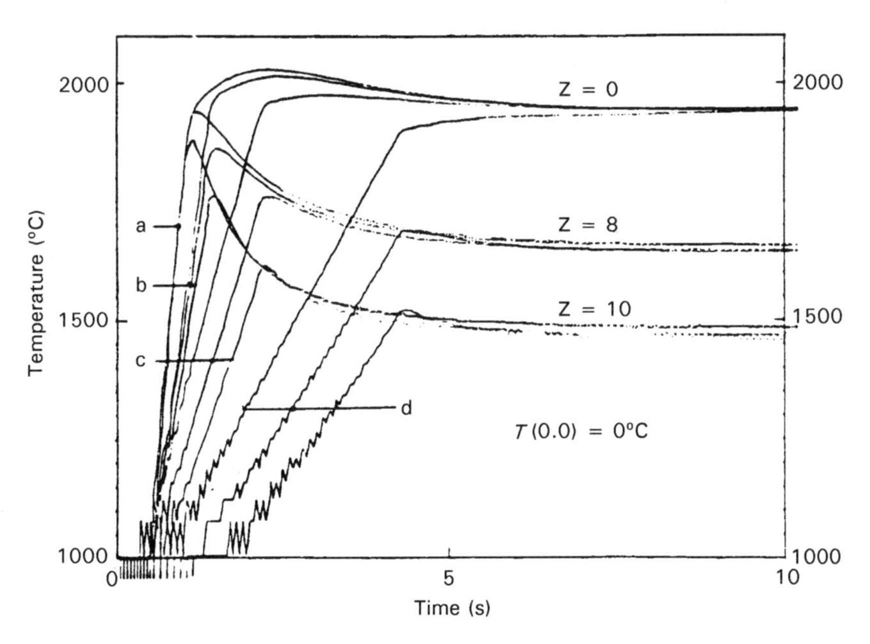

Figure 3.12 Temperature at three tube positions versus time for different heating rates, starting at room temperature, for a standard Perkin-Elmer HGA 500 pyrolytically coated graphite tube. Heating rates: (a) full power, (b) $1000\,K\,s^{-1}$, (c) $500\,K\,s^{-1}$, (d) $250\,K\,s^{-1}$; z = distance from tube center (mm) [6]. (Reprinted from *Spectrochim. Acta*, Vol. **40B**, H. Falk, A. Glisman, L. Bergann, G. Minkwitz, M. Schubert and J. Skole, Time-dependent temperature distribution of graphite tube atomizers, pp. 533–542, 1985, with kind permission of Elsevier Science – NL, Sara Burgerhartstraat 25, 1055 KV Amsterdam, The Netherlands.)

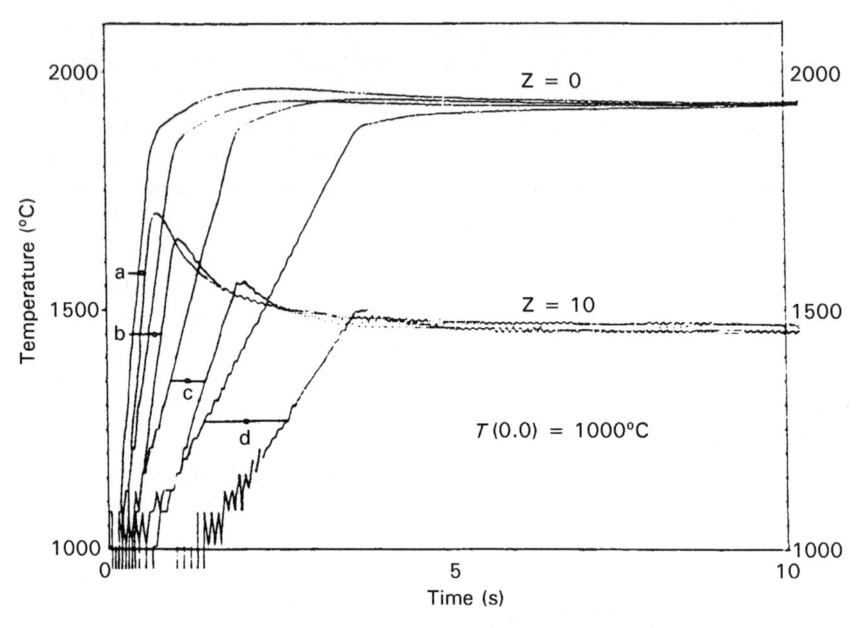

Figure 3.13 Temperature at two tube positions versus time for different heating rates, starting at 1000°C, for a standard Perkin-Elmer HGA 500 pyrolyically coated graphite tube. Heating rates: (a) full power, (b) 1000 K s^{-1}, (c) 500 K s^{-1}, (d) 250 K s^{-1}; z = distance from tube center (mm) [6]. (Reprinted from *Spectrochim. Acta*, Vol. **40B**, H. Falk, A. Glisman, L. Bergann, G. Minkwitz, M. Schubert and J. Skole, Time-dependent temperature distribution of graphite tube atomizers, pp. 533–542, 1985, with kind permission of Elsevier Science – NL, Sara Burgerhartstraat 25, 1055 KV Amsterdam, The Netherlands.)

especially during the atomization portion of the cycle. This is indicated by the time delay in the appearance of the absorption signal, and shows that sample atomization is delayed relative to that observed for vaporization from the tube wall. Intuitively, one would expect this delay to result in vaporization into a furnace atmosphere that is at a higher temperature and a more nearly constant temperature, as initially suggested.

3.1.9 Delayed temperature (atomization) techniques

The degree of dissociation of the analyte into atoms and the diffusion rate of analyte atoms depend on its temperature. Additionally, in the Massmann furnace, with its associated temperature gradient, these factors will depend on time and the location within the tube. Ediger [19] showed that the presence of nickel as a chemical modifier in certain samples caused a marked shift in the onset of vaporization of selenium and arsenic. The addition of nickel binds the element in a compound with a different volatility. The temperature at which analyte atomization occurs is therefore altered and experimental conditions may thus be chosen so that the sample enters the vapor phase at a time more conducive to efficient atomization, for example a vapor isothermal in time and space.

L'vov *et al.* [20] proposed the well known technique of vaporizing the sample from a graphite platform inside the furnace. As the platform and sample are heated by radiation from the wall of the tube, it was suggested that there is time for the tube and the vapor in the tube to reach a stable final temperature before the platform reaches a temperature where the sample is volatilized. They proposed that, as the temperature of the platform always lags behind that of the wall, a nearly constant temperature environment in the furnace would be achieved when the analyte vaporizes from the platform. As will be seen later, a radial temperature gradient persists and, in most cases, the wall and platform temperature are still increasing when vaporization occurs. However, the objective of analyte vapors entering a higher gas phase temperature is achieved (see Section 1.5.1). The temperature environment so created at the time of atomization reduces analytical interferences associated with variations in atomization conditions, due to differences in sample matrix, by promoting molecular dissociation in the more elevated temperature of this region above the platform. As mentioned, the ideal goal is to delay sample atomization until the temperature of the tube and the inert gas have stabilized at a temperature higher than the platform. The theoretical basis for L'vov's platform method rationalized that, if the vapor temperature does not change while the sample is vaporized from the platform, each atom will be in the vapor phase for the same length of time regardless of the rate of vaporization. Thus, each atom will contribute equally to the absorbance signal. As shown previously, the integrated absorbance signal is relatively independent of the rate of evolution of the atomic vapor. Although variations in the matrix may produce a variable rate of evolution of the atomic vapor, this should not significantly alter the integrated absorbance signal if the furnace is at its final constant temperature during atomization. Thus, integration of the absorbance signal should show a much smaller dependence on changes in the rate of vaporization when compared with peak height measurements.

3.1.10 Models

The temperature of the graphite tube plays a decisive role in most of the mechanisms by which it operates, and ultimately its analytical performance, Hence, knowledge of the temperature of the furnace with respect to time and position is necessary for an understanding of these processes. It is therefore not surprising that attempts were made early on to predict, by way of calculation, the temperature of the furnace atomizer with respect to time and position in the tube. Tube shape changes (i.e. contoured tubes) have been made to eliminate or minimize the spatially nonuniform temperature along the length of the tube. An analytical model that makes possible the calculation of the dynamic temperature distribution along the length of the furnace wall with accuracy would therefore have been useful in analyzing a variety of furnace tube designs directed towards reducing the temperature gradient along the tube. It is equally beneficial to predict the time-dependent platform temperature that might result from alterations in the tube design. In applying such a model, there are several important design questions that can be answered concerning contoured tubes:

- Which tube design will ensure optimum temperature–time characteristics (for example, the maximum tube length having a constant temperature profile for as long a time as necessary)?
- Which material should be used, what is the influence of the pyrolytic coating, and how thick should it be?
- Within what temperature range will the temperature profile have the desired characteristics?
- How can mechanical strength be maintained to ensure maximum tube life without adversely affecting the required temperature–time characteristics?
- How can the optimum temperature–time characteristics be maintained while maintaining economic fabrication?

Cresser and Mullins [21] proposed the theoretical basis that may be used to calculate the temperature of a filament atomizer. Slavin *et al.* [22] developed and used an analytical model to calculate the steady state temperature of a graphite tube furnace, aimed at shaping the graphite tube in an attempt to eliminate interferences. These first efforts to calculate the temperature distribution were of little value, because a steady-state temperature was calculated, and atomization usually occurs early in the atomization cycle, whereas the temperature and the temperature distribution over the tube length is changing rapidly. Rademeyer and Human [23] proposed a method of calculating the dynamic temperature characteristics of a heated graphite tube, and presented results for different tube designs. For the purpose of the calculations, the furnace tube was considered to consist of 30 ring segments. When an electric current is passed through such a ring segment, the temperature increase over a period of time Δt is given by

$$\Delta t = m \int_{t_1}^{t_2} \frac{C\,\mathrm{d}t}{P} \tag{3.4}$$

where m is the mass of the segment, C is the heat capacity of the graphite, and P is the net power input by the electric current, i.e. the sum of the energy input by resistive heating, absorption of radiation from the walls and conduction from adjacent segments, minus the energy lost by radiation and conduction.

Equation (3.4) can be converted to a finite sum and rearranged to calculate the time Δt required to increase the temperature of a segment of the graphite tube by a temperature ΔT:

$$\Delta t = m\sum_{T_i}^{T_f} \frac{C\Delta T}{P} \tag{3.5}$$

where T_i is the initial and T_f is the final temperature. This calculation was performed for all the tube segments, using iterative techniques to determine the temperature distribution over the tube length as a function of time. Results of such calculations were used to determine the shape of atomizers that have the targeted temperature distributions. Using this model, it was shown that tubes with contoured shapes can be designed that allow containment of the atomic vapor in the analysis volume for longer times.

Chakrabarti *et al.* [24] proposed an even more sophisticated model. Their calculations indicated that, even in the absence of a forced convective flow of a protective gas through the tube, the gas temperature is

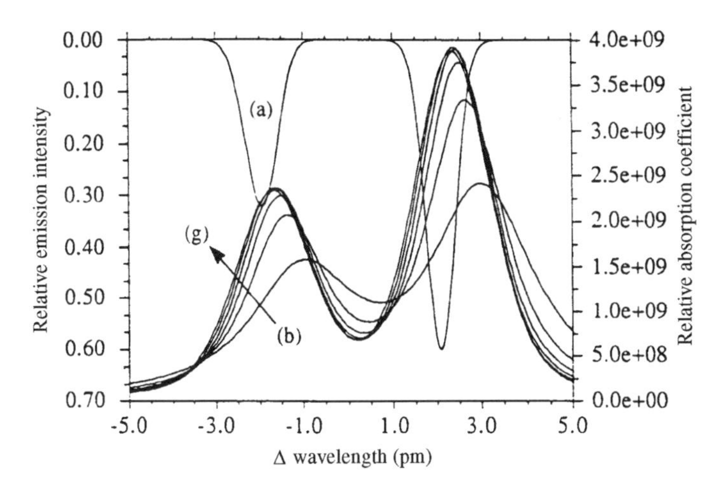

Figure 3.14 Comparison of (a) the hollow cathode lamp emission profile to the absorption profile for copper in 1 atm at various temperatures: (b) 500 K; (c) 1000 K; (d) 1500 K; (e) 2000 K; (f) 2500 K; (g) 3000 K [25]. (Reprinted by permission of the Society for Applied Spectroscopy from, O.A. Guell and J.A. Holcombe, Monte Carlo study of the relationship between atom density and absorbance in electrothermal atomizers, *Appl. Spectrosc.*, **45**, 1171–1176 (1991).)

lower than the wall temperature until the wall has attained a steady-state temperature.

Monte Carlo simulations were used by Guell and Holcombe [25] to obtain time and spacial gas-phase atom density and temperature data. These data were used to investigate the assumption that the absorbance signal is proportional to the gas-phase atom density throughout the transient absorbance profile. This approach simulated the position and local temperature of every particle in the absorption volume at any given time, permitting absorption profiles to be constructed as a function of temperature, as shown in Figure 3.14. Very informative results were obtained, supporting the assumption of direct proportionality between gas density and absorbance.

3.2 CONTROLLING THE TEMPERATURE

From the previous discussion, it is clear that the temperature of the graphite furnace should be carefully controlled to provide an optimal reproducible temperature of the tube wall or the platform to ensure efficient atomization. Theoretically, this may be accomplished by introducing the analyte into a well-defined volume at an already sufficiently high and constant temperature. The temperature of such an environment (such as the inside of a graphite tube) should be chosen to allow complete volatilization of the analyte. The widely used Massmann furnace does not approach such an ideal case. However, it offers an operationally practical system, and the majority of commercial systems presently in use are derivations of that design. Other systems have been described [10,24,26] and employ special power supply and control systems.

It is essential, therefore, to maintain accurate temperature control of the sample-laden atomization surface (i.e. tube wall or platform) while drying, pyrolyzing, and atomizing the sample. This control promotes reliable, reproducible and interference-free results. In commercial and research Massmann instruments, this is accomplished by applying a time-dependent potential to the tube, in order for the tube to reach a predetermined temperature at the desired rate. Control is accomplished by means of a triac or silicon controlled rectifier (SCR). These are semiconductor devices able to control relatively large currents by means of small control potentials applied to the device. The temperature of the tube and the rate at which it increases can, therefore, be controlled by varying the relatively small control potential, (typical of the order of a few volts).

Figure 3.15 shows a typical electronic circuit diagram employing a triac as controlling device. Effective control of the temperature of the graphite furnace requires continuous monitoring while the temperature is increasing (or, less important, while it is decreasing). A feedback system to the controlling circuit is, therefore, essential for measuring the temperature of the graphite tube. Frequently, optical pyrometry is used for this

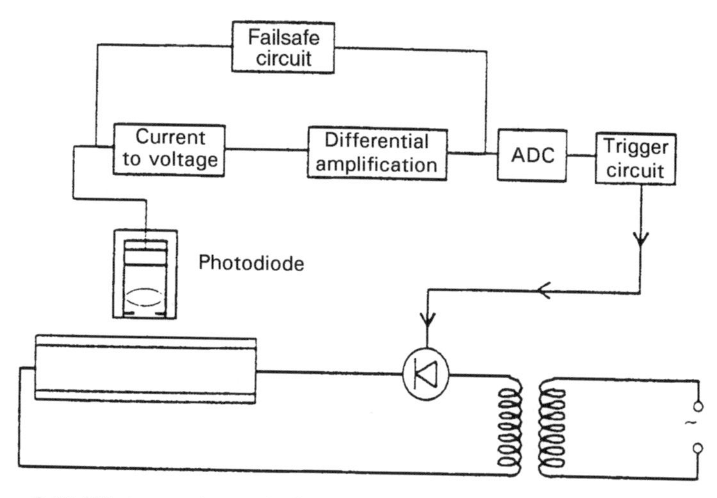

Figure 3.15 Diagram of a typical power supply and controlling circuit for ETA.

purpose [27]. An optical pyrometer is placed close to the graphite tube, as shown in Figure 3.15, in order to receive radiation from it. This device consists of a photodiode and a filter to cut off short wavelengths (below 650 nm) to avoid ambiguity resulting from the combination of the sensitivity curve of the diode and the wavelength variation of the black body radiation. The signal from the photodiode is fed to circuitry that creates the control signal to the triac. The signal may, of course, be derived from a computer system, allowing elegant control [28].

Such a controlling system should measure the temperature of the graphite furnace accurately, and Section 3.3 discusses the problems associated with measuring the temperature of the furnace (or platform) surface. Additionally, while determining this temperature for platform atomization, it should be remembered that the platform temperature always lags behind the wall temperature, especially during the thermal ramp. When the potential is applied, at the onset of the cycle, the temperature of the atomization surface will increase with time until the heat loss through radiation and conduction balance the supplied electrical power. The heating rate, the final temperature, and the time allowed for each step might thus be selected to obtain the optimum analytical result for a given element. Guidelines for these optimum conditions are supplied with instruments, but should be established by the analyst. More detailed information may be found in the literature.

3.3 TEMPERATURE MEASUREMENT

When determining physical parameters, such as the degree of atomization, diffusion coefficients and other thermochemical constants, it is necessary to know the temperature of the graphite surface and atomic vapor. Consequently, an ability to measure these quantities is necessary. The temperature of the surface of a solid, such as graphite, may be measured by a thermocouple. The large temperature range in ETAAS may necessitate the use of different thermocouples for different temperature ranges; for example, temperatures up to 750°C may use an iron–constanton couple, whereas platinum–platinum/10 per cent rhodium thermocouples can be used for temperatures up to 1700°C. Voltages are measured by high-impedance voltmeters, and standard tables supplied with the thermocouples are used to convert measured voltages to temperatures. It is necessary to ensure that good thermal contact is maintained between the junction and the graphite surface whose temperature is measured, and that the heat capacity of the thermocouple is sufficiently small to follow the rapid temperature change during a heating ramp. Furthermore, it is necessary to ensure that the thermocouple itself does not interfere with the measurement. Measurements should be calibrated, and using the melting point of pure metals is one way of doing this. However, this is not the ideal method, as the metals themselves may act as heat sinks, and contact with the graphite surface may be poor. These metals should be chosen so that temperatures may be calibrated over the entire range. For such measurements, small particles of the pure metal are placed on the surface of the graphite at the point where temperature is to be measured. The graphite tube is then progressively heated, with periodic visual inspections to establish when melting has occurred. Alternatively, the shiny surface of metals, such as platinum may be used to reflect a laser beam. Onset of melting of the metal is indicated by the loss of reflection. Apart from metals, use may also be made of the melting point of pure compounds or salts, such as sodium chloride.

Thermocouples have mostly been used at low temperatures, and the temperature of a hot graphite surface is generally measured with an optical pyrometer by measuring thermal radiation. A body radiates energy at all temperatures, and this is visibly noticeable at temperatures above 800°C. This radiant energy and its spectral distribution changes with temperature according to the Planck radiation law for a black body radiator:

$$I_b(\lambda, T) = 2C_1\lambda^{-5}\left[\exp\left(\frac{C_2}{\lambda T}\right) - 1\right]^{-1} \tag{3.6}$$

where I_b is the intensity (W cm^{-2}), λ is the wavelength (nm), T is the temperature (K), C_1 is 5.95×10^{-17} m^2 W and C_2 is 1.44×10^{-2} mK. Because the radiation from the body is dispersed over all wavelengths, the color of the heated body is affected and this wavelength dependence will also affect the measurement. If the measured temperature of a body is unaffected by the wavelength at which the measurement is made, it is termed a gray body. If measurements are made through a very small hole into a cavity in an opaque material, it would appear perfectly black if all radiation is absorbed. If the construction of the cavity were such that very little light can be reflected back through the hole, the body would be a near-perfect absorber. Conversely, a small hole in a cavity is the most perfect emitter and is termed a black body emitter. The emissivity ε of such a body is one. The emissivity $(0 < \varepsilon < 1)$ indicates how close a real object is to a perfect blackbody $(\varepsilon = 1)$. The emissivities of real bodies are strongly dependent on the material from which the body is made, the surface structure, the geometry of the system, the wavelength of observation, and the body's temperature. In practice the emissivity of the surface of a graphite furnace changes continuously throughout its lifetime. As a consequence, the emissivity of a flat strip of pyrolytic graphite will vary between 0.85 and 1.00, thus influencing any temperature measurements. Calibration of a system for temperature measurements is therefore essential. Thus, even comparisons of relative temperature measurements should only be made between tubes that have been used to the same degree, unless $\varepsilon = 1.0$ as a consequence of the system geometry.

The distinction between *brightness temperature* and the *true temperature* is clear from the above discussion. The *brightness temperature* of a surface is defined as the temperature of a black body with the same radiant intensity as that surface at a fixed wavelength. The relationship between true temperature T and brightness temperature S of a surface is given by

$$I(\lambda, S) = \tau, \varepsilon(\lambda, T)I(\lambda, T) \tag{3.7}$$

where $I(\lambda, S)$ is the radiant intensity of a black body with brightness temperature S, $I(\lambda, T)$ is the radiant intensity of a tungsten ribbon with true temperature T, $\varepsilon(\lambda, T)$ is the emissivity of tungsten, τ is the transmission factor of the window striplamp $(0 < \tau < 1)$, and λ is the wavelength of radiation (nm). Substituting Wien's radiation law:

$$\frac{1}{S} - \frac{1}{T} = -\frac{\lambda}{C_2}\ln[\tau, \varepsilon(\lambda, T) \tag{3.8}$$

where $\varepsilon(\lambda, T) = Ke(k/\lambda)$, are K and k are independentof wavelength. In colorimetry and in problems concerning color properties of light sources, the color temperature is the important quantity. The color temperature of a body is defined as that of a black body having the same spectral distribution of energy in the visible region as that of the body under consideration. These aspects are discussed here, but can be found in various texts [29].

Radiation from a black body depends only on the temperature of the body. Radiation from a Massmann furnace, or any of its modifications, does not represent black body radiation, because of the radiation lost through the relatively large open tube ends. The two most important deviations from black body radiation result from the dependence of the radiation on the surface characteristics, and on the wavelength at which the measurements are made. Therefore, when performing pyrometric measurements on graphite surfaces in a graphite furnace atomizer, the important consideration is how to make corrections for deviations from black body radiation. Falk [30] outlined the necessary corrections when making measurements of the graphite surface. They are valid for the following experimental conditions: (1) emission from the sample introduction hole located in the middle plane of the tube is observed in a radial direction; (2) the diameter of the sample introduction hole is small compared to the inner diameter of the tube; (3) the tube temperature is a function of the tube length only; (4) the inner surface of the tube obeys the cosine law of diffuse reflection; and (5) the emissivity depends only on l and T.

The relative deviation of the measured temperature T_h from the true temperature T_0 of the furnace wall for $z = 0$ (tube center) is given by

$$\Delta T_h = \frac{T_0 - T_h}{T_0} \tag{3.9}$$

T_0 and T_h can be calculated for the case where $T_z = T_0$ as

$$T_h = \frac{C_2}{\lambda}\left\{\ln\left[\frac{\exp(C_2/\lambda T_0) - 1}{f(z_0)} + 1\right]\right\}^{-1} \tag{3.10}$$

and

$$T_0 = \frac{C_2}{\lambda}\left\{\ln\left[\exp\left(\frac{C_2}{\lambda T_h} - 1\right)f(z_0) + 1\right]\right\}^{-1} \tag{3.11}$$

where

$$f(z_0) = \varepsilon_0 + \frac{1 - \varepsilon_0}{\sqrt{8\pi}} J_0(z_0) \tag{3.12}$$

The expressions hold for the case where the temperature is constant. Obviously, a more complicated relationship applies when the temperature is changing in the z direction (along the length of the tube):

$$T_h = \frac{C_2}{i}\left(1 - \ln\left\{\varepsilon_0\left[\exp\left(\frac{C_2}{iT_0}\right) - 1\right]^{-1} + \frac{(1 - \varepsilon_0)i^5}{\sqrt{8\pi C_1}} J(i, z_0, T_z)\right\}\right)^{-1} \tag{3.13}$$

Shown in Figure 3.16 are the relative error ΔT_h and the temperature T_h, measured pyrometrically through the sample introduction hole as a function of the tube length with constant temperature T_0. The corrections given in Eqn (3.14) (below), are applied. The influence of the wavelength of observation is clearly shown. It is clear from the figure that the longer

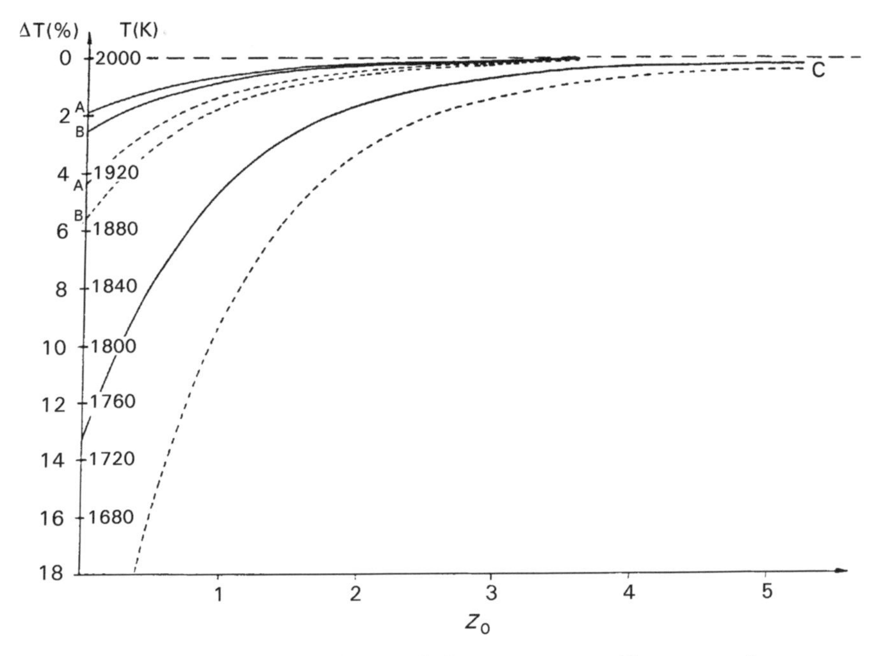

Figure 3.16 The relative error ΔT_h and the temperature T_h measured pyrometrically in the sample introduction hole as a function of the tube length with constant temperature T_0: (—) $\varepsilon = 0.8$; (---) $\varepsilon = 0.6$; (A) $\lambda = 0.65\,\mu m$; (B) $\lambda = 0.85\,\mu m$; (C) $\lambda = 8\,\mu m$; $T_0 = 500\,K$ [30]. (Reprinted from *Spectrochim. Acta*, Vol. **39B**, H. Falk, On the pyrometric temperature measurements in graphite furnaces: theoretical approach, pp. 387–396, 1984, with kind permission of Elsevier Science – NL, Sara Burgerhartstraat 25, 1055 KV Amsterdam, The Netherlands.)

the wavelength of observation, the greater the deviation. Therefore, systematic errors are minimized by making the operational wavelength of the pyrometer as short as possible. When the tube has a constant temperature plateau, which has a length of four or more times the inner diameter of the tube, there is the maximum amount of reflected intensity in the tube, and therefore the temperature error caused by the finite diameter of the sample injection hole can be neglected.

The corrections mentioned above cannot be applied to tubes with specular reflecting surfaces, such as polished metals. Graphite furnace atomizers with pyrolytic coatings do not show considerable deviations from cosine reflection, because such surfaces are not microscopically smooth. The tube furnace with diffuse reflecting inner surfaces, observed through the injection hole, can be considered as blackbody radiators when the wall temperature is constant for a tube length more than three times the diameter. In such cases, for optical pyrometers working in the visible spectral range, the error can be kept to less than 1 per cent at temperatures higher than 900°C. When the temperature of the tube furnace is maximum at the center of the tube and decreases monotonically to the ends, then blackbody radiance cannot be reached even for long tubes. Consequently, the error in the pyrometric temperature measurement is higher, compared with a tube of the same length but at constant temperature. In the case of a L'vov platform, the situation is more complicated. The emissivity of the platform determines its rate of heating through absorption of thermal radiation from the tube. It also determines the systematic error of the pyrometric temperature measurement that is caused by the reflected part of the wall radiation. All pyrometric measurements of the platform should therefore be corrected for the emissivity of the platform material and for reflected radiation from the tube wall using Eqn (3.14), as proposed by Falk [30] and applied by Welz *et al.* [31]:

$$T_{\mathrm{P_{corr}}} = \frac{C_2}{\lambda_{\mathrm{m}}}\left\{\ln\varepsilon_p - \ln\left[\exp\left(\frac{-C_2}{\lambda_{\mathrm{m}}T_{\mathrm{p}}}\right) - (1-\varepsilon_{\mathrm{p}})g_0\exp\left(\frac{-C_2}{\lambda_{\mathrm{m}}T_{\mathrm{r}}}\right)\right]\right\}^{-1} \quad (3.14)$$

where $T_{\mathrm{P_{corr}}}$ is the corrected platform temperature, C_2 is Wien's second radiation constant, ε_{p} is the emissivity of the platform, T_{p} is the measured platform temperature, g_0 is the fetching factor for reflected radiation, T_{r} is the corrected tube temperature, and λ_{m} is the wavelength of the pyrometric measurement. Sperling *et al.* [32] provided detailed results of experimental pyrometrically-measured platform temperatures, including measurements of the side-heated integrated contact furnace (see Figure 1.11). As an example, Figure 3.17 shows the increase in temperature of the integrated platform as a function of time.

3.3.1 Measurement of gas temperature inside a graphite furnace

As indicated earlier, the effective vapor temperature in the graphite furnace directly influences atomization and other important factors such as the magnitude of interferences and matrix effects. Knowledge of the vapor temperature is, therefore, essential for an understanding of the different mechanisms of the operation of the atomizer. This permits a more systematic approach towards the design of the optimum atomizer. Several experimental techniques may be used to measure the temperature of vapors in flames, including optical methods based on the line reversal phenomenon, brightness and emissivity, color temperature, and spectral line intensity ratio. Other methods include measurement of the effective translational temperature by Doppler broadening, the effective rotational temperature, and the vibrational temperature.

Other techniques that may be employed include measurement of the refractive index of higher energy (α-particles and X-rays) species, sonic methods, hot wire methods, and ionization methods. However, none of

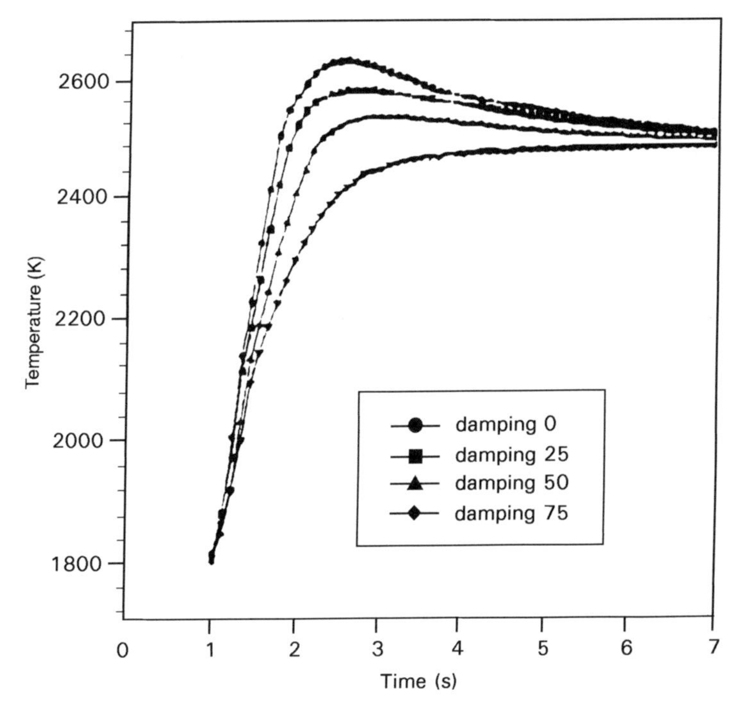

Figure 3.17 Surface temperature of the integrated platform inside the integrated contact furnace, from room temperature to 2473 K, with maximum power and gas-stop, viewed through the sample injection hole with different damping parameters of the power supply [32]. (Reprinted from *Spectrochim. Acta*, Vol. **51B**, M. Sperling, B. Welz, J. Hertzberg, C. Rieck and G. Marowsky, Temporal and spatial temperature distributions in transversely heated graphite tube atomizers and their analytical characteristics for atomic absorption spectrometry, pp. 897–930, 1996, with kind permission of Elsevier Science – NL, Sara Burgerhartstraat 25, 1055 KV Amsterdam, The Netherlands.)

these is directly applicable for use in electrothermal atomizers for various physical and chemical reasons. For example, the vapors produced in electrothermal atomizers are nonluminous, precluding methods based on brightness and emissivity. The vapor temperature may be measured by observing the vibrational and rotational Boltzmann distribution of emission bands of C_2 and CN produced within a graphite atomizer, as described by Massmann and Grucer [33]. L'vov *et al.* [34] used the atomic vapor itself as a spectroscopic thermometric species. They measured the effective electronic excitation temperature of an absorbing layer of tin in an isothermal graphite cuvette by a two-line atomic absorption method using a sharp-line source. The thermometric species was produced by volatilizing in the same way as any ordinary sample. However, such a procedure allows the vapor temperature to be determined only for the finite period during which the sample atoms are present in the analysis volume.

The energy ($E = hC/\lambda$) of the radiation emitted by a species in the vapor phase is dependent on the nature of the emitting species and the energy state of the species during emission. Considering a cloud of emitting species in a graphite furnace, the probability that a given species has a given energy state depends on a number of factors, of which the energy of the state relative to the ground state and the temperature of the species are the most important. The well known Boltzmann distribution describes the relative population of the different energy states as a function of temperature. Browner and Winefordner [35] described the two-line method that may be used for vapor temperature measurement in an ETA. Radiation from an uncalibrated continuum source is passed through the atomic vapor and the fraction of radiation absorbed a is measured for two atomic lines having the same upper energy level but different lower energy levels, and also similar wavelengths (e.g. the indium 410.18 nm and 451.13 nm lines). One wavelength involves a resonance transition, but the other involves a transition from a nonground-state level that is populated thermally. They showed that the temperature for low absorbances is given by

$$T = \frac{E_1 - E_0}{k \ln\left[\left(\frac{g_1 f_1}{g_0 f_0}\right)\left(\frac{\lambda_1}{\lambda_0}\right)^2\left(\frac{a_0}{a_1}\right)\right]} \qquad (3.15)$$

and, for high absorbances,

$$T = \frac{E_1 - E_0}{k \ln\left[\left(\frac{g_1 f_1}{g_0 f_0}\right)\left(\frac{\lambda_1}{\lambda_0}\right)^2\left(\frac{a_0}{a_1}\right)^2\left(\frac{\Delta\lambda_{L^1}}{\Delta\lambda_{L_0}}\right)\right]} \qquad (3.16)$$

where E is the energy of transition, the subscripts 0 and 1 represent the electronic states of the two transitions involved, k is the Boltzmann constant, g is the statistical weight, f is the absorption oscillator strength, λ is the wavelength, and $\Delta\lambda_L$ is the Lorentz halfwidth. If the growth curves (log a versus log analyte solution concentration) for the two wavelengths are parallel, and if the intersection point of the low and the high absorbance asymptotes occurs at the same concentration, then Eqn (3.16) reduces to Eqn (3.15). Use of these equations should be limited to low absorbances, because high absorbances will lead to systematic errors. Furthermore, a monochromator with spectral bandwidth of less than 0.1 nm should be used to ensure adequate absorption at the wavelengths of interest. With α values less than about 0.03, the wings of the absorption line extend beyond the spectral bandwidth of the monochromator and the approximations inherent in the derivation of the temperature equations no longer hold. Elements should be chosen so that absorption oscillator strengths are comparable.

The principle problem with this type of procedure is that a line-of-sight measurement is usually taken along the length of the tube, and therefore through an inhomogeneous temperature region. Temperature conditions are dynamic, and this temporal behavior governs the atomic population by the rate of introduction of atoms into and their rate of dissipation from the analysis volume. Any thermometric species employed for such measurements, which is optically detected, continues to increase in population with time, and ultimately reaches a maximum value only to decay as the atomizer continues to increase in temperature. The temperature of the vapor will, therefore, be nonuniform and any measurement will be a mean temperature weighted towards the higher temperature. Furthermore, the method is based upon the assumption that the emission line from the light source is narrower than the absorption line width and that the peak wavelengths coincide. Neither assumption is met, with the result that accurate absorbance is not measured.

All results obtained using this method indicate that the gas temperature closely follows the wall temperature during the initial stages of the heating cycle. Thereafter, it lags behind by about 100–200 K. One reason for this phenomenon may be found in earlier explanations. With time, the distribution of the atoms in the tube becomes more uniform. This results in the relative proportion of the atoms at the cooler ends of the tube increasing. Hence, the weight of those cooler atoms in determining the effective vapor temperature increases, thus producing a lower effective temperature. Terui *et al.* [36] proposed that a correction factor be used in two-line method calculations. This correction factor is experimentally determined by comparing the temperature of the vapor in the furnace with that of an air–acetylene flame. This method still seems to have some limitations, although it could be useful with further refinement.

Recently, coherent anti-Stokes Raman scattering (CARS) was employed to measure the temperature of the gas phase in the graphite tube. As a species- and state-selective technique, it allows spatially and temporally precise measurements of concentration and temperature of molecular species. The simultaneous determination of population densities of several rotational and vibrational states of a given species allows a description in terms of a population distribution, and hence, the determination of a temperature, provided the conditions of a Boltzmann distribution pertain. This concept of spectroscopic temperature determination was introduced as early as 1976 by Nibler *et al.* [37]. State-selective measurements using the CARS technique do not perturb the population distribution, provided the optical input of the diagnostic laser radiation is kept below certain power-density limits. These limiting intensities depend on environmental conditions, such as total ambient pressure, temperature and molecular relaxation behavior. Wenzel *et al.* [4], and later Welz *et al.* [31], used the CARS technique to investigate the longitudinal and radial gas-phase temperature distribution within the absorption volume of a graphite tube furnace, and its change with time during periods of rapid heating. Gas-phase temperatures were correlated with pyrometrically measured tube wall and platform temperatures, in order to find out how rapidly the gas temperatures follow changes of the wall temperature and how well temperatures agree under equilibrium conditions. They measured the steady-state gas temperature over the length of a pyrolytically coated graphite tube without a platform in the gas-stop mode. The results are depicted in Figure 3.18. The development of the temperature at the center as a function of time is shown in Figure 3.19. From these results it was concluded that, under steady state conditions, the wall and gas temperatures are very similar. This is in agreement with the models of van den Broek *et al.* [38] and of Rademeyer and Human [23], who found that there are only a few degrees difference between the wall and gas

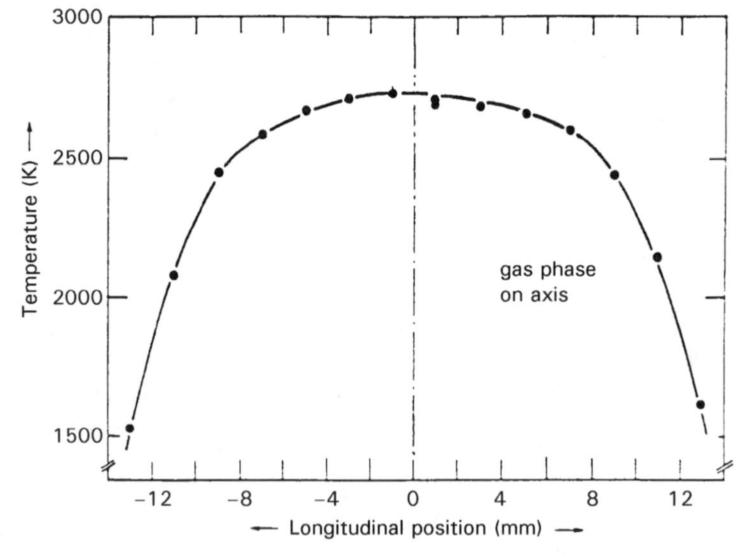

Figure 3.18 Longitudinal distribution of CARS-monitored gas temperature on the tube axis ($r = 0$) measured in a graphite tube under steady-state conditions [31]. (Reprinted from *Spectrochim. Acta*, Vol. **43B**, B. Welz, M. Sperling, G. Schlemmer, N. Wenzel and G. Marowsky, Spatially and temporally resolved gas phase temperature measurements in a Massmann-type graphite tube furnace using coherent anti-Stokes Raman scattering, pp. 1187–1207, 1988, with kind permission of Elsevier Science – NL, Sara Burgerhartstraat 25, 1055 KV Amsterdam, The Netherlands.)

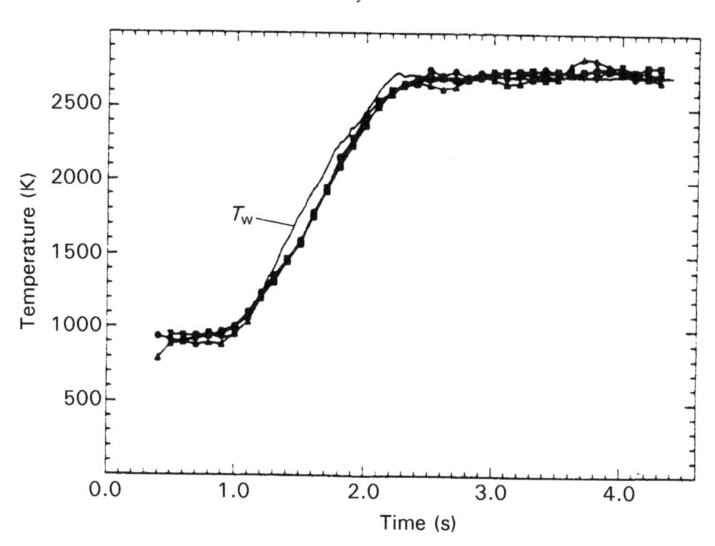

Figure 3.19 Development of gas temperature in a graphite tube without platform under gas-stop conditions at location $z = 0$ (tube center) and four radial locations: (■) $r = -1$; (●) $r = 0$; (△) $r = +1$; (▽) $r = +2$. Also shown is the pyrometrically measured tube wall temperature (Tw) [31]. (Reprinted from *Spectrochim. Acta*, Vol. **43B**, B. Welz, M. Sperling, G. Schlemmer, N. Wenzel and G. Marowsky, Spatially and temporally resolved gas phase temperature measurements in a Massmann-type graphite tube furnace using coherent anti-Stokes Raman scattering, pp. 1187–1207, 1988, with kind permission of Elsevier Science – NL, Sara Burgerhartstraat 25, 1055 KV Amsterdam, The Netherlands.)

temperature at each point along the furnace length. From their results, Welz *et al.* were able to conclude that the gas escapes predominantly through the tube ends, and that the effective extension of the hot gas volume may be greater than previously anticipated. This is in agreement with the model of Holcombe [39].

REFERENCES

[1] W. Slavin and D.C. Manning, *Anal. Chem.* **51**, 261 (1979).
[2] C.L. Chakrabarti, S. Wu, R. Karwowska, J.T. Rogers, L. Haley, P.C. Bertels and R. Dick, *Spectrochim. Acta, Part B* **39B**, 415 (1984).
[3] D.C. Manning and W. Slavin, *Anal. Chem.* **51**, 2375 (1979).
[4] N. Wenzel, B. Trautmann, H. Grosse-Wilde, G. Schlemmer and B. Welz, *Opt. Comm.* **68**, 75 (1988).
[5] C.J. Rademeyer, H.G.C. Human and P.K. Faure, *Spectrochim. Acta, Part B* **41B**, 439 (1986).
[6] H. Falk, A. Glisman, L. Bergann, G. Minkwitz, M. Schubert and J. Skole, *Spectrochim. Acta, Part B* **40B**, 533 (1985).
[7] C.J. Rademeyer and H.G.C. Human, *Prog. Anal. Spectrosc.* **9**, 167 (1986).

[8] J.M. Ottaway, J. Carroll, S. Cook, S.P. Corr, D. Littlejohn and J. Marshall, *Fresenius' Z. Anal. Chem.* **323**, 742 (1986).
[9] C.L. Chakrabarti, H.A, Hamed, C.C. Wan, P.C. Bertels, D.C. Gregoire and S. Lee, *Anal. Chem.* **52**, 167 (1980).
[10] W. Frech, D.C. Baxter and B. Hütsch, *Anal. Chem.* **58**, 1973 (1986).
[11] W. Frech and S. Jonsson, *Spectrochim. Acta, Part B* **37B**, 1021 (1982).
[12] O.O. Ajayi, D. Littlejohn and C.B. Boss, *Anal. Proc.* **25**, 75 (1988).
[13] R.J. Lovett, *Appl. Spectrosc.* **39**, 778 (1985).
[14] D.C. Gregoire, C.L. Chakrabarti and P.C. Bertels, *Anal. Chem.* **50**, 1730 (1978).
[15] J. Zsako, *Anal. Chem.* **50**, 1105 (1978).
[16] C.W. Fuller, *Electrothermal Atomization for Atomic Absorption Spectrometry*, The Chemical Society, London, 1977.
[17] C.L. Chakrabarti, C.C. Wan, R.J. Teskey, S.B. Chang, H.A. Hamed and P.C. Bertels, *Spectrochim. Acta, Part B* **36B**, 427 (1981).
[18] N.G. Zhou, W. Frech and L. de Galan, *Spectrochim. Acta, Part B* **39B**, 225 (1984).
[19] R.D. Ediger, *At. Absorpt. Newsl.* **14**, 127 (1975).
[20] B.V. L'vov, L.A. Pelieva and A.I. Sharnopolsky, *Zh. Priki. Spectrosk.* **27**, 395 (1977).
[21] M.S. Cresser and C.E. Mullins, *Anal. Chim. Acta* **68**, 377 (1974).
[22] W. Slavin, S.A. Myers and D.C. Manning, *Anal. Chim. Acta* **117**, 267 (1980).
[23] C.J. Rademeyer and H.G.C. Human, *J. Anal. At. Spectrom.* **4**, 393 (1989).
[24] C.L. Chakrabarti, S. Wu, R. Karwowska, J.T. Rogers and R. Dick, *Spectrochim. Acta*, **40B**, 1663 (1985).
[25] O.A. Guell and J.A. Holcombe, *Appl. Spectrosc.* **45**, 1171 (1991).
[26] G. Lundgren, L. Lundmark and G. Johansson, *Anal. Chem.*, **46**, 1028 (1974).
[27] B.V. L'vov, *Spectrochim. Acta Part B* **33B**, 153 (1978).
[28] D.K. Eaton and J.A. Holcombe, *Anal. Chem.* **55**, 1821 (1983).
[29] G.A.W. Rutgers and J.C. De Vos, *Physica* **20**, 715 (1954).
[30] H. Falk, *Spectrochim. Acta, Part B*, **39B**, 387 (1984).
[31] B. Welz, M. Sperling, G. Schlemmer, N. Wenzel and G. Marowsky G, *Spectrochim. Acta, Part B*, **43B**, 1187 (1988).
[32] M. Sperling, B. Welz, J. Herzberg, C. Rieck and G. Marowsky, *Spectrochim. Acta, Part B* **51B**, 897 (1996).
[33] H. Massmann and S. Grucer, *Spectrochim. Acta, Part B* **29B**, 283 (1974).
[34] B.V. L'vov, D.A. Katskov and L.P. Kruglikova, *Zh. Priki. Spektrosc.* **14**, 784 (1971).
[35] R.F. Browner and J.D. Winefordaer, *Anal. Chem.* **44**, 247 (1972).
[36] Y. Terui, K. Yasuda and K. Hirokawa, *Anal. Sci.* **7**, 599 (1991).
[37] J.W. Nibler, J.R. McDonald and A.B. Harvey, *Opt. Commun.* **18**, 371 (1976).
[38] W.M.G.T. van den Broek, L. de Galan, J.P. Matousek and E.J. Czobik, *Anal. Chim. Acta*, **100**, 121 (1978).
[39] J.A. Holcombe, *Spectrochim Acta, Part B* **38B**, 609 (1983).

4

Instrumentation

James M. Harnly

4.1 INTRODUCTION

The evolution of graphite furnace atomic absorption spectrometry (GFAAS), from its original design into the instrument so widely used today, can be traced through a series of discrete instrumental developments. It is a field that has relied primarily on instrumental innovation, rather than chemical methodology, to overcome the problem of accurate measurement of atom species in samples with complex chemical matrices. Two distinct areas of development can be identified: first, the development of the graphite furnace as an atomizer that provides reproducible and matrix-independent gaseous populations of the analyte in the atomic state; and second, the development of the detection system, the optical, and electronic subsystems necessary to rapidly, accurately, and precisely measure the analyte absorption.

Flame atomic absorption spectrometry (FAAS), as first described by both Walsh [1] and by Alkemade and Milatz [2], used an optical design that, conceptually, was not significantly different from conventional ultraviolet–visible absorption spectrometers of the time. A hollow cathode lamp (HCL) was substituted for the continuum source and a flame atomizer, located between the source and the disperser, replaced the absorption cell. These new components comprised the advantages of FAAS–the specificity and low noise of the HCL, and the energetic conversion of aqueous compounds to gaseous atoms by the flame. The static nature of the absorption signal and the concept of single and double beam absorption measurements, however, were not new. Consequently, the existing optical and electronic technology were adequate for accurate determinations.

Electrothermal Atomization for Analytical Atomic Spectrometry. Edited by K. W. Jackson
© 1999 John Wiley & Sons Ltd

When L'vov [3] developed the graphite furnace atomizer as an alternative to flame atomization, he initiated a revolution in instrument design. Although the goal, accurate measurement of absorption, is the same, the rapid, transient nature of graphite furnace signals make accurate measurements a much more complicated process. The rapidly heated graphite furnace efficiently atomizes close to 100 per cent of the sample, analyte and matrix, into a small confined space, typically $0.8\,cm^3$ or less. As the loss of the gas-phase atoms from the furnace is theoretically dependent on the rate of diffusion, the residence time of the analyte in the optically critical measurement volume is orders of magnitude greater than for systems employing pneumatic nebulizers [4,5] with continuous gas flow. With pneumatic nebulization, the analyte is convectively swept through the flame or plasma. Temporally, GFAAS signals are short lived, generally having a peak half-width of 0.5–2 s. Thus, accurate characterization of the height or area of the peak, in absorbance, requires data acquisition intervals that are much shorter than the duration of the peak. Additionally, the combination of high volatilization efficiency and the relatively long residence time in the small volume of the furnace results in high analyte and matrix concentrations in the gas phase. This produces high analyte sensitivity but also the potential for high levels of nonspecific background attenuation and extended interaction between the analyte and the matrix.

It was eventually realized that direct substitution of a graphite furnace for a flame atomizer did not yield accurate results. The need for rapid and accurate background-corrected absorbances and the concurrent developments in solid-state electronics led to a dramatic evolution in design of the detection system. Four different background correction methods, each based on a different principle, were developed within a 10-year span. Sophisticated optical designs, employing either a double beam configuration or single beam configuration with double beam performance, were developed to provide the best signal-to-noise ratios. Automated sampling systems were necessary to reproducibly deliver microliter samples to the furnace. Digital data acquisition systems were developed to permit the measurement of thousands of intensities per second in complex patterns. Finally, microprocessors were incorporated into the instruments to permit the efficient processing and storage of the tremendous volume of data.

Of course, graphite furnace designs also improved. Increased understanding of the complex physical and chemical processes taking place in the furnace and discovery of the importance of temporal and spatial isothermality (see Chapter 3) led to the development of new furnace designs such as the L'vov platforms, the integrated contact cuvettes, and the two-step furnaces (see Section 1.4). These changes in furnace design, in turn, placed new demands on the detection system. As a result, the designs of today's atomic absorption spectrometers bear little resemblance to the instruments originally described by Walsh [1] and Alkemade and Milatz [2].

The only appropriate yardstick for evaluating current instrument designs are the analytical figures of merit, i.e. the accuracy, sensitivity, precision, and signal-to-noise ratio (SNR) or limits of detection. Every instrumental development has produced an improvement in one or more of these defining parameters; for example, the graphite furnace has been a success because it improved the sensitivity and detection limits for AAS by 1000-fold. Background correction, the L'vov platforms, contoured tubes, and the two-step furnaces improved the accuracy of furnace determinations, and autosamplers improved the precision. In this chapter, all aspects of instrumentation are discussed on the basis of the analytical figures of merit.

4.2 FUNDAMENTAL CONSIDERATIONS

4.2.1 Analytical signal

The conventional measurement parameter for AAS is absorbance A, which is computed as

$$A = \log_{10} \frac{I_0}{I} \tag{4.1}$$

where I_0 is the intensity of the incident radiation and I is the intensity of the transmitted radiation. The absorbance also has the relationship

$$A = abc \tag{4.2}$$

where a is the absorptivity, b is the path length, and c is the concentration of the absorbing species. This classic equation defines a linear relationship between absorbance and concentration that is accurate if: (1) the incident radiation is monochromatic; (2) the absorbing medium does not scatter the radiation (i.e. there is no loss of incident intensity due to nonspecific background attenuation; and (3) the absorbing medium is homogeneous [6]. This equation and its underlying assumptions have some very interesting implications for the optical design of GFAAS instruments.

Absorbance, as computed in Eqn (4.1), is the logarithm of the ratio of the transmitted intensity to the incident intensity. The incident intensity serves as an internal reference. As a result, absorbance is dimensionless and independent of the source intensity. In principle, every AA instrument with the same path length and the same concentration of atoms c in the light path will provide the same absorbance. Thus, absorbance is a more fundamental measurement (i.e. less dependent on instrument parameters) than intensity. The price of this more fundamental value is the added complexity of data acquisition and processing for two intensity measurements. The measurement of the incident intensity is of equal importance as the measurement of the transmitted intensity, and the two intensities must be converted to the absorbance domain prior to further calculations.

Absorbance measurements can be made using either a single beam or double beam optical configuration Figure (4.1). The basic detection unit, consisting of a dispersing device and a photoelectric detector, is known as a spectrometer. With a single beam spectrometer (Figure 4.1(A)), the incident radiation I_0 is measured prior to the introduction of the sample (or standard or blank) to the atomization cell. The stored value of I_0 is then used to calculate an absorbance for each measured value of I during the atomization of the sample. With a double beam spectrometer (Figure 4.1(B)), the source radiation is split prior to the atom cell and I_0 and I are measured either simultaneously with two separate spectrometers (double beam in space) or sequentially with the same spectrometer (double beam in time). Thus, double beam spectrometers allow measurement of I_0 and I either simultaneously or in a rapid sequential manner; the time interval between the I_0 and I measurements is much shorter than it is for a single beam spectrometer. A double beam spectrometer with accompanying electronics and detector for the determination of absorbance is a spectrophotometer. Most commercially available GFAAS instruments are double beam in time. Although all double beam instruments must be

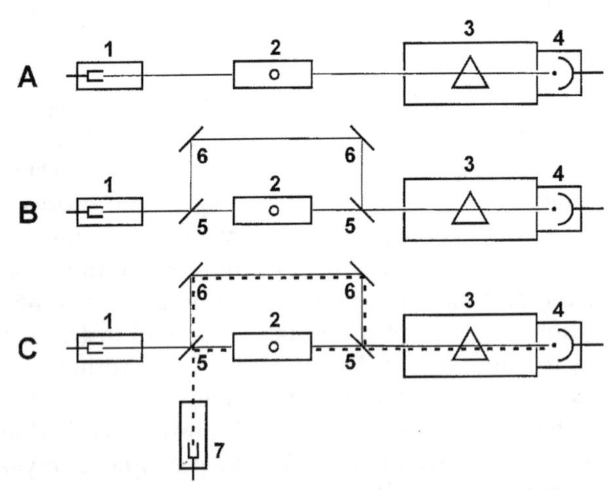

Figure 4.1 Optical arrangements of spectrometers: (A) single beam; (B) double beam; (C) dual double beam. The components are: (1) source; (2) graphite furnace atomizer; (3) spectrometer; (4) detector; (5) beam combiner/splitter or chopper with reflecting surfaces; (6) mirrors; and (7) secondary source.

spectrophotometers to produce absorbance signals, they are almost exclusively referred to as spectrometers.

The first assumption of Eqn (4.2), that the incident radiation is monochromatic, has always been reasonable for GFAAS because HCLs and electrodeless discharge lamps (EDLs) have been used exclusively for commercial instruments. These sources, known collectively as line sources (LSs), emit radiation composed of discrete electronic transitions, or lines. As LSs are operated at reduced pressures and temperatures (see Section 4.3.1.1) compared to the flame or furnace, the emitted lines have a full width at half height (FWHH) which is 3–5 times less than that of the absorption profiles. Although LSs are not truly monochromatic, their relatively narrow width, with respect to the absorption profiles, makes the assumption of monochromaticity valid to a first approximation. The operation of LSs at excessive currents or power, however, can significantly increase the width of the emission profiles and reduce sensitivities. This problem is discussed in more detail in Section 4.3.1.1.

The second assumption of Eqn (4.2), that the absorbing medium (the vaporized sample in the furnace volume) does not scatter the radiation, is problematic for GFAAS. The high volatilization efficiency of the graphite furnace results in a high probability of nonspecific background attenuation from reflection, refraction, and broad band molecular absorption by the sample matrix. Thus, indiscriminate application of Eqn (4.2) will result in an absorbance that is the sum of the analytical absorbance and the background absorbance. Computation of an accurate background corrected absorbance A_{bc}, requires subtraction of the background absorbance A_b from the total absorbance (Eqn (4.1)):

$$A_{bc} = A - A_b \tag{4.3}$$

Historically, the first approach to background correction required a separate source and measurement to determine A_b. For Eqn (4.3) to be valid, the second source must be insensitive to the analyte absorbance and only measure the background absorbance. By convention, the source measuring the analyte absorbance (a line source for all commercial instruments) is referred to as the primary source, and the source measuring only the background absorbance is referred to as the secondary source. If double beam optics are employed for both the primary and secondary source (Figure 4.1, arrangement C), the instrument is a dual double beam spectrometer.

Combining Eqn (4.1) and (4.3) shows that two intensity measurements for each of the sources are required to compute A_{bc}. Thus,

$$A_{bc} = \log_{10}\frac{I_0}{I} - \log_{10}\frac{I_{0,b}}{I_b} \tag{4.4}$$

where I_o and $I_{o,b}$ are the intensities of the incident radiation of the primary and secondary source, respectively, and I and I_b are the intensities of the transmitted radiation of the two sources. If the two sources are balanced by adjusting the operating currents, such that $I_o = I_{o,b}$, the Eqn (4.4) becomes

$$A_{bc} = \log_{10} \frac{I_b}{I} \tag{4.5}$$

Equation (4.5) is identical to Eqn (4.1), except that the incident radiation of the primary source has been replace by the transmitted radiation of the secondary source. It is now appropriate to redefine I as the analytical radiation and I_b as the reference radiation. The reference radiation, unlike the previously defined incident radiation, must pass through the furnace. The need for suitable reference radiation has led to the development of four uniquely different background correction designs. These four designs make use of: (1) a line source with a secondary continuum source (LS/CSAAS); (2) Zeeman splitting of the source or analyte absorption (ZAAS); (3) self-reversal of the source (SRSAAS); and a primary continuum source with off-line correction (CSAAS). The first three designs can be found in commercially available instruments, whereas the fourth design is found only in prototype instrumentation. Each of these designs is discussed in detail in section 4.5.

The two most useful values for quantifying the transient analytical signals of the graphite furnace atomizer are the peak height, or peak absorbance A_p and the peak area, or integrated absorbance, A_i [7]. The peak absorbance is the maximum absorbance observed for the analytical peak, whereas the integrated absorbance is the sum of absorbances over the peak normalized by the time interval between absorbance calculations, t_A.

$$A_i = t_A \int A dt \tag{4.6}$$

The integrated absorbance must be normalized by the time interval of absorbance computation to render the integrated absorbance independent of the measurement frequency. Although both values demand the same frequency response of the electronics (Section 4.2.4), integrated absorbances are fundamentally more accurate. Provided that the analyte loss mechanism is constant (i.e. the temperature is constant as a function of time and space), the integrated area is independent of the analyte supply function. With optimized integration intervals, integrated absorbances also offer better detection limits (Section 4.2.3).

Because peak and integrated absorbance are independent of the source intensity the signal generated per unit of analyte concentration, the sensitivity, is a useful diagnostic tool. The sensitivity is dependent on the furnace parameters (temperature, length, diameter, and rate of heating), and the width ratio of the HCL and the absorption profile. Consequently, with a lamp current equal to or less than the maximum specified by the manufacturer, the sensitivity can be used as an indicator of the analytical performance of the furnace. Significant deviation of the sensitivity from previously obtained performance or from the published value of the manufacturer is an indication of improper operation or a malfunction. Sensitivity for GFAAS is preferably expressed in terms of the characteristic mass [8]. The characteristic mass for peak absorbance m_p is the analyte mass necessary to give a peak absorbance of 0.0044:

$$m_p = \frac{0.0044m}{A_p} \tag{4.7}$$

where m is the mass of a standard whose peak absorbance A_p falls in the linear range. The characteristic mass for integrated absorbance m_0 is the analyte mass necessary to give an integrated absorbance of 0.0044 s:

$$m_0 \frac{0.0044m}{A_i} \tag{4.8}$$

Table 4.1 lists the elemental characteristic masses for several commercially available instruments described in Section 4.6. It has recently been shown that for Zeeman AAS, the characteristic mass is dependent on the design and operating currents of the line source and the magnet field strength [14,15]. Su *et al.* [15] have suggested that, for these instruments, a new measure of sensitivity be defined as corrected characteristic mass, ${}^{c}m_o$. The corrected characteristic mass is obtained by correcting characteristic mass by the ratio of sensitivities with the magnet off and on R and the rollover absorbance, A_r. Still another definition of sensitivity has been suggested for absorbances that are integrated with respect to wavelength as well as time [16]. Using a continuum source and an array detector, absorbance is integrated with respect to wavelength, for each absorption profile, and then with respect to time in the normal manner (Section 4.7). This doubly integrated absorbance must then be normalized with respect to time (between absorbance computations) and wavelength (pixel width in picometers) [16]. Consequently, this measure of sensitivity, the intrinsic mass m_i, is the analyte concentration that produces an absorbance of 0.0044 pm s. Intrinsic mass is independent of all instrument operating parameters except the furnace atomization efficiency. Conventional characteristic masses can be converted to intrinsic masses by normalizing by the spectral width of the HCL emission line and the ratio of the HCL and absorption profile widths [16].

Table 4.1 Characteristic mass (pg).

		Perkin–Elmer				Varian	
Element	*Wavelength (nm)*	*HCL/D_2*	*Transverse Zeeman*	*Longitudinal Zeeman*	*Longitudinal Zeeman*	*HCL/D_2*	*Transverse Zeeman*
Ag	328.1	1.5	1.5	4.5		0.72	0.70
Al	309.3	10	10	31		5.0	5.0
As	193.7	15	15	40	49	12	10
Au	242.8	8.5	10	18		4.7	4.4
B	249.7	700	1000	600		1200	860
Ba	553.6	6.5	6.5	15		17	17
Be	234.9	0.5	1.0	2.5		0.78	0.50
Bi	223.1	19	31	60		10	9.0
Ca	422.7	0.80	0.80	1.0		0.64	0.60
Cd	228.8	0.35	0.35	1.3	1.8	0.23	0.20
Co	240.7	6.0	7.0	17		4.1	4.0
Cr	357.9	3.0	3.3	7.0	7.0	1.5	1.5
Cs	852.1	5.0	5.3	12		19	11
Cu	324.7	4.0	8.0	17	20	7.1	6.0
Dy	421.2	40	40	40		46	46
Er	400.8	70	70	71		100	100
Eu	459.4	20	20	18		26	26
Fe	248.3	5.0	5.0	12		1.2	1.2
Ga	287.4	12	20	50		5.8	4.6
Ge	265.1	26	30	25		9.0	9.0
Hg	253.7	72	100	220		220	150
In	303.9	14	15	80		7.0	7.0
Ir	264.0	250	250	230		140	136
K	766.5	0.80	1.1	2.0		0.44	0.40
Li	670.8	1.4	1.4	5.5		8.2	4.0

Mg	285.2	0.30	0.35	0.40		0.27	0.20
Mn	279.5	2.0	2.2	6.3	5.4	0.65	0.60
Mo	313.3	9.0	9.0	12	15	7.3	7.0
Na	589.6	1.0	1.0	1.2		0.11	0.10
Ni	232.0	13	13	20		4.9	4.8
P	213.6	4200	5500	21000		3200	2200
Pb	283.3	10	12	30	60	6.1	5.6
Pd	247.6	22	24	52		8.6	8.6
Pt	265.9	95	115	220		85	70
Rb	780.0	2.3	2.4	10		1.1	1.0
Rh	348.5	10	10	24		8.4	8.0
Ru	349.9	31	35	45		15	15
Sb	217.6	22	22	55		10	10
Se	196.0	23	28	45	43	15	14
Si	251.6	40	40	120		15	15
Sn	224.6	10	24	90		11	10
Sr	460.7	1.4	1.4	4		2.1	2.0
Tb	432.6					4.0	3.6
Te	214.3	17	17	50		9.7	9.0
Ti	365.3	45	45	70		50	50
Tl	276.8	13	19	53	54	24	15
V	351.5	30	40	42	51	28	22
Yb	398.8	2.5	2.5	3.0		3.1	3.0
Zn	213.9	0.40	0.45	1.00		0.16	0.15
Reference		9	10	11	12	13	13

4.2.2 Noise

Noise can be defined as the random variation of a measured signal around the average value. The noise is generally characterized by determining the standard deviation σ of n measured signals:

$$\sigma = \sqrt{\frac{\sum_{i=1}^{n}(x_i - \bar{x})^2}{n-1}} \tag{4.9}$$

For GFAAS, the noise of interest is the variation of the intensity measurements. The fundamental noise component of all absorption spectrometric techniques is the source photon shot noise, or simply shot noise. Shot noise arises from the random arrival of photoelectrons at the detector and has its origins in the quantum nature of light. Shot noise σ_s is proportional to the square root of the signal:

$$\sigma_s = \xi_s \sqrt{I}, \tag{4.10}$$

where ξ_s is a proportionality constant. For detection with a photomultiplier tube (PMT), the proportionality constant ξ_s is dependent on the PMT again. For photon counting, ξ_s is equal to unity. For solid state detectors, ξ_s is equal to unity when I is expressed in terms of electrons, i.e. the signal taken directly from the diode prior to amplification. As the description of shot noise implies, there is no elimination of this noise component. It is a law of physics. The relative contribution of the shot noise, however, can be reduced. The relative noise, σ_s/I decreases inversely as the signal I increases. Thus, more intense radiation sources or longer integration times result in improved signal-to-noise ratios (Section 4.2.3).

The radiation source(s) may also contribute source fluctuation noise σ_f to the intensity measurement. Source fluctuation noise is a low-frequency noise arising from variations in the power and temperature of the source. Types of fluctuation noise are periodic noise, $1/f$ noise and source drift. Periodic noise is a repetitive variation with one or more discrete frequencies. The $1/f$ noise has a noise power spectrum (variance) which is inversely proportional to frequency. Extremely low-frequency intensity fluctuation is commonly known as drift. The amplitude of source fluctuation noise is proportional to the signal:

$$\sigma_f = \xi_f I \tag{4.11}$$

where ξ_f is the fluctuation factor. Consequently, the relative noise will not decrease with greater intensities or longer integration times. Source $1/f$ and drift noise are usually eliminated by the use of a double beam spectrometer.

Other sources of noise are the atomizer, the detector, and the electronics. The atomizer may give rise to analyte fluctuation noise ($\sigma_{a,f}$), analyte shot noise ($\sigma_{a,s}$), background shot noise ($\sigma_{b,s}$), background fluctuation noise ($\sigma_{b,f}$), and furnace shot noise ($\sigma_{f,s}$). The analyte fluctuation noise arises from the nonrandom movement of the gaseous analyte atoms in the furnace and from variations in the position of the sample on the wall or platform at the start of volatilization. Nonrandom movement of the gaseous atoms in the furnace can be attributed to convective components from gas expansion during the heating process or from temperature differences within the furnace after the maximum temperature has been reached. Variations of the position of the dried sample on the wall or platform can come from random movement during the drying and pyrolysis stages. At analytical concentrations, well above the detection limit, the limiting uncertainty of the furnace absorption signal is usually the analyte fluctuation. Analyte shot noise arises from radiation emitted by the analyte. Thermal and photon excitation of the analyte can induce emitted radiation.

Furnace shot noise occurs as a result of the black body emission of the furnace wall. Proper imaging of the radiation source onto the entrance slit of the spectrometer can minimize this noise. Furnace and analyte shot noises are primarily problems at wavelengths greater than 400 nm; both sources are very weak in the ultraviolet region of the spectrum. Both noises tend to be more troublesome in the presence of high background attenuation when the source intensity is significantly reduced.

Double beam operation will eliminate background fluctuation noise in the same way it eliminates source fluctuation noise. Background fluctuation noise is similar to source fluctuation noise, because the changing background attenuation produces a change in the source intensity. Both fluctuation noises can be eliminated provided the time interval between the measurement of I_0 and I is small compared to the rate of change of the reference intensity. For GFAAS, the interval between the I_0 and I is dictated by the background fluctuation noise, not the $1/f$ noise of the HCL. The fluctuation noise of an HCL generally occurs at a very low frequency and can usually be more accurately characterized as drift noise. Background fluctuation can be rapid and, in general, an interval of less than 2 ms between the I_0 and I measurements is required to minimize this noise (Section 4.2.4).

The detector noise σ_d depends on the type of detector. The multiplication process of the PMT does not produce a measurable noise component. A dark current noise, however, arises from the random arrival of electrons at the anode, which have leaked into the dynode cascade (Section 4.3.4.1). This noise is usually very small and is a function of the temperature and the gain of the PMT. Cooling of the PMT can dramatically reduce the dark current, but it is usually sufficiently low compared to the electron flux generated by normal line source radiation that this is never done. Two major types of detector noise are associated with solid-state detectors and the specific levels depend on the type of detector: a linear photodiode array (LPDA), a charge coupled device (CCD), or a charge injection device (CID). Each of these detectors is discussed in detail in Section 4.3.4. The noise sources for these detectors are read noise (σ_{read}) and dark current shot noise (σ_{ds}). The read noise can be reduced with correlated double sampling. The read noise for the LPDA is typically 3000 e^- and that for the CCD is 5–50 e^-, depending on the design of the device and the area of the pixels. In both of these cases the read is a destructive process. The read noise of the latest CIDs is around 200 e^-, but nondestructive reads are possible. Consequently, the relative read noise can be decreased by the square root of the number of nondestructive reads. The dark current shot noise is primarily a concern when long integration times are used and can be significantly reduced by cooling of the array. For the short integration intervals necessary for GFAAS (below 20 ms), the dark current shot noise is generally not significant.

Electronic noise σ_e is the variation in the signal which is introduced by the electronic components. For well-designed circuits, this noise is insignificant compared to other noises. One electronic noise component that can become significant is the quantization noise σ_q. In contemporary instruments, the signal from the PMT, in volts, is converted to a digital number for processing in the computer. Suitable amplification is chosen such that the measured voltage is equal to the full scale of the analog-to-digital converter (ADC). High background attenuation, however, may reduce the source intensity to the point that the intensity noise is small compared to the voltage of a single digital step of the ADC, i.e. the resolution of the ADC. The quantization noise is

$$\sigma_q = \frac{r}{\sqrt{b}} \tag{4.12}$$

where r is the resolution (in volts) and b is the number of bits of the ADC [17]. For 10 V full scale quantization, noise is obviously less problematic with a 16 bit ADC (0.000 04 V) than with a 12 bit ADC (0.006 V).

Each of the noise sources just discussed may contribute uncertainty to the intensity measurement. The contribution of these noise sources will add quadratically. Thus the total incident intensity noise σ_{I_0} is

$$\sigma_{I_0} = \sqrt{\sigma_s^2 + \sigma_f^2 + \sigma_{f,s}^2 + \sigma_d^2 + \sigma_q^2} \tag{4.13}$$

and the transmitted intensity noise σ_I is

$$\sigma_I = \sqrt{\sigma_s^2 + \sigma_f^2 + \sigma_{a,s}^2 + \sigma_{a,f}^2 + \sigma_{b,f}^2 + \delta_{f,s}^2 + \sigma_d^2 + \delta_q^2} \tag{4.14}$$

This assumes that the incident radiation is measured either prior to the sample atomization or passes around the furnace. With background correction, noises comparable to those shown in Eqns (4.13) and (4.14) are expected for $\sigma_{I_{a,b}}$ and σ_{I_b} respectively, i.e. the noises for the transmitted and incident radiation of the secondary source (Eqn (4.4)). Each of the noise components in Eqns (4.13) and (4.14) are constantly present, but not necessarily significant. For conventional GFAAS using an HCL, the shot noise σ_s is dominant at low analyte concentrations. As the analyte concentration increases, so does the analyte fluctuation noise $\sigma_{a,f}$ and it will eventually dominate. With high and/or rapid background attenuation, the background fluctuation noise $\sigma_{b,f}$ may become important, depending on the time interval between I_0 and I measurements of the double beam spectrometer. Also in the presence of high background attenuation, furnace emission shot noise $\delta_{f,s}$ and quantization noise σ_q may become significant. Thus, the most significant noise sources are highly dependent on the circumstances of the determination.

The precision of the computed absorbance is directly related to uncertainty of the measurement of the transmitted and incident intensities. Using the mathematics established for the propagation of errors [18], it can be shown that the absorbance noise σ_A is

$$\sigma_A = 0.4343\sqrt{\left(\frac{\sigma_I}{I}\right)^2 + \left(\frac{\sigma_{I_0}}{I_0}\right)^2} \tag{4.15}$$

where σ_I and S_{I_0} are defined in Eqns (4.13) and (4.14). If two sources are used, as is the case with line source/continuum source (LS/CS) background correction, the uncertainty of the intensity measurements of the secondary source must also be taken into account. Thus, the background corrected absorbance noise $\sigma_{A_{bc}}$ is

$$\sigma_{A_{bc}} = 0.4343\sqrt{\left(\frac{\sigma_I}{I}\right)^2 + \left(\frac{\sigma_{I_0}}{I_0}\right)^2 + \left(\frac{\sigma_{I_b}}{I_b}\right) + \left(\frac{\sigma_{I_{0,b}}}{I_{0,b}}\right)^2} \tag{4.16}$$

where the subscript b is used to denote the incident and transmitted intensity and noise of the secondary source. When no analyte or sample matrix is present and the primary and secondary sources have equal intensities, the baseline absorbance noise, from Eqn (4.15) becomes

$$\sigma_{A_{bc}} = \frac{(0.434)(\sqrt{2})\sigma_{I_0}}{I_0} = \frac{0.614\sigma_{I_0}}{I_0} \tag{4.17}$$

and the baseline for equation 4.16 becomes

$$\sigma_{A_{bc}} = \frac{(0.434)(2)\sigma_{I_0}}{I_0} = \frac{0.868\sigma_{I_0}}{I_0} \tag{4.18}$$

The baseline absorbance noise is a factor of $\sqrt{2}$ less in Eqn (4.17), as only half as many intensity measurements are involved. Equations (4.15) and (4.16) assume that the uncertainty of both the analytical and reference intensities of the primary and secondary sources contribute to the absorbance noise. As a result, Eqns (4.17) and (4.18) deviate by factors of 2 and $\sqrt{2}$, respectively, from conventional statements of this relationship [19], which assumes that only the variation of the transmitted intensity I affects the noise level. It must be emphasized that Eqns (4.17) and (4.18) specify the mathematical relationship between the absorbance noise and the intensity noise and do not depend on the relationship of σ_{I_0} to I_0, i.e. do not depend on whether photon shot noise or fluctuation noise is dominant.

Equations (4.17) and (4.18) also point out an irrevocable effect of background attenuation. Even though the accuracy of the analytical measurement can be restored through the use of the various background-correction techniques, the precision of the computed absorbance, in the presence of background attenuation, will be worse because of the reduced intensity. The only way to return to the original precision levels (with no background attenuation) is to eliminate the background attenuation. If photon shot noise is limiting for both sources, then Eqn (4.10) can be used to modify Eqns (4.17) and (4.18), to give

$$\sigma_{A_{bc}} = \frac{0.614}{\sqrt{I_0}} \tag{4.19}$$

and

$$\sigma_{A_{bc}} = \frac{0.868}{\sqrt{I_0}} \tag{4.20}$$

respectively, for the case of one and two sources. The decrease in absorbance noise with increasing intensity runs counter to intuition, but arises from the ratioing of intensities for the absorbance computation.

Equations (4.19) and (4.20) represent the minimum noise levels that can be expected for an absorption spectrometer where source intensities are equal, the sources are shot noise limited, and analyte absorption and background attenuation are nonexistent or very low. In practice, the noise levels may be greater. For example, if either of the sources has a significant (comparable to or greater than the shot noise) flicker noise component, then Eqns (4.19) and (4.20) are no longer valid, and either Eqns (4.17) or (4.18) must be used with noise computed from Eqn (4.13).

Another common problem is that the secondary source is less intense than the primary source, i.e. the secondary source cannot be operated at a sufficiently high current to give a comparable intensity. Usually, the operating current of the primary source is reduced so that the intensities of the two sources are equal. Equation (4.20) shows that $\sigma_{A_{bc}}$ will now be greater because the value of I_0 has decreased. Equations (4.19) and (4.20) provide some perspective as to the variation in intensity levels that are expected of a radiation source for AAS. An absorbance noise level of 0.1 per cent requires the relative intensity noise (σ_{I_0}/I_0) to be 0.16 per cent. This is, in general, the maximum allowable instability that is desirable for a radiation source. Assuming a shot noise-limited source, a relative noise of 0.16 per cent corresponds to an I_0 value of 10^6 photons. A stable HCL will have a relative precision closer to 0.03 per cent or $\sigma_A = 0.0002$, corresponding to an I_0 of 10^7 photons.

The time integrated absorbance is simply the summation of a specified number of individual absorbances n over a specified time interval T. The variables T and n are related by the relationship $n = T/t_A$ where t_A is the period, in seconds, between absorbance computations. Further, $t_A = 1/f$ where f is the frequency of absorbance computation. Using the rules for the propagation of errors, it can be shown that the baseline standard deviation of the integrated absorbance is

$$\sigma_{A_\mathrm{i}} = \sigma_A \sqrt{n} \tag{4.21}$$

As the integrated absorbance must be normalized for frequency by multiplying by $1/f$ or t_A, and n can be replaced by T/t_A,

$$\sigma_{A_\mathrm{i}} = \sigma_A \sqrt{T t_A} \tag{4.22}$$

This relationship should be verified on every instrument before being used routinely. Deviations occur (σ_{A_i} is greater than predicted) if there is a drift in the baseline absorbance. Upon verification, however, this relationship offers the analyst the opportunity to extrapolate the baseline-integrated absorbance noise from the standard deviation of the individual absorbances without performing 20 or more atomizations to obtain a statistically reliable estimate of the integrated noise.

This discussion of noise has dealt strictly with instrumental considerations. The variance of the analytical result will also include factors for sample handling and preparation. To most analysts, with sample concentrations well above the instrumental detection limits and with well engineered, commercial instruments, the instrumental noise will be insignificant and the uncertainty of the sample blank will be limiting and, consequently, of greater concern. The concern of this chapter, however, is instrumentation.

4.2.3 Signal-to-noise ratio

The term signal-to-noise ratio (SNR) is commonly used in atomic spectrometry to refer to the ratio of the analytical signal to a noise component of the signal or the blank. The SNR is a useful concept because optimal instrument performance cannot be achieved by simply maximizing the sensitivity or minimizing the noise. Optimal instrument performance is achieved only by maximizing the SNR. The usual measure of the SNR is the detection limit. IUPAC [20] states that 'the limit of detection, expressed as a concentration c_L (or mass q_L), is derived from the smallest measure x_L that can be detected with reasonable certainty for a given analytical procedure'. Mathematically, this definition is expressed as

$$x_L = x_B + k\sigma_B \tag{4.23}$$

where x_B is the mean value of the blank response, k is a numerical factor chosen to give the desired confidence level, and σ_B is the standard deviation of the blank response. This expression is valid regardless whether peak absorbance or peak area is measured. By definition,

$$q_L = \frac{x_L - x_B}{S} = \frac{m(x_L - X_B)}{0.0044} \tag{4.24}$$

where S is the analytical sensitivity (absorbance/mass, i.e. the slope of the calibration curve) and m has been previously defined as the characteristic mass. Substitution of Eqn (4.23) into (4.24) yields

$$q_L = \frac{k\sigma_B m}{S} \tag{4.25}$$

Long and Winefordner [21] have shown that $k = 3$ allows a confidence level of 99.86 per cent that the signal is not due to a random fluctuation of the blank (assuming a normal distribution). They concluded that a value of $k < 3$ should never be used.

The definition of the blank response x_B can vary. In considering instrument operation, the blank response may be measured without furnace atomization or with furnace atomization, but without a blank solution. These measures characterize the response of the absorption measurement system with and without the furnace power system engaged. For a well-engineered instrument, the two blank responses should be the same. A detection limit based on the blank with the furnace power engaged is the instrumental detection limit. The instrumental detection limit is a useful measure of instrument performance. Like the characteristic mass, it can be used to evaluate day-to-day instrument performance and to compare performance between instruments. Whereas the characteristic mass reflects differences in the atomizer efficiency, the instrument detection limit will reflect total instrument performance. A listing of detection limits for several commercially available instruments is shown in Table 4.2.

A detection limit based on the blank response obtained when an appropriate series of method blanks are atomized is the method detection limit. This detection limit accounts for variation due to background

Table 4.2 Detection limits (pg).

		Perkin-Elmer				Varian		Leeman Labs
Element	*Wavelength (nm)*	*Model 5000 transverse Zeeman*	*Model 4100ZL longitudinal Zeeman*	*SIMAA 6000 multielement longitudinal Zeeman*	*SIMAA 6000 single element longitudinal Zeeman*	*Spectra 880 HCL/D_2*	*Spectra 880Z transverse Zeeman*	*Analyte 5 self-reversed*
Ag	328.1	0.5	1					
Al	309.3	4	5			4	4	
As	193.7	20	10	8	6	5	5	30
Au	242.8	10	8					
B	249.7	1000	900					
Ba	553.6	10	16					
Be	234.9	1	0.4					
Bi	223.1	10	12					
Ca	422.7		0.6					
Cd	228.8	0.4	0.4	0.2	0.1	0.2	0.2	
Co	240.7	2	8					
Cr	357.9	1	2	1	0.4	2	2	
Cs	852.1							
Cu	324.7	1	5	7	4			
Dy	421.2							
Er	400.8							
Eu	459.4							
Fe	248.3	2	6					
Ga	287.4							
Ge	265.1							
In	303.9							
Ir	264.0		140					
K	766.5		0.4					
Li	670.8	2	3					
Mg	285.2	0.5	0.2					

(continued)

Table 4.2 *(continued)*

		Perkin-Elmer				*Varian*		*Leeman Labs*
Element	*Wavelength (nm)*	*Model 5000 transverse Zeeman*	*Model 4100ZL longitudinal Zeeman*	*SIMAA 6000 multielement longitudinal Zeeman*	*SIMAA 6000 single element longitudinal Zeeman*	*Spectra 880 HCL/D_2*	*Spectra 880Z transverse Zeeman*	*Analyte 5 self-reversed*
Mn	279.5	1	2	1	0.6			
Mo	313.3	4	4	2	1			
Na	589.6		1					
Ni	232	10	16			10	10	
P	213.6		6500					
Pb	283.3	5	3			4	4	15
Pd	247.6	25	40	8	4			
Pt	265.9	50	100					
Rb	780	5	2					
Rh	348.5							
Ru	349.9		60					
Sb	217.6	15	8					60
Se	196	30	15	22	9			60
Si	251.6	40	50					
Sn	224.6	20	10					
Sr	460.7	2	1					
Tb	432.6							
Te	214.3	10	20					
Ti	365.3	50	18			15	15	
Tl	276.8	10	6	12	9			15
V	351.5	20	6	4	2	10	10	
Yb	398.8							
Zn	213.9	0.1	6					
Reference		22	23	12	12	24	24	25

attenuation and instrumental noise, as well as analyte contamination and variations arising from the sample preparation procedure. As stated at the end of the previous section, the method detection limit is usually of more concern to the analyst.

The precision of instrumental detection limits is highly dependent on the number of measurements on which the value is based, i.e. the number of measurements used to determine σ_B and m (Eqn (4.25)). A common approach to computing the detection limit is to make 10 repeat atomizations of a low concentration standard, use the average signal to compute m (assuming $x_B = 0$), and use the standard deviation for σ_B. It is assumed that, at a low concentration, the precision of the measurement is dominated by the shot noise and will be the same for the blank or a low concentration standard. Williams [26] and Stevens and Winefordner [27] showed that for 10 repeat measurements, differences between detection limits must exceed a factor of 2 to be significant at the 95 per cent confidence level. With fewer measurements, the differences between detection limits can be even greater without being significant. Williams compiled confidence limit tables for the comparison of detection limits.

The most accurate estimate of the detection limit is computed using separately determined values for the characteristic mass and the baseline absorbance noise that have been averaged for an extended period of operation. For peak absorbance

$$q_L = \frac{k\sigma_B m_p}{0.0044} \tag{4.26}$$

where σ_B is the standard deviation of the individual absorbances and, like m_p and 0.0044, has no units. For integrated absorbance,

$$q_L = \frac{k\sigma_B m_0}{0.0044} \tag{4.27}$$

where σ_B is the standard deviation of integrated absorbance for the desired interval (either computed directly from repeat atomizations or from Eqn (4.22) and, like m_0 and 0.0044, has units of seconds. The major difficulty in using this algorithm is the measurement of the baseline absorbance noise. Although many commercial instruments make it easy to compute characteristic mass, it is not easy to compute the standard deviation of individual or integrated absorbances.

A long-standing controversy has existed as to whether the peak height or area provide the best detection limits. As discussed with respect to the analytical signal (Section 4.2.1), there are valid reasons for employing peak area, instead of peak height, to characterize the transient signal. Although, in the past, analysts have found that peak height routinely gives the best detection limits, this is, in general, the result of using too large an integration interval. Voigtman [28] showed that gated peak integration is optimum when absorbances are summed from approximately one-third height on the leading edge to one-third height on the trailing edge. There is some variation in the optimum integration interval with the peak shape. Voigtman also showed that a matched filter, based on the noise power distribution and the peak shape, provided the best SNR possible. Harnly [29] showed experimentally that gated integration from the peak appearance (when absorbance exceeded 3σ of the baseline absorbance noise) to the point where the absorbance fell to $1/e$ or $1/e^2$ (approximately 91 per cent and 94 per cent, respectively, of the peak area) provided detection limits superior to those computed from peak height.

Another useful representation of the SNR is the relative concentration error, specifically a plot of the relative concentration error (σ_c/c) *vs* c. This value is most appropriate for evaluating the useful working range of a calibration curve, where 'useful' can be defined in terms of relative concentration error. The calibration curve is used to translate A and σ_A into c and σ_c. Thus, the relative concentration error is the inverse of the absorbance SNR corrected by the slope of the calibration curve. As the slope of the calibration curve deviates from linearity, the absorbance error σ_A will intercept a larger concentration range. Consequently, although σ_A/A may become small at higher concentrations, σ_c/c will grow as the calibration curve bends towards the concentration axis.

4.2.4 Accuracy

The *accuracy* of a measurement or result indicates how close the experimental result comes to the true value. An interference is a systematic deviation of the measurement or result from the true value. A more pragmatic definition of an interference is a systematic deviation of the signals of the calibration standard from those of the sample. By this definition, deviations of the analytical absorbance from the true absorbance are not an interference as long as the standards and samples are affected equally. It is the author's opinion, however, that an interference will ultimately produce a less than optimum instrumental response and, therefore, should not be ignored.

In general, there are three classes of interference for GFAAS, chemical, spectral and instrumental. Chemical interferences are a result of interaction of the analyte with the sample matrix. In most cases, interferences arise because the analyte diffuses from the furnace before atomization takes place, i.e. the analyte is lost as a molecule. The use of delayed atomization (with a L'vov platform or contoured tube) and chemical modification reduces many of these interferences, as discussed in Section 1.5.2.

Spectral interferences can be classified as broad-band background attenuation, structured interferences, and stray radiation. Broad-band background attenuation, as previously discussed, is very common in GFAAS and arises from molecular absorption and scattering of the primary source. Molecular absorption in the ultraviolet and visible regions of the spectrum is well documented. In GFAAS, molecular absorption arises from those molecules that are still intact at the atomization temperature of the analyte. An example of the absorption spectra of some of the more stable halide compounds is shown in Figure 4.2. Scattering interferences also result in decreased radiation intensity at the detector. Scattering by atoms or small molecules (particles smaller than the wavelength) is called Rayleigh scattering and is proportional to λ^{-4}. Scattering by larger molecules and particles is classified as Debye (particles of the same size as the wavelength) and mie (particles larger than the wavelength) scattering. Both scattering and molecular absorption occur frequently in GFAAS, especially for more complex sample matrices. This constant potential for background attenuation has spurred the development of four separate background correction systems (Section 4.5).

Structured interferences, or line overlap interferences, occur when the absorption profile of a nonanalyte element or narrow molecular structure overlaps the bandwidth of the primary or secondary radiation source. Line overlap interferences are rare for HCL emission lines (operated at currents within the specifications of the manufacturer) because of their narrow width. The spectral bandwidth (SBW) viewed by the secondary

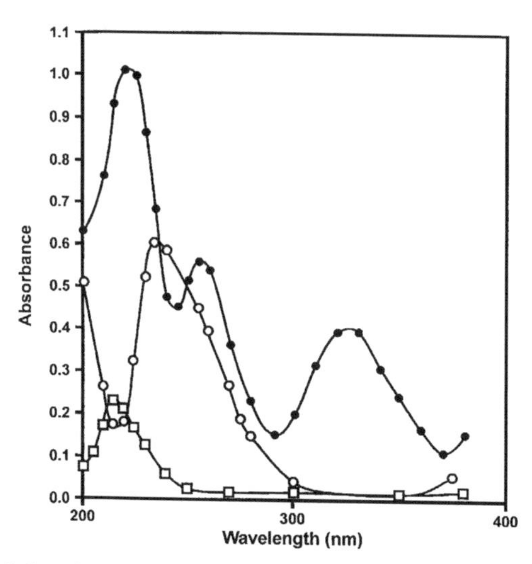

Figure 4.2 Molecular absorption spectra of 5 μL of 0.1 per cent (w/v) solutions of (○) NaCl, (□) NAF, and (●) NaI in a graphite furnace atomizer [30]. (Reprinted with permission from B.R. Culver and T. Surles, Interferences of molecular spectra due to alkali halides in non flame atomic absorption spectrometry, *Anal. Chem.* **47**, 920–921. Copyright 1975 American Chemical Society.)

continuum source used for background correction, however, may exceed the of width of the HCL line by 10–1000 times and, hence, is far more susceptible to line overlap interferences (Section 4.5.2). A compilation of line overlap interferences is listed in Table 4.3. In each case, the wavelength of the analytical line, the interferent, the wavelength difference, and the relative sensitivity of the analytical line are shown. Only those interferents falling within 1 nm of an analyte line have been listed. A relative sensitivity of 1.0 can be expected for the most sensitive, most frequently used analytical lines. In most cases, the interferent must be present at 10–100 times the concentration of the analyte before a significant interference is observed. No attempt has been made to establish the concentration of the interferent necessary to provide comparable absorption to that of the analyte, as it will vary according to the specific atomization conditions.

Stray radiation, or stray light, is explicitly defined as the occurrence of radiation outside the spectral bandwidth of the spectrometer that is passed to the detector. By this definition, stray radiation can only arise from flaws in the grating (ghosts) and unmasked reflective surfaces in the spectrometer (far stray radiation), and it cannot be absorbed by the analyte. For this reason, stray radiation has sometimes been termed unabsorbable stray light. With AAS, however, radiation found within the spectral bandwidth, other than the 'monochromatic' primary source line, will also influence the accuracy of the absorbance of the measurement. This extraneous radiation, sometimes described as polychromatic radiation, has a finite probability of being absorbed by the analyte. Absorptivity is at a maximum at the line center and decreases systematically as the distance from the line center increases. Polychromatic radiation has also been called absorbable stray light. Polychromatic radiation originates from the HCL, secondary analyte lines, lines from elemental contaminants of the cathode, or fill gas lines. In some cases, the broadened wings of the primary analytical line can also qualify as polychromatic radiation. The effect of stray radiation on the computed absorbance can be shown as

$$A + \log\left(\frac{I_0 + I_S}{I + I_S}\right) \tag{4.28}$$

where I_S is the stray radiation. As stated earlier, the definition for stray radiation requires that the absorptivity of the stray radiation is zero. The primary effect of stray radiation is to reduce sensitivity and introduce nonlinearity into the calibration curves; for example, in the presence of 10 per cent stray radiation, the maximum absorbance is 1.0 ($\log[1.0/0, 1] = 1.0$). Stray radiation does not affect detection limits because, at low absorbances, a similar decrease in the absorbance and the baseline absorbance noise is observed. The mathematical description of the influence of polychromatic radiation on the computed absorbance is much more complex. An accurate model requires convolution of the

Table 4.3 Atomic and molecular spectral interferences[a]

Analyte element	*Wavelength (nm)*	*Relative sensitivity*[b]	*Interfering element*	*Wavelength (nm)*	*Wavelength difference*
Al	308.215	0.67	V	308.211	0.004
Ag	328.068	1.00	Rh	328.055	0.013
			Cu	327.396	0.672
			Yb	328.937	0.869
			Zn	328.233	0.165
			PO	–	–
	338.289	0.49	Ni	338.085	0.204
As	193.696	1.00	Ag	193.200	0.496
			Al	193.691	0.005
			P_2	–	–
Au	242.795	1.00	Co	242.493	0.302
	267.595	0.58	Co	267.598	0.003
			BaO	–	–
B	208.893	0.48	Ir	208.882	0.011
Be	234.861	1.00	As	234.984	0.123
Bi	223.061	1.00	Cu	223.008	0.053
	227.658	0.07	Re	227.525	0.133
			Sn	226.891	0.767
	306.772	0.35	OH	–	–
Ca	422.673	1.00	Ge	422.567	0.106
			CuH	–	–
Cd	228.802	1.00	As	228.812	0.010
			PO	–	–
	326.106	0.01	PO	–	–
Co	240.725	1.00	Ru	240.272	0.453
	242.493	0.80	Au	242.795	0.302
			Sn	242.949	0.456
	252.136	0.43	In	252.137	0.001
			Fe	252.285	0.149
			V	251.962	0.174
	384.547	–[c]	CN	–	–
	388.187	–[c]	CN	–	–
Cu	324.754	1.00	Eu	324.753	0.001
			Pd	324.270	0.484
			Ni	324.306	0.448
	327.396	0.45	Ag	328.068	0.672
			Rh	328.055	0.659
	216.509	0.16	Fe	216.677	0.168
			Pb	216.999	0.490
			Pt	216.517	0.008
	217.894	0.15	Fe	217.809	0.085
			Pb	216.999	0.895
			Sb	217.581	0.313
Cr	357.869	1.00	Fe	358.120	0.251
			Nb	358.027	0.158
			CN	–	–
	359.349	0.78	Ni	359.770	0.421

(continued)

Table 4.3 *(continued)*

Analyte element	*Wavelength (nm)*	*Relative sensitivity*[b]	*Interfering element*	*Wavelength (nm)*	*Wavelength difference*
	360.533	0.56	Ni	361.046	0.513
	425.435	0.39	CuH	–	–
Dy	419.484	0.75	Tm	418.762	0.722
	421.172	1.38	Rb	421.556	0.384
Er	248.747	–[c]	Cu	249.215	0.468
			Pt	248.717	0.030
			Fe	248.327	0.420
Fe	246.264	0.09	PO	–	–
	248.327	1.00	Cu	249.215	0.888
			Pt	248.717	0.390
			PO	–	–
	250.113	0.11	PO	–	–
	252.285	0.56	PO	–	–
	271.903	0.29	Pt	271.904	0.001
			Ga	271.965	0.062
			Ta	271.467	0.436
Ga	287.424	1.00	Pb	287.332	0.092
			Fe	287.417	0.007
	294.418	1.18	Fe	294.788	0.370
			V	294.149	0.269
	403.298	0.39	Mn	403.307	0.309
			CuH	–	–
Gd	368.413	1.00	V	368.807	0.394
			Pb	368.348	0.065
	405.937	0.86	Nb	405.894	0.043
			Pb	405.783	0.154
Hg	253.652	1.00	Co	253.649	0.003
			Fe	253.560	0.092
			Po	–	–
Ho	410.384	1.00	In	410.176	0.208
			Tm	410.584	0.201
			Y	410.238	0.146
In	303.936	1.00	Fe	303.739	0.197
			Ni	303.794	0.142
			Sn	303.412	0.524
	325.856	0.95	PO	–	–
Ir	208.882	1.00	B	208.893	0.011
	284.972	0.80	Mg	285.213	0.241
K	404.414	0.01	CuH	–	–
	404.721	0.01	CuH	–	–
Lu	356.784	0.46	Ni	356.637	0.147
			Ru	357.059	0.275
Mg	383.826	–[c]	CN	–	–
Mn	279.482	1.00	Mg	279.553	0.071
			Pb	280.199	0.717
			AlBr	–	–
	279.827	0.78	Mg	279.553	0.071

(continued)

Table 4.3 (*continued*)

Analyte element	*Wavelength (nm)*	*Relative sensitivity*[b]	*Interfering element*	*Wavelength (nm)*	*Wavelength difference*
			Pb	280.199	0.372
	280.106	0.47	Pb	280.199	0.193
	403.307	0.08	Ga	403.298	0.009
Mo	313.259	1.00	Ni	313.411	0.152
			Ir	313.332	0.073
Ni	232.003	1.00	Sn	231.723	0.280
	341.476	0.35	Co	341.263	0.213
			Pd	342.124	0.648
	352.454	0.36	Co	352.685	0.231
	378.353	–[c]	CN	–	–
Os	378.220	–[c]	CN	–	–
P	213.618	1.00	Fe	213.859	0.241
Pb	216.999	2.36	Cu	216.509	0.490
			Sb	217.581	0.582
			Ni	216.555	0.444
			Fe	217.809	0.810
			AlO	–	–
	261.418	0.41	Co	261.413	0.005
	283.306	1.00	Pt	283.030	0.276
			Sn	283.999	0.693
			Os	283.863	0.557
			S_2	–	–
Pd	244.300	–[c]	PO	–	–
	244.791	1.00	Cu	244.164	0.627
			Ru	245.553	0.762
	247.642	0.88	Pb	247.638	0.004
			Fe	247.978	0.335
			PO	–	–
Pr	492.495	0.49	Nd	492.453	0.042
Pt	265.945	1.00	Ir	266.479	0.534
			Sn	266.124	0.179
			Ru	265.962	0.017
	271.904	0.10	Fe	271.902	0.002
	299.797	0.24	Fe	299.443	0.354
			Ni	300.249	0.452
			Pd	300.265	0.468
Re	345.188	0.39	Co	345.350	0.162
	346.046	1.00	Co	345.523	0.323
	346.473	0.58	Co	346.580	0.103
Rh	343.489	1.00	Ni	343.356	0.133
			Ru	343.674	0.185
Ru	349.894	1.00	Co	350.228	0.334
			Ba	350.111	0.217
			Ni	349.296	0.558
Sb	206.833	0.75	PO	–	–
	217.023	–[c]	Pb	216.999	0.024
			Co	217.460	0.437

(*continued*)

Table 4.3 (*continued*)

Analyte element	*Wavelength (nm)*	*Relative sensitivity*[b]	*Interfering element*	*Wavelength (nm)*	*Wavelength difference*
	217.581	1.00	Pb	216.999	0.582
			Co	217.460	0.121
			Cu	217.894	0.313
			Fe	217.809	0.228
	231.147	0.5	Ni	231.097	0.050
	306.800	–[c]	Bi	306.772	0.028
Sc	327.363	0.08	Ag	328.055	0.069
			Cu	327.396	0.033
			Rh	328.055	0.692
	391.181	1.00	Eu	390.710	0.471
Se	196.026	1.00	Bi	195.948	0.078
			Co	197.385	1.359
			Pd	196.011	0.015
			Fe	195.949	0.077
			NO,PO	–	–
	203.985	0.10	Cr	203.900	0.085
Si	250.690	0.36	Fe	250.113	0.577
			V	250.690	0.000
	251.432	0.30	Fe	251.083	0.349
	251.611	1.00	Fe	252.285	0.774
			Co	252.136	0.525
			V	251.962	0.351
	252.411	0.30	Co	252.136	0.175
			Fe	252.285	0.126
Sn	224.605	1.00	Cu	224.426	0.179
			Pb	224.688	0.083
	235.484	0.77	As	234.984	0.500
	286.333	0.55	InBr	–	–
Te	214.281	1.00	Zn	213.856	0.425
			Sn	214.873	0.592
			PO	–	–
Ti	364.268	1.00	Pb	363.958	0.310
Tl	276.787	1.00	Pd	276.309	0.478
			CS	–	–
Tm	371.791	1.00	Fe	371.994	0.203
V	306.638	0.41	Bi	306.772	0.134
			Pt	306.471	0.167
Y	410.238	1.00	Ho	410.384	0.146
			In	410.176	0.062
			Tm	410.584	0.346
Yb	398.799	1.00	Ti	398.976	0.177
Zn	213.856	1.00	Fe	213.859	0.003
			Te	214.281	0.425
			NO,PO	–	–

[a]Data taken from References 31, 32, and 33.
[b]Sensitivity of the analyte line as a fraction the most sensitive analyte line.
[c]No reference for line sensitivity.

instrument slit function, the source intensity function, and the analyte absorption function. There is no obvious solution to the problem of stray radiation. It will be minimized in higher quality spectrometers with better-ruled gratings and improved baffling. Polychromatic radiation can be eliminated by reducing the spectral bandwidth or using a higher-resolution spectrometer. A high-resolution echelle spectrometer, however, can introduce other sources of stray radiation. Radiation from overlapping orders qualifies as stray radiation and may exceed 50 per cent at higher wavelengths (above 400 nm) depending on the slit height and the between-order resolution of the echelle. In addition, the coarsely ruled echelle gratings give rise to unusually large amounts of stray radiation in the far-ultraviolet region (below 250 nm).

Instrumental interferences are systematic biases arising from the inability of the instrument to accurately measure the atomic absorption of the analyte. The rapid, transient nature of the analytical signal and the background attenuation make the accuracy of the computed absorbance in Eqn (4.4) extremely dependent on the timing associated with the instrument electronics [34]. To obtain accurate absorbances, the time interval between the analytical and reference intensity measurements should be minimal, the time constant for the two intensity measurement circuits should be small compared to the rate of change of the analytical signal or the background, the time constant for the two measurement circuits should be perfectly matched, and the time constant for the computed absorbance should be small with respect to the width of the analytical peak.

In general, the greatest accuracy for transient signals is achieved if the time intervals and time constant are kept as small as possible, i.e. the data acquisition frequency is maximized. With a shot noise-limited instrument, however, better SNRs are obtained with longer integration times, or time constants, which require longer intervals between measurements. Consequently, there is a conlict between accuracy and SNR when selecting the optimal instrument parameters. Obviously, parameters suitable for accurate determinations must take first priority. Unfortunately, the selection of parameters that will guarantee accuracy means trying to anticipate the worst possible condition–conditions that are not routinely encountered in the laboratory.

The choice of suitable time intervals and time constants must start with consideration of the rate of change of the analytical absorption and the background attenuation. For intensity measurements, the rate of change of the background attenuation is most critical. In the presence of a large matrix component, the background attenuation has been reported to change by as much as 7–10 s^{-1} [35,36]. Thus, in a 2 ms time interval, the measured absorbance can change by 0.014 to 0.020. This possibly worst case example was obtained by atomizing solutions of 1 per cent sodium chloride solution and sea water from the furnace wall at the maximum heating rate. Ideally, the analytical and reference intensities would be measured simultaneously (double beam in space). With a double beam in time instrument, they should be measured within as short a time interval as possible, or mathematical extrapolation should be used to predict the reference intensity at the time of the analytical intensity measurement. In general, 2 ms is considered the maximum allowable time interval without using mathematical extrapolation.

Time constants for the intensity measurement circuit(s) are similarly restricted. As shown in Eqns (4.1) and (4.4), absorbance, not intensity, is linear with respect to concentration. Time constants that are large with respect to the rate of intensity change will result in integration in the intensity domain and produce nonlinearity and shape dependency [37]. With a double beam in space instrument, the time constants for the paired detection circuits must be perfectly matched to avoid inaccuracies in the computed absorbance [38]. With a double beam in time instrument, the maximum allowable time constant is restricted either by the time interval between measurements or the rate of change of the intensities. If a time interval of 2 ms is used, then the analog time constant must be less than 0.4 ms, because response time is roughly 5 times the time constant [39]. If the time interval is greater than 2 ms, then longer time constants are possible, but nonlinearity will be introduced with rapid changes in intensity. Consequently, the maximum allowable time constant for the intensity measurements is approximately 0.5 ms.

Suitable time constants for filtering of the transient absorbance versus time signals depend on whether peak or integrated absorbance is being measured and the nature of the filter. It has been shown that large time constants will reduce the peak absorbance [34]. The degree of reduction is dependent on the length of the time constant and the temporal width of the analytical signal. It has been suggested that an analog time constant should be 10 times less than the half-width of the analytical peak. With digital filtering, the problem becomes more complex. The reduction of the peak absorbance will depend on the type of filter, the width of the filter compared to the half-width of the peak, and the degrees of freedom of the filter [40]. Inaccuracies can arise if filtering of the standards and samples introduces relative differences in the peak absorbances. This is generally a result of subtle differences in the shapes of the peaks, in which case peak absorbance is not an appropriate measure.

In general, the time constant, or filter, will have less influence on the integrated absorbance [34]. As a filter moves systematically across the peak, every data point is eventually provided with the same weighting as all the other data points. For example, in the case of the sliding average filter, it can be shown mathematically that the integrated absorbance is identical for the filtered and unfiltered signal. Thus, for integrated absorbance, the contribution of each absorbance is not changed significantly, unlike the case for the peak absorbance.

4.3 INSTRUMENTAL COMPONENTS

4.3.1 Sources

Only two types of source have been used with GFAAS, line sources and continuum sources. Line sources, such as HCLs, EDLs, lasers, and laser diodes, emit electronic excitation spectra that are much more intense than predicted by their kinetic temperature. Continuum sources, such as deuterium lamps, hydrogen lamps, metal halide lamps, and xenon arc lamps, are black body emitters which emit recombination spectra and whose intensities are predictable from their operating temperatures. In commercial instruments, HCLs or EDLs are used without exception as the primary source, and continuum sources are used strictly as a secondary source to measure the background attenuation (Section 4.6). Continuum sources have been used as primary sources for GFAAS only in prototype instruments (Section 4.7).

4.3.1.1 Line sources

Line sources, specifically the HCL, are almost solely responsible for the success of AAS and the lack of success of multielement AAS. In general, line sources provide intense and stable spectra specific to the element of interest. The high intensity of these lamps is due to the lack of local thermodynamic equilibrium at the low operating pressures. High electron excitation temperatures are achieved at low kinetic temperatures and input power. The intensity and stability of these sources result in low relative noise levels; short-term baseline standard deviations, in terms of absorbance, are typically 0.0005 or less. A source spectrum specific to the element of interest means that the spectrometer is easily aligned and, barring background attenuation, there is no ambiguity in the analytical determination. The narrow width of the spectral emission line, relative to the width of the analyte absorption profile (Figure 4.3), serves to approximate a monochromatic source, an assumption on which the validity of Eqn (4.2) is based. As a result, medium resolution (approximately 0.03 nm) monochromators are adequate, and calibration curves for AAS are linear for 1.5 to 2.5 orders of magnitude (up to absorbances of 0.3 to 0.5).

Hollow cathode lamps

An HCL employs a high voltage (150–200 V), low current (1–25 mA) glow discharge within a cylindrical cathode [41]. A self-sustaining glow discharge is established at reduced pressures by placing a d.c. electrical

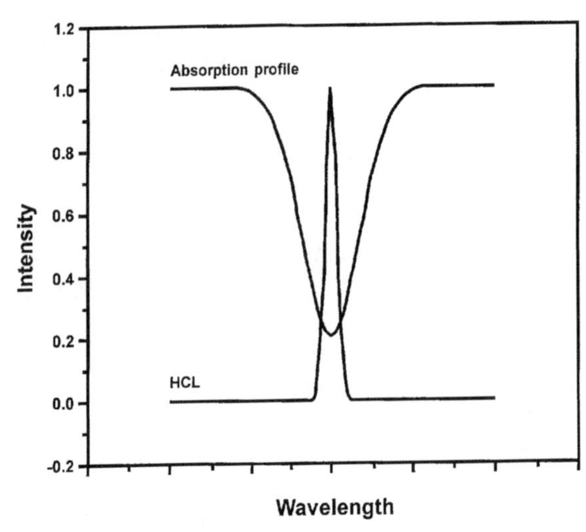

Figure 4.3 Illustration of atomic absorption process for conventional GFAAS. The atomic absorption profile has been drawn 5 times wider than the HCL emission line.

potential between two electrodes. For HCLs, argon and neon are the most commonly used fill gases at pressures of 1 to 5 torr. Construction of cathodes of the appropriate material results in emission sources that are intense, stable, and have a very pure spectrum for the fill gas and the cathode material. The dominant noise is photon shot noise; there is almost no fluctuation noise. As the lamps grow older, a long-term drift component may be observed, but it is readily eliminated using double beam operation.

The difficulty in preparing intense, multielement HCLs arises from the compatibility of cathode materials and the reduction of the solid concentration of each of the components. Consequently, although HCLs make excellent single-element sources, they have been found inadequate as multielement sources. This necessitates a large inventory of HCLs for the analyst who wishes to determine a large number of elements.

The low operating pressures of HCLs and EDLs produce spectral emission lines that are narrow compared to the width of the absorption profile at atmospheric pressure (Figure 4.3). At low pressures Doppler, resonance (Holtzmark), and natural broadening effects (see Section 2.7.1) are the primary factors in determining the widths of the lines. As a result the spectral radiations of HCLs have a FWHH ranging from 0.50 to 2.0 pm for the most commonly determined elements between 200 and 600 nm. Larkins [42] confirmed these widths with experimental measurements for a series of HCLs and EDLs at wavelengths ranging from 194 to 307 nm. At atmospheric pressure, the width of the absorption profiles is dominated by collisional (Lorentzian) broadening (see Section 2.7.1). Few experimental results are available, but calculated absorption widths [43] range from 2 to 8 pm. Cooke *et al.* [44] suggested that these widths are too narrow, and that the assumed collisional cross-sections were too small. They proposed collisional broadening effects that are a factor of 4 larger. Neither set of calculations, however, took hyperfine splitting of the absorption profile into account. For some elements this could be significant. In general, the widths of the absorption profiles range from a factor of 3 to 5 larger than the HCL line widths (Figure (4.3).

The number of neutral atoms sputtered from the cathode, and the intensity emitted from the cathode are roughly proportional to the square of the operating current. At low currents (currents less than the maximum specified by the manufacturer) the sputtered neutral atoms are, in general, restricted to the confines of the cathode. At higher currents, a cloud of ground-state neutral atoms forms immediately in front of the cathode and absorbs some of the emitted radiation from the cathode. This self-absorption process decreases the lamp peak intensity and broadens the line; the FWHH is greater as the height of the emission profile is reduced. As the lamp current is further increased, the population of atoms in front of the cathode increases until the absorption process results in significant absorption in the center of the HCL line (Figure 4.4) and the line is said to be self–reserved Thus, operation of HCLs at currents exceeding the manufacturers suggested maximum can lead to self-absorption and self-reversal of the spectral lines and shortened lamp lifetimes. Self-absorption and self-reversal produce a redistribution of the intensity of a HCL emission line. The region of maximum line intensity is shifted away from the region of maximum absorptivity of the absorption profile. Thus, at high currents, the source is less monochromatic and the average absorptivity decreases. As a result, the sensitivity decreases and the nonlinearity of the calibration curve increases. The nonlinearity introduced by a polychromatic source is superficially similar to that produced by stray radiation (Section 4.2.4 and Eqn (4.28)) but is mathematically different. Differences in the shapes of the calibration curves become more obvious as wider calibration ranges are examined.

Although sensitivity decreases with increasing lamp current, the baseline absorbance noise also decreases. Consequently, the SNR will increase with increasing current, albeit with diminishing improvement. The decreased absorbance noise and the increased SNR obtained with increasing lamp current results in a strong tendency on the part of analysts to operate HCLs at too high a current. Differences in HCL line widths with current, age, manufacturer, and between individual lamps can lead to significant variations in the sensitivity, or characteristic mass. In a recent study, Shuttler *et al.* [14] documented a factor of 2 range in the computed characteristic masses for common elements determined by the users of the Perkin-Elmer Zeeman GFAAS. L'vov *et al.* [45] have shown that this variation in characteristic masses correlates well with the width

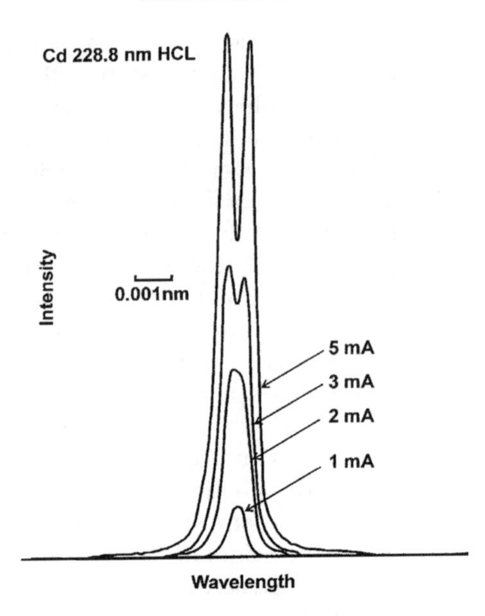

Figure 4.4 Measured profiles showing the variation of line shape and intensity with current for the 228.8 nm line from a Cd HCL [42]. (Reprinted from Spectrochim. Acta, Vol. Vol. **40B**, P.L. Larkins, Atomic line profile measurements on hollow cathode and electrodeless discharge lamps using a high-resolution echelle monochromator, pp. 1585–1598, 1985, with kind permission of Elsevier Science – NL, Sara Burgerhartstraat 25, 1055 KV Amsterdam, The Netherlands.)

of the HCL line as determined by the rollover of the background-corrected absorbance (Section 4.5).

The self-reversal phenomenon described in the preceding paragraphs is the basis for the self-reversed source background-correction method, first applied to conventional HCLs by Baranova *et al.* [46] and later by Smith and Hieftje [47]. Operation of the HCLs at extremely high currents for short periods of time produces emission lines that are dramatically self-reversed. The degree of self-reversal depends on the volatility of the individual element and its sputtering efficiency. Although there is still considerable overlap between the self-reversed lines and the absorption profile, the sensitivity is reduced by factors of 2–10. This process is discussed in more detail in Section 4.5.4.

Electronic modulation or chopping may be used to allow the HCL radiation to be differentiated (through synchronous detection) from the furnace emission component. With electronic modulation, pulse current levels and duty cycles are selected such that the average currents do not exceed the recommended maxima for the lamps. Under these conditions, little broadening of the spectral lines is observed and the stability of the lamp appears unaffected, i.e. no increase in the fluctuation noise is observed. It is not known if this is also true of very high current pulses, which produce severely self-reversed lines. An increase in the intensity variation between pulses would result in an increase in the fluctuation noise component. Although the loss in precision may not be large, if the relative precision of the pulsed intensity exceeds that of d.c. lamp operation, then the result is a noisier baseline absorbance. As discussed earlier (Section 4.2.2), the pulse precision must be close to 0.1 per cent to achieve a baseline absorbance noise of 0.001.

Electrodeless discharge lamps

These lamps were investigated extensively in the 1970s as high-intensity sources for atomic fluorescence spectrometry and have now found a place as AAS sources, primarily for volatile elements and elements for which it is difficult to make cathodes. The metal, metal halide, or the metal plus iodine is sealed in a quartz bulb with several torr of an inert gas (usually argon or krypton). The bulb is mounted inside the coil of a radiofrequency generator which induces a discharge inside the bulb, volatilizing and exciting the metal, and producing intense electronic excitation spectra. Initially, EDLs were restricted to elements with high vapor pressures, such as arsenic, cadmium, selenium, and zinc. With improved technology and understanding of the chemistry, EDLs are being made for elements for which HCLs are difficult to prepare, such as cesium, phosphorus, and rubidium. For the nonvolatile elements, iodine is usually used to enhance the vapor pressure. In each of these cases, the EDL offers significant improvements in intensity and detection limits over an HCL; for example, the arsenic EDL provides almost an order of magnitude improvement in detection limits oyer an HCL.

As EDLs, like HCLs, are low-pressure line sources, the widths of the emission lines are comparable and yield comparable analytical sensitivities. Significant reductions in sensitivity are observed if the lamps are operated at power levels exceeding the maximum recommended by the manufacturer. This reduction in sensitivity is a result of broadening of the emitted lines due to self-absorption. The SNR, however, continues to improve as the power is increased. Larkins [42] used a high-resolution spectrometer to measure the emission profiles of HCLs and EDLs as a function of the operating current and power, respectively. He found that EDLs with the elements present as the metal (arsenic, cadmium, selenium, and zinc) showed rapid increase in line width with increasing power and self-reversal of the resonance lines at low power settings. Those EDLs that appeared to contain iodine (bismuth, lead, antimony, tin, and tellurium) did not show broadening until considerably higher power levels. In most cases, the EDL line widths, at recommended operating power levels, were greater than those for HCLs operated at the recommended current levels.

Like HCLs, EDLs are quiet sources and photon shot noise is dominant. A long-term drift component is sometimes present which is generally tolerable on single beam systems and easily corrected on double beam instruments. Like HCLs, EDLs are primarily single-element sources. Multielement EDLs have not been successful. A separate lamp is required for each element to be determined.

Lasers and laser diodes.
Lasers are, potentially, an ideal source for AAS. They provide intense beams of radiation with bandwiths less than those of HCLs. Unfortunately, they are also quite expensive, large, unwieldy, and lack the reliability and wavelength range necessary for application to the field of AAS.

In recent years, laser diodes have been developed which are dramatically smaller, less expensive, and more reliable [48,49]. Unfortunately, wavelength coverage in the ultraviolet region is still problematic. Semiconductor laser diodes of InGaAsP and AlGaAs provide intense radiation at narrow bandwidths but are restricted primarily to the red and infrared (above 630 nm) spectral range. Lower wavelengths can be reached using second harmonic generation (SHG) but significant intensity (a factor of 100–1000) is lost. The high intensity of the laser diodes in the primary and SHG modes, results in extremely low absorbance noise (Eqn (4.20)) and makes the fluctuation noise of the source limiting. Wavelength modulation (obtained by adding a small, sinusoidally modulated current to the diode current) has been used in both modes to significantly improve detection limits. With SHG and wavelength modulation, detection limits for aluminum (396.2 nm) have been achieved that are comparable to those for LSAAS (with a HCL at 309.3 nm) [48]. At this time, laser diodes do not appear ready to replace HCLs. As solid-state technology continues to improve, however, these sources may eventually become ideal line sources.

4.3.1.2 Continuum sources

The use of a continuum source for AAS was first suggested by Walsh [1]. With a polychromatic continuum source, only one lamp is necessary rather than a collection of HCLs. Substitution of a continuum source for HCLs, however, has generally proved unsuccessful. This is because of the nature of the emitted radiation of the two sources (pseudomonochromatic versus polychromatic), the noise characteristics of the lamps, and the tendency to use continuum sources in the same optical arrangement as HCLs. Compared to line sources, continuum sources are nonspecific, unstable, and insensitive. Commercially, continuum sources have only been used to provide highly polychromatic radiation for background correction (Section 4.5.2). The continuum sources generally used for AAS are black body emitters; hence the emitted intensities are predictable from their operating temperature. With input power comparable to that used for HCLs, the intensity for continuum sources in the ultraviolet region are, at best, comparable to that of HCLs. Intensities exceeding those of HCLs can be obtained through the use of high-pressure arc lamps and high operating powers. Unfortunately, high-pressure arc lamps have a high fluctuation noise component and are thus inherently much less stable than HCLs. At high intensities, wavelength modulation is necessary to reduce the fluctuation noise of the lamps.

Analytical sensitivities are greatly reduced when continuum sources are substituted for HCLs for use with a medium resolution monochromator. From Figure 4.5, it can be seen that a considerable amount of radiation from the continuum source will fall outside the absorption profile and have no possibility of being absorbed. This polychromatic radiation will act like stray radiation and reduce the sensitivity (Eqn (4.28)). The amount of polychromatic radiation reaching the detector is dependent on the SBW of the spectrometer. The SBW of a medium resolution (0.03 nm) spectrometer (routinely used for LSAAS) is approximately 10 times the width of the absorption profile and 30–50 times the width of a HCL emission line. Consequently, use of a continuum source requires a high-resolution spectrometer in order to obtain the sensitivities and calibration curve linearities achieved with HCLs.

Given the general characteristics of continuum sources it is easy to understand their lack of use in modem AAS. With proper design of the

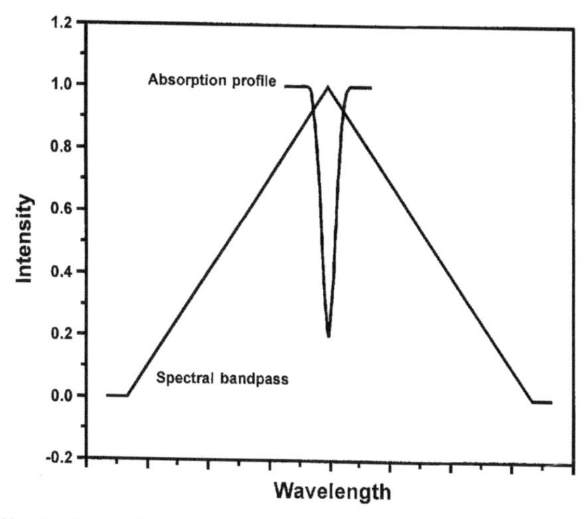

Figure 4.5 Illustration of atomic absorption of a continuum source. The spectral bandwidth of the spectrometer limits the transmitted intensity and has been drawn to be 12 times the width of the absorption profile.

instrument, however, use of continuum sources has the potential for considerable impact, as discussed in Section 4.7.

Deuterium lamps

Deuterated hydrogen discharge (D_2) lamps are the most commonly used secondary continuum source for background correction in commercially available spectrometers. They emit continuum spectra between 185 and 370 nm. They are, however, very weak at the higher wavelengths and considerable reduction of the intensity of the primary source is required to balance the intensities. In operation, deuterium lamps provide roughly three times the intensity of hydrogen lamps, and consequently hydrogen lamps are seldom used. Deuterium lamp designs consist or either a d.c. arc discharge in deuterium at a pressure of several torr or an HCl with deuterium as the fill gas. Deuterium HCLs are significantly weaker than the arc discharge lamps, but they are also significantly quieter. A d.c. arc lamp experiences some wander of the discharge, and the noise power spectrum shows a significant fluctuation noise component. When used in a double beam system with sampling frequencies in excess of 50 Hz, they are shot noise limited.

The strong ultraviolet component (below 400 nm) of the D_2 lamps produces ozone in the light path. Purging the interior of the instrument with an inert gas is therefore desirable, to prevent corrosion of the optics. In addition, the ozone must be removed from the environment to protect the analyst.

Metal halide lamps

To obtain continuum radiation at wavelengths greater than 300 nm, metal halide or quartz halogen lamps are used. These lamps employ high-pressure mercury or xenon discharges with added metal halides. The combination of pressure-broadened electronic emission, molecular emission, and recombination emission provide a pseudotcontinuum spectrum. Commonly used metal salts are iodides or bromides of sodium, scandium, lithium, dysprosium, thallium, cesium, holmium, thulium, indium, and gallium. The noise power spectra of the metal halide lamps are similar to the D_2 lamps, although the fluctuation noise component may be stronger due to the higher pressures and arc wander.

Xenon arc lamps

These lamps have been used quite extensively in prototype instruments where a continuum source is the primary source. They are by far the most intense of the continuum sources. Traditionally, they employ a d.c. arc in 5–20 atm of xenon and are classic, black body emitters. As the operating temperature of the lamps goes up, the wavelength of maximum emission is shifted to a lower wavelength and short-wavelength emission is enhanced. Interest in continuum sources with intense emission from 190 nm to 600 nm is a direct result of interest in the development of multielement GFAAS. The lamps are available in a wide variety of operating powers, ranging from tens to thousands of watts. From a spectroscopic point of view, the higher power lamps are ineffective. As power increases, the arc gap increases, and the brightness (intensity per stereradian) decreases. As the lamp radiation must ultimately be passed through the small entrance aperture of the spectrometer, the diffuse source produced at higher powers is of little practical use.

Of greatest spectroscopic interest is a subset of xenon arc lamps known as short-arc lamps. The very small arc gap of these lamps results in a very intense, highly defined ultraviolet emission region at the tip of the cathode. This region is more easily imaged onto the entrance slit of the spectrometer, and thus gives higher intensities at wavelengths below 300 nm. Xenon short arc lamps, operated at 150–500 W are the most intense continuum sources available for spectroscopic applications. There are two major limitations of xenon arc lamps. First, they are subject to arc precession and wander, and thus have a strong fluctuation noise component (Figure 4.6). Short-arc lamps traditionally demonstrate larger fluctuation noise components than those with larger gaps. This large noise component necessitates double beam operation. Second, radiated intensity in the far ultraviolet region (below 280 nm) is weak. As the wave-

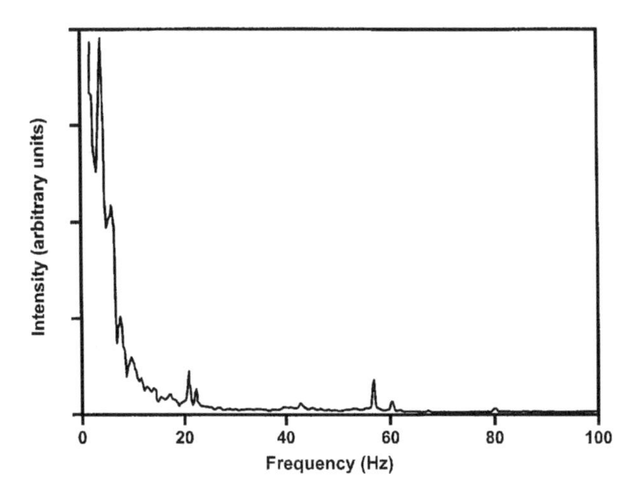

Figure 4.6 Noise power spectrum of 150 W Cennax Xenon arc lamp [50]. (Reprinted with permission from R.L. Cochran and G.M. Hieftje, Spectral and noise characteristics of a 300-watt Eimac arc lamp, *Anal. Chem.* **49**, 2040–2043. Copyright 1977 American Chemical Society.)

length decreases from 300 nm to 200 nm, the intensity diminishes by almost an order of magnitude.

Two types of xenon short-arc lamp are commercially available. The more conventional variety employs electrodes sealed in a glass envelope with an external reflector, and the second variety uses an internal reflector and a single, circular window as the sealed lamp housing. The former are generally operated with the electrode axis in a vertical position, perpendicular to the optical path. The latter has electrodes located on the axis of a parabolic reflector. The anode lies at the apex of the parabola and the cathode is held in place just in front of the window by three struts. The window is made of sapphire (to transmit radiation below 400 nm) and the reflector is a silicon-based ceramic with an aluminum coating. The matching composition of the window and housing makes it possible to operate the lamp with only air cooling. With this design, the lamp is operated with the electrode axis parallel to the optical path. The internal reflector lamp has been used extensively as primary sources in prototype multielement AAS instruments in conjunction with wavelength modulation and high-resolution spectrometers (Section 4.7).

The emitted intensity of the 300 W internal reflector lamp described above has been shown to be more intense than commercially available Xe arc lamps with external reflectors operated under comparable conditions. The 300 W internal reflector lamp has also been shown to be slightly more intense, at 200 nm, than the 500 W version of the same lamp [51]. This illustrates the importance of the size of the arc gap. The bright spot of the 300 W lamp is slightly more intense than that of the 500 W, despite having 40 per cent less power. The 300 W internal reflector lamp has also been shown to be more intense than HCLs, operated at the manufacturers maximum recommended current (Fig, 4.7). Using the same dispersion optics and detector, with mathematical correction for the difference in the convoluted profiles arising from the bandwidth differences between the continuum and line sources, the 300 W xenon arc lamp was found to be 3–100 times more intense than HCLs between 193.7 nm and 589.6 nm [52]. The lack of intensity associated with the use of the 300 W xenon arc lamp for multielement GFAAS comes from slit height and transmission limitations of the echelle spectrometer. The parabolic reflector of the internal reflector lamp produces a collimated beam of light which is passed through the furnace. Radiation emerging from the furnace is focused onto the entrance slit of the monochromator using a lens one focal length from the spectrometer. The lens is chosen so that the ratio of the focal length to 6 mm (the internal diameter of the furnace) is equal to the F-number of the spectrometer. Maximum intensity is achieved by locating the lamp as close as possible to the furnace and adjusting the vertical and horizontal position until maximum intensity is achieved, i.e. locating the 'hot spot' of the lamp.

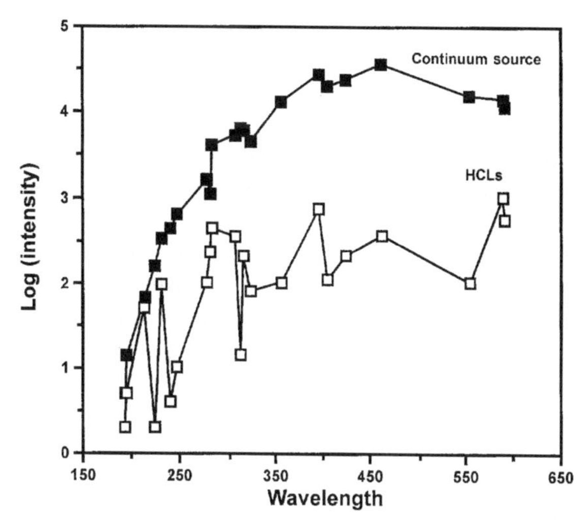

Figure 4.7 Comparison of intensities for HCLs and a 300 W Cermax xenon arc lamp at commonly used wavelengths, with the same spectrometer, detector, and entrance and exit apertures [52].

4.3.2 Atomizers

The function of the atomizer is to volatilize and atomize the analyte in a reproducible manner, independent of the sample or standard matrix. Any dependence of the atomization efficiency on the matrix may potentially result in a variation in detectability between the standards and the samples, i.e. an interference. The generation function of the analyte is difficult to totally dissociate from the sample matrix. If, however, the generation function is fast compared to the loss function and if the loss function is constant over the duration of the analytical signal, the residence time of the individual atoms will be constant and the analyte absorbance signal will be independent of the analyte matrix and the generation function.

The dominant loss function of most enclosed furnaces, with no inner flow of inert gas, is gaseous diffusion. Under nonideal conditions, the loss function is dependent on convective movement of the internal gases. If the convective flow is due to the sample matrix, then this may also result in an interference, a variation in detectability between the standards and the samples. Under ideal conditions, diffusion is limiting. The diffusion coefficient will be constant, provided the temperature is constant. Thus, in principle, the temperature of the furnace should be uniform along its length (spatially isothermal) and for the duration of the presence of the analyte (temporally isothermal). In addition, the furnace must be heated rapidly. This ensures that the generation function will be rapid compared to the diffusion loss function.

Restriction of the atomization process to a small, confined volume with a long path length maximizes the gaseous analyte concentration, the residence time, and the analyte absorbance. Guell and Holcombe [53] showed that theoretically, the ideal atomizer, yielding the best SNR, would be approximately 40 mm long and 5 mm in diameter. However, 80 per cent the optimum SNR could be achieved with furnaces 25 mm in length. In reality, furnace design is determined by the practical limits of the power supply, the optics, the need to put a magnet around the furnace, and the need for temporal and spatial isothermality. Still, the atomization efficiency and the long residence time in the observation volume (compared to other analytical methods) are the strengths of a furnace atomizer.

There are many ways to construct atomizers with the characteristics just described. Designs have varied with respect to heating (resistive and inductive), material (carbon, metal, and nonconductive ceramics), configuration (tubes, cups, and filaments), direction of current flow (longitudinal and transverse), tube shape (uniform and contoured), and number of components (single- and two-step atomizers). This section does not present a comprehensive review of every atomizer ever developed, nor is it restricted to commercially available designs. Designs are considered based on their utility and impact in the field.

4.3.2.1 Graphite furnace atomizers

The graphite furnace or graphite tube is by far the most commonly used of the furnace atomizers. The design and size of the graphite furnaces, as shown in Table 4.4, are almost the same for all manufacturers. In general, graphite furnaces are made of polycrystalline electrographite with a pyrolytic graphite coating. All graphite furnaces are resistively heated. The longitudinally heated furnaces are hollow cylinders or tubes with a 6 mm inner diameter, a 9 mm outer diameter, and a 28 mm length (Figure

Table 4.4 Dimensions for commercially available graphite furnaces [54]

Atomizer	*Tube radius (mm)*	*Tube length (mm)*	*Injection port radius (mm)*
HGA[a]	2.950	28.00	1.000
THGA[a]	2.892	17.55	0.900
GTA-96[b]	2.500	25.00	1.000
SP-9[c]	2.500	30.00	1.000
PU 9390[c]	3.250	30.00	0.293
ICC[d]	2.800	19.00	0.850

[a]Perkin–Elmer Corporation.
[b]Varian Associates.
[c]Pye Unicam Ltd.
[d]Ringsdorff-Werke GmbH.

4.8(a)). Two varieties of transversely heated furnaces are commercially available. Both have a 6 mm inner diameter and a 19 mm length. The primary difference between the two is the robustness of the exterior, as shown in Figure 4.8 (b and c). The more massive furnace, the transversely

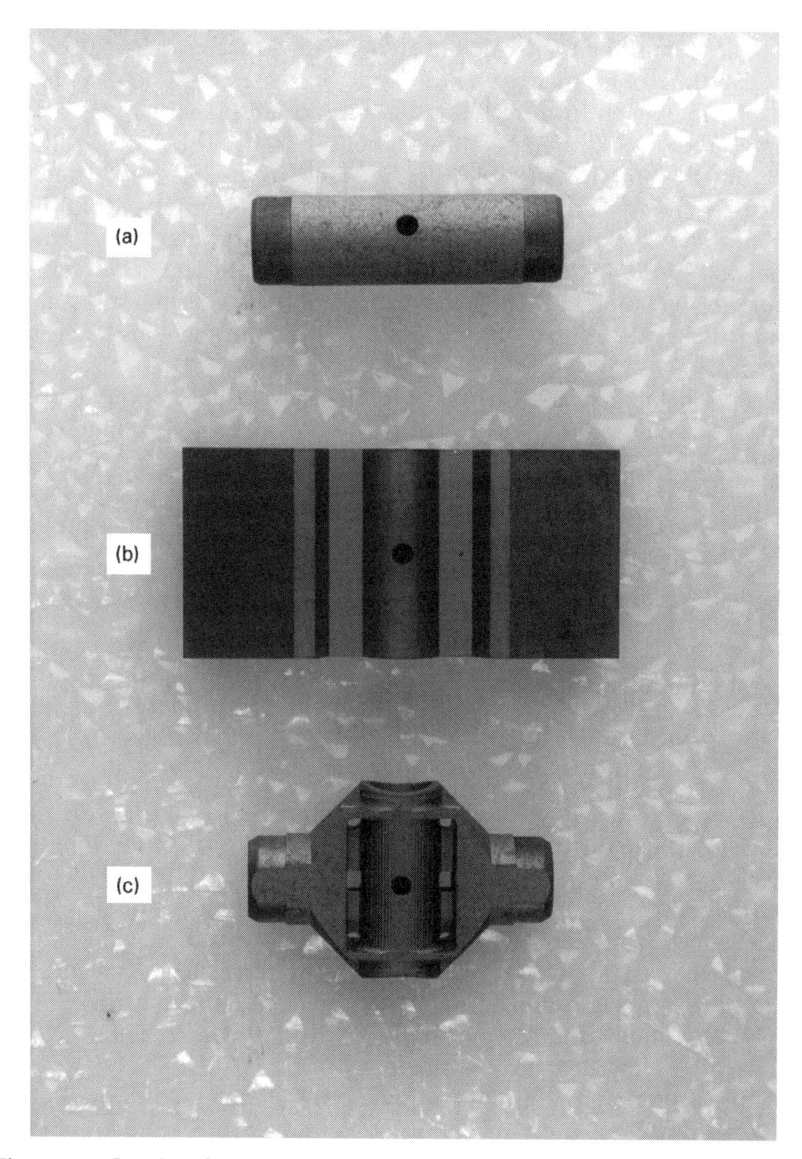

Figure 4.8 Graphite furnace atomizers: (a) An end-heated tube similar to that used in most commercially available instruments, (b) a transversely heated graphite atomizer, and (c) an integrated contact cuvette.

heated graphite atomizer (THGA), is designed to resist compression from the sides, as the furnace is pneumatically clamped between the two electrical contacts (Figure 4.8(c)). The more delicate furnace, the integrated contact cuvette (ICC), is designed to have each wing separately clamped from above and below in a vice-like apparatus (Figure 4.8(b)).

In every furnace, a sample injection hole is located midway between the ends. The furnace is either sealed in an inert atmosphere or enveloped in a protective flow of an inert gas. Both approaches are designed to prevent atmospheric oxygen from reaching the furnace and destructively attacking the carbon at high temperatures (above 1800 K). The furnace life is a direct function of the efficiency with which oxygen is kept from the furnace. In nonsealed furnaces employing a protective gas flow, an inner and outer flow are usually employed. The outer flow, around the outside of the furnace, is continuous, whereas the inner flow, inside the furnace, is stopped during the atomization cycle. Thus, the analyte loss function is determined by diffusion rather than the convective flow of the gas.

Resistive heating of the furnaces permits easy control of the temperature for thermal pretreatment, volatilization, and atomization of the analyte. Most manufacturers allow a minimum of five steps with programmable temperatures, ramp times, hold times, gas flows, and alternate gas selections for each step. In this manner the furnace heating program can be fine tuned to provide the optimum conditions (drying, pyrolizing or oxidizing, volatilizing, and atomizing) for the analyte and the matrix. The atomization temperature is selected to provide efficient atomization and maximum residence time of the analyte in the furnace. Although the temperature must be high enough to ensure that complete atomization is achieved, excessively high temperatures will increase the diffusion rate and decrease the residence time. During the atomization stage, the furnace is heated as rapidly as possible to ensure that the generation function is rapid compared to the loss function. The temperature during this rapid heating step is usually monitored using optical, electrical, or thermometric feedback systems to maximize the heating rate and minimize the overshoot. The optical feedback method matches the furnace emission to the predetermined emission level of the desired atomization temperature. Electrical feedback systems use the resistance of the furnace, current flow, and/or power consumption to maintain a constant temperature. With the thermometric method, a thermocouple in physical contact with the tube provides a temperature measurement.

The desire for temporal isothermality has been addressed through three separate approaches: graphite platforms, contoured tubes, and graphite or tungsten probes. In all three approaches, the concept is to delay volatilization of the analyte until the furnace atmosphere has reached a stable temperature. This concept is discussed in Sections 1.4 and 1.5.

Graphite Platforms

Graphite platforms are made from pyrolytic graphite, and sit in the center of the furnace immediately below the sample injection hole. The sample is deposited onto the platform where the thermal pretreatment and volatilization take place. At the start of the atomization cycle, the furnace wall heats rapidly to a constant temperature. The gas temperature lags behind the wall temperature during the rapid heating step and then stabilizes at a slightly lower temperature than the wall. As the platform is heated radiatively, the platform temperature lags behind that of the wall and the gas. Ideally, the platform reaches a temperature sufficient to volatilize/atomize the analyte only after the gas temperature has stabilized. Consequently, the analyte will experience a constant temperature and will exhibit a constant diffusion coefficient (see Section 1.5). The heating process of the platform is dependent on how it is mounted in the furnace. Falk and Glissman [55] demonstrated that if the platform contacts the tube wall at different distances along the current path, the platform will experience some resistive heating. Currently there are four general types of platforms: captured, free, forked, and pedestal. The captured platforms are held in place by parallel grooves etched into the pyrolytic graphite. The free platforms sit loosely on the wall surface. After several firings, however, the platforms may become attached to the wall as a result of the pyrolized carbon generated during atomization. Forked platforms are held in position through the spring-like action of the fork in one end of the platform. The other end rests lightly against the wall. Again, the free end is likely to become firmly attached to the wall after several atomizations. The pedestal platform, as its name implies, is attached to the wall at a single contact point, although it is not in the center of the platform. This last platform has a theoretical advantage, but from a performance standpoint there is little difference between it and the forked platform.

Contoured furnaces

These have been machined so that the wall at the ends of the furnace is thinner than it is at the center. The heating rate of the graphite is inversely related to the thickness – the larger the cross-section, the less the current density, and the less the resistive heating. Thus, the furnace ends will heat most rapidly to a constant temperature, followed by the gas temperature. The middle region of the furnace, where the dried analyte is located, will heat at a slower rate and the analyte will be volatilized into a temperature-stabilized gas. In practice, this approach is not as successful as the platform at retarding the volatilization of the analyte and providing temporal isothermality.

Graphite probes

Probes made of graphite (and also tungsten) are seemingly an ideal tool for achieving temporal isothermality. The analyte is deposited through the injection port onto the probe while it is inserted in the furnace. While still in the furnace, the sample can be subjected to thermal pretreatment. The probe is then withdrawn and the furnace is heated to the desired temperature. After a stable gas temperature has been achieved, the probe is re-inserted into the furnace for volatilization/atomization of the analyte under temporally isothermal conditions. Two approaches have been used for introducing the probe into the furnace – through a hole in the side of the furnace and through the end of the furnace. Two major disadvantages of the first approach are the disruption of the spatial isothermality and the reduction of the analyte residence time by the hole in the side of the furnace.

Inserting the probe through the end of the furnace avoids the disadvantage of reduced analyte residence time encountered with the side entry approach, but spatial isothermality remains a problem. The cold mass of the probe serves as an optimal site for condensation of the analyte. Sensitivities for nonvolatile elements are significantly reduced. In addition, there is no discernible advantage of probe atomization over platform atomization in reducing interferences [56]. Reductions in sensitivity of the nonvolatile metals were also encountered with tungsten and tantalum probes due to dissolution of the analyte in the probe metal. With atomization temperatures in excess of 2500 °C, dramatically reduced signals were observed for molybdenum. At these high temperatures, the probes became slightly malleable. It appears that the molybdenum was amalgamated with the probe and never volatilized.

The concern for spatial isothermality has been addressed through the use of transversely heated furnaces, the THGA and the ICC. Longitudinally heated furnaces may experience temperature drops of 500–1000 °C between the middle and the ends of the furnace. These cooler ends can be sites for condensation of the analyte as well as matrix components. These cooler ends demand elevated atomization temperatures to produce satisfactory analytical signals and can be a source of residual analyte (memory) between successive atomizations. Frech *et al.* [57] showed that transversely heated furnaces have spatially uniform temperatures, permit lower atomization temperatures, allow the use of platform atomization for even the most nonvolatile elements, and dramatically reduce the memory effects. This shorter furnace (19 mm versus 28 mm for the longitudinally heated furnace) also gave poorer sensitivities. Recently, Frech *et al.* [58] have shown the existence of a convective gas flow at the stabilized temperature. Heated gases leaving the furnace through the injection port tend to pull cooler gases in through the ends of the furnace. The presence of this cooler gas near the tube ends can produce condensation of the analyte and the matrix as microdroplets. With the use of a chemical modifier such as palladium, the microdroplets of the modifier serve as a condensation site for the analyte. This phenomenon is called matrix trapping. Frech and L'vov [59] proposed a design modification that minimizes matrix trapping and enhances the sensitivity of the THGA. They inserted small graphite rings (end-caps) into both ends of the furnace. These rings restrict the openings at both ends and conduct current. As a result, matrix trapping is dramatically reduced, the characteristic mass is reduced by an average of approximately 2, and the overshoot during the rapid heating step is reduced [60]. Use of end-caps reduces the solid angle of light that can pass through the furnace but does not appear to have a significant effect on the SNR. The reduced solid angle makes furnace emission more problematic and requires careful alignment of the instrument optics.

Alternative carbonaceous materials have been used for the construction of furnaces, but none have shown significant advantages over pyrolytically coated electrographite. Furnaces composed of glassy carbon had the perceived advantage of greater inertness and longer lifetime. The electrical conduction properties of glassy carbon furnaces, however, were significantly different than those of the conventional furnaces. Glassy carbon furnaces heated more slowly and required a higher potential from the power supply. Consequently, they were not readily substituted for conventional furnaces. Most importantly, there was no significant reduction in any of the interference levels.

4.3.2.2 Nongraphitc atomizers

The less-attractive features of graphite atomizers (carbide formation, low heating rate, and diffusion of the analyte into the graphite substrate) have led to the investigation of nongraphite surfaces and atomizers. These investigations have taken two approaches, modification of the existing graphite tube, and construction of metal atomizers.

Modifications of conventional graphite tubes have included impregnating or coating the surface with metal, lining the tube with foil or metal liners, and depositing the sample on metal platforms or in metal cups. These approaches have removed interferences associated with specific elements and matrices. In general, the metal surface reduces carbide formation, enhances sensitivity, reduces tailing of the analytical peaks, and reduces interferences. Each of these approaches, however, has limitations, which has negated its general application.

The metal atomizers have consisted of metal strips, filaments, coils, crucibles, and tubes [61,62]. The metals used have been molybdenum, tantalum and tungsten. For some elements, metal atomizers provide enhanced sensitivities and lower atomization temperatures. In general, peak height detection limits are comparable to those for conventional graphite furnaces, but peak area detection limits are generally worse. Almost all the research on metal atomizers has come from prototype furnaces, although at least one tungsten strip atomizer and one metal tube atomizer has been offered commercially (see Section 1.4).

Totally metal atomizers (strips, crucibles, and microtubes) can achieve higher heating rates (as much as 3000 K s^{-1}) with much lower power than graphite atomizers. Heating rates must be carefully controlled, however, to prevent distortion of the atomizer. This is particularly true of the metal strip atomizers where the height of the observation zone above the surface is critical to sensitivity. Distortion can also be a problem for metal tubes with a smaller internal diameter (less than 3 mm). The maximum achievable temperatures of the metal atomizers are also problematic. The bulk of the experimental data is for volatile and intermediately volatile elements, although some results for vanadium have been reported. Besides the structural integrity of the atomizer, the problem of the analyte dissolving in the atomizer, i.e. amalgamation, must be considered. There is currently a shortage of data with which to evaluate these questions. The analytical signals for the totally metal atomizers are much more rapid than those for graphite atomizers. Typically, the FWHH is around 0.1 s for intermediately volatile elements such as copper, manganese and nickel. Usually the analyte appears during the rapid heating of the furnace and convective losses are a problem. It can be seen that the extremely narrow FWHH of the analytical peaks would require optimized integration intervals in order for the peak area detection limits to be comparable to the peak height detection limits.

Recently, a commercially available metal furnace has been described that resembles a transversely heated graphite atomizer. The metal leads have been split to provide forked contacts [62]. This design provides an 'autoplatform effect'. Pyrometric studies showed that the ends of the furnace heated faster than the center of the furnace and that the atomization of the analyte was delayed in relationship to the wall temperature. It was shown that, as with graphite furnaces, palladium served to stabilize the analyte and allowed the use of higher temperatures for pyrolysis and atomization. With the palladium modifier, characteristic masses approached those of graphite atomizers. The peak area detection limits were generally worse than those for graphite atomizers, with some notable exceptions such as selenium.

4.3.2.3 *Two-step atomizers*

The two-step furnace is described in Section 1.4. It is a double furnace consisting of an integrated contact cuvette (ICC) mounted directly above a carbon cup [63]. A hole in the bottom of the ICC (directly below the sample introduction hole) admits the sample vapor volatilized from the cup. The sample is introduced into the cup through the sample introduction hole. In the original design, the top of the cup was firmly inserted into the hole in the ICC, thus eliminating any gap through which the analyte might escape. This physical contact of the two components meant that the temperature control of the cup was not totally independent of the ICC temperature. Alternate designs have employed a 1–2 mm gap between the cup and the ICC without reducing the transport efficiency of the analyte. This design allows more autonomous temperature control of the two components. The two-step furnace is designed to separate the volatilization and atomization processes and to obtain spatial and temporal isothermality. In operation, the ICC is first heated to the necessary temperature to ensure atomization of the element of interest. The cup is then heated to volatilize the sample. As the ICC is spatially and temporally isothermal, the rate of volatilization of the analyte from the cup is not important. The ICC and cup temperature programs can be optimized independently. It has been shown that lower atomization temperatures can be used with the two-step furnace than are used for conventional furnaces.

4.3.2.4 Sample introduction devices

The material to be analyzed is usually introduced into the graphite furnace in a liquid form. Volumes of 1–100 μl have been used, although most platforms typically use 20–40 μl. The extreme sensitivity of the instrument and the small sample sizes necessitate the use of an autosampler. Conventional commercially available autosamplers deliver discrete volumes of 1–100 μl with analytical precisions of approximately 1 per cent relative standard deviation (RSD), for aqueous solutions. Alternatives to a discrete volume delivery are aerosol depostion, flow injection (FI), in situ trapping for cold vapor generation systems, direct solid sampling, and slurry sampling.

With aerosol deposition, a pneumatic nebulizer is used to generate an aerosol that is sprayed through the dosing hole to impact on the furnace wall [64]. As with all pneumatic nebulizers, only 10 per cent (at the most) of the solution will actually be nebulized. The rest of the solution is lost down the drain of the spray chamber unless a recirculating system is used. Maximum deposition efficiency is achieved with a furnace temperature of slightly greater than 100 °C. Using this technique, analyte can be deposited for any length of time in the furnace. This has the advantage of improving concentration detection limits. Thus, this technique is best suited for samples with low analyte concentrations where sample volume is not a constraint. Flow injection (also discussed in Section 1.5.2.2) is a versatile technique, which allows the sample to be treated in a variety of ways prior to injection into the furnace. The most popular approach uses solid sorbents to concentrate the analyte [65]. The most concentrated fraction of the eluate is then deposited in the furnace (eluate zone sampling). Using this methodology, enhancement factors from 20- to 25-fold have been achieved with loading times of 60 s. Obviously, greater enhancements can be achieved with longer loading times provided the analyte concentration is low enough and/or the column capacity is high enough. In addition, separation of the analyte from the sample matrix and separation of the analyte by oxidation state have been reported. A wide variety of FI-GFAAS prototypes have been reported, many custom designed for specific problems. At least one company has developed an FI accessory for use with their furnace. Although numerous Fl stand-alone systems are available, coordination of timing between the Fl and the GFAAS can be a major obstacle.

In situ trapping has been used very successfully with cold vapor techniques to deposit volatile elemental species (e.g. hydrides and carbonyls) directly into the furnace [66]. The overall approach is very similar to that of aerosol deposition, except the aerosol generation mechanism is replaced by the cold vapor generation process. Maximum deposition efficiency is element dependent but the furnace temperature generally ranges from 200 to 500°C.

Introduction of dry solids into the furnace has been achieved using a variety of sampling boats or cells [67]. A number of these are commercially available. This process avoids the sometimes difficult step of digesting the sample. It does, however, place considerable emphasis on the homogeneity of the sample. This technique is discussed in detail in Chapter 6. Introduction of solids as finely ground particles carried in liquid medium, or slurry, is more easily automated [68]. With slurry sampling, conventional autosamplers can be used to inject samples into the furnace. An agitation device is needed to keep the particles evenly dispersed in the liquid phase until the sample can be extracted by the autosampler. This can be a magnetic stirrer, a vortex mixer, an ultrasonic agitator, or a gas bubbler. Of these devices, only the ultrasonic agitator is commercially available. Slurry sample introduction, like solid sampling, places emphasis on the homogeneity of the sample. This technique is also discussed in detail in Chapter 6.

4.3.3 Spectrometers

The requirements of the spectrometer for conventional GFAAS are relatively undemanding. The sensitivity, specificity, and linearity of the absorbance measurement are dictated by the stability and the narrow width of the HCL emission line (Section 4.3.1). The role of the spectrometer is to isolate the HCL emission line of interest and limit the amount of stray light reaching the detector. For this purpose a 'medium' resolution spectrometer, with spectral bandwidth of approximately 0.03 nm, is adequate. For routine operation, spectral bandwidths ranging from 0.1 to 1.5 nm are usually employed.

When a continuum lamp is used as the primary source, the specifications for the spectrometer are more demanding. The spectral bandwidth of the spectrometer replaces the HCL line in determining the width of the spectral region over which the absorption measurement is made. Thus, the spectral bandwidth of the spectrometer determines the analytical sensitivity and the linearity of the calibration curve. The most commonly used high-resolution spectrometers use echelle gratings and Littrow prisms to provide a highly resolved two-dimensional spectrum. For a typical echelle spectrometer, the spectral bandwidth ranges from approximately 0.002 nm (at 200 nm) to 0.005 nm (at 500 nm) with 25 μm slit widths. Thus, the spectral bandwidth is comparable to the width of the analyte absorption profile and a factor of 3–5 broader than typical widths of HCL emission lines.

4.3.3.1 Monochromator characteristics

A spectrometer disperses polychromatic light with respect to wavelength, isolates the wavelength region(s) of interest, and measures the transmitted intensity using a photoelectric detector. If the spectrometer uses a single exit slit in the focal plane, it is called a monochromator. If multiple exit slits are employed, it is a polychromator. If the spectrometer is employed to ratio two intensities, the reference intensity and the sample intensity, to determine an absorbance, the system may be called a spectrophotometer. A spectrometer that employs photographic film or a photographic plate as a detector in the focal plane is a spectrograph. Spectrometry refers to the measurement process using a spectrometer, and spectroscopy refers to the measurement process using a spectrograph. There are numerous excellent sources, which provide full mathematical descriptions of spectrometers. A brief summary of these relationships is provided here as a basis for comparison of conventional and echelle spectrometers and their effect on the SNR.

A spectrometer most commonly consists of an entrance aperture, a collimating mirror, a diffraction grating, a focusing mirror, and an exit aperture. The use of separate mirrors for collimation and focusing is called a Czerny–Turner configuration. The most common parameters used to characterize a spectrometer are dispersion, spectral bandwidth, resolving power, and luminosity.

A diffraction grating is a planar or concave surface that is ruled with closely spaced grooves (Figure 4.9(a)). The wavelength dependence of the dispersion of radiation by the grating is described by the general grating

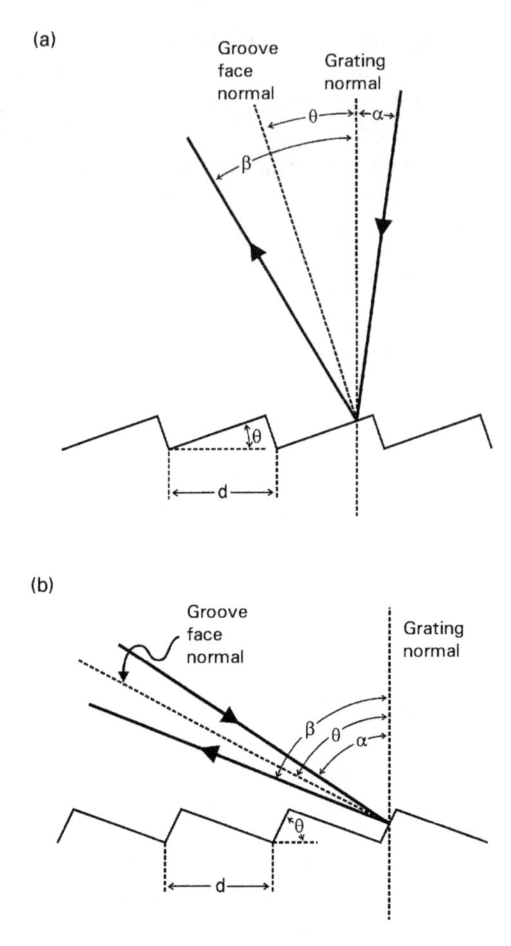

Figure 4.9 Ray paths for: (a) transmission grating used in a Fastie–Ebert or Czerny–Turner configuration; (b) transmission grating used in a Littrow configuration with echelle spectrometer.

equation (Eqn (4.29) in Table 4.5). A grating is characterized by its area, the number of grooves per unit length, and the blaze angle Θ. The blaze angle is the angle the groove face makes with the plane of the grating (Figure 4.9(a)), or the angle made by a line normal to the groove face with the grating normal. The blaze angle can be translated into a blaze wavelength λ_b using Eqn (4.29) in Table 4.5 with $\alpha = 0$ and $\beta = 2\Theta$. The blaze wavelength is the wavelength of maximum spectral efficiency, i.e. maximum diffracted intensity compared to a blank (no grooves) with the same coating. The grating efficiency falls off to 50 per cent of the maximum efficiency at $2/3\lambda_b$ and $3/2\lambda_b$. The dispersion of the grating is characterized by the angular dispersion D_a, defined as $d\beta/d\lambda$ (in radians per nanometer), where $d\beta$ is the angular separation of two wavelengths of separation $d\lambda$. Taking the derivative of Eqn (4.29) with respect to λ yields Eqn (4.30), in Table 4.5. Thus, the angular dispersion is proportional to the distance between ruled lines on the grating. The more lines per unit length, the greater the angular dispersion. Angular dispersion is translated into linear dispersion D_l at the focal plane, by multiplying D_a by the focal length (Eqn (4.31) in Table 4.5). A more frequently used value is the reciprocal linear dispersion R_d the inverse of D_l (Eqn (4.32) in Table 4.5).

The spectral bandwidth $\Delta\lambda_s$ or SBW is the product of the slit width W (in mm) and the reciprocal linear dispersion (Eqn (4.33) in Table 4.5). This assumes equal widths of the entrance and exit slit and perfect optical alignment of the spectrometer. The resultant slit function is triangular with a full width at half-maximum equal to $\Delta\lambda_s$. When the widths of the slits are unequal, the slit function is a trapezoid. The width of the trape-

Table 4.5 Equations for spectrometer parameters

Parameter	*Symbol*	*Conventional monochromator*[a]	*Eqn*	*Echelle monochromator*	*Eqn*
Grating equation	–	$m\lambda = d(\sin\alpha + \sin\beta)$	(4.29)	$m\lambda = 2d\sin\Theta$	(4.61)
Angular dispersion	D_a	$\frac{m}{d\cos\beta}$	(4.30)	$\frac{2\tan\Theta}{\lambda}$	(4.62)
Linear dispersion	D_l	$\frac{fm}{d\cos\beta}$	(4.31)	$\frac{2f\tan\Theta}{\lambda}$	(4.63)
Reciprocal linear dispersion	R_d	$\frac{d\cos\beta}{fm}$	(4.32)	$\frac{\lambda}{2f\tan\Theta}$	(4.64)
Spectral bandwidth	$\Delta\lambda_s$	$\frac{Wd\cos\beta}{fm}$	(4.33)	$\frac{W\lambda}{2f\tan\Theta} + \frac{\lambda^2}{2W_g\sin\Theta}$	(4.65)
Diffraction limited	$\Delta\lambda_d$	$\frac{\lambda d}{W_g m}$	(4.34)	$\frac{\lambda^2}{2W_g\sin\Theta}$	(4.66)
Resolving power	R	$\frac{\lambda_{ave}}{\Delta\lambda_s}$	(4.35)	$\frac{\lambda_{ave}}{\Delta\lambda_s}$	(4.67)
Theoretical resolving power	R_0	$\frac{W_g m}{d}$	(4.36)	$\frac{2W_g\sin\Theta}{\lambda}$	(4.68)

α =angle of incident radiation with grating normal;
β =angle of diffracted radiation with grating normal;
Θ =blaze angle;
d =distance between grooves on the grating;
m =order, in integers;
λ =wavelength;
W =width of matched entrance and exit slit;
W_g =width of the grating;
λ_{ave} =average of wavelength of two lines that are just resolved;
f =focal length.

zoid at half height is given by Eqn (4.33) when the width of the widest slit is used for W. The baseline width for unequal slits is also given by Eqn (4.33) if the sum of the slit widths is used for W. The slit width cannot be decreased infinitely. A diffraction-limited slit width is reached when the slit image turns into a Fraunhofer diffraction pattern, whose central fringe has the angular half-width λ/W'_D, where W'_D is the width of the beam emerging from the disperser. The half-width of the central fringe at the exit plane is $\lambda f/W'_D$. In general, gratings are square or rectangular with a width of W_g. Any rotation of the grating from a position normal to the incident radiation diminishes the effective width of the grating. Thus, the effective grating width is $W_g \cos\beta$, and the half-width of the central Fraunhofer fringe is $\lambda f/(W_g \cos\beta)$. For conventional spectrometers, $\cos\beta$ is close to unity in the first order. The diffraction-limited spectral bandwidth $\Delta\lambda_d$ (Eqn (4.34) in Table 4.5), is obtained by substituting the width of the central fringe of the diffraction pattern for the slit width into Eqn (4.33). To ensure negligible effects of diffraction on the slit function, $\Delta\lambda_s$ should be much larger than $\Delta\lambda_d$. In general, the experimentally determined spectral bandwidth should be approximately equal to the sum of the dispersion and the diffraction widths. The degree of agreement of the experimental and predicted spectral bandwidths can be used as an indicator of the quality of the optical components. Spherical aberration, coma, and astigmatism can result in slit image mismatch which degrades (increases) the spectral bandwidth.

The resolving power R of a spectrometer is defined in Eqn (4.35), where λ_{ave} is the average wavelength of two closely spaced lines with peak centers λ_1 and λ_2, that are separated by at least one bandwidth $\Delta\lambda_s$, as determined by the slit width W. The theoretical resolving power of a spectrometer, as shown in Eqn (4.36), is defined as the average wavelength divided by the diffraction-limited spectral bandwidth.

Typical parameters for a medium resolution monochromator are given in Table 4.6. These values were taken from a commercially available monochromator that employs a separate grating for the ultraviolet and the visible regions. Consequently, there are two values listed for each parameter. These values serve to provide a general characterization of the type of monochromator used in conjunction with an HCL.

The light-gathering capability of a spectrometer is the F-number, defined as

$$\text{F-number} = f/D \qquad (4.37)$$

where f is the focal length and D is the diameter of the optical component. The larger the solid angle that the spectrometer intercepts, the greater the intensity at the detector, and the smaller the F-number. The F-number specifies the characteristics of the optical components external to the spectrometer necessary to provide maximum transmission. For GFAAS, a 1:1 image of the HCL is first projected into the furnace by a lens, or mirror,

Table 4.6 Typical specifications for conventional and echelle spectrometers

Parameter	*Conventional Zeeman 5000*[a]	*Echelle spectraspan V*[b]
Focal length	408 mm	750
Blaze wavelength: UV	210 nm	–
Visible	580 nm	–
Blaze angle: 210 nm	18° 37′	63° 26′
580 nm	28° 19′	
Grating size: 210 nm	84 mm×84 mm	46 mm×96 mm
580 nm	84 mm×84 mm	
Grooves per millimetre: 210 nm	2880	79
580 nm	1440	
D_a: 210 nm	0.0036 nm^{-1}	0.019 nm^{-1}
580 nm	0.0026 nm^{-1}	0.0069 nm^{-1}
D_l: 210 nm	1.47 nm mm^{-1}	14.3 nm mm^{-1}
580 nm	1.06 nm mm^{-1}	5.18 nm mm^{-1}
R_d: 210 nm	0.68 nm mm^{-1}	0.070 nm mm^{-1}
580 nm	0.94 nm mm^{-1}	0.193 nm mm^{-1}
$\Delta\lambda_s$ (with 25 μm slits): 210 nm	0.017 nm	0.0021 nm
580 nm	0.024 nm	0.0070 nm
$\Delta\lambda_d$: 210 nm	0.000 87 nm	0.000 26 nm
580 nm	0.004 79 nm	0.002 2 nm
R_0: 210 nm	242 000	817 000
580 nm	121 000	296 000

[a]Perkin-Elmer Corporation [69].
[b]Fisons Instruments/ARL [70].

a distance $2f$ from the source and from the furnace. A 1:1 image of the image of the HCL in the furnace is then projected onto the entrance slit by a second lens, or mirror, $2f$ distant from both the furnace and the spectrometer. To completely fill the collimating mirror, a lens should be chosen such that $2f/D$ equals the F-number of the spectrometer. The limiting solid angle for all commercially available GFAAS instruments is in the monochromator. A furnace 6 mm in diameter and 28 mm long provides an F-number of 2.3. Shorter furnaces, 19 mm long, with the same diameter will permit the collection of a larger solid angle of radiation, and provides an F-number of 1.6. An end-capped furnace, with half the aperture diameter and a 19 mm length, yields an F-number of 3.2. The F-numbers for the furnaces are sufficiently low that the economics of constructing a high transmission monochromator are limiting.

Another factor that quantifies the light gathering and transmission qualities of the spectrometer is the monochromator throughput factor or étendue $Y(\lambda)$. The étendue is defined as the product of the source area viewed, the solid angle collected by the spectrometer, the transmission of the spectrometer optics, and the slit function:

$$Y(\lambda) = WH\Omega\tau t(\lambda) \tag{4.38}$$

where H is the slit height, τ is the dimensionless transmission factor, $t(\lambda)$ is the slit function, and Ω is the solid angle of light collected by the spectrometer. The solid angle is defined as A/f^2, where A is the effective grating area (width adjusted by the angle the grating makes with the normal of the plane of the limiting aperture). The transmission factor τ reflects the grating efficiency (distance of the diffraction angle of the wavelength of interest from the blaze angle), and the quality of the optical components. The radiant power Φ passing through the exit slit is

$$\Phi = B_\lambda WH\Omega\tau t(\lambda) \tag{4.39}$$

where B_λ is the source spectral radiance. Equation (4.39) can be evaluated for two cases: a monochromatic source (an HCL) and a continuum source. For a monochromatic source, the slit function is equal to unity and the radiant power Φ_L becomes

$$\Phi_L = B_L WH\Omega\tau \tag{4.40}$$

where B_L is the line radiance. This same simplification applies to Eqn (4.38). Thus, for a monochromatic source, the étendue becomes

$$Y(\lambda) = \frac{WHA\tau}{f^2} = \frac{A_s A_g \tau}{f^2} \tag{4.41}$$

where A_s the area of the entrance slit and A_g is the effective area of the grating. For a continuum source, the slit function is equal to the spectral bandwidth and Eqn (4.39) becomes

$$\Phi_C = B_\lambda WH\Omega\tau(\Delta\lambda_s) \tag{4.42}$$

Substituting Eqn (4.39) and $\Omega = A/f^2$ into Eqns (4.40) and (4.42), respectively, gives

$$\Phi_L = \frac{B_L H D_a A\tau(\Delta\lambda_s)}{f} \tag{4.43}$$

and

$$\Phi_C = \frac{B_\lambda H D_a A\tau(\Delta\lambda_s)^2}{f} \tag{4.44}$$

Equations (4.43) and (4.44) demonstrate that, for matched entrance and exit slit widths, the intensity for an HCL increases linearly with the spectral bandwidth, and the intensity for a continuum source increases with the square of the spectral bandwidth.

Equations (4.43) and (4.44) can be used as a basis for evaluating optimum monochromator parameters for the photon shot, fluctuation, and read noise-limited cases. Snelleman [71] was the first to do this with a continuum source and a shot noise-limited system. This approach, however, can also be applied to line sources and other noise sources. With an HCL, the fraction of absorbed radiation α_{HCL} is independent of the spectrometer (assuming that the spectral bandwidth is greater than the width of the HCL) and proportional to N, the concentration of the analyte in solution. At the detection limit, the fraction of absorbed radiation by the minimum detectable concentration N_{min} is proportional to the relative precision of the radiant power Φ:

$$\alpha_{HCL} = k_{HCL} N_{min} = \frac{\sigma_{\Phi}}{\Phi} \tag{4.45}$$

where k_{HCL} is a proportionality constant and σ_{Φ} is the standard deviation of the measured radiant power, Φ. This provides a minimum detectable signal of

$$N_{min} = \frac{\sigma_{\Phi}}{k_{HCL}\Phi} \tag{4.46}$$

Table 4.7 presents the limiting noise (σ_{Φ}; Eqns (4.47)–(4.49)) and the minimum detectable signals (N_{min}; Eqns (4.50)–(4.52)) for an HCL source in terms of the monochromator parameters. For the shot noise-limited case, the noise (Eqn (4.47)) is proportional to the square root of the radiant power (Eqn (4.43)). Substituting Eqns (4.43) and (4.47) into (4.46) provides the minimum detectable concentration, Eqn (4.50). It can be seen that the monochromator parameters that maximize Φ will minimize N_{min}. Thus, a short focal length, a large slit height, grating area, spectral bandwidth, and a high source radiance will improve the detection limit.

There are limitations to the optimization of these parameters. The slit height is limited by the physical size of the cathode of the HCL and the diameter of the graphite furnace. In general, a height of 2 mm is the maximum limit. The spectral bandwidth is limited by spectral interferences. A wide spectral bandwidth is more likely to allow inclusion of stray light, resulting in nonlinear calibration curves with shortened dynamic ranges. These equations, however, show clearly how the spectrometer parameters can affect the detection limits.

Table 4.7 Limiting noise and minimum detectable concentration for an HCL

Type of noise	*Limiting noise, σ*	*Eqn*	*Minimum detectable signal, N_{min}*	*Eqn*
Photon shot	$\sqrt{\dfrac{B_L H D_a A \tau(\Delta\lambda_s)}{f}}$	(4.47)	$\dfrac{1}{k_{HCL}}\sqrt{\dfrac{f}{B_L H D_a A \tau(\Delta\lambda_s)}}$	(4.50)
Fluctuation	$\xi_f\left(\dfrac{B_L H D_a A \tau(\Delta\lambda_s)}{f}\right)$	(4.48)	$\dfrac{\xi_f}{K_{HCL}}$	(4.51)
Read (n detectors)	$\sigma_{read}\sqrt{\dfrac{\Delta\lambda_s}{\Delta\lambda_{pix}}}$	(4.49)	$\dfrac{f\sigma_{read}}{k_{HCL} B_L H D_a A \tau\sqrt{\Delta\lambda_s \Delta\lambda_{pix}}}$	(4.52)

In the case where the source fluctuation noise is limiting, the precision of the radiant power is proportional to the total power ($\sigma_\Phi = \xi_f \Phi$), where ξ_f is the fluctuation factor (Eqn (4.11)). As a result, the minimum detectable signal is dependent on the proportionality constant (Eqn (4.51)). The smaller the fraction the noise is of the radiant power, the lower the detectable signal.

Finally, N_{min} is defined for the read noise limited case (Eqn (4.52)). This case arises when the read noise of a solid-state detector, such as a linear photodiode array (LPDA), is limiting. Several prototype multielement systems have used such detectors. The total read noise is dependent on the read noise for a single pixel σ_{read} and the square root of n, the number of pixels used to cover the SBW ($n = \Delta\lambda/\Delta\lambda_{\text{pix}}$), where $\Delta\lambda$ is the spectral width of the image of the entrance slit at the focal plane and $\Delta\lambda_{\text{pix}}$ is the spectral width of a pixel. As shown in Eqn (4.52), the best detection limits are obtained by maximizing $\Delta\lambda$ and the spectral width of the pixels ($\Delta\lambda_{\text{pix}}$). The limitations for the optimization of the parameters in Eqn (4.52) are the same as those previously discussed for Eqn (4.50).

A similar evaluation of the optimum spectrometer parameters can be made for a continuum source. The fraction of absorbed radiation with a continuum source α_c is inversely dependent on the spectral bandwidth:

$$\alpha_c = \frac{A_t}{\Delta\lambda_s} = \frac{k_c N}{\Delta\lambda_s} \tag{4.53}$$

where A_t is the total absorption k_c is the proportionality constant. As the spectral bandwidth is increased, more unabsorbed radiation is transmitted and the fraction of absorbed radiation decreases. The total absorption is proportional to the concentration of the analyte in solution. It is assumed in this case, that the spectral bandwidth of the spectrometer is greater than the width of the absorption profile. The minimum detectable concentration is achieved when the fraction of absorbed radiation equals the relative precision of the radiant power:

$$N_{\text{min}} = \frac{\Delta\lambda_s \sigma_\Phi}{k_c \Phi} \tag{4.54}$$

Table 4.8 presents the limiting noise, Eqns (4.55)–(4.57), and the minimum detectable signal, Eqns (4.58)–(4.60), for the shot, fluctuation, and read noise-limited cases using a CS. This table is analogous to that presented for the HCL (Table 4.7). Equation 4.58 shows that, for the shot noise-limited case, the minimum detectable signal is independent of the $\Delta\lambda$. For the fluctuation noise-limited case (Eqn (4.59)), the detection limit is proportional to the fluctuation factor and the spectral bandwidth and, for the read noise-limited case (Eqn (4.60)), the noise dependence on the spectrometer parameters is identical to that for the HCL.

Table 4.8 Limiting noise and minimum detectable concentration for a continuum source

Type of noise	*Limiting noise, σ*	*Eqn*	*Minimum detectable signal, N*$_{\mathrm{min}}$	*Eqn*
Photon shot	$\Delta\lambda_s\sqrt{\frac{B_\lambda H D_a A\tau}{f}}$	(4.55)	$\frac{1}{k_c}\sqrt{\frac{f}{B_\lambda H D_a A\tau}}$	(4.58)
Fluctuation	$\frac{\xi_f B_\lambda H D_a A\tau(\Delta\lambda_s)^2}{f}$	(4.56)	$\frac{\xi_f \Delta\lambda_s}{k_c}$	(4.59)
Read (n detectors)	$\sigma_{\mathrm{read}} = \sqrt{\frac{\Delta\lambda_s}{\Delta\lambda_{\mathrm{pix}}}}$	(4.57)	$\frac{f\sigma_{\mathrm{read}}}{k_c B_\lambda H D_a A\tau\sqrt{\Delta\lambda_s \Delta\lambda_{\mathrm{pix}}}}$	(4.60)

4.3.3.2 Echelle characteristics

One of the most commonly used high-resolution spectrometers is the echelle polychromator. The echelle employs a coarsely ruled echelle grating with a large blaze angle (Table 4.6) for wavelength dispersion and a cross-disperser, a prism or grating, to separate the orders. The result is a two-dimensional spectrum at the exit plane of the spectrometer with the orders dispersed vertically and the wavelength dispersed horizontally (Figure 4.10). All echelles employ a Littrow configuration (Figure 4.9(b)) where $\alpha = \beta$. Thus the general grating equation, Eqn (4.29), can be expressed as Eqn (4.61) (Table 4.5). With the simplification of the grating equation, the general equations for angular, linear, and reciprocal linear dispersion, bandwidth and resolving power can also be restated as shown in Eqns (4.62)–(4.68). In each case, the order, m has been replaced by $(2d\cos\Theta/\lambda$, obtained from Eqn (4.61). Thus, the dispersion characteristics of the spectrometer are now wavelength dependent. This is a result of the systematic change in order with increasing wavelength. Equation (4.65) for spectral bandwidth of the echelle assumes that the diffraction-limited spectral bandwidth is significant compared to the geometric spectral bandwidth. Thus, the spectral bandwidth is equal to the sum of the geometric and diffraction-limited bandwidths. If the diffraction-limited spectral bandwidth is not significant, the second term can be dropped.

The design of the echelle produces a short free spectral range for each order that requires cross-dispersion to avoid severe spectral overlaps. However, the spectral efficiency of each order, is very high, ranging from a maximum at the center of the free spectral range to approximately half this value at each end of the range. A typical spectrum is shown in Figure 4.10 for a Spectraspan III (Spectrometrics; now ARL). On this spectrometer, orders 118 through 28 are used to cover the wavelength range from

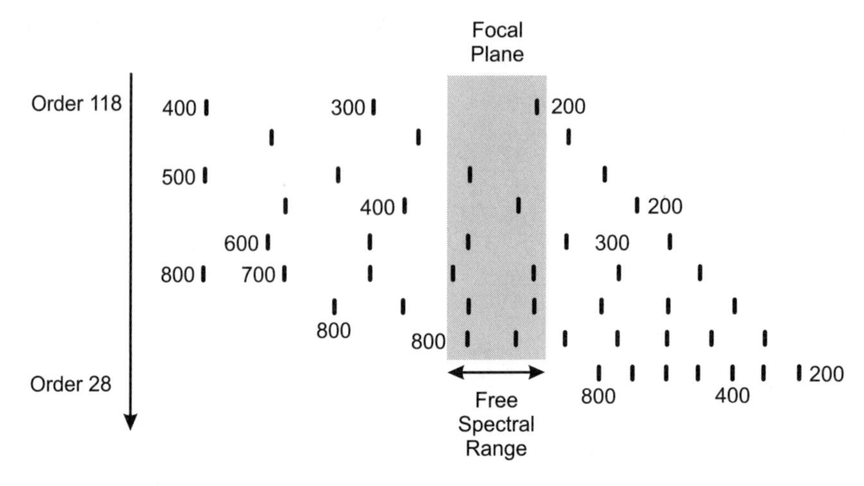

Figure 4.10 Illustration of the two-dimensional spectrum obtained from an echelle spectrometer. Dispersion on the horizontal axis is obtained with a grating and dispersion on the vertical axis is obtained with a prism.

190 to 800 nm, with the highest dispersion (from the highest orders) occurring for the shortest wavelengths. In many cases, a wavelength will fall on more than one order. Usually there is a consistent bias in spectral efficiency towards one end of the free spectral range. This generally determines which order is used. It has been documented that the Spectraspan III echelle has a rapid fall-off in maximum order efficiency below 250 nm [72]. From a maximum efficiency of approximately 50 per cent at 230 nm the efficiency drops to approximately 20 per cent at 200 nm. This characteristic limits the utility of the Spectraspan for the determination of elements with wavelengths below 230 nm. There is no theoretical reason for the echelle to have such a dramatic drop in efficiency. It can only be assumed that the loss of efficiency is due to the design and quality of the optical components. A more recent echelle, designed for use as a disperser for an inductively-coupled plasma (ICP) source, has order efficiencies as high as 50 per cent at 200 nm [73].

The throughput factor for the echelle is mathematically defined in the same manner as it is for a conventional monochromator (Eqn (4.38)). It thus follows that the derived equations for the transmission of radiant power (Eqns (4.39)–(4.43)), and the limiting noise and detection limits for an HCL source (Eqns (4.47)–(4.52) in Table 4.7) and a continuum source (Eqns (4.55)–(4.60) in Table 4.8) are also the same. The general comments offered for a conventional monochromator also apply for the echelle spectrometer. The only special consideration is for those cases where it was assumed that the spectral bandwidth was larger than the absorption profile. With the smallest slit widths of the echelle, this assumption is no longer valid. At larger slit widths this assumption will hold and the equations will be valid.

4.3.4 Detectors

The photomultiplier tube (PMT), based on vacuum tube technology, has long been the detector of choice for AAS. The high gain and low noise of these detectors make them practically transparent to the measurement process (i.e. contributing no significant noise to the signal) and the standard against which all other detectors are judged. With a PMT and a HCL source, the absorption detection system is shot noise limited. Until recently, PMTs were the only detectors found in commercial AAS instruments. In 1994, the first commercial GFAAS instrument with non-PMT detection was announced.

The alternatives to PMTs are solid-state detectors. The interest in solid-state detectors has been driven by their small size and their inherent suitability for combination into linear and two-dimensional arrays. Solid-state detectors are constructed of semiconductor materials, either as photodiodes or metal–oxide semiconductor (MOS) capacitors. Mounted in the focal plane of a conventional or echelle spectrometer, these solid state arrays can provide multiwavelength detection. Solid-state array detectors have been used extensively with ultraviolet-visible spectrophotometers, with ICP emission spectrometers, and for a variety of prototype AAS instruments with either HCL or continuum sources. The driving force in all three cases has been the demand for multiwavelength, or multi-element, detection. The biggest hindrance to the development of multi-element GFAAS has not been the detector, but the lack of a suitable source. Consequently, use of solid state array detectors for GFAAS has been restricted to prototype instruments. To date, no array detectors have been employed with commercial GFAAS. In 1994 a multielement GFAAS instrument (SIMAA 6000, Perkin-Elmer Corp.) was introduced that employed a detector consisting of an array of single photodiodes [74,75]. The photodiodes are used as direct substitutes for PMTs.

4.3.4.1 Photomultiplier tubes

A PMT converts the energy from a photon into an amplified electron current. A PMT consists of a photosensitive cathode, a series of dynodes, and a collection anode. The cathode is constructed of a photoemissive material that ejects an electron when struck by a photon. The potential for each dynode becomes increasingly positive (as compared to the cathode) in uniform, discrete steps. Depending on the type of PMT, the positive electrical bias of the anode (compared to the cathode) may range from 500 to 2500 V. Impact of each photoelectron on a dynode releases multiple electrons. Each electron in turn releases multiple electrons at the next dynode, and so on. Thus, each photon striking the cathode surface initiates a cascade of electrons that produce a pulse of electrons (current pulse) at the anode. The number of electrons generated at the anode by a single photon at the cathode can range from 10^4 to 10^7, depending on the voltage drop between the anode and cathode. A minimum, or threshold, energy is required of a photon to liberate an electron from the cathode. This threshold energy corresponds to the maximum wavelength at which the PMT is useful. The threshold wavelength limit of the PMT is dependent on the material of the photocathode. At lower wavelengths photons will liberate electrons with an efficiency which is wavelength dependent. The lower wavelength limit is also ultimately determined by the material of the photocathode. In most cases, however, the practical lower limit is determined by either transmission through air, the transmission efficiency of the PMT window, or the transmission efficiency of the spectrometer [72]. In general, PMTs, like solid-state detectors, are indiscriminate with respect to the energy of the photons. The PMT is dependent on the spectrometer to isolate the wavelength region of interest.

The quantum efficiency $K(\lambda)$ is the ratio of photoelectrons generated to the number of photons striking the cathode surface. The rate at which photoelectrons are generated is the product of the quantum efficiency and the incident radiant flux Φ_p (photons per second). The radiant cathodic responsivity $R(\lambda)$ of the PMT (in amps per watt) is the efficiency with which photons are converted to photoelectrons:

$$R(\lambda) = \frac{K(\lambda)e}{h\nu} = (8.06 \times 10^{-4})K(\lambda)\lambda \qquad (4.69)$$

where e is 1.6×10^{-19}, h is Planck's constant, and ν and λ are the frequency and wavelength, respectively, of the radiation. The spectral response is a plot of the responsivity as a function of wavelength. The gain m is the number of electrons generated at the anode by a single photoelectron. The range over which the signal is linear is the linear dynamic range. The time necessary for the signal to rise from 10 per cent to 90 per cent of its final value is the rise time and is determined by the spread in the transit time of the electrons and the electron optics of the PMT. The anodic current i_{ap}, induced by photons at wavelength λ, is

$$i_{ap} = m\eta\Phi R \qquad (4.70)$$

where η is the anodic collection efficiency. Figure 4.11 gives the quantum efficiency of a broad response and a solar blind (unresponsive above 400 nm) PMT. The maximum quantum efficiency at 200 nm is about 28 per cent and 40 per cent, respectively. The maximum values will vary considerably, depending on the photocathode material and between individual PMTs of the same type.

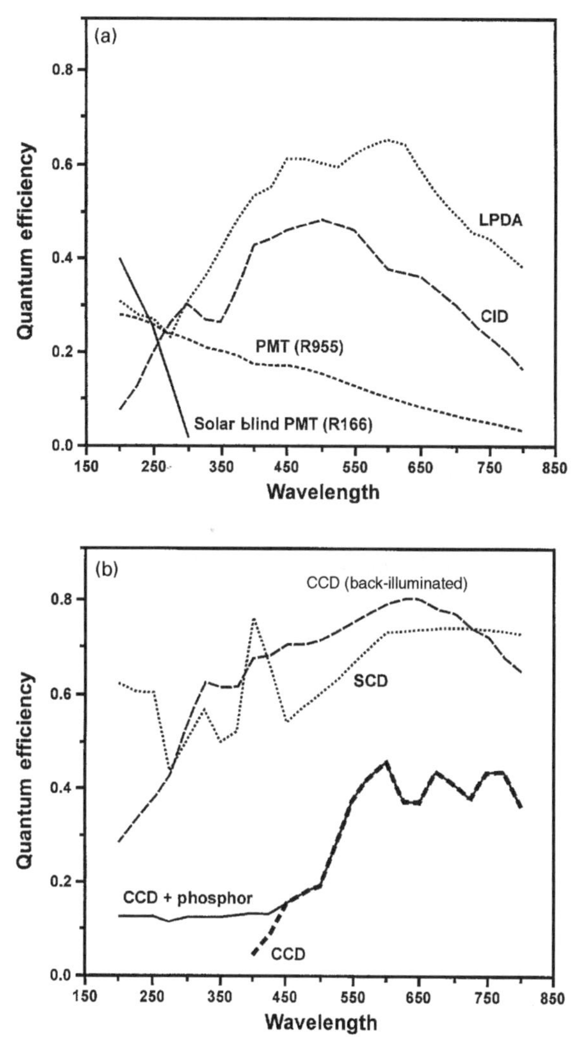

Figure 4.11 Quantum efficiencies of: (a) a broad-band response PMT (R955), a solar-blind PMT (R166), an LPDA, and a CID; (b) a CCD, a CCD with a phosphor coating, a thinned CCD lit from the backside, and an SCD.

In general, the PMT is used to measure the average current generated by a radiation flux, i.e. the response time of the electronics is slow compared to the duration of the current pulses. With fast electronics and electron optics tuned to provide a rapid rise time, pulses of electrons at the anode can be counted to provide the equivalent of photon counting. This technique is best suited for measuring low radiation levels and has proved more useful for emission and fluorescence measurements than for absorption measurements. With AAS, the best detection limits and the measurement of lowest analyte levels are achieved with the highest possible source radiation. Consequently, photon counting has had little impact on AAS.

The limiting noise of the PMT is dark current shot noise. The dark current i_{ad} arises from thermal emission of the cathode and dynodes. Although the emission from any component may be small, the large gain factor m can produce dark currents of 10^{-7}–10^{-11} A at the anode, depending on the voltage. This compares with currents of 10^{-4}–10^{-8} A using HCLs. As the lowest absorbance signals are obtained when the sample intensity is almost equal to the reference intensity (when the radiant flux is the highest), the dark current shot noise is not a significant factor for routine GFAAS determinations. In cases of high background attenuation, the dark current shot noise increases relatively, but is still less than the source shot noise. The dark current shot noise can be reduced by cooling the PMT and thus reducing the thermal emission. The relative contribution of the dark current shot noise has always been sufficiently low that cooling the PMT was never justified.

With respect to the SNR, the PMT is an almost ideal detector. Arriving photons are converted to a current pulse and amplified by a factor of 10^4–10^7 with no significant noise contribution. At extremely low photon fluxes, the dark current shot noise of the PMT may contribute to the overall noise level, but these occasions are rare. Thus the PMT is the ideal transparent detector/amplifier.

4.3.4.2 Solid state detectors

The three types of solid state detector that have been adapted for atomic spectrometry are the photodiode (PD), the charge coupled device (CCD), and the charge injection device (CID). All three of these detectors employ a silicon matrix which is doped (impregnated with low concentrations of impurities) to render them electron acceptors (*p*-doped) or electron donors (*n*-doped). The PDs are solid state diodes that use the *p–n* junctions as a charge storage device [76], whereas the CCDs and CIDs are metal–oxide semiconductor (MOS) capacitors, and are more accurately described as charge transfer devices [77,78]. Historically, photodiode arrays (PDAs) have seen more widespread use with prototype systems, because they were the first array detectors developed and because of the greater simplicity of the read process. Unfortunately, the LPDA read noise is also much greater than that for the other two devices, resulting in read noise being limiting. The quieter CCDs and CIDs, tend to be photon shot noise limited, even at the lowest photon fluxes, and have been developed as detectors for emission spectrometry with an ICP source [73,79,80]. As mentioned previously, it is only recently that a commercially available

GFAAS instrument has been developed that uses an array of individual PDs. There is still no commercial GFAAS instrument that employs linear or two-dimensional solid-state arrays.

Photodiodes and photodiode arrays

Photodiodes were the first solid-sate detectors and PDAs (linear) were the first of the new generation of solid-state array detectors available for spectrometric applications. In general, the greatest spectroscopic interest in PDs has been in the array format. Historically, linear PDAs have consisted of linear arrays of 128–4096 photodiodes, or pixels, which were read in a sequential manner. They are suitable in a linear (rather than two-dimensional) format, because their height to width ratio, or aspect, can be as high as 100:1 (Table 4.9). Today, PDAs are also available with smaller aspects, in two-dimensional arrays, and with random access capability. The parameters for a commonly used PDA are provided in Table 4.9. It is interesting that the demand for better PDAs has led to improved PDs which are now suitable as replacements for PMTs. Table 4.9 also shows the specifications for the PD used commercially as a detector for GFAAS. Simplistically, a photodiode is a junction of p-doped and n-doped silicon to which a reverse bias is applied (negative voltage to the p side and positive voltage to the n side). This bias voltage draws the electrons to the n side and the electron holes to the p side, creating a depletion region and a potential between the two halves, i.e. a charged capacitor. When the bias is removed, photons striking the PD generate electrons and holes, which provide a current that discharges the capacitor. Either the decrease in potential or the current necessary to restore the potential is measured. Most commercially available PDs function primarily in the second mode.

For a PDA there are four noise sources of consequence: fixed pattern noise, dark current shot noise, read noise, and source carried noise. The fixed pattern noise arises from a fixed intensity pattern for the pixels, which is readily observed at low intensities. This pattern arises from the clocking sequence used in the interrogation process and produces a repetitive pattern. The pattern is roughly constant at a constant temperature and pixel read frequency, is small compared to the full scale of the PDA, and is additive in nature. Consequently, the pattern is easily removed by subtraction of a reference profile. Variation in the pattern between reads gives rise to a noise component that is insignificant compared to the other noise sources. Dark current arises from the thermal generation of electrons and holes in the photodiode. It usually ranges from 0.2 to 5 pA. For short exposures or integration times, variation of the current is usually negligible (150–700 e^- for a 16 ms integration time) compared to the other noise sources, depending on the array and the radiation intensity.

Table 4.9 Characteristics for selected solid-state detectors

Type	*Photodiode array* [81]	*Photodiode* [75]	*Charge coupled device* [82]	*Charge coupled device* [73]	*Charge injection device* [79]
Manufacturer	Hamamatsu Corp.	Perkin-Elmer Corp.	Photometrics Corp.	Perkin-Elmer Corp.	CID Technology Corp.
Size ($\mu m \times \mu m$)	25×2500	1000×2000	23×23	25×170	23×27
Format	1×256	1	384×576	1×64	512×1512
Read noise (e^-)	2800	1000	6–15	15	180
Quantum efficiency (200 nm)	30%	70%	22%[a]	60%	35%[a]
Dark current (e^- s)	8×10^5[b]	2.7×10^7[b]	8[c]	120[d]	5×10^6[b]
Saturation charge (e^-)	1.25×10^8	7.5×10^6	2.5×10^5	9.8×10^5	1.07×10^6

[a]Detector surface coated with Metachrome II, a patented phosphor.
[b]Detector at room temperature (≈ 21 °C).
[c]Detector cooled to −45 °C.
[d]Detector cooled to −40 °C.

As the dark current is an exponential function of the temperature, dark current shot noise can be dramatically reduced by cooling the PDA. This is not usually necessary for GFAAS applications where 10–20 ms integration times are necessary, but is frequently done for emission spectrometry where considerably longer integration times are used. In general, the dark current is reduced by a factor of 2 for every 6° the detector temperature is lowered.

Read noise is the noise that arises from reading a pixel. It is proportional to $(kTC)^{1/2}$, where k is Boltzman's constant, T is the absolute temperature, and C is the capacitance associated with the pixel. The large surface area of the pixels for the PDA listed in Table 4.9 results in a high capacitance and, hence, a very large read noise of 2800 e^-. Smaller pixels have correspondingly lower read noises. The read noise is usually the dominant noise source for the array and is reduced by approximately 12 per cent by cooling the array to −40°C.

Finally, source carried noise arises from the photons arriving at the PD. For an HCL, the only source carried noise is photon shot noise, the fundamental noise arising from the random arrival of photons at the detector. As discussed in Section 4.2.2 (Eqn (4.10)), the shot noise for a solid state detector, expressed in electrons, is equal to the square root of the measured charge. For the PD listed in Table 4.9, the photon shot noise is 10 000 e^- at 90 per cent saturation and 7900 e^- at 50 per cent saturation. The source carried noise for a continuum source can also include source fluctuation noise. Fluctuation noise is usually eliminated through double beam operation. With a PDA, pixels to either side of the absorption profile can provide a reference intensity. Thus, use of a PDA with a continuum source provides double beam operation and eliminates fluctuation noise.

To compare the noise contribution of the various sources, Table 4.10 was constructed, assuming a PDA with the same specifications shown in Table 4.9, a quiet, variable gain amplifier between the LPDA and the ADC, an ADC range of 10.24 V, and 16 ms (60 Hz) integration time. For photon shot noise to be equal to the read noise, 6.3 per cent saturation is necessary. For the shot noise to be limiting (90 per cent of the noise), approximately 20 per cent saturation is needed. In the far ultraviolet region (around 200 nm), HCLs and continuum sources yield approximately 250 000 e^-s, or approximately 0.2 per cent saturation. Thus, read noise is usually dominant. Amplification of the signal will not change the relative contribution of the read noise. Amplification will, however, reduce the contribution of the quantization. This can be important at high analytical concentrations or with high background when the source is significantly attenuated. Obviously, a 16 bit ADC will offer lower quantization noise than a 12 bit ADC.

The quantum efficiency of the PDA is compared to a PMT and the other solid state detectors in Figure 4.11. It can be seen that the PDA offers an

Table 4.10 Noise Sources for a photodiode array.

	Amplifier gain			
Source	×1	×5	×10	×20
Read noise	2 800	14 000	28 000	56 000
Dark current	140	700	1 400	2 800
Quantization noise				
12 bit ADC	15 000	15 000	15 000	15 000
16 bit ADC	800	800	800	800
Photon shot noise				
5% saturation	2 500	7 500	25 000	50 000
10% saturation	3 500	17 500	35 000	–
20% saturation	5 000	25 000	–	–
50% saturation	7 900	–	–	–
90% saturation	11 000	–	–	–

All values given in e^-, where $2000\,e^- = 0.096$ mV. The voltage range of the ADC is 10.24 V. Saturation of the PDA is 6.0 V. Values for the photon shot noise are not listed when the amplification would have resulted in a signal exceeding saturation.

advantage over the PMT. The exact advantage will depend on specific PDAs and PMTs that are used.

Charge coupled devices

These devices typically have much smaller surface areas (the same width but with an aspect of 1) than LPDAs (Table 4.9) and have been used almost exclusively in large, two-dimensional arrays. The CCD is a series of MOS capacitors that store charges generated by photons striking the detector (Figure 4.12). The photosensitive region of a CCD is the epitaxy layer which is 10–20 μm thick and is composed of lightly p-doped silicon. The epitaxy is mounted on the substrate, consisting of highly p-doped silicon for support. The epitaxy is separated from the polysilicon electrodes, or gates, by thin ($\approx 0.02\,\mu$m) insulating layers of silicon oxide and silicon nitride. The polysilicon gates are not metal as MOS implies, but highly doped polycrystalline silicon. A positive potential on the gates creates a depletion layer below the electrode. During the exposure time, photons produce electron–electron hole pairs in the epitaxy; the electrons are stored under the positively biased gates while the electron holes are drawn off through the substrate.

The strength of the CCD is its extremely low read noise of 5–25 e^- (Table 4.9). If this value is substituted into Table 4.10 it can be seen that with a 16 bit ADC and with cooling of the detector to minimize dark current shot noise, the fundamental photon shot noise is always dominant. Each pixel consists of 1–4 gates. By systematically varying the potential on the gates the electron charge can be shifted sequentially from

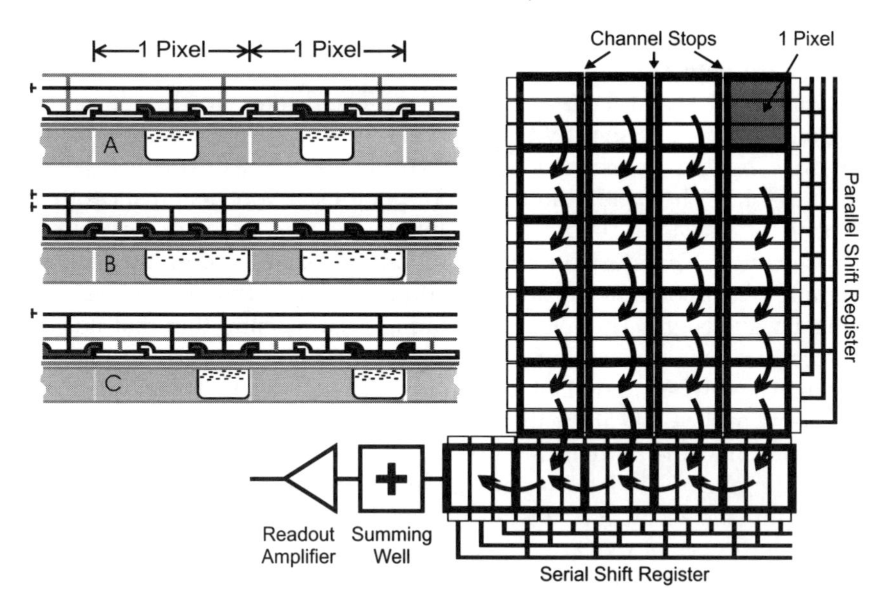

Figure 4.12 Diagram of a CCD, illustrating how multiple gates are used to shift the charge sequentially down the CCD.

pixel to pixel, as shown in Figure 4.12. The common gates of every pixel are tied together so that the charge coupling occurs in phase for all the pixels. This charge coupling is the concept which gives the detector its name. The size of a pixel is dictated by the width of the gates, which are in turn restricted in size and spacing, by the need for a charge transfer efficiency close to unity.

A two-dimensional array can be thought of as multiple columns of linear arrays where each column is electrically isolated, but shifts charges in phase with all the other columns, i.e. parallel registers. Each shift of the parallel registers deposits new charges on the serial register. The charges on the serial register are then sequentially shifted to the readout amplifier. For very large arrays several million shifts may be necessary to read the entire array. As a result the charge transfer efficiency is an important specification. For most two-dimensional CCD arrays, the transfer efficiency is very close to unity (99.99–99.999 per cent). It can be seen that even a slight deviation from unity can cause smearing of the image. Like PDAs, cooling of a CCD array reduces the dark current increasing the possible integration time and reducing the dark current shot noise. Excessive cooling of the array, however, can impede the charge transfer efficiency. Consequently, the CCD arrays are only cooled to around −40 °C. CCD arrays are also vulnerable to blooming; excessive build-up of charge under one gate (from extremely intense illumination) spilling over to the next gate.

Usually, CCDs suffer from low quantum efficiency in the ultraviolet region. This lack of sensitivity is due to absorption and reflection of the photons by the polysilicon gate electrodes and the insulating layers (Figure 4.13). The penetration depth of photons at wavelengths less than 400 nm is less than 1 μm. Two solutions to this problem have been implemented. The most common solution is to coat the detector with an organic phosphor that absorbs the ultraviolet light and fluoresces at a longer wavelength. The phosphor coating is sufficiently thin that no significant reduction in resolution occurs. As the fluorescence is omnidirectional, the highest possible quantum efficiency is half the quantum efficiency of the CCD at the fluorescing wavelength. The second approach is to illuminate the array from the back side which requires removal of the substrate layer. This latter approach offers the best quantum efficiency. Quantum efficiencies as a function of wavelength are shown in Figure 4.11 for several types of CCDs.

The read noise for the CCD arises primarily from the capacitance associated with the readout amplifier. Depending on the circuit design, the read noise can range from 5 to 25 e^-. Fluctuation noise associated with the read process can be eliminated by correlated doubling sampling, i.e. sampling the potential prior to and after transferring the charge. With correlated double sampling the limiting noise is proportional to $(kTC)^{1/2}$ as described in the previous section. The read process for the CCD is problematic for spectroscopic applications. The inherent charge coupled

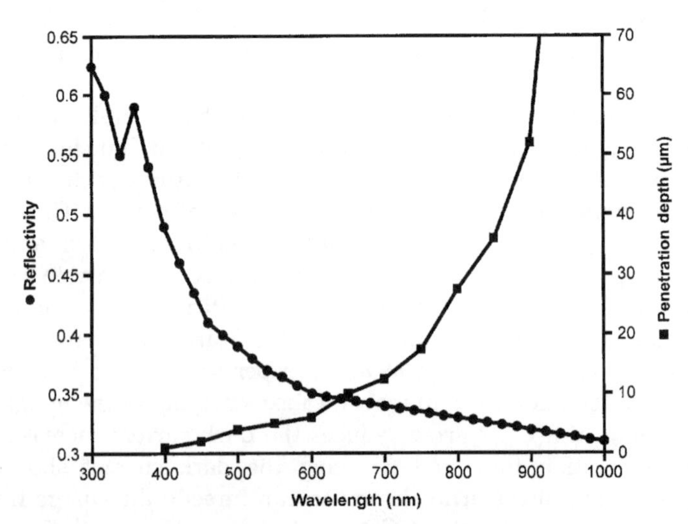

Figure 4.13 Photon penetration depth (■) (thickness necessary to absorb 90 per cent of the radiation) and reflectivity of silicon (●) as a function of wavelength.

transfer character necessitates a sequential read process and does not permit random access of the pixels. This means that a considerable time and number of reads may be necessary to acquire the data for the entire array, even though the desired spectral regions may only constitute a small fraction of the viewed spectrum. One solution to this problem has been to use a series of CCD arrays imbedded in a silicon chip [73]. This segmented CCD array detector (SCD) consists of linear arrays of 16–32 pixels (Table 4.9), placed at strategic positions in the focal plane, and offers rapid access to the spectral regions of interest. In addition, the quantum efficiency of the segmented array CCD is 60 per cent at 200 nn, a factor of two better than the two-dimensional arrays with a phosphor. This improvement comes from the lack of polysilicon gate over the epitaxy. Blooming between arrays is prevented by a protective barrier to drain off excess charge. Blooming between pixels within an array is still a concern.

Charge injection devices

These devices are similar to CCDs with respect to basic structure, but quite different functionally. Structurally, the CIDs are identical to the CCD, except that an *n*-doped epitaxy is used. As a result, electron holes are collected under the gate electrodes when the array is exposed to light as shown in Figure 4.14 (a). Each pixel consists of two gates, the collection and the sensing gates. The charge on any pixel can be selectively read by adjusting the potential of the two gates. The charge on the sensing electrode is read with (Figure 4.14, (b)) and without (c) the contribution of the potential of the collected electron holes. The signal is the difference between the two integrated values. This version of correlated double sampling serves to eliminate fluctuation read noise. The charge on the sensing gate can be cleared by injecting the electron holes into the substrate (d) or reread by transferring the charge back to the collecting gate (c). A read followed by injection of the charge into the substrate is called a destructive read, and a read followed by a transfer of the charge back to the collecting gate is called a nondestructive read. An unlimited number of nondestructive reads is possible without compromising the integrity of the signal. This technique can be used to monitor the signal while the array is exposed, or to improve the SNR.

The read process for the CID (at the time this was written) can be best described as pseudorandom access. Although the pixels themselves can be randomly accessed, the x and y axis address registers must be clocked to the correct address and the clocking is not reversible nor can the register be cleared. The read noise for the CID, after correlated double sampling, is approximately the same as that of the CCD. The gain of the CID (usually 30–50 nV per e^-)is less, however, so the relative read noise (Table 4.9) of the CID is higher. Multiple nondestructive reads, can be used to improve the SNR. The signal will increase linearly with the number of

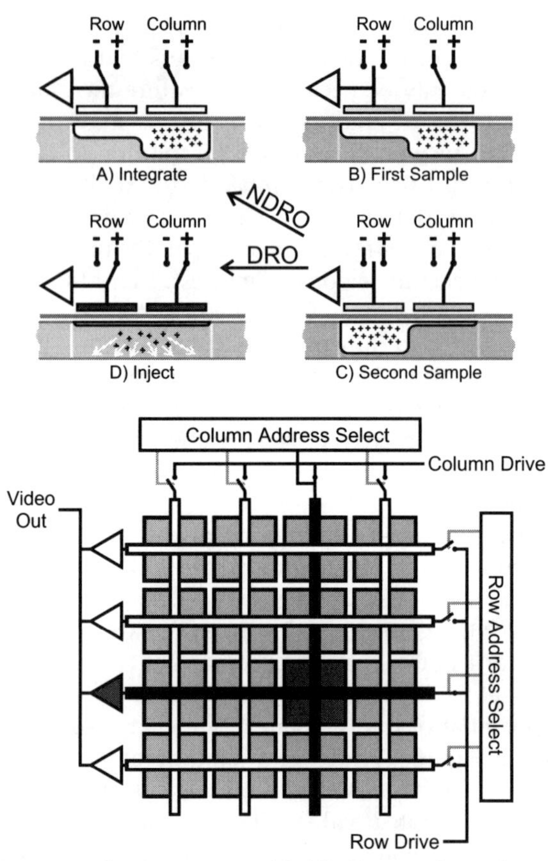

Figure 4.14 Diagram of a CID. Inserts (a)–(d) illustrate how the row and column (sense and collection) gates collect and transfer charge for destructive (DRO) and non-destructive (NDRO) reads.

reads, whereas the noise increases with the square root of the number of reads. For the CID specified in Table 4.9, 100 nondestructive reads is considered standard. Thus, the read noise is effectively $18\,e^-$. An examination of Table 4.10 shows that with 100 nondestructive reads, cooling with liquid nitrogen, and a 16 bit ADC, the CID, like the CCD, is almost always photon shot noise limited.

Charge injection device arrays have several advantages over CCD arrays. Blooming is less severe for the CID, as the overflow of charge for an intensely illuminated pixel will be drained off by the *p*-doped substrate layer. Nondestructive reads can be used to actively monitor the signal while the array is exposed and to improve the SNR. Finally, the read process of the CID makes random access of individual pixels possible. The disadvantage of the CID array is the inherently poorer SNR that makes multiple nonrestructive reads necessary.

4.4 ELECTRONICS, COMPUTERS AND SOFTWARE

The field of analytical instrumentation has progressed dramatically since the mid-1980s as a result of the tremendous advances in solid-state electronics and computer technology. GFAAS is no exception. Every instrument is under microprocessor control. All aspects of the analytical process can be automated–from instrument set-up, to sample introduction, data acquisition, data processing and report generation. Most of the analytical capabilities that are taken for granted today are a direct result of the advances in solid state electronics.

Today, the analyst can run a GFAAS almost completely from the computer keyboard. Certain physical chores, such as inserting the correct HCLs into the turret and loading samples, standards, and blanks into the autosampler, are still required, but the rest is automated. Given the element to be analyzed, the computer can set up the instrument using a default program for the desired element or prompt the analyst for specific parameters for a custom program. In each case the computer checks to make sure the correct lamp is in position, sets the required current, and optimizes the analytical wavelength. The computer controls all aspects of the furnace operation: calibrating the furnace for the desired atomization temperature; signaling the autosampler to deposit the sample; and controlling the selected temperature ramps, hold times, and temperatures for multi-step programs. Computer control of the autosampler provides flexibility in delivering the desired sample to the furnace. Accurate programming of the pipetted volume and random access to the autosampler tray permit an almost limitless selection of sample volumes, dilution factors, chemical modifiers, sampling order, and repeat determinations. Complex quality assurance protocols can be implemented for running blanks, standards, samples, spiked samples, and quality assurance standards. The method of additions can easily be implemented on most instruments. Thus, a significant fraction of the sample preparation can be allocated to the computer and autosampler.

Data acquisition has been revolutionized by solid-state electronics. Logic for complex timing schemes involving multiple sources, pulsing at different frequencies, and gating of the detector is readily implemented. Analog intensity measurements, made with a PMT or a solid state detector, are immediately converted to the digital domain where all the data processing is conducted. In the digital domain, it is possible to use algorithms that were previously far too slow and cumbersome to be applied to real-time measurements. The data can be displayed in many different formats and formats can easily be changed at the analyst's request. Calibration curves can be constructed, standards added or dropped, and complex nonlinear fits made to the data. The data can be manipulated in almost every conceivable manner to meet the requirements of the analyst. As the acquired data and all the auxiliary information (sample identification, dilution factors, etc.) are stored in appropriate files in the computer, final reports can be generated in almost any format for analyte concentration, spike recoveries, and reference materials. With the low cost of computer memory and the proliferation of peripheral storage devices, vast amounts of data can be stored in an easily retrievable manner.

The computer hardware for GFAAS should be transparent to the analyst. Generally, it consists of a microprocessor with an ADC, a digital-to-analog converter, serial and parallel digital input and output, real-time clocks, and large amounts of memory. The extremely low cost of memory makes it possible to employ megabytes of random access memory and thousands of megabytes for hard disks. Obviously, the specific hardware will vary between manufacturers and instrument models.

The computer programs, or software, are highly variable between manufacturers, highly proprietary, and very expensive. Today, the software must be versatile and user friendly. Consequently, the cost and development time of the software for a GFAAS can equal or exceed that of the hardware. In general, manufacturers are very protective of their software and are reluctant to furnish the analyst with a copy (source listing) of the programs. This can be problematic, as it makes it impossible to verify the algorithm that the computer is using. Thus, the accuracy of data processing can only be tested by examining the final result. This places great emphasis on the use of reference materials to ensure that the instrument is operating properly and producing valid results. Still, the full capability of computer technology has yet to reach GFAAS. Economic considerations and the speed of technology development produce a lag between the capabilities found in AAS instrumentation and state-of-the-art technology. There are many aspects of instrument operation and quality assurance that have yet to be addressed. Time and the development of more sophisticated multielement instruments, however, will demand expanded computer power. Many manufacturers are already using off-the-shelf computers as instrument controllers rather than developing their own microprocessor system. Thus, the instrument manufacturers are getting out of the computer hardware business and very much into the development of computer software.

4.5 BACKGROUND CORRECTION

The importance of background correction for the computation of accurate absorbance signals has been emphasized in preceding sections of this chapter. It was pointed out that the need for background correction, combined with the transient nature of the GFAAS signals, has been the driving force in the development of new instrument designs. As stated previously, four separate background correction methods have been developed and will be considered in this section: line source with a secondary continuum source (LS/CSAAS), Zeeman splitting of the source or analyte absorption (ZAAS), self-reversal of the source (SRSAAS), and primary continuum source with off-line correction (CSAAS). The first three methods are commercially available and the fourth method is found only in prototype instrumentation. Six configurations of the four background correction methods listed above are considered in this section. Characteristics of these instruments are summarized in Table 4.11. In keeping with the goals of this chapter, the SNRs of these techniques are evaluated based on known attenuation of the source(s) and loss of analytical sensitivity. These SNRs assume similar performance for lamp operation, source intensities, transmission efficiency of the spectrometers and external optics, and data acquisition parameters. In practice, unique designs are used for all these instruments and the assumption of comparable operation is not valid. Instrument designs will compensate for known weaknesses. Thus, the SNR is intended only as a tool for evaluating potential advantages and weaknesses for specific methods.

4.5.1 General considerations

4.5.1.1 Accuracy

It must be emphasized that background correction only restores the accuracy of an absorbance measurement, not the loss in the SNR. Background correction cannot restore the intensity lost due to background attenuation. Consequently, accurate background corrected absorbances can be computed, but the SNR will reflect the attenuated intensity. For complex sample matrices with significant attenuation, precisions comparable to those obtained with a full strength source can only be achieved by removing the source of the attenuation, either chemically or physically. Every background correction method employs a minimum of two measurements (Eqn 4.54)): the analytical intensity, which is diminished by the analyte absorbance and the background attenuation; and the reference intensity, which is diminished only by the background attenuation. In principle, a single source can be used for both measurements. This, however, requires additional instrumentation to differentiate between the analytical and reference intensities.

The accuracy of the background correction is dependent on the temporal, spectral, and spatial matching of the radiation used for the analytical and reference intensities. Maximum accuracy is achieved if the two

Table 4.11 Background methods

Method	*LS/CSAAS transverse a.c. field*	*Direct ZAAS transverse a.c. field on source*	*Inverse ZAAS transverse a.c. field on furnace*	*Inverse ZAAS longitudinal a.c. field on furnace*	*SRSAAS*	*CS-AAS*
Sources	2	1	1	1	1	1
Optimum optical design	DDB[a]	DDB[a]	SB[b]	SB[b]	DDB[a]	SB[b]
Reference source	D_2 unaltered	LS magnetically shifted	LS unaltered	LS unaltered	LS broadened and self-reversed	CS (Xe arc) unaltered
FWHH (pm)	100–1000	22–28	0.5–2.0	0.5–2.0	3–10	26–80
Transmission	0.5	0.5	0.5	1.0	1.0	1.0
Sample source	LS unaltered	LS unaltered	LS unaltered	LS unaltered	LS unaltered	CS (Xe arc) unaltered
FWHH (pm)	0.5–2.0	0.5–2.0	0.5–2.0	0.5–2.0	0.5–2.0	9–26
Transmission	0.5	0.5	0.5	1.0	1.0	1.0
Analyte absorption	unaltered	unaltered	magnetically shifted	magnetically shifted	unaltered	unaltered
FWHH (pm)	2–8	2–8	24–36	24–36	2–8	2–8
Sensitivity	1.0	0.4–0.5	0.85–1.0	≈ 0.48	≈ 0.6	1.0
Susceptible interval for spectral Interferences (pm)[c]	100–1000	12–18	14–21	14–21	3–10	13–40
SNR (compared to double beam instrument)[d]	0.5	0.28–0.45	0.60–0.71	≈ 0.48	≈ 0.6	1.0

[a]DDB=dual double beam instrument.
[b]SB=single beam instrument.
[c]Interval to either side of line center within which line overlaps will cause an interference. Compare to $\Delta\lambda$ in Table 4.3.
[d]SNR as compared to a standard double beam instrument with comparable source intensity, optics, and data acquisition but no background correction.

intensities can be measured at the same time (temporal accuracy), at the same wavelength and with the same bandwidth (spectral accuracy), and over the same space, i.e. the radiation beams intercept an identical volume of space in the furnace (spatial accuracy). Significant deviations from these ideal conditions reduce the possibility of accurate background correction. Temporally, the analyte and background absorbance may change very rapidly during atomization, with each increasing by as much as 7–10 per second. Ideally, the analytical and reference intensities are measured simultaneously. For double beam in time optical designs (there are no double beam in space designs for commercial systems), this is not possible. To obtain temporal accuracy, two methods are generally employed. Either the measurements are made in a very rapid, sequential manner, or the intensity of one of the sources (usually the reference source) is mathematically extrapolated to provide an accurate estimate of the intensity at the time of the measurement of the second source.

It is generally acknowledged that 2 ms is the maximum allowable interval between the analytical and reference measurement. It has been shown that computation of background absorbances using only a single reference measurement, either preceding or trailing the analytical measurement, produces errors that have a first derivative function [83]. If the reference measurement precedes the analytical measurement, then, with increasing background attenuation, the measured reference intensity will be biased high (Figure 4.15) and the computed absorbance will be biased high. With diminishing background attenuation, the measured reference intensity will be biased low, as will the absorbance. Measurement of the reference intensity after the analytical measurement will reverse the preceding logic; absorbance will be biased low as attenuation increases and high as it decreases. The maximum error will depend on the rate of change of the background attenuation and the time interval between measurements. Thus, the absorbance computation error can be minimized by minimizing the time interval between measurements. With a 2 ms time interval, an error of 0.0036 is obtained for a background absorbance changing at a rate of 1.8 per second. In Figure 4.15, the error using a 9 ms time interval with a background absorbance changing at 1.8 per second is about 0.016. An important aspect of the first derivative nature of the error function is that the sum of the errors will be zero if absorbances are summed over the duration of the background signal. Thus, if a sufficiently long integration interval is used, integrated absorbance, unlike peak absorbance, is inherently less sensitive to temporal background correction errors.

The method of mathematical extrapolation is conceptually simple. Reference intensities preceding and following the analytical intensity are used to estimate the reference intensity at the time of the analytical measurement. Minimally, one reference intensity before and one reference

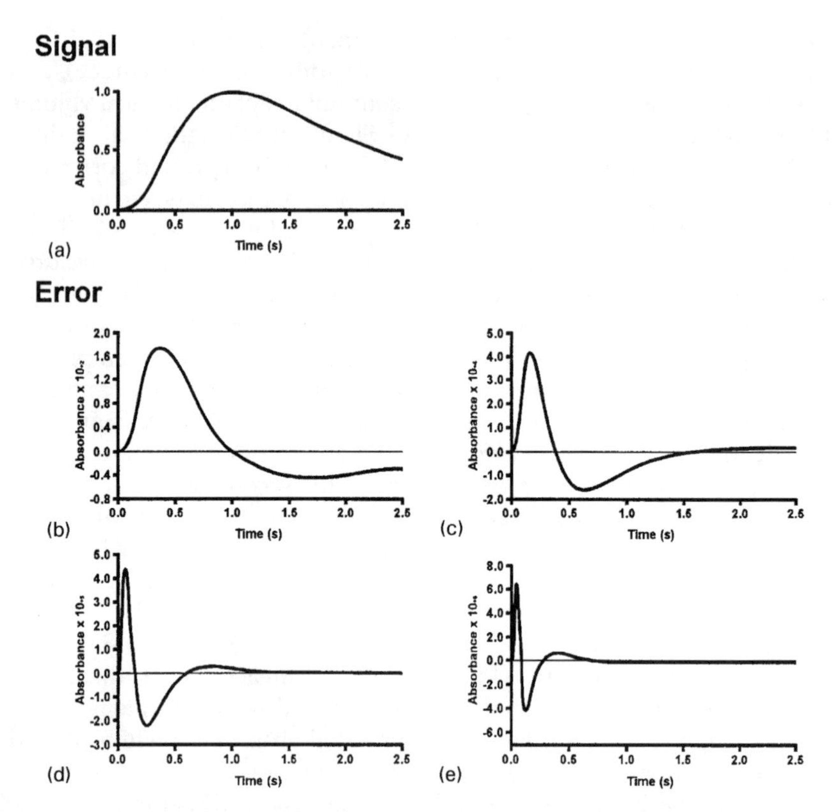

Figure 4.15 (a) Transient absorbance signal, and temporal absorbance computation errors when (b) one reference measurement, (c) two reference measurements, (d) three reference measurements, and (e) four reference measurements are used to compute the reference intensity at the time of the analytical measurement [84]. Note the differences in absorbancescales for (b)–(d).

intensity after the analytical measurement can be averaged to provide a linear extrapolation of the reference intensity. A nonlinear extrapolation can be implemented by using three or more reference intensities and a quadratic least-squares fit [84]. For both correction methods (linear and non-linear), the error functions are theoretically predictable. The maximum error will decrease dramatically as more reference intensities are used in the calculation (Figure 4.15). The error functions of the extrapolations for linear, nonlinear with three points, and nonlinear with four are proportional to the 2nd, 3rd, and 4th derivative of the background absorbance–time function. The decrease in the maximum error is dependent on the time between the measurements and the rate of change of the background absorbance. For a 56 Hz sampling frequency and a 1.8 per second change in background absorbance (Figure 4.15), improvement factors of 40, 320, and 2300 were obtained for the three extrapolation methods, respectively, compared to the error for nonextrapolated absorbances.

Spectrally, it is desirable for the analytical and reference measurements to be made at the same wavelength with identical spectral bandwidths. The ideal solution would be to make identical measurements and simply remove the analyte or the background attenuation from one of the measurements. Less than ideal, but still very practical, is the measurement of the background attenuation off line (shifted away from the analytical wavelength) but with the same spectral bandwidth. The least-ideal case is the use of significantly different spectral bandwidths for the two measurements. The last case allows an increased possibility of the background component being different for the two measurements, i.e. at structured background affecting the measurement with the larger bandwiths. Off-line background correction is standard for emission spectrometry. However, with high temperature sources, such as the ICP, the surrounding spectra may be complex and the exact position of the off-line reference measurement is critical. The surrounding spectra must often be carefully examined to find a wavelength region without structure. This problem is dramatically reduced for GFAAS but does occur. The lower temperature of the graphite furnace (compared to a source such as the ICP) results in much simpler spectra. The incidence of structured interferences, or interfering line overlaps, is low (Table 4.3). This is fortunate, as conventional AAS is operating blind. Without a broad band source, it is not possible to obtain the absorption spectra of the surrounding wavelength region.

For broad band background interferences, the four background correction methods mentioned at the start of this section offer accurate absorbance measurements. For these interferences, attenuation changes little over a range of tens of nanometers. By comparison, the spectral bandwidths and the wavelength shifts of the background correction methods for GFAAS are small. Consequently, the probability for accurate correction of broad band attenuation is the same for all four methods. This is not the case for structured interferences. Accurate background correction is assured only if the analytical and reference measurements are made at the same wavelength with the same spectral bandwidth. With off-line measurements, the probability of an inaccurate correction increases, because a different spectral region is being examined and because the effective bandwidth is doubled if measurements are made to both sides of the analytical wavelength. Fortunately, even with off-line measurements, the simplicity of the spectra at furnace temperatures means that the occurrence of line overlap interferences is still quite rare (Table 4.3). The probability of a line overlap interference for off-line measurements is independent of the distance of the wavelength shift if the spectral bandwidth is constant, i.e. the probability of an interfering absorption line falling within the spectral bandwidth is the same, whether the measurement is made 0.1 nm or 1.0 nm away from the profile center.

Spatially, maximum accuracy is achieved if the radiation beams employed for the analytical and reference measurements have identical and exactly overlapping spatial intensity distributions. The spatial resolution of GFAAS optical systems has only recently come under careful consideration. With the technique of shadow spectral filming, Gilmutdinov *et al.* [85] have shown that the radial distribution of the analyte and matrix is highly irregular. They have also shown that the optical image of an HCL is highly dependent on the lamp current [86] and that the spatial intensity distributions of HCLs, EDLs, and D_2 lamps are dramatically different [87] (see Chapter 2). These observations have demonstrated the potential importance of the spatial positioning of the radiation beams for the analytical and reference measurements. Research is continuing to quantify the errors arising from spatial mismatches.

4.5.1.2 *Sensitivity and calibration*

Background correction frequently introduces changes in the measurement of the analytical absorbance that affect the sensitivity and the shape of the calibration curve. These changes in the absorbance generally arise from optical changes that result in the reference intensity being sensitive to the analyte concentration. As a result, the reference intensity, like the analytical intensity, decreases as a function of concentration, although at a slower rate. The degree of loss of sensitivity and the shape of the calibration curve is a function of the background correction method.

Classical flame AAS employs reference beams that are totally unresponsive to the analyte concentration. This lack of response is assured by employing a reference intensity measured prior to the start of the aspiration (single beam) or by passing the reference beam around the flame (double beam). As a result the absorption coefficient of the analytical beam is the sole determinant of the analytical sensitivity. With increasing analyte concentration, the analytical intensity asymptotically decreases to a minimum value determined by the stray light. The computed absorbance, as a function of the analyte concentration in the radiation beam, asymptotically approaches a maximum value. Despite further increases in concentration, the computed absorbance remains constant. Accurate background correction for GFAAS makes it necessary for the reference beam to pass through the atomizer. All four background correction methods do this by employing radiation close to the analytical wavelength for the reference measurement. In each case, physical limitations restrict the distance the reference wavelength can be moved from the line center. A small, but finite absorption coefficient at the reference wavelength means that the reference intensity, like the analytical intensity, will decrease with increasing concentration, albeit at a slower rate. Figure 4.16 shows the effect of the gaseous analyte concentration in the absorption cell (flame or furnace) on the analytical and reference intensities, the computed absorbance and the relative concentration error. The reversal in the computed absorbance occurs because the analytical intensity has reached a minimum, but the reference intensity can continue to decrease with increasing concentration leading to decreasing absorbances.

The point of maximum absorbance is called the roll-over absorbance, A_r. The roll-over absorbance is determined by the fraction of stray light and the sensitivity ratio, R of the two wavelengths:

$$R = 1 - \frac{m_I}{m_{I_0}} \tag{4.71}$$

where m_I is the characteristic mass of the analytical intensity and M_{I_0} is the characteristic mass of the reference intensity. As R decreases the roll-over absorbance decreases. Some manufacturers express R as a fraction (1.0 is maximum), and some as a percentage (100 per cent is maximum). The roll-over of the absorbance as a function of gaseous analyte concentration produces unique analytical peak shapes at high concentrations for GFAAS. As the analyte is atomized in the furnace, the gaseous analyte concentration in the furnace will increase rapidly and then decrease giving the characteristic transient absorbance peak. As the maximum gaseous concentration increases, absorbance will increase until the roll-over

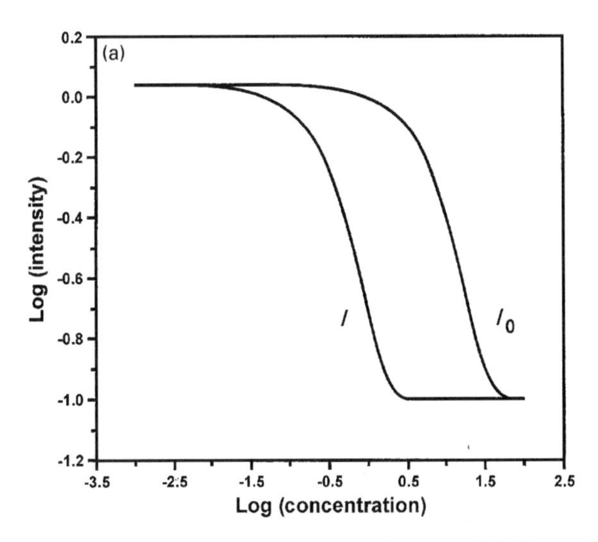

Figure 4.16 Modeled data for Zeeman AAS showing: (a) the analyte intensity I and the reference intensity I_0. Stray light α is 10 per cent and the reltive sensitivity R is 0.90

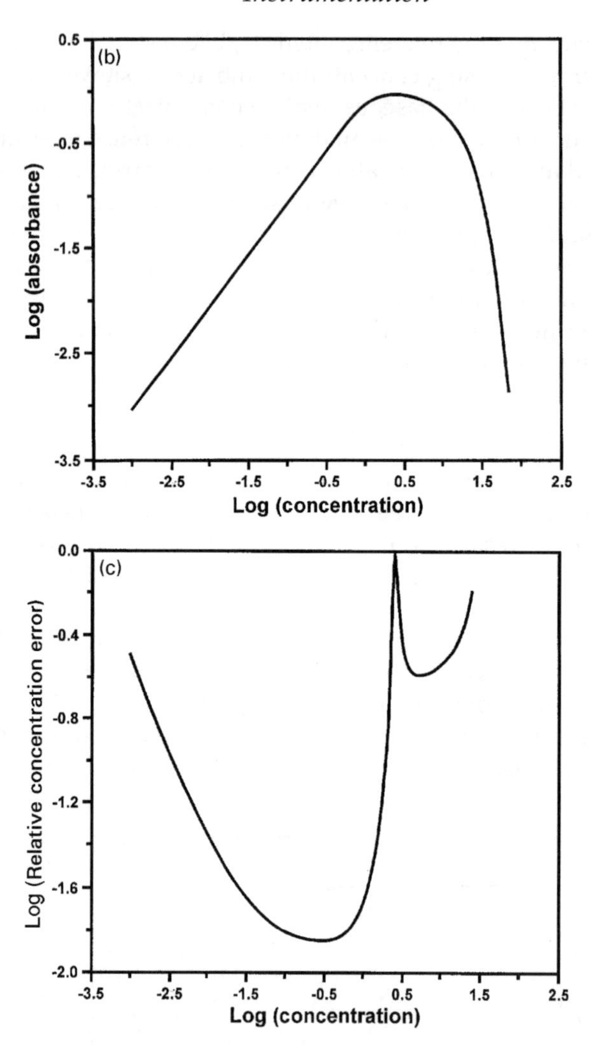

Figure 4.16 Modeled data for Zeeman AAS showing: (b) absorbance; (c) δ_c/C as a function of concentration. Stray light α is 10 per cent and the relative sensitivity R is 0.90.

absorbance is reached. At higher concentrations, the absorbance decreases. This decrease will continue until the gaseous analyte concentration has reached its maximum. As the concentration decreases, the absorbance will first increase and then decrease as the absorbance passes through the roll-over absorbance a second time. Thus, as higher masses of analyte are injected into the furnace, the analyte peaks develop a double maximum as a dip or reversal appears (Fig. 4.17). These dips become deeper as the injected mass increases. The minimum of the dip will eventually return to the baseline when the reference intensity reaches the stray light level (Fig. 4.16). The depths of the reversal can be used for a peak height calibration with a negative slope.

As a result of the absorbance roll-over, calibration curves for peak absorbance and integrated absorbance asymptotically approach a maximum value at high concentrations and may even show a slight reversal. For peak height, the maximum value is A_I^r, and is reduced compared to the maximum absorbance for nonbackground-corrected calibration curves. The minimum of the dip in the peak maximum and the distance between the twin peaks can both be used for calibration. All three measurements (peak absorbance and width and depth of the dip) are dependent on the peak shape (i.e. the convolution of the analyte appearance and loss functions), and are less accurate than the peak area. For integrated absorbance, the existence of a maximum value is in sharp contrast to the monotonically increasing (although not linear) absorbances observed for nonbackground-corrected measurements.

4.5.2 Line source and secondary continuum source

The use of a secondary continuum source for background correction was first described by Koirtyohann and Pickett [88], and was the first background correction system to be commercially developed. This method (LS/CSAAS), which is still commercially available today, employs a line source as the primary source and a continuum source, usually a D_2 lamp, as the secondary source. Ideally, both sources are operated in the double beam mode (dual double beam in Figure 4.1(c)). The principle, as embodied in Figure 4.18, is simplistic. The line source is used for the analytical measurement whereas the continuum source is used for the reference measurement. Although the analyte will also absorb the continuum

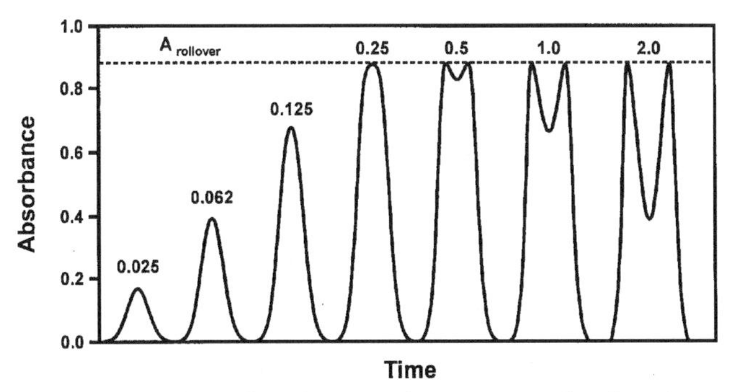

Figure 4.17 Modeled graphite furnace absorption peaks showing reversal due to close proximity of the reference intensity measurement to the line center. Stray light α is 10 per cent and the relative sensitivity R is 0.90.

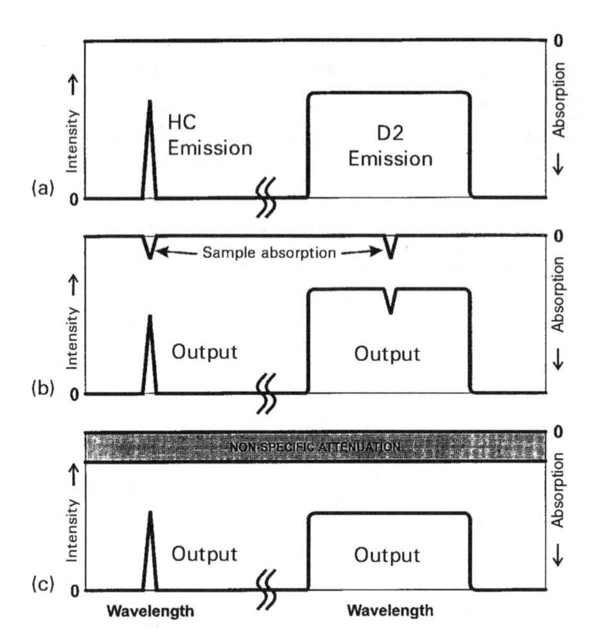

Figure 4.18 Operation of secondary continuum source background correction showing the effect of (a) no absorption, (b) atomic absorption, and (c) nonspecific background attenuation on the transmitted intensity of the HCL and the continuum source [90]. (Reprinted from H.L. Kahn, Background compensation system for atomic absorption, At. Absorpt. Newsl., 1968, **7**, 40–43, with kind permission from The Perkin-Elmer Corporation, Norwalk, CT, USA.)

radiation at the analytical line, the spectral bandwidth is so large that analyte absorption is insignificant compared to background attenuation. The absorbance of the continuum source is then subtracted from the line source absorbance to provide a background corrected absorbance (Eqn (4.3)). The characteristics of LS/CSAAS are summarized in Table 4.11.

Potential inaccuracies exist due to lack of temporal, spectral, and spatial matching of the sources, as considered in Section 4.5.1.1. Temporal errors can be remedied by reducing the time interval between measurements, by appropriate mathematical algorithms, or a combination of the two approaches. As separate sources are used, the first method is easily implemented by pulsing each source within the 2 ms interval. Spectral and spatial errors, unfortunately, are inherent in the method. A wide spectral bandwidth is needed for LS/CSAAS to ensure that the fraction of the continuum radiation absorbed by the analyte is not significant. As shown in Eqn (4.53) for a continuum source, the fraction of analyte absorption is inversely proportional to the spectral bandwidth. The fraction of nonspecific background attenuation of the continuum source (over a short wavelength region) and the fraction of analyte absorption of the line source are constant, independent of the spectral bandwidth. Thus, opening the entrance slit will decrease the analyte absorption for the continuum source and have no effect on the analyte absorption for the line source. Conventional monochromators with bandwidths of 0.1–1.0 nm, compared with 0.002–0.008 nm widths of the absorption profiles, easily minimize the continuum absorption of the analyte. As a result, the sensitivity ratio remains close to 1.0 and no reversal of the analyte peaks are seen. The large spectral bandwidth of the monochromator, as compared to the line source, dramatically increases the probability of structured absorbance interferences. All the interferences listed in Table 4.3 could potentially contribute inaccuracies for background correction using a secondary continuum source. The effectiveness of the background correction and its susceptibility to spectral interferences will be dependent on the element and the spectral bandwidth. In retrospect, the occurrence of line overlap interferences are not frequent, but they do exist. The large spectral bandwidth used with the secondary continuum source means that the problem cannot be ignored.

Spatial errors are also inherent in LS/CSAAS. To obtain accurate background correction, the radiation beams from two structurally different sources must be superimposed within the furnace. This necessitates placing masks in front of the sources and precisely aligning the two beams–a difficult task. As shown by Gilmutdinov *et al.* [87], despite these precautions, the spatial intensity distribution will still be different. Consequently, there is almost always a variation in the spatial sampling of the two beams. The success of LS/CSAAS over the years, however, suggests that the error in the background-corrected absorbance arising from the variation in spatial sampling is not significant for most elements and sample matrices.

The SNR of LS/CSAAS can be expected to be approximately 71 per cent of that for a double beam instrument. This reduction in the SNR is a result of the two sets of measurements being required for background corrections (compare Equations 4.19 and 4.20) and of the attenuation of both sources by the beam combiner. A half-silvered mirror is the only means of superimposing the two source beams and achieving reasonable spatial accuracy. Between 200 and 300 nm, where the D_2 lamp is most intense, the intensities of the sources are comparable. Attenuation of the sources by the beam combiner, however, still reduces the intensity of each source by 50 per cent. Above 300 nm the SNR for secondary continuum source background correction will deteriorate further as the D_2 lamp loses intensity, (Eqn 4.18). It must be remembered that the SNR in Table 4.11 is a theoretical estimate based on comparable lamp operation, intensities, optics, and data acquisition parameters. In practice, instrument designs will not be comparable.

4.5.3 Zeeman effect

Placing an atom in a magnetic field will cause splitting of an emission line or absorption profile into multiple components, based on orbital and spin angular momentum. This phenomenon is called the Zeeman effect and is the basis of the most common background correction method used for commercial instruments. Placing the magnetic field around the source (emitting atoms) is called the direct Zeeman effect, whereas placing the field around the furnace (absorbing atoms) is called the inverse Zeeman effect.

The Zeeman effect is predictable from quantum mechanics. Simplistically, the transition of an electron between orbits is characterized by a single emission or absorption line with a finite width. The natural width of these lines (natural broadening) arises from Heisenberg's uncertainty principle, and is primarily dependent on the finite lifetime of the excited state (when the lower energy level is the ground state). The energy states of the electrons within each orbital are almost identical, and their differences are insufficient to produce noticeable broadening or shifting of the profile. When a magnetic field is placed around the atom, the energy states within the orbitals are enhanced or decreased by the contribution of the electron orbital angular momentum. The allowed transitions permit an electronic transition to occur at approximately the same energy (as observed with no magnetic field), the π component, or at discretely higher and lower energies, the σ component. Thus the normal Zeeman effect splits the spectrum of a single transition into three components, which are distinguishable by a wavelength shift and difference in polarization.

The magnitude of the wavelength shift of the σ components away from the line center is dependent on the strength of the magnetic field and is approximately 10 pm for a field of 8–12 kG. The π component is polarized parallel to the magnetic field and the σ components are polarized perpendicular to the magnetic field. In some cases, the electron spin angular momentum also contributes to the energy levels within the orbit and produces internal splitting of the π and σ components, as shown in Figure 4.19. These anomalous Zeeman effect splitting patterns increase the width of the π and σ components and increase their overlap. Like the shift of the σ components away from the line center, the internal splitting of the π and σ components is dependent on the strength of the magnetic field. The magnetic field can be either perpendicular or parallel to the optical path. If the magnetic field is perpendicular or transverse to the optical path, then both the π and σ components are present, either emitted by the source or absorbed by the analyte. If the magnetic field is parallel or longitudinal to the optical path (by drilling holes in the magnets to pass the radiation), only the σ component is present. The π component cannot be observed by the detector.

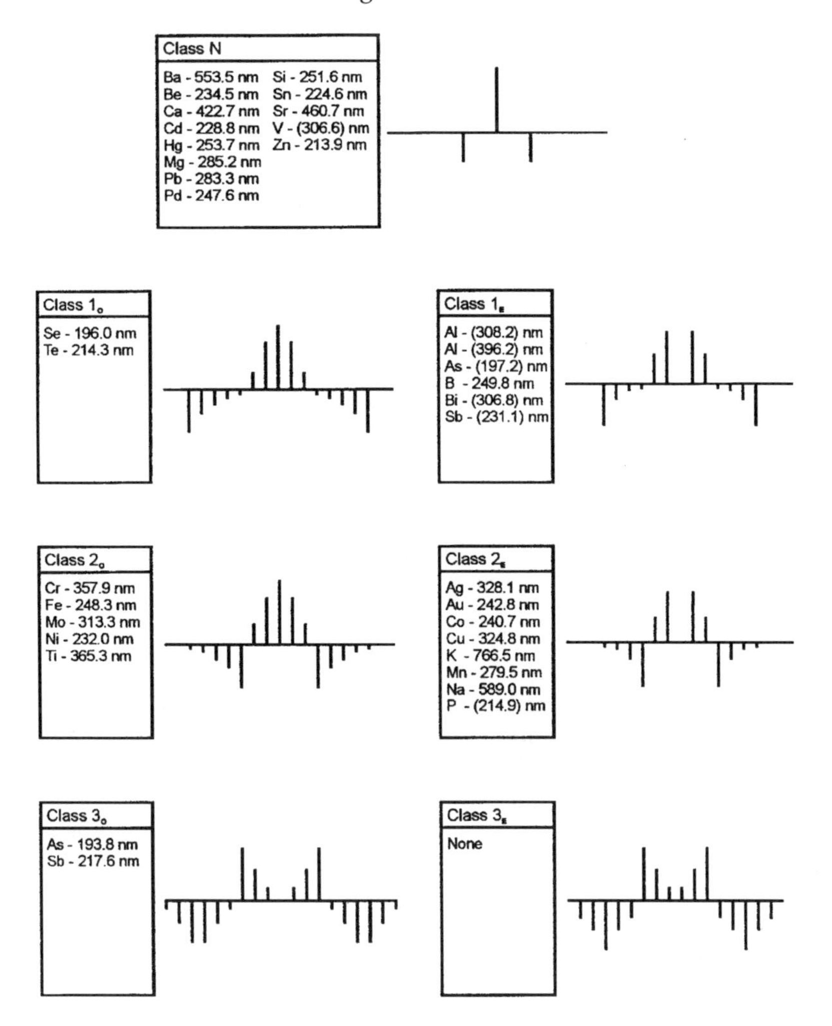

Figure 4.19 Zeeman splitting patterns: one normal (Class N) and 3 abnormal (Classes 1, 2 and 3) patterns with odd (O) and even (E) groups noted by the subscripts [90]. (Reprinted from B. Welz, Atomic Absorption Spectrometry, 2nd Completely Revised Edition, 1985, p.141, with kind permission from Wiley-VCH Verlag GmbH, Weinheim, Germany.)

The key to the success of ZAAS is the orthogonal (right angle or uncorrelated) polarization of the π and σ components. This is depicted in Figure 4.19 by the positive and negative direction of the π and σ components, respectively, from the horizontal axis. Despite their close spectral proximity (about 10 pm), the two components can be determined inde-

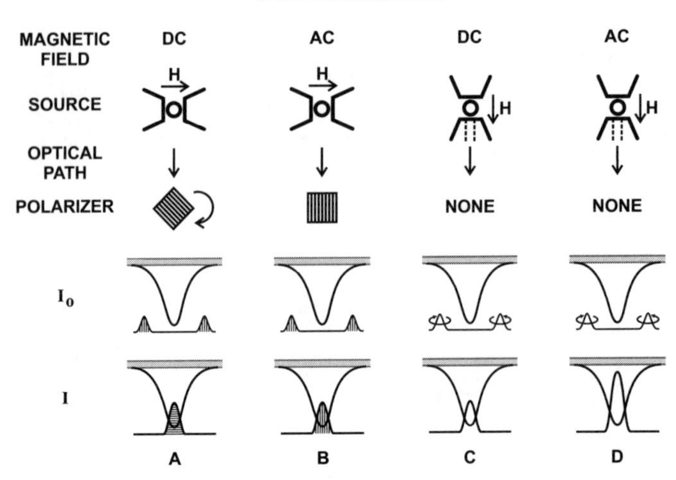

Figure 4.20 Possible configurations for a direct Zeeman effect background-correction instrument: (a) fixed transverse magnetic field with rotating polarizer; (b) alternating transverse field with fixed polarizer; (c) fixed longitudinal field; (d) alternating longitudinal field. The arrow denotes the direction of the magnetic field, H.

pendently on a medium-resolution spectrometer (a SBW of 0.03) using a polarizing filter. The principle of ZAAS is relatively simple–the magnetic field is used to either shift the absorption profile away from the analytical line or shift the analytical line away from the absorption profile to allow an intensity measurement that reflects only the background attenuation. This measurement satisfies the requirements for a reference intensity, I_b in Eqn (4.5), and allows computation of a background-corrected absorbance. The possible combinations of the direct and inverse Zeeman effect, transverse and longitudinal magnetic fields, and d.c. and a.c. operation are shown in Figures (4.20) and (4.21). The three ZAAS instruments found commercially (at the time of writing this chapter) are: (1) the transverse a.c. field on the source (Figure 4.20(c)); (2) the transverse a.c. field on the furnace (Figure 4.21(b)); and (3) the longitudinal a.c. field on the furnace (Figure 4.21(d)).

4.5.3.1 Transverse AC magnetic field on the source

The first conceptualization of ZAAS in the mid 1970s employed a d.c. magnet around an HCL (direct Zeeman effect) transverse to the optical path (Figure 4.20(a)). In this configuration both the π and σ components were present simultaneously and continuously. They were measured in a rapid sequential manner by placing a rotating polarizer in the radiation beam. The radiation of the π component served as the analytical intensity I

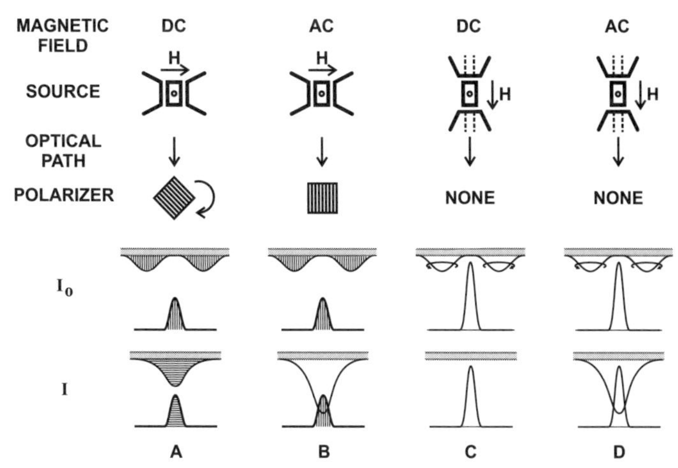

Figure 4.21 Possible configurations for an inverse Zeeman effect background correction instrument: (a) fixed transverse magnetic field with rotating polarizer; (b) alternating transverse field with fixed polarizer; (c) fixed longitudinal field; (d) alternating longitudinal field. The arrow denotes the direction of the magnetic field, H.

was diminished by analyte absorption and background attenuation. The radiation of the σ component served as the reference intensity I_b and was diminished only by the background attenuation. Use of these two intensities to compute an absorbance (Eqn (4.4)) resulted in a background-corrected absorbance.

Compared to a double beam instrument with no background correction, the SNR for a transverse d.c. field on the source is reduced by 29 per cent. This estimate is based on the known reduction of the analytical and reference intensities by a factor of two, a result of splitting the emitting atoms evenly between the π and σ components. This estimate is also based on the assumption that photon shot noise is limiting and that all source, magnetic field, and data acquisition parameters are identical for the two instruments. In practice, three difficulties were encountered: the magnetic field destabilized the HCL; the analytical sensitivity was extremely dependent on the strength of the magnetic field and was considerably reduced compared to sensitivities with no background correction; and the rotating polarizer was found to introduce fluctuation noise which increased the baseline absorbance noise. As a result of these problems, the SNR for a transverse d.c. field around the source was degraded by more than 29 per cent.

It is easy to understand why a magnetic field can have a deleterious effect on an HCL. As discussed in Section 4.3.1.1, the emitting atoms are sputtered from the cathode surface and are excited by high energy electrons. The sputtering process and the high excitation temperature of the plasma within the cathode lead to a relatively large number of cations. Both cations and electrons will be influenced by a strong magnetic field. Conceivably, the positioning of a magnetic field around the cathode can lead to increased fluctuation noise as well as shortened lamp life. Some of these difficulties have been overcome through changes in the lamp design. The dependence of the analytical sensitivity on the strength of the magnetic field is primarily a result of the anomalous Zeeman effect splitting patterns of many of the elements (Figure 4.19). For the normal Zeeman effect, the larger the magnetic field strength, the greater the shift of the σ components away from the center of the absorption profile, the less the analyte absorption of the reference intensity, and the closer the analytical sensitivity approaches that obtained without background correction.

With anomalous Zeeman effect splitting patterns, increasing the strength of the magnetic field increases the average shift of the σ components away from the absorption profile but also increases the internal splitting of both the π and σ components. Too strong a magnetic field causes the internal lines of the π component to be shifted away from the absorption profile. This leads to lower sensitivity (by decreasing the effective absorption coefficient) and increased nonlinearity (by increasing the nonmonochromaticity of the source). The balancing of these two demands–a strong field to increase the shift of the σ components and a weak field to reduce the width of the π component–leads to a high dependence of the analyte sensitivity on the magnetic field strength. Maximum sensitivities were found at intermediate field strengths, which varied with the element and ranged from 3 to 6 kG [31]. The maximum sensitivities averaged 60–70 per cent less than those obtained with no background correction and background attenuation. Thus, the SNR of the transverse d.c. field on the source was significantly degraded compared to operation with no background correction.

The problems of analyte sensitivity and fluctuation noise introduced by the d.c. magnetic field with a rotating polarizer were eliminated by using an a.c. field and a fixed polarizer (Figure 4.20(b) and Table 4.11). With this modification, the analytical intensity is measured with the magnet off, the reference intensity is measured with the magnet on, and the polarizer is held fixed, perpendicular to the magnetic field. The fixed polarizer contributes no fluctuation noise and a magnetic field strength of approximately 9 kG is sufficient to shift the σ components away from the absorption profile. Consequently, sensitivities were obtained which were only 10–30 per cent [31] less than those obtained with no background correction.

With respect to temporal errors, the design of an instrument based on a transverse a.c. magnetic field around the source, like all ZAAS instruments, is less flexible than LS/CSAAS in reducing the time interval between the measurement of the analytical and reference intensities and, hence, is more dependent on correction through the use of mathematical algorithms. The data acquisition frequency is limited by the frequency of the a.c. magnetic field. The field cannot be changed rapidly enough to allow measurement of the analytical and reference intensities within a 2 ms interval. The time intervals are usually symmetrical, i.e. the time interval between the reference and analytical measurement, and the interval between the analytical and the following reference measurement are equal. Thus, a nonlinear extrapolation is the preferred method of computing the appropriate reference intensity at the time of the analytical measurement (Section 4.5.1.1).

Spatially, the potential for errors for ZAAS with a transverse a.c. field around the source would appear to be less than those for LS/CSAAS. As the same source is used for the analytical and reference intensities, the instrument has a single beam design and inherently the same spatial region of the furnace is sampled for both measurements. The possibility, however, exists that the total intensity and the spatial intensity distribution may change systematically in response to the oscillation of the magnetic field. If the discharge is limited to inside the hollow cathode (which is normal for HCL operation), the problem would seem to be negligible, as the image size and path through the furnace would be unaffected. If the magnetic field draws the discharge out of the cathode, then the spatial intensity distribution and the spatial overlap of the radiation in the furnace, with and without the magnetic field, are significant problems.

Spectrally, the potential for errors for ZAAS with a transverse a.c. field around the source is dramatically reduced compared to LS/CSAAS. As with LS/CSAAS, the analytical intensity is only measured over the width of the HCL line (approximately twice the spectral bandwidth, or 1–4 pm). Unlike LS/CSAAS, the reference intensity is measured at two wavelengths, just ± 10 pm from the line center. The FWHH of the split lines will cover 21–24 pm (20 pm difference between the σ components plus twice the 0.5–2.0 pm width of the lines). With the anomalous Zeeman effect, the width of each of the σ components is roughly doubled, and the FWHH is 22–28 pm, whereas the baseline interval is approximately 24–36 pm. Compared to broad band nonspecific background attenuation, this interval is extremely small and accurate background correction is not a problem. Line overlap interferences will be a problem if they overlap the unsplit or split source. Thus, line overlaps are potential interferences if they fall within 12–18 pm of the line center. An examination of Table 4.3 reveals that very few of these overlaps are a potential problem.

It is interesting that no commercial version of a longitudinal a.c. field around the source has been developed (Figure 4.20(d)). This approach has the advantage that no polarizer is required. The reference intensity is reduced by a factor of two, as only half the atoms emit σ radiation when the magnet is on, but the analytical intensity is not attenuated by a polarizer in the optical beam. As a result the SNR for a longitudinal a.c. field around the source is theoretically superior to that for the use of a transverse a.c. field, but still not as good as that obtained with no background correction. Unfortunately, HCLs would have to be redesigned in order to fit between the poles of the magnet.

4.5.3.2 Transverse a.c. magnetic field on the furnace

The next major step, commercially, was the development of an instrument based on the inverse Zeeman effect. In principle, both approaches, direct and inverse, produce equally accurate background correction. Placing the magnet field around the furnace provides simpler optical designs and avoids complications arising from source instability introduced by the magnetic field. This design, however, requires the magnets to be extremely close to the furnace to achieve the desired field strength. When this chapter was written, the market was dominated by instruments employing the inverse Zeeman effect. The inverse Zeeman effect (Figure 4.21) is analogous to the direct Zeeman effect, except the analyte atoms in the furnace are being polarized. As the source is not in a magnetic field, the polarization of each emitted photon is random, so approximately 50 per cent of the total source intensity will contribute to each orientation (perpendicular or parallel). Analyte atoms polarized parallel to the magnetic field around the furnace (the π component) and analyte atoms polarized perpendicular to the magnetic field (the σ component) will absorb source radiation with the same polarization. As the background attenuation is unaffected (not polarized) by the magnetic field, it will attenuate all of the source intensity equally.

When based on a transverse a.c. field around the furnace (Figure 4.21(b) and Table 4.11), ZAAS measures the analytical intensity with the magnet off and the reference intensity with the magnet on. With the magnetic field off, 100 per cent of the analyte atom population is capable of absorbing the source radiation. With the magnet on, the absorption by either component (π or σ) will be half the value expected with no background correction, because the analyte population is divided evenly between the two polarities. This is immaterial to the measurement process, however, as the purpose of the magnetic field is to shift the analyte absorption away from the HCL line. The analytical sensitivities for a transverse a.c. field around the furnace are 85–100 per cent of those obtained with no background correction. The source intensity is attenuated by 50 per cent by the fixed polarizer, which further reduces the SNR. As a result, the SNR is reduced by 29 per cent or more, compared with the case of no background correction, and is comparable to that obtained with a.c. transverse field around the source.

The potential for temporal errors for ZAAS with a transverse a.c. field around the furnace is identical to that for a transverse a.c. field around the source. The frequency of data acquisition for the analytical and reference intensity is limited by the frequency of the a.c. field. A nonlinear mathematical extrapolation is necessary to compute the appropriate reference intensity at the time of the analytical measurement (Section 4.5.1). Spatially, the potential for errors with ZAAS is close to nonexistent. With inverse ZAAS, radiation from the same source and of the same intensity is used for the analytical and reference measurements. The only possible source of spatial inaccuracy would be a systematic shift of the position of analyte atoms in the furnace in response to the magnetic field. Such an interference could take two possible forms: shifting of the absorbing atoms in response to the a.c. field; or varied degrees of shifting of the analyte as a function of the matrix. Neither of these interferences is likely, however, because the magnetic field will primarily affect ions and it is well documented that the furnace is a very poor ionization source. Spectrally, the potential for errors for ZAAS with a transverse a.c. field around the furnace can be shown to be slightly greater than that for a transverse a.c. field around the source. As the source (for transverse a.c. field) is not shifted by the magnetic field, correction for broad-band attenuation is made at the wavelength of interest over the measurement interval of 1–4 pm (as discussed for a.c. transverse source around the source). With the magnet on, overlap interferences can arise if the σ components of an interfering element overlap the HCL line. As the shift of the a components is ± 10 pm, and the FWHH of the absorption profiles range from 2 to 8 pm, the influence of an interfering line can extend 12–18 pm from the line center with a FWHH of 24–36 pm. For anomalous Zeeman splitting patterns, the width of the σ components is roughly doubled and the interval sensitive to line overlap interferences is 28–42 pm. This is twice the interval computed for a transverse a.c. field around the source. An examination of the wavelength differences in Table 4.3 shows that there are very few potential interferences.

4.5.3.3 Longitudinal a.c. magnetic field on the furnace

The most recent commercial ZAAS instrument is based on a longitudinal a.c. magnetic field around the furnace (Figure 4.21(d) and Table 4.11). The σ components, which are polarized perpendicular to the magnetic field, appear circularly polarized when viewed parallel to the magnetic field. The π component, which is polarized in the direction of the magnetic field, is invisible to the detector when viewed parallel to the field. As a result, the measurement scheme for a longitudinal a.c. magnetic field around the furnace is identical to that for a transverse a.c. field around the furnace, except that no polarizer is required. As there is no attenuation of the source by a polarizer, the SNRs are comparable to those with no background correction. The use of a longitudinal field requires that the length of the furnace be reduced to maintain the desired magnetic field strength and that the electrical contacts for the furnace be made from the side. Both of these requirements have resulted in a shorter (19 mm versus 28 mm, see Table 4.4), transversely heated furnace. The shorter furnace length results in shorter residence times and, consequently, reduced sensitivities. Transverse heating offers better spatial isothermality (than the longitudinal heating), more efficient atomization of the analyte, lower atomization temperatures, and reduced memory between atomizations. In all, the shorter furnace results in a factor of 2.5 decrease in sensitivity.

The SNRs for ZAAS with a longitudinal a.c. magnetic field around the furnace can be expected to be only 35–40 per cent of those for a double beam instrument if it is assumed only the furnace has been changed. This decrease in the SNRs arises primarily from the shorter furnace. The a.c. magnetic field gives 85–100 per cent of the sensitivity obtained for operation of the shorter furnace without a magnetic field. Fortunately, neither the analytical nor reference intensities nor the analyte population are reduced by the magnetic field. This provides an SNR advantage compared to the other background correction methods. In general, detection limits for the commercialized instrument using ZAAS with a longitudinal a.c. magnetic field around the furnace are comparable to those for other ZAAS instruments and other background correction methods. This implies that other optical improvements have been made to reduce the noise level in order to offset the decrease in sensitivity. This is indeed the case as discussed in Section 4.6.3.

The temporal error considerations are identical to those for instruments using a transverse a.c. field around the source and a transverse a.c. field around the furnace. Spatially, the potential for errors for ZAAS with a longitudinal a.c. field around the furnace is identical to those for ZAAS with a transverse a.c. field around the furnace. As the instrument has a single beam design, the analytical and reference beams are spatially identical. Spatial inaccuracy arising from a systematic shift of the position of analyte atoms in the furnace in response to the magnetic field are unlikely, as discussed for a transverse a.c. field around the furnace. Spectrally, the potential for errors from interferences is identical to that discussed for ZAAS with a transverse a.c. field around the furnace. Background correction is implemented using intensities measured at the line center. Correction for broad band interferences is expected to be spectrally accurate. As the absorption profile is broader than line source emission lines, ZAAS with a.c. longitudinal field around the furnace is slightly more susceptible to line overlap interferences than a transverse

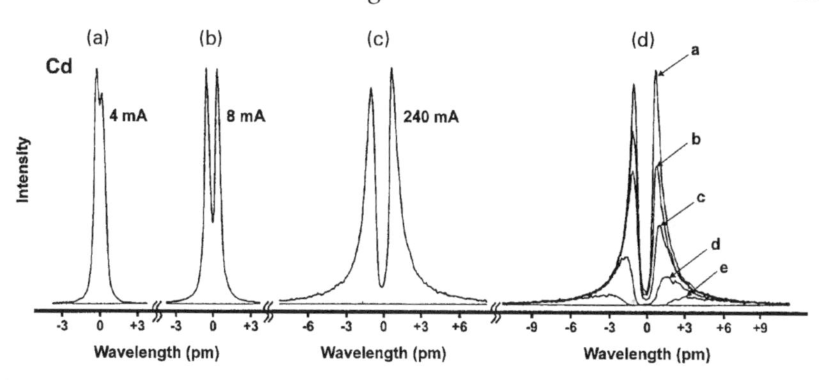

Figure 4.22 High resolution spectral scans in an air–acetylene flame showing the shape of the Cd (228.8 nm) HCL line at (a) 4 mA, (b) 8 mA, (c) 240 mA; (d) 240 mA with absorption by 0, 2, 5, 20, and 100 μg ml^{-1} solutions (curves a–e, respectively) [91]. (Reprinted from *Spectrochim. Acta*, Vol. **43B**, P.L. Larkins, The effect of spectral line broadening on the shape of analytical curves obtained using pulsed hollow-cathode lamps for background correction, pp. 1175–1186, 1988, with kind permission of Elsevier Science – NL, Sara Burgerhartstraat 25, 1055 KV Amsterdam, The Netherlands.)

a.c. field around the furnace, but the list of potential interferences within 14–21 pm is very small (Table 4.3).

4.5.4 Self-reversed source

A description of the use of a self-reversed source for background correction (SRSAAS) was first published by Baranova *et al.* [46], and was soon followed by the work of Smith and Hieftje [47]. As the name implies, background correction is accomplished by pulsing an HCL to a very high current (200–400 mA) to produce self-reversal of the HCL emission line. At high operating currents, a cloud of ground state atoms is formed, just in front of the cathode, which absorbs the radiation emitted from the cathode. The resultant self-reversed profile of the emitted line, shown in Figure 4.22 [91], is similar to that of the σ components for the Zeeman effect, although the wavelength shift is not as great. Thus, the analytical measurement is made with the HCL operated at normal currents and the reference measurement is made with the HCL pulsed to a high current.

The degree of self-reversal, or the extent of splitting of the source, is dependent on the magnitude and length of the high current pulse and the volatility of the element. In general, the shift of the source is sufficient to produce a 40 per cent decrease in the analyte absorption, although there are several elements for which self-reversal is poor, even at the highest currents. The fraction of absorption by the reference radiation is comparable to that of ZAAS with a transverse a.c. field around the source.

Thus, the sensitivity of the self-reversed source method ranges from 20 per cent to 90 per cent of that obtained without background correction, and averages about 60 per cent. The high analytical sensitivity of the reference radiation, coupled with normal amounts of stray light, predicts roll-over of the calibration curve, as shown in Figure 4.16. In practice, however, roll-over for SRSAAS does not appear to be as severe as that encountered with all forms of ZAAS. It was suggested by de Galan and de Loos-Vollebregt [92] that the lack of roll-over was a result of the increased stray light arising from the high current pulse. More accurately, the lack of roll-over was due to the large increase of nonmonochromatic radiation within the spectral bandwidth. Larkins [91] confirmed this experimentally. Figure 4.22 shows the emission profile of cadmium (228.8 nm) at 4, 8, and 240 mA, and with absorption by 0, 2, 5, 20, and 100 μg ml^{-1}. With a high current pulse, the HCL line width (FWHH) is dramatically broadened as compared to operation at normal currents. More importantly, at these high currents the wings of the HCL profile will provide an increasing fraction of the total emission. As a result, there is a large increase in radiation at wavelengths that are not efficiently absorbed by the analyte. The results in Figure 4.16 were obtained using a model that assumed 10 per cent stray light (light not absorbable by the analyte). If the simplifying assumption is made that the polychromatic radiation in Figure 4.22 acts like stray radiation, then the model yields the analytical and reference intensities and computed absorbance shown in Figure 4.23. In this case, it is hypothesized that the 10 per cent stray radiation associated with normal HCL operation increases to 40 per cent with a high current pulse. The roll-over of the absorbance in Figure 4.23 is much less severe than that observed in Figure 4.16. The roll-over could be easily transformed into a plateau by assuming more stray light.

Physically, the design of a SRSAAS instrument is single beam in nature. The beam of the source is altered (self-reversed), however, to produce a suitable reference radiation for background correction. The effect of the high current on the lamp image, the spatial intensity distribution, and the precision of operation are unknown. Of primary concern for the SNR, is the reproducibility of the high current pulse. Such a high current pulse could introduce a fluctuation noise component that will increase the absorbance noise. Commercially, a dual double beam instrument and single beam instrument have been developed. The dual double instrument uses a reflecting mirror with a hole in the center to reflect a portion of the source to a second PMT–double beam in space. In this manner, a single absorbance is based on four intensity measurements. The single beam instrument ratios the analytical and reference intensities to values acquired prior to atomization [47]; each absorbance computation is based on two intensity measurements. This approach assumes that the precision of the self-reversed radiation is comparable to a conventional HCL operated in the DC mode or in the normal pulsing mode (Section 4.3.1.1).

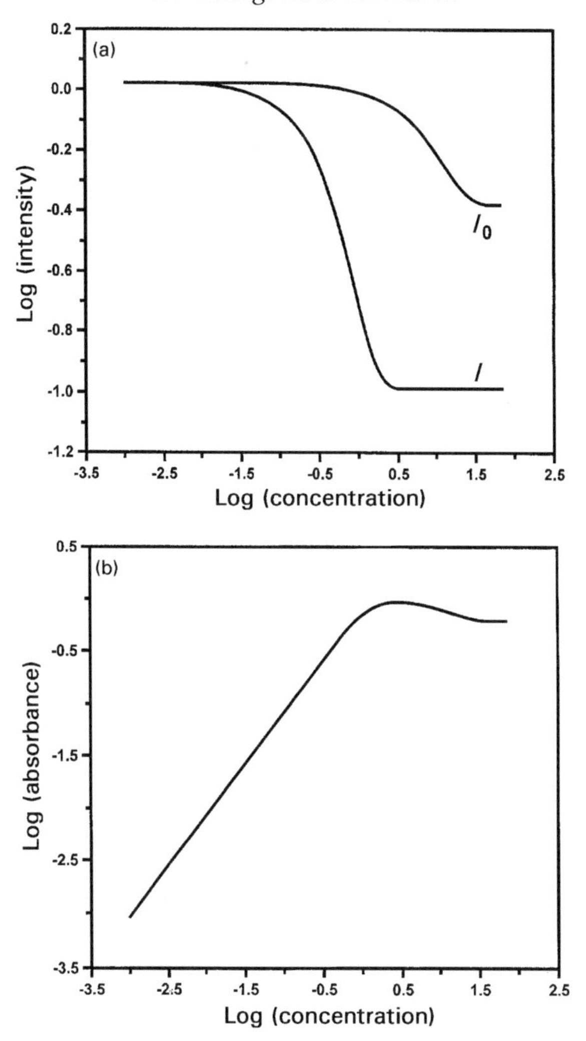

Figure 4.23 Modeled data for self-reversed AAS showing: (a) the analyte intensity I, and reference intensity I_0; and (b) absorbance. Stray light (α) was 10 per cent at the normal lamp current and 40 per cent with the pulsed current, and the relative sensitivity R is 0.90.

Assuming that shot noise is limiting, the SNR for SRSAAS will be worse than that for a double beam instrument. The background correction process does not attenuate the source intensity but the overall sensitivity is reduced by 40 per cent since the reference measurement is made close to the absorption file. The pulsing of the lamp for background correction if it does not introduce a fluctuation noise, may actually reduce the baseline absorbance noise (Eqn (4.18)) and bring the SNR closer to that of a double

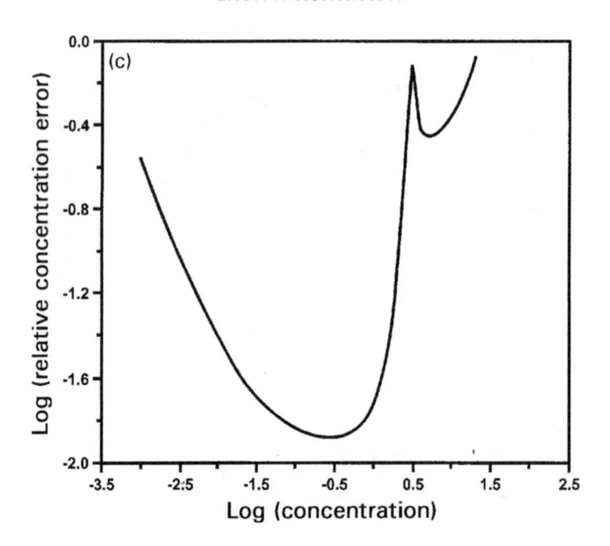

Figure 4.23 Modeled data for self-reversed AAS showing: (c) σ_c/C as a function of concentration. Stray light (α) was 10 per cent at the normal lamp current and 40 per cent with the pulsed current, and the relative sensitivity R is 0.90.

beam instrument. However, the reduction in the analytical sensitivity, resulting from the small shift of the reference intensity, results in a 40 per cent loss in the SNR (Table 4.11). The spatial accuracy of SRSAAS is problematic. The high current pulsing of the HCL can change the source image, the spatial intensity distribution, and the spatial sampling volume within the furnace. Recent data show that modest pulsing of a HCL changes the shape of the image and the spatial distribution of the intensity. It must be emphasized that there are no data with respect to high current pulsing. It would be reasonable to assume, however, that the distortion of the image and the intensity distribution is more extreme than that observed at lower pulse currents. Temporally, SRSAAS is easily adapted to a short time interval between the analytical and reference measurement. In most designs, the lamp is turned on at a normal current and the analytical measurement is made. The lamp is then pulsed to a very high current and the reference measurement is made. These measurements are generally made within a 2 ms interval. It must be assumed that the cloud of ground-state atoms forms very rapidly in front of the cathode. The relaxation time of the HCL, or the time necessary for the cloud of ground-state atoms to disperse from in front of the cathodes, is dependent on the design of the HCL and the amplitude of the high current pulse. Although there are no published data the pulsing frequency of various instruments suggest that it is between 20 and 50 ms. Although this does not affect the temporal accuracy in terms of correcting for background attenuation, it has serious implications for the capability of the instrument to accurately characterize the transient absorbance signal of the graphite furnace. Operation at 20 Hz or less is marginal for accurate measurement of the analytical signal.

The SRSAAS method is less susceptible to spectral errors than any of the other background correction methods. As illustrated in Figure 4.22, the FWHH for the self-reversed lamp line ranges from 3 to 10 pm with a baseline range of approximately 6 to 20 pm. These data suggest that line overlap interferences that fall within 3–10 pm of the line center could cause problems.

4.5.5 Continuum source/off-line correction

Continuum source/off-line background correction (CSAAS) has only been employed on prototype instruments. Conceptually, it is analogous to off-line background correction used with emission spectrometry. As the source intensity is available over the spectral interval of interest, reference intensities attenuated only by the background can be made to either side of the profile, whereas analytical intensities are measured over the absorption profile (Figure 4.24). Absorbance is then computed using Eqn (4.45). With a high-resolution echelle spectrometer, the reference intensity can be measured within a few picometers of the line center. In this manner, CSAAS is similar to direct ZAAS. The continuum source, however, does not have to be altered for the reference measurement. As for inverse ZAAS, the source is operated normally for both the analytical and reference intensity measurements.

Historically, CSAAS progressed through a series of designs that reflected existing state-of-the-art in technology. Initially, mechanical wavelength modulation was used with PMT detection and a lock-in amplifier to quantify the analytical signal [93]. Wavelength modulation was implemented using a quartz refractor plate mounted on a galvanometer. The wavelength shift induced by the refractor plate is described by Snell's law and is dependent on the refractive index of the quartz, and the angle of rotation and thickness of the refractor plate. As the refractor plate was oscillated, rotated first one direction and then the other, the spectral region viewed by the PMT was scanned across the absorption profile (Figure 4.24(a)). Scanning across the profile (from one extreme of the profile to the other) produced an analytical signal at twice the refractor plate frequency. The furthest distance the wavelength could be shifted from the line center was determined by the rotational limit of the galvanometer ($\pm 12^{\circ}$) on which the refractor plate was mounted. The speed with which the wavelength could be shifted was determined by the mass of the armature of the galvanometer. At higher frequencies (28 Hz for the galvanometer, 56 Hz for the spectrum) the modulation waveform was restricted to modified sine waves.

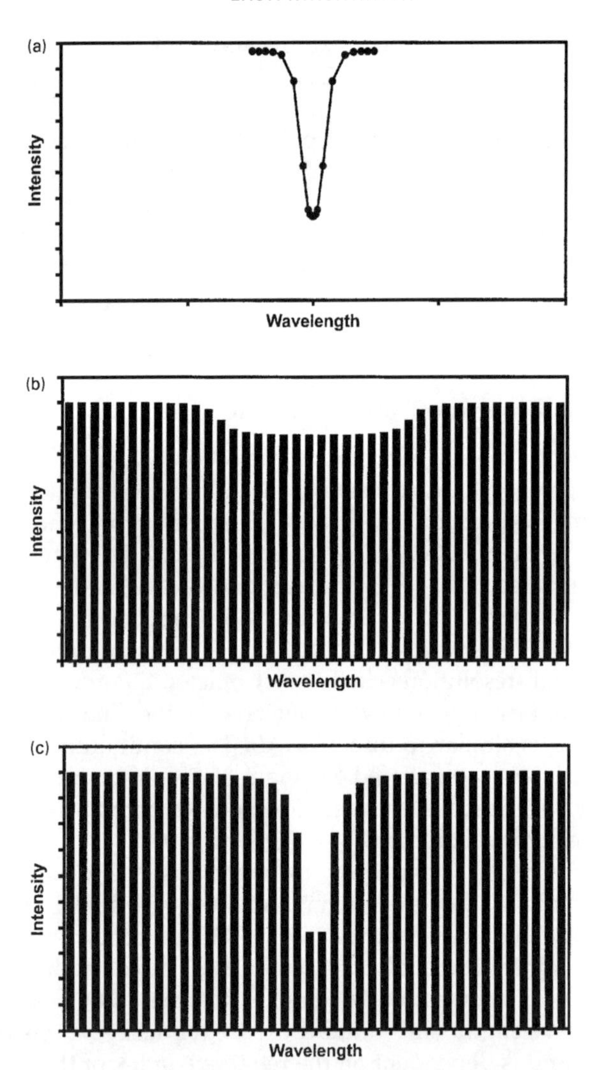

Figure 4.24 Absorption profile measured using a continuum source and: (a) PMT detection with wavelength modulation; (b) LPDA detection with a 500 μm entrance slit width; (c) CCD detection with 25 μm entrance slit width.

The versatility of the measurement process and the SNR of CSAAS were improved significantly with the substitution of a linear photodiode array (LPDA) for the refractor plate and the PMT [94]. This method of detection allows ease in selecting the portions of the spectrum to be used in the absorbance calculation, and requires no mechanically moving parts. In the initial design, a 256 pixel LPDA was dedicated to viewing the spectra around a single absorption profile. Figure 4.24(b) presents a model of the spectrum observed for cadmium (228.8 nm) with a 500 μm entrance slit width and a 25 μm pixel width [95]. At 228.8 nm, the cadmium absorption profile has a width of 2.7 pm, and the spectral width of a 25 μm pixel is 2.2 pm. Improved SNRs were achieved with the LPDA, due to the multiplex advantage of simultaneous detection at all wavelengths and the higher quantum efficiency of the LPDA (as compared to a PMT) below 220 nm. The read noise of the LPDA exceeded the shot noise in the ultraviolet region, but the relative contribution of the read noise is inversely dependent on the square root of the spectral bandwidth (Eqn (4.60)). With a 500 μm entrance slit width (30–60 pm for the echelle spectrometer between 200 μm and 400 nm), the LPDA provided a significant improvement in the detection limits. For the first time, detection limits for CSAAS were comparable to those for LSAAS (Section 4.7.1). Most recently, CSAAS has been further improved by using a charge coupled device (CCD) instead of the LPDA [96]. The low read noise of the CCD resulted in photon shot noise being the dominant noise source at all wavelengths. With shot noise, the minimum detectable signal is independent of the spectral bandwidth (Eqn (4.58)). Consequently, state-of-the-art SNRs can be obtained with a resolution of 3–6 pm (Figure 4.24(c)).

Absorbances for CSAAS were computed for all three detectors (PMT, LPDA, and CCD) in basically the same manner; reference intensities were chosen off the line on both sides and absorbances were then computed using the analytical intensities as shown in Eqn (4.54). With the PMT, the resultant analytical signals were time-integrated absorbances, had units of seconds, and were similar to those computed for LSAAS. For the LPDA and CCD, however, the absorbances were integrated with respect to wavelength and time, had units of picometer-seconds, and were not directly comparable to the absorbances computed for conventional AAS. The wavelength-integrated absorbances represent a more fundamental analytical signal, since it is independent of the width of the source emission line. The differences in the computation method for the absorbances, and the noise characteristics of the detectors, produced fundamental differences in the shapes of the calibration curves and the relative concentration errors for the continuum source and the three detectors. The characteristics of all three continuum source detectors are discussed in detail in Section 4.7.

Spatially, the accuracy of CSAAS is comparable to inverse Zeeman effect background correction. The analytical and reference measurements are made using the same radiation beam. Thus, the source image and the radiation path through the furnace are unchanged throughout the measurement process. Temporal errors are inherently absent with CSAAS. With array detection, the analytical and reference intensities are measured simultaneously. The only possibility for the introduction of temporal errors is in the read process. If the array is read sequentially, the pixel intensities can be influenced differentially by the fluctuation noise of the xenon arc lamp. These errors can be eliminated by simultaneously shifting the pixel charges vertically to a linear array of storage registers. The charges can then be read sequentially without the introduction of temporal errors. This approach also offers the advantage of separating the integration and interrogation process. The pixels can be integrating the source intensity while the previous data are read from the storage register.

The probability of spectral interferences for CSAAS is extremely small, but is difficult to quantify, because it can vary with the specific mode of operation of each experiment and with the analytical concentration range that is anticipated. Continuum source/off-line correction is most similar to direct ZAAS, because the reference intensity is measured just off the absorption profile. The profiles in Figure 4.24(c), show that pixels 10–12 pm from the line center can be used for reference intensities. If high analytical concentrations are anticipated, the pixels used for the intensity measurement may be chosen as far as 40 pm from the line center. The probability for line overlap interferences, however, will be more dependent on the number of pixels used for the reference measurement than on their position. If two pixels are used on each side of the profile, the spectral coverage is about 9 pm (four pixels × 2.2 pm per pixel). With more pixels, the spectral coverage increases, and so does the possibility of a line overlap interference. In theory, the probability of a line overlap is only dependent on the spectral coverage, not the distance between the analytical and the reference pixels.

Based on the data in Figure 4.24(c), a reasonable algorithm for the computation of absorbance would be to use the middle four pixels to measure the analytical intensity, and to select two pixels on either side of the analytical profile to determine the average reference intensity. If a two-pixel gap is left between the analytical and reference pixels, then the analytical and reference pixels cover 9 pm (four pixels × 2.2 pm per pixel). The wavelength range susceptible to line overlap interferences is 26 pm (12 pixels × 2.2 pm per pixel). At higher wavelengths, and lower orders, the dispersion is less and wavelength covered by an equivalent number of pixels. At 550 nm, the wavelength range would be close to 80 pm. In addition, larger wavelength intervals would also be used to allow the determination of very high analyte concentrations, an option not available to the other background correction methods.

The spectral region covered by CSAAS exceeds that for all the other methods except LS/CSAAS. Unlike all the other methods, however, line overlap interferences can be readily detected. Identification of spectral interferences and the shapes of the calibration curves are discussed more completely in Section 4.7.

4.6 COMMERCIAL INSTRUMENTS

Previous sections of this chapter have discuss the individual components of AA spectrometers in detail. The purpose of the next section is to illustrate the numerous ways these components have been combined to meet the challenge of achieving the best possible accuracy and SNR for absorbance measurement. The purpose of this section is not to be a critical comparison of the instruments. Such a comparison is not practical for several reasons. First, although the instruments discussed were commercially available at the time this book was published, newer and better designs will continue to evolve as technology improves. Second, some of the instruments can be operated with variable optical parameters which will influence their performance. Finally, a truly critical evaluation of the instrumental SNR would require more detailed information than is available. Most importantly, the SNR cannot be evaluated without knowing the transmission efficiency of the spectrometers (τ in Eqn (4.38)).

4.6.1 Single-element AAS

As previously discussed, the use of the graphite furnace atomizer has led to a variety of designs to meet the demands of accurately characterizing the rapid, transient analyte absorbance in the presence of high, rapidly changing nonspecific background attenuation. Tables 4.12 and 4.13 list the characteristics of four AA spectrometers that are based on three different principles of background correction. Of the background correction techniques considered in Section 4.5, only the primary continuum source has not been incorporated into a commercially available instrument.

4.6.1.1 Line source/secondary continuum source

The instrument considered in this section is the Spectra 880 with the GTA-96 graphite furnace atomizer (Varian Associates, Wood Dale, IL, USA). This instrument employs an HCL or EDL as the primary source and a secondary continuum source for background correction (Section 4.5.2, Figure 4.18, and Table 4.12) with a dual double beam optical design (Figure 4.1, arrangement C). The instrument characteristics are summarized in Tables 4.12 and 4.13. The two sources are optically aligned along two paths using a beam combiner. This is a half-silvered mirror, which splits 50 per cent of each source along two different optical paths–the first path passes through the furnace, whereas the second path passes around the furnace. Pulsing of the HCL and D_2 lamp, and use of a rotating sectored mirror, permits the two sources to be alternately viewed by a single PMT. Thus, as shown in Tables 4.11 and 4.13, both sources are attenuated by 50 per cent for all measurements.

Absorbances are computed at a frequency of 60 Hz (50 Hz in Europe). Each absorbance is based on six intensity measurements. These intensity measurements correspond to the four source intensities, described in Eqn (4.4), plus a PMT dark current and a furnace emission intensity. With the sources directed around the atomizer, the PMT dark current is measured

Table 4.12 Commercial single-element AAS

Background correction method	*Optical design*	*Optical performance*	*Source*	*Furnace*	*Spectrometer and detector*	*Absorbance computation frequency*	*Time between I and I_0*	*Extrapolation between I_0 points*
LS/CSAAS, (Spectra 880), Varian Associates	dual double beam in time	Double beam (each source)	HCL and deuterium or metal halide lamp	Longitudinal current	Czerny–Turner, PMT	60	2 ms	none
Inverse ZAAS, transverse a.c. field on furnace (Spectra 880 Zeeman, Varian Associates)	Single beam	Double beam	HCL	Longitudinal current	Czerny–Turner PMT	120	4 ms	three point quadratic
Inverse ZAAS longitudinal a.c. field on furnace (4100ZL, Perkin-Elmer)	Single beam	Double beam	HCL	Transverse current	Czerny–Turner PMT	60	8 ms	linear
SRSAAS, (AA Scan, Thermo Jarrell-Ash)	Single beam	Pseudo double beam	HCL	Longitudinal current	Czerny–Turner PMT	20	< 2 ms	none

Table 4.13 Characteristics of commercial single-element AAS

Parameter	*LS/CSAAS (Spectra 880, Varian)*	*Inverse ZAAS transverse field (Spectra 880 Z, Varian)*	*Inverse ZAAS Longitudinal field (4100 ZL, Perkin-Elmer)*	*SRSAAS (AA-Scan, Thermo Jarrell-Ash)*
Focal length (mm)	330	330	267	330
Blaze wavelength (nm)	240	240	236	250
Blaze angle	12° 48′	12° 48′	12° 16′	8° 44′
Grating size (mm×mm)	30×32	30×32	62×72	40×40
Grooves per millimetre	1800	1800	1800	1200
D_a (mm nm^{-1})	0.0020	0.0020	0.0018	0.0012
D_l (mm nm^{-1})	0.659	0.659	0.492	0.415
R_d (nm mm^{-1})	1.52	1.52	2.03	2.41
$\Delta\lambda_s$ (25 μm slits) (nm)	0.038	0.038	0.051	0.060
Luminosity (mm^2)[a]	0.0088 $A_s\tau$	0.0088 $A_s\tau$	0.034 $A_s\tau$[b]	0.014 $A_s\tau$
Optical path transmission	0.5[c]	0.5[d]	1.0	1.0
Absorption sensitivity	1.0	0.85–1.0	0.48	0.6
Integration interval (ms)	1.4	0.7	2.5	11 (0.28)[e]
Measurements per cycle	4	2	2	1 (1)[e]
Cycles per second	60	120	60	20 (20)[e]
Total integration interval (ms)	336	168	300	440

[a]Calculated using Eqn (4.41), A_s = entrance slit area.
[b]Includes additional factor to account for double blazed grating.
[c]50 per cent attenuation by beam combiner.
[d]50 per cent attenuation by fixed polarizer.
[e]Values outside parenthesis are for intensity measurements of lamp at normal current and values in parenthesis are for intensity measurements of lamp with a high current pulse.

with both sources off, the primary source reference intensity is measured when the HCL is pulsed, and the secondary source reference intensity is measured when the D_2 lamp is pulsed. The integration intervals of the three intensities are approximately 1.4 ms with interval centers 2 ms apart. A half cycle (8.3 ms) later, the three measurements are repeated with the source beams passing through the atomizer. The furnace emission is measured with both sources off, the primary source analytical intensity (reduced by atomic absorption and nonspecific attenuation) is measured when the HCL is pulsed, and the secondary source analytical intensity (reduced by nonspecific attenuation) is measured when the D_2 lamp is pulsed. Again, the integration intervals are 1.4 ms with centers 2 ms apart. Subtraction of the PMT dark current from the reference intensities and subtraction of the furnace emission from the analytical intensities yields I_0, $I_{0,b}$, I, and I_b, respectively. These values are used with Eqn (4.4) to compute the background corrected absorbance. Measurement of the two source intensities within 2 ms of each other minimizes the possibility of a change in the background attenuation within the interval of the measurements. Consequently, no extrapolation of the reference intensities is used to achieve a more accurate estimate of the background attenuation. Fluctuation or drift of either source is eliminated through the double beam operation.

The SNR could be computed from the data in Table 4.13 if the line radiance B_L, the area of the entrance slit A_s, and the instrument transmission factor τ, were known (Eqn (4.38)). The first two values are variable, depending on the lamp current and the entrance slit width and height, respectively. The last value can only be obtained by direct measurement, and was not available for this instrument. The use of a secondary continuum source for background correction reduces the SNR of the Spectra 880 by a factor of $\sqrt{2}$. The SNR is reduced by another factor of $\sqrt{2}$ by the attenuation of the sources by the beam combiner. As discussed previously (Section 4.5.2 and Table 4.11), there is no deterioration of the analytical sensitivity. The secondary continuum source background correction gives characteristic masses comparable to those obtained for instruments without background correction.

4.6.1.2 *Inverse Zeeman effect – transverse a.c. field*

The instrument considered in this section is the Spectra 880 Zeeman with the GTA-96 Zeeman graphite furnace atomizer (Varian Associates, Wood Dale, IL, USA). This instrument uses the inverse Zeeman effect for background correction with a transverse a.c. field around the furnace (Section 4.5.3.2, Figures 4.16 and 4.21(b), and Table 4.11). The instrument characteristics are summarized in Tables 4.12 and 4.13. Optically, this is a single beam instrument, with the exception of a fixed polarizer in the optical beam, which provides double beam performance. The intensity reaching the detector, however, is half that of the single beam instrument, because the fixed polarizer will attenuate approximately 50 per cent of the intensity from the unpolarized HCL. The transverse magnetic field is compatible with the longitudinally heated graphite furnace.

Absorbances are computed at a frequency of 120 Hz (100 Hz in Europe), the same frequency as the oscillating magnetic field. During each cycle, the furnace emission is measured with the HCL off and the magnet on and off, the reference intensity I_b is measured with the magnetic field and the HCL on, and the analytical intensity I measured with the HCL on and the magnetic field off. The reference and analytical intensities are measured 4 ms apart with integration intervals of approximately 0.7 ms. As the interval between intensity measurements is too large to allow the assumption that the background is constant, the reference intensity at the time of the analytical intensity is extrapolated using a quadratic equation fit to three reference intensities which bracket the analytical intensity (Section 4.5.1.1). After subtracting the intensity of the furnace emission from the analytical and extrapolated reference intensities, absorbance is computed as shown in Eqn (4.5).

The double beam performance eliminates source flicker noise and provides a photon shot noise-limited instrument. It is difficult to compare this instrument to the Spectra 880, because with different optical components and a different number of mirrors, the optical transmission factor τ can be different. The fixed polarizer, like the beam combiner in the Spectra 880, attenuates the source by 50 per cent (Tables 4.11 and 4.13) leading to a reduction by $\sqrt{2}$ in the SNR, compared with double beam instrument without background correction. The source intensity is further reduced by the shorter integration interval (0.7 ms versus 1.4 ms) dictated by the oscillating magnetic field. The shorter integration interval, however, can be compensated for by using higher lamp currents. Unfortunately, the increased lamp Current will result in a slightly broader line source and a reduction in sensitivity, but an improved SNR. Thus, the exact ratio of source intensities and the absorption sensitivities of the two instruments are difficult to predict, but it is possible to make the SNR of the Spectra 880 Zeeman comparable to the Spectra 880 by increasing the HCL intensity. Varian reports comparable detection limits for the two instruments.

4.6.1.3 Inverse Zeeman effect – longitudinal a.c. field

The instrument considered in this section is the model 4100ZL with the THGA transversely heated graphite atomizer (Perkin-Elmer Corporation, Norwalk, CT, USA). This instrument uses the inverse Zeeman effect for background correction, with a longitudinal a.c. field around the furnace

(Section 4.5.3.3, Figure 4.21(d), and Table 4.11). The instrument characteristics are summarized in Tables 4.12 and 4.13. This instrument has a single beam optical design, which provides double beam performance. An alternating magnetic field is used to shift the absorption profile away from the source. No polarizer is required as there is no measurable component at the line center.

Absorbances are computed at a frequency of 54 Hz, the same frequency as the oscillating magnetic field. With the magnet off; the HCL is turned on and the analytical intensity I is measured. With the magnet on, the HCL is turned on again and the reference intensity I_b is measured. The furnace emission is measured prior to the analytical and reference measurements during the transition period. After subtracting the furnace emission intensity, the analytical and reference intensities are used to compute absorbance, as shown in Eqn (4.5). The integration intervals are each 2.5 ms and are separated by 8.3 ms. The reference intensity at the time of the analytical measurement is computed using a linear extrapolation, this being the average of the reference intensities before and after the analytical measurement.

It is not possible to compare the SNR of this instrument with that of either of the Varian instruments without further information. For the 4100ZL, the full intensity of the HCL is transmitted to the detector because no polarizer is required. The integration interval is 2.5 ms instead of 2.8 ms, which leads to a 5 per cent loss in the SNR (compared to a double-beam instrument without background correction). A slight loss in sensitivity (5–15 per cent) occurs, as the absorption profile cannot be shifted sufficiently far from the line center and a fraction of the intensity of the reference beam is absorbed. It can be seen in Table 4.13 that the luminosity of the 4100ZL will be approximately four times greater than the Spectra 880 with a comparable slit area. Values for τ have been reported for the 4100ZL [74], but values are not known for the Varian instruments. In addition, the SNR will also be dependent on the spectral bandwidth and the HCL operating current, which can be varied on all three instruments.

4.6.1.4 Self-reversed source

The instrument considered in this section is the AA-Scan with the CTF 755 constant temperature furnace (Thermo Jarrell Ash Corporation, Franklin, MA, USA). This instrument uses a self-reversed source for background correction (Section 4.5.4, Figures 4.22 and 4.23, and Table 4.11). The instrument characteristics are summarized in Tables 4.12 and 4.13. The instrument design and operation is that of a single beam instrument, but the absorbance calculation is identical to that for a double beam instrument (Eqn (4.5)). Reference intensities for the normal lamp intensity (operation of lamp at normal currents) and the pulsed intensity are acquired prior to the atomization cycle. Achievement of true double beam performance is dependent on the reproducibility of the high intensity pulse. The HCL is operated in a three-step cycle at a frequency of 20 Hz. Prior to turning on the HCL, the furnace emission intensity is measured. The HCL is then turned on at a normal operating current for 14 ms during which an 11 ms integration interval is used to measure the analytical intensity of the primary source I. The HCL is then pulsed to 200 mA for 0.4 ms. After allowing the source to stabilize, the analytical intensity of the secondary source I_b is measured over an interval of 0.28 ms. The lamp then remains off for approximately 35 ms until the start of the next cycle. After subtracting the furnace emission intensity from the two measurements, absorbance is computed as shown in Eqn (4.5). The values for the reference intensities I_0 and $I_{0,b}$ are obtained from a series of measurements prior to the atomization cycle.

In Table 4.12, the performance of this instrument is listed as pseudo-double beam. This evaluation is based on the lack of information concerning the reproducibility of the intensity from the high-current pulse of the HCL. Available data for HCLs, operated in the d.c. mode or pulsed to moderate currents, show that the emitted intensity is very stable, i.e. there is no significant fluctuation noise component. There are no published data, however, for the stability of HCLs pulsed at the high currents necessary to achieve self-reversal. Leeman Laboratories (see next section) also employ self-reversal for background correction, and have chosen to split the source beam and to measure I_0 and $I_{0,b}$, simultaneously with I and I_b. It must be remembered that the relative precision of the self-reversed intensity must be around 0.03 per cent, be comparable to accepted performance for a HCL. Consequently, there is some question as to the precision of the intensities of the high-current pulses and as to whether double beam performance is achieved.

Again, it is impossible to compute the SNR of the AA-Scan or to compare it with the other instruments without further information. As shown in Table 4.11, the analytical sensitivity is approximately 60 per cent of that of an identical instrument without background correction, because of the finite analytical sensitivity of the reference beam (the reference measurement is even closer to the line center than the Zeeman measurement). The luminosity of the spectrometer falls between that of the 4100ZL (Perkin-Elmer) and the Spectra 880 (Varian). The source intensity is difficult to compare because τ is not known, because it is not known if the pulsed intensity (measured for 0.28 ms) is comparable to the normal intensity (measured for 11 ms), and because several different spectral bandwidths may be employed. If comparable lamp currents are used with normal operation, then the AA-Scan, with its longer integration time will provide greater intensity. However, integration of intensity, and not absorbance, over the 11 ms interval can lead to nonlinearity. In general, the

AA-Scan can be expected to have slightly poorer SNRs than the other instruments. The loss of absorption sensitivity is only slightly offset by the increased source intensity.

4.6.2 Multielement AAS

Development of multielement AAS has been limited primarily by the lack of an appropriate multielement source. The four commercially available multielement systems (each described here) employ combinations of LS and, as a result, are restricted to 4–6 elements (Tables 4.14 and 4.15). They all use either inverse Zeeman effect or self-reversal of the source for background correction. Background correction with a secondary continuum source requires too complex an optical arrangement to be suitable for multielement determinations. A multielement instrument using a primary continuum source has yet to be commercialized.

4.6.2.1 Inverse Zeeman effect–transverse d.c. field

The instrument considered in this section is the Z-9000 (Hitachi Instruments, Inc., Naperville, IL, USA). This instrument uses the inverse Zeeman effect with a transverse d.c. magnetic field around the furnace (Figure 4.16 and 4.21(a)). The characteristics of this instrument are summarized in Tables 4.14 and 4.15. Optically, this instrument can be visualized as four single beam instruments, as shown in Figure 4.25(a): four sources, a common furnace, four separate monocchomators, and four separate detectors (PMTs). The beams of four HCLs are focused through a common point in the center of the furnace using four separate mirror systems. The beams do not overlap and only converge at the furnace center. To accomplish this, the F-number of each mirror system is a factor of two less than that for an optimized single-element instrument, and the light throughput is reduced by a factor of four. The diverging beams emerging from the furnace are then focused by a concave spherical mirror onto a common entrance slit for a quadruplet of Czerny–Turner monochromators. The beams from the four exit slits are then passed through a Wollaston prism and a rotating beam selector (a chopper) which alternately selects the σ and π components for imaging onto a PMT. The Wollaston prism and rotating beam selector are used in place of a rotating polarizer. There are four lamp positions, and four elements can be determined simultaneously. The sources are modulated at either 1200 or 1800 Hz and are detected by appropriate, phase-sensitive electronics associated with each PMT. Modulation of the sources permits the elimination of the furnace emission component. The π (reference intensity) and σ (analytical intensity) components are measured at a frequency of 60 Hz. The reference intensity at the time of the analytical measurement is computed as the average of the preceding and following reference

Table 4.14 Commercial multielement AAS

Background correction method	*Elements*	*Optical design*	*Optical performance*	*Sources and combiner*	*Spectrometer and detector(s)*	*Absorbance computation frequency*	*Time between I and I_0*	*Extrapolation between I_0 points*
Inverse ZAAS, transverse d.c. field on furnace (Z 9000, Hitachi Instruments)	4, selectable	single beam	double beam	HCLs, mirrors	quad Czerny–Turner, PMTs	60	8 ms	linear
Inverse ZAAS, longitudinal a.c. field on furnace (SIMAA 6000, Perkin-Elmer)	4, selectable	echelle poly-chromator	double beam	HCLs, mirrors	echelle, array of diodes	54	8 ms	linear
SRSAAS (AA Scan 4, Thermo Jarrell-Ash)	4, selectable	single beam	Pseudo double beam	HCLs, none (rapid sequential)	Czerny-Turner, PMT	20	< 2 ms	none
SRSAAS (Analyte 5, Leeman Labs)	5, (As, Pb, Sb, Se, and Tl)	double beam in space	double beam	HCLs, reverse Rowland circle	none, PMT	50	< 2 ms	none

Table 4.15 Characteristics of commercial multielement AAS

Parameter	*Inverse ZAAS, transverse field (Z 9000, Hitachi Instruments)*	*Inverse ZAAS longitudinal field (SIMAA 6000, Perkin-Elmer)*	*SRSAAS, conventional (AA Scan 4, Thermo Jarrell-Ash)*	*SRSAAS fixed elements (Analyte 5, Leeman Labs)*
Focal length (mm)	450	501	330	350
Blaze wavelength (mm)	200	none[a]	250	None[b]
Blaze angle	10° 33′	63° 24′	8° 44′	12° 2′[b]
Grating size (mm×mm)	32×32	74×143	40×40	52×52
Grooves per millimetre	1800	79	1200	1700
D_a (nm^{-1})	0.0019	0.017	0.0012	0.0019
D_l(mm nm^{-1})	0.772	8.32	0.415	0.652
R_d(nm mm^{-1})	1.30	0.12	2.41	1.53
$\Delta\lambda_s$ (25 μm slits)	0.032	0.004	0.060	0.038
Luminosity (mm^2)[c]	0.0050 $A_s\tau$	0.019 $A_s\tau$	0.014 $A_s\tau$	0.022 $A_s\tau$
Optical path transmission	0.5[d]	0.25[e]	1.0	1.0
Absorption sensitivity	0.85–1.0	0.48	0.6	0.29
Integration interval (ms)	8.0	6.0	11.0 (0.28)[f]	1.0 (0.08)[f]
Measurements per cycle	2	2	1 (1)[f]	2 (2)[f]
Cycles per second	60	60	5 (5)[f]	50 (50)[f]
Total integration interval (ms)	960	720	110	200

[a]Echelle spectrometer, all values computed for 241 nm, center of order 47.
[b]Holographic grating, all values computed for 240 nm.
[c]Calculated from Eqn (4.41), A_s =entrance slit area.
[d]75 per cent attenuation by reduced F-number of optics and 50 per cent attenuation by Wollaston prism.
[e]75 per cent attenuation by four-faceted mirror.
[f]Values outside parenthesis are for intensity measurements of lamp at normal current and values in parenthesis are for intensity measurements of lamp with high current pulse.

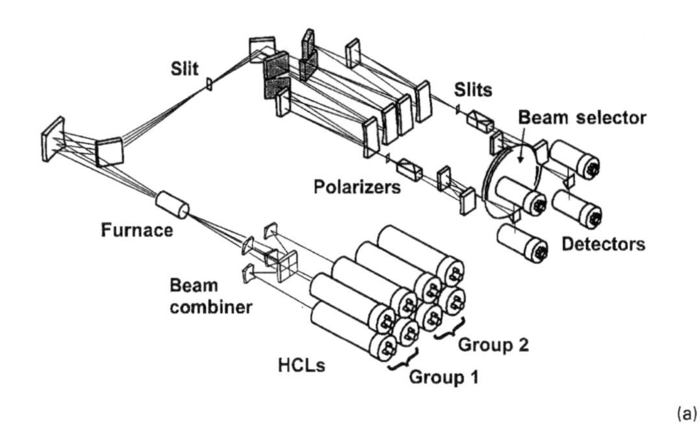

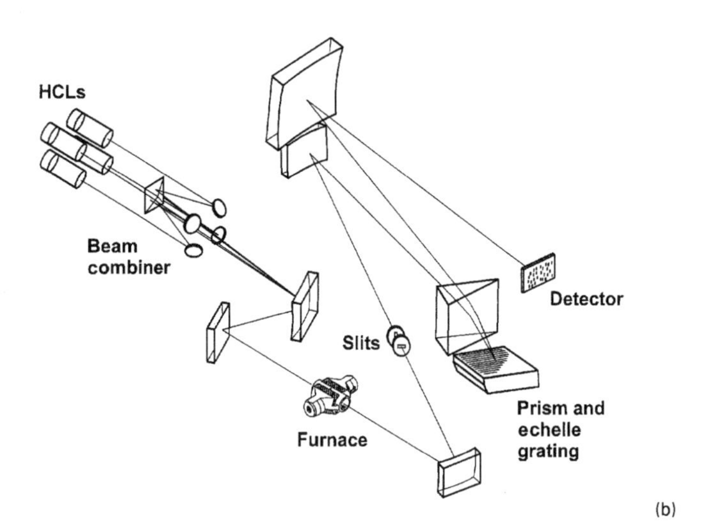

Figure 4.25 Diagrams of four multielement AAS spectrometers. (a) Z-9000 (Hitachi Instruments) [97] (Reprinted from *American Laboratory*, Volume 21, number 12, page 24, 1989 by permission of International Scientific Communications Inc.). (b) SIMAA 6000 (Perkin-Elmer Corporation) [74] (Reprinted by permission of the Royal Society of Chemistry from B. Radziuk, G. Rodel, H. Stenz, H. Becker-Ross, and S. Florek, Spectrometer system for simultaneous multi-element electrothermal atomic absorption spectrometry using line sources and Zeeman-effect background correction, *J. Anal. At. Spectrom.*, **10**, 127–136 (1995))

intensities. An absorbance is computed from the analytical and reference intensity, as shown in Eqn (4.5), at a frequency of 60 Hz. This quadruplicate single beam system produces double beam performance for each of the four channels.

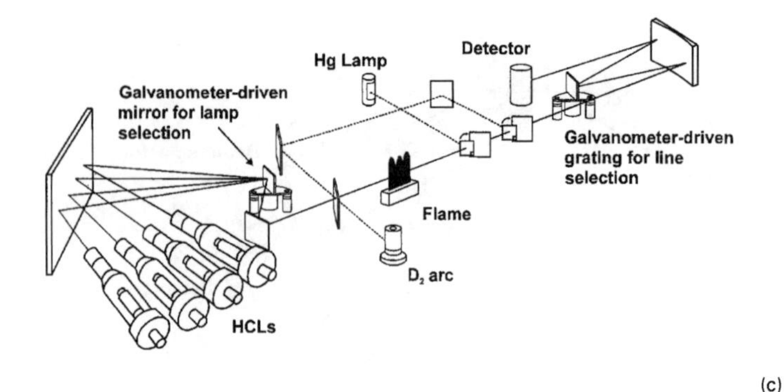

Figure 4.25 Diagrams of four multielement AAS spectrometers. (c) AA Scan 4 (Reproduced by permission of Thermo Jarrell Ash Corporation). (d) Analyte 5 [25] (Reproduced by permission of Leeman Labs, Inc., *Leeman Letter*, 1994, **33**, 3).

In this instrument, the F-number of the optics has been reduced by a factor of two to permit the four beams to be focused through the furnace. The source is attenuated by an additional 50 per cent by the fixed Wollaston prism. Additional noise arises from the shot noise contribution of one of the other sources. Source pairs passed through the same Wollaston prism are modulated at different frequencies to allow separation of the signals. The noise contribution, however, cannot be separated. The total shot noise from both sources will determine the absorbance noise for each channel; thus, in Eqn (4.16), I_0 will be the intensity of a single source, whereas σ_{I_0} will be the total shot noise for both sources.

4.6.2.2 Inverse Zeeman effect–longitudinal a.c. field

The instrument considered in this section is the SIMAA 6000 (Perkin-Elmer Corporation, Norwalk, CT, USA) [74,75]. This instrument uses the inverse Zeeman effect for background correction with a longitudinal a.c. field around the furnace (Section 4.5.3.3, Figures 4.16 and 4.21(d) and Table 4.11). The instrument characteristics are summarized in Tables 4.14 and 4.15. The beams from four HCLs are combined using a four-faceted mirror (Figure 4.25(b)), which transmits one quarter of the image and the intensity of each HCL. The F-number of the optical system is the same as that for an optimized single-element instrument. After passing through the furnace, the radiation is dispersed using an echelle spectrometer, and individual intensities at specific wavelengths are measured simultaneously using a monolithic detector consisting of a series of individual photodiodes. Each photodiode is 1 mm high (cross-disperser axis) and 2 mm wide (wavelength axis). Photodiodes are located at the primary resonance wavelengths of the 39 elements most commonly determined by AAS and at 16 secondary lines. The lamp turret holds four lamps, but up to six elements can be determined simultaneously using multielement lamps. With inverse Zeeman effect background correction, double beam performance is achieved for each element with a single beam optical configuration. The entrance slit height stop (1.0 mm) and width stop (2.3 mm) are separated by approximately 8 mm to offset the astigmatism introduced by the optical design of the echelle spectrometer. Despite the large slit width, the high resolution of the echelle provides spectral bandwidths of 0.2 nm and 0.55 nm at 200 nm and 550 nm, respectively. Thus, the echelle provides high luminosity and resolution comparable to that of the conventional monochromators usually employed for GFAAS. The specifications for the photodiode detectors are provided in Table 4.9 [75].

Absorbance is computed for each channel, according to Eqn (4.5), at a frequency of 54 Hz. The reference and analytical intensities are measured as described in Figure 4.21(d) at 9.5 ms intervals. As this time interval is large compared to the rate of change of the background, the reference intensity used in Eqn (4.5) is computed as the average of the reference intensities preceding and following the analytical intensity. The intensity measured for each channel of this instrument is diminished by 75 per cent by the four-faceted beam combiner. In addition, the furnace sensitivity is reduced by 52 per cent by the shorter furnace, the THGA, used with the longitudinal magnetic field. On the positive side, the high luminosity (with a 2.3 mm slit width) increases the transmitted intensity and reduces the absorbance noise. In general, the detection limits for the SIMAA 6000 (Table 4.2), in the multielement mode, are comparable to those for the 4100ZL (see section 4.6.1.3) in the single-element mode. These instruments are compared in detail in Section 4.6.3.

4.6.2.3 *Self-reversed source*

The instrument considered in this section is the AA-Scan 4 with the CTF 755 (Thermo Jarrell Ash Corporation, Franklin, MA, USA). This instrument uses a self-reversed source for background correction (Section 4.5.4, Figures 4.22 and 4.23, and Table 4.11). The instrument characteristics are summarized in Tables 4.14 and 4.15. This instrument is identical to the one described for single-element operation in Section 4.6.1.4. Multielement capability is achieved by placing a mirror in front of a planar array of four HCLs (Figure 4.25(d)). The mirror and the spectrometer grating are each mounted on galvanometers to select the appropriate lamp and wavelength, respectively. Thus, each element is determined sequentially within a 50 ms interval. Data acquisition and processing is identical to that described in the preceding section, except that the frequency is four times slower; i.e. absorbances are computed for each channel at a frequency of 5 Hz. The lower, frequency of data acquisition and absorbance computation is dictated by the time necessary to accommodate the movement and settling of the galvanometers on which the mirror and grating are mounted.

In theory, the SNR for the AA-Scan 4 is 50 per cent that of the AA Scan, because the total integration interval and the total integrated intensity are reduced by a factor of four. However, Thermo Jarrel Ash has reported that the SNR is degraded by more than a factor of two–by a factor closer five or six [98]. This unexpectedly large degradation is attributed to the settling time of the galvanometers. They also reported that an absorbance computation frequency of 5 Hz is too slow to accurately characterize the rapid transient signal from the graphite furnace.

4.6.2.4 *Self-reversed source–Rowland circle*

The instrument considered in this section is the Analyte 5 (Leeman Labs, Lowell, MA, USA). This instrument uses self-reversed sources for background correction (Section 4.5.4, Figures 4.22 and 4.23, and Table 4.11). The instrument characteristics are summarized in Tables 4.14 and 4.15. As shown in Figure 4.25(c), this instrument has a dramatically different optical design to any of the other single or multielement instruments described in this section. The Analyte 5 was developed specifically for the measurement of five elements (arsenic, lead, antimony, selenium, and thallium) in ground water, required by the United States Environmental Protection Agency. The beams from the five HCLs are combined using a reverse spectrometer. That is, the lamps are positioned at their wavelength in the exit plane of a conventional Rowland circle spectrometer, and their beams are combined by the grating and focused through the entrance slit [99]. The furnace is located just in front of the entrance slit. Radiation emerging from the entrance slit is focused onto a PMT. Not shown in Figure 4.25(d) is a mirror with a hole in it which sits just in front of the furnace. The bulk of the radiation passes through the hole, but some of the peripheral radiation is reflected onto a second PMT. Optically this provides a double beam in space configuration. The furnace used is an integrated contact cuvette (ICC in Section 4.3.2.1 and Fig. 4.8). The data for each source are obtained at 50 Hz. In sequence, each lamp is turned on at a normal current for approximately 2 ms and then pulsed to a high current for 100 μs. Prior to turning the lamp on, the furnace emission intensity is measured. The analytical intensity is measured for 1 ms during the normal current operation, and the reference intensity is measured for 80 μs during the high current pulse. During the normal and high current phases, intensity measurements are also made by the second PMT. Intensities measured for radiation passed through the furnace correspond to land I and $I_{0,b}$ respectively, while intensities measured for the radiation reflected from the mirror correspond to I, and $I_{0,b}$. After subtracting the furnace emission from I and I_b, absorbance is computed as shown in Eqn (4.4).

The SNR for the Analyte 5 is difficult to predict. The ICC is approximately two-thirds the size of a conventional graphite furnace (Table 4.4), and therefore has only approximately half the sensitivity. The integration time for the analytical intensity measurement is short (1 ms) compared to the other instruments, but the luminosity of the spectrometer is the highest. It is not possible to determine if the integrated intensity (over 0.08 ms) for the high current pulse I_b is comparable to that of the analytical intensity I. It is also not known whether the reflected intensifies I_0 and $I_{0,b}$ are comparable to the intensities passed through the furnace. More information is required for a rigorous comparison of the SNR of the Analyte 5 with that of the other instruments. Detection limits reported by the manufacturer are shown in Table 4.2.

4.6.3 A specific comparison

In the preceding sections, a rigorous comparison of the SNRs of the instruments was not possible because of a lack of published information. Suitable information has been published by one manufacturer and allows the evolution of a multielement instrument to be traced through the various intermediate stages. Table 4.16 summarizes the relative SNRs for the Zeeman 5000, 4100ZL, and SIMAA 6000, in the single and multi-element mode (all instruments manufactured by Perkin-Elmer Corp., Norwalk, CT, USA). These instruments were introduced in 1987, 1991, and 1994, respectively. These three instruments make several interesting transitions: (1) from inverse Zeeman effect background correction with a transverse a.c. field and a longitudinally heated furnace (Zeeman 5000) to a longitudinal a.c. field with a transversely heated furnace (SIMAA 6000);

Table 4.16 Computed signal-to-noise ratios

			SIMAA 6000	
Parameter	*Zeeman 5000*	*4100ZL*	*Multielement*	*Single element*
Étendue (mm^2)	$0.040\ A_s\tau^a$	$0.034\ A_s\tau^b$	$0.019\ A_s\tau^c$	$0.019\ A_s\tau$
Slit area, A_s (mm^2)	0.25	0.25^d	2.3^d	2.3^d
Optical transmission factor, τ	0.24	0.24^d	0.18^d	0.18^d
Optical path transmission	0.5^e	1.0	0.25^f	1.0
Integration interval (ms)	2.5	6.0	6.0	6.0
Measurements per cycle	2	2	2	2
Cycles per second	60	60	60	60
Total intensity (ms mm^2), i.e. product of above parameters	0.36	1.47	1.42	5.66
Noise, i.e. (total intensity)$^{-1/2}$	1.66	0.82	0.84	0.42
Absorbance signal	1.0	0.48	0.48	0.48
SNR (noise/absorbance signal)	0.60	0.58	0.57	1.14

[a]Calculated from Eqn (4.41) and specifications given in Table 4.6.
[b]Calculated from Eqn (4.41) and specifications given in Table 4.13.
[c]Calculated from Eqn (4.41) and specifications given in Table 4.15.
[d]From reference 75.
[e]50 per cent attenuation by fixed polarizer.
[f]75 per cent attenuation of intensity of each lamp by four faceted mirror.

(2) from PMT detection (Zeeman 5000) to photodiode detection (SIMAA 6000); (3) from a conventional, medium resolution monochromator (Zeeman 5000) to an echelle polychromator (SIMAA 6000); and (4) from single-element determinations (Zeeman 5000) to multielement determinations (SIMAA 6000).

The introduction of the transversely heated furnace in the 4100ZL is a key step to multielement determinations, but results in an inherently less sensitive atomizer. The shorter furnace results in approximately a factor of two decrease in sensitivity. To obtain the same relative SNR as the Zeeman 5000, the integrated intensity of the 4100ZL must be increased by a factor of four. Table 4.16 shows that the intensity increase came from the elimination of the polarizer from the optical path transmission and from the increased integration interval. As a result, the SNRs are almost identical for the two instruments. These values are based on two very reasonable assumptions, that the LSs are operated at the same current on both instruments, and that the optical transmission factor for the Zeeman 5000 is the same as the published value for the 4100ZL [74]. A comparison of the detection limits in Table 4.2 supports the validity of these assumptions and the accuracy of the computed SNRs in Table 4.16.

In the multielement mode, the SIMAA 6000 loses a factor of four in intensity to the combination of the LSs (Table 4.16). This intensity loss is compensated by the increased luminosity of the echelle spectrometer. With an entrance slit area of 2.3 mm^2, the echelle provides four times the luminosity of the Zeeman 5000 or 4100ZL, and a spectral bandwidth ranging from 0.2 at 200 nm to 0.55 at 550 nm. The resulting SNR for the SIMAA 6000, in the multielement mode, is approximately equal to that of the 5000 and 4100ZL. The SNR for the SIMAA 6000 will improve slightly in the far ultraviolet region (below 220 nm), where the quantum efficiency of the photodiode will be better than that of a PMT. The computed SNRs in Table 4.16 are generally accurate in predicting the ratios of the detection limits in Table 4.2. In the single-element mode, the SIMAA 6000 gains a factor of four in source intensity, as only a single line source and a flat mirror (not a four-faceted mirror) are used. As a result, the relative SNR of the SIMAA 6000 in the single-element mode is a factor of two better than that of the multielement mode or the 5000 or 4100ZL. These results are again supported by the listed detection limits in Table 4.2.

The trend in the evolution of the multielement instrument is clear. The reduction in sensitivity due to the shorter, transversely heated furnace and the combination of the LSs, is offset by increased transmission of the optical system. This was accomplished by elimination of the polarizer, increasing the integration interval, and replacing a medium-resolution monochromator with an echelle polychromator. In each case, the newer instrument provided an SNR similar to that of the previous instrument.

4.7 PROTOTYPE INSTRUMENTS WITH PRIMARY CONTINUUM SOURCES

The use of a continuum source for AAS has always been appealing. A continuum source, with radiation at all wavelengths, eliminates the need for a large inventory of line sources and offers the potential for multielement determinations. The use of CSAAS for multielement determinations was first investigated by Walsh [1] and later by Fassel *et al.* [100]. They found that the sensitivities and the detection limits were dramatically worse with a continuum source than with a line source. In retrospect, this is not surprising. The much reduced fraction of absorbed light (Figure 4.5) for the continuum source (using the same spectral bandwidth employed with line sources) results in a reduced absorbance. In addition, short arc lamps (Section 4.3.1.2), the most commonly used continuum sources, are inherently less stable than line sources, due to the strong flicker component arising from arc wander and the schlieren effect found at the high internal operating pressures. Thus, it is not possible to substitute a continuum source directly for a line source and achieve comparable results.

Successful use of a continuum source requires a new design philosophy for the instrument. In general, the new design includes a high-resolution spectrometer, systematic off-line intensity measurements, and high-speed data acquisition. The high resolution is necessary to maximize the sensitivity of the absorbance measurement. The small spectral bandwidth of a high-resolution spectrometer, like the narrow line source line, limits the spectral region over which the absorption measurement is made. Instrumental spectral bandwidths similar in width to that of the line source provide similar sensitivities. Off-line intensity measurements are mandatory for the elimination of the source fluctuation noise and for accurate background correction. The off-line intensity measurements satisfy the requirements for a reference intensity measurement that reflects the background attenuation of the source (Eqn (4.5)). The logarithm of the quotient of the intensity measured off the analytical line and the intensity measured on the line provides a background corrected absorbance, minimizes the fluctuation noise, and produces the desired shot noise-limited condition. These characteristics are achieved provided the intensities are measured at a higher frequency than the intensity fluctuations of the lamp. Thus, high-speed data acquisition is also necessary for the success of CSAAS.

A series of prototype instruments has been developed, based on the unique requirements of the continuum source. Initially, an echelle polychromator was used with PMT detection [93,101,102]. Off-line intensity measurements were made by mechanically shifting the viewed spectral region with a quartz refractor plate. More recently, radiation detection has been accomplished using solid-state array detectors, such as LPDAs,

CCDs, and CIDs. The multiwavelength detection capability of these arrays permits simultaneous measurement of the off-line intensities. Array detectors offer improved SNRs, compared with PMTs, due to the multiplex advantage of simultaneous multi-wavelength detection and, in some cases, due to their higher quantum efficiency in the far ultraviolet region. Using solid-state detectors, detection limits were obtained for graphite furnace CSAAS for the first time which were comparable to or better than those for graphite furnace LSAAS.

Historically, the primary attraction of array detectors, for both absorption and emission spectrometry, has been simultaneous multielement determinations. In both fields, the major obstacle has been the compromise between array size (wavelength coverage), pixel size (resolution), and read time. Arrays that provide wavelength coverage and high resolution can take several seconds to read, and are not compatible with graphite furnace atomization. Arrays that provide wavelength coverage and rapid read times will consist of too few pixels, and will not have the resolution necessary for CSAAS. Arrays that provide adequate resolution and rapid read times do not have the necessary wavelength coverage. An obvious solution is to use a series of arrays, each dedicated to an analytical wavelength of interest. The ideal array detector, for use with a continuum source and graphite atomization, should have a pixel width equal to or less than the width of the entrance slit, a rapid read time, and low read noise. If the width of a pixel equals the width of the entrance slit, then the spectral width of a pixel will be equal to the spectral bandwidth, regardless of the resolution of the spectrometer. Consequently, the absorption profile will fall on two or more pixels. In practice, the smallest pixel widths range from 10 to 25 μm, close to the smallest slit widths used for most spectrometers. A fast read time is 20 ms or less, as 50 Hz is the lowest recommended frequency for characterizing graphite furnace signals. The read noise should be small enough to provide a shot noise-limited instrument under most conditions.

The GFAAS requirement for fast read times means that short arrays, or arrays with random access capability, will be the most useful. Thus, short arrays with 10–25 μm pixels can be used with high-resolution spectrometers to provide high resolution, and still maintain a rapid read time. A multi-element instrument therefore requires multiple short arrays, or a very large array with random access capability. In this section, prototype instruments with LPDA, CCD, and CID detectors are considered. As shown in Section 4.3.4.2, these detectors display a wide range of characteristics. Each prototype instrument has been customized to suit the detector.

4.7.1 Continuum source – linear photodiode array detection

The most frequently used solid-state array detectors for prototype spectroscopic applications have been LPDAs. This is because they have been available longer, are relatively inexpensive, are fairly easy to interface to a computer, and have a large height to width aspect for each pixel. The large aspect makes linear arrays feasible and the read process simpler and faster. The major restrictions for usage of LPDAs are the read noise and the compromise between the array size and the read time. In general, two approaches have been used: large arrays covering the maximum possible wavelength for multielement detection [104]; and short arrays dedicated to the measurement of intensities in the region of the wavelength of a single element [94]. With the use of a high-resolution spectrometer, however, even the large arrays (4096 pixels) do not provide the necessary wavelength coverage. Multiple arrays are necessary for true multielement GFAAS.

The CS-LPDA-AAS prototype, with a short array [94], has only been implemented in the single-element mode, but was developed based on the premise that a multielement instrument would require a separate array for each analytical wavelength. In this instrument, the quartz refractor plate was removed from the echelle spectrometer, and the exit cassette was modified to mount the LPDA and driver/amplifier board (MOS series S3904-256Q and C4070, Hamamatsu Corp., Bridgewater, NJ, USA). The LPDA, with 256 pixels on 25 μm centers, provided a window of 0.4–1.1 nm (for values of R_d ranging from 0.059 nm mm^{-1} at 200 nm to 0.177 nm mm^{-1} at 600 nm) depending on the order of the echelle. In addition, a second LPDA (MOS series S3904-256Q) with 128 pixels on 50 μm centers was also tested. With a slit width of 500 μm (the largest available on the echelle), the 256 pixel array required approximately 60 pixels (0.09–0.26 nm windows at 200 and 600 nm, respectively) to cover the absorption profile and the off-line reference intensities. A 64 pixel array (on 25 μm centers) would have been sufficient, but was not commercially available. All exposure times were 20 ms and the read frequency was 50 Hz. A wavelength-integrated absorbance was computed for each read of the 256 pixels of the LPDA at 50 Hz [94]. At the start of each experiment, the center of the absorption profile was determined, the locations specified for the analytical pixels covering the absorption profile, and an equal number of reference pixels, distributed to both sides of the absorption profile, specified for determining the average reference intensity I_b. Usually, a range of 32 analytical pixels was selected with 16 reference pixels, to both sides of the profile, used to compute I_b:

$$I_b = \frac{\sum\limits_{p=-35}^{-20} I_p + \sum\limits_{p=19}^{34} I_p}{32} \tag{4.72}$$

Absorbances were computed for each of the analytical pixels and then summed and normalized to give a wavelength-integrated absorbance:

$$A_{\lambda} = \Delta\lambda_{\text{pix}} \sum_{p=-16}^{15} \log\left(\frac{I_{\text{b}}}{I_{p}}\right) \tag{4.73}$$

where $\Delta\lambda_{\text{pix}}$ is the normalization constant, the spectral width of a pixel in picometers. Without normalization, the wavelength integrated absorbance for the same spectrometer would increase as the pixel width decreased or, for the same detector, would increase as the resolution of the spectrometer increased. Wavelength-intergrated absorbance was then summed with respect to time in the normal manner over the duration of the furnace signal to provide wavelength and time integrated absorbance, $A_{\lambda,\text{t}}$:

$$A_{\lambda,\text{t}} = t_{\text{A}} \sum A_{\lambda} \tag{4.74}$$

where t_{A} is the time between absorbance computations or the inverse of the array scan frequency. The resultant normalized absorbance has units of picometers-seconds.

Wavelength-integrated absorbance is independent of the source width and the entrance slit width. As shown in Figure 4.26, the absorption profile becomes broader and shallower as the entrance slit is widened. The integrated absorbance, however, remains constant if a sufficient number of pixels is used to cover the absorption profile, as shown in Figure 4.27. If the absorbance was computed after summing the intensity of the analytical pixels (wavelength integrated intensity instead of absorbance), then the absorbance would be dependent on the entrance slit width and would decrease linearly with an increasing width.

The sensitivity for CS-LPDA-AAS is defined as intrinsic mass [12], which is the mass necessary to provide an absorbance of 0.0044 pm s (Section 4.2.1). Intrinsic masses are shown in Table 4.17 for a range of elements between 190 nm and 330 nm. The intrinsic mass provides a measurement of the instrument sensitivity which is independent of the resolution of the spectrometer or the width of the pixels of the array. It can be argued that the intrinsic mass is a more fundamental measurement than the characteristic mass. The characteristic mass is dependent on the width ratio of the line source and the absorption profile. When operated at the manufacturers recommended currents, line source widths for each element are approximately the same and no normalization for the spectral bandwidth has been used. It has been shown, however, that line width variation between lamps, between lamp designs, and with age can produce a factor of two to three variation in the characteristic mass. The intrinsic mass is independent of the measurement system, the source, the spectrometer, and the detector. The intrinsic mass is only dependent on the atomization efficiency of the graphite furnace. Intrinsic and characteristic masses cannot be compared directly. It is possible, however, to

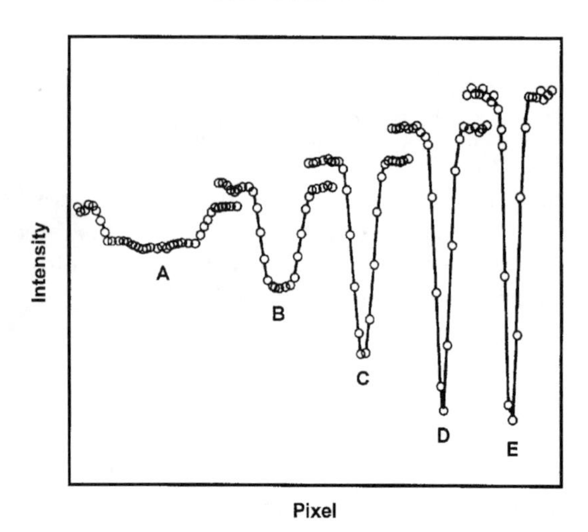

Figure 4.26 Absorption profiles obtained by ensemble averaging the first 2 s of atomization for Cu (324.7 nm). Entrance slit widths were: curve a 500 μm, b 200 μm, c 100 μm, d 50 μm, and e 25 μm. The intensities of each profile were scaled approximately inversely by the entrance slit width to yield comparable values [94].

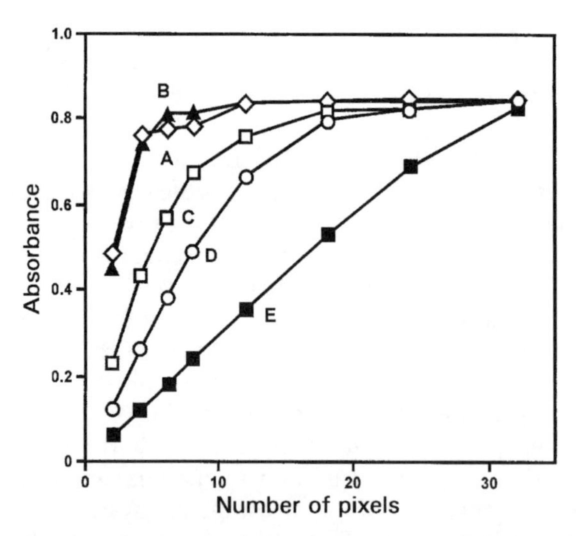

Figure 4.27 Wavelength integrated absorbance A_λ as a function of the number of pixels used to cover the absorption profile for 20 μl of 10 ng ml^{-1} of Cd. Entrance slit widths were: curve A 25 μm, B 50 μm, C 100 μm, D 200 μm, and E 500 μm [95].

mathematically convert characteristic mass to intrinsic mass if the width of the line source and the absorption profile are known. In Table 4.17, the characteristic masses obtained with line sources have been converted to intrinsic mass by normalizing the characteristic mass by the width of the line source, and then applying a correction factor for the ratio of the width of the absorption profile to the width of the line source. Unfortunately, the normalization and correction factors must be based on limited experimental data for the line source widths and modeled absorption widths. Under these circumstances, the agreement of intrinsic mass for CS-LPDA-AAS and conventional LSAAS, shown in Table 4.17, is very reasonable.

It was determined that, under optimum conditions, the LPDA read noise was approximately 2800 e^- [94], and that read noise was the dominant noise source for all the elements between 200 nm and 350 nm. The relative read noise will decrease inversely with increasing intensity. Using the propagation of errors, it was shown that the standard deviation of the wavelength integrated absorbance σ_{A_λ}, with no analyte present, is

$$\sigma_{A_\lambda} = \frac{0.43\sigma_{\text{read}}\sqrt{n_{\text{pix}}+1}}{I_b} \tag{4.75}$$

Table 4.17 Continuum source intrinsic mass

Element	Wavelength (nm)	HCL Source m_0 [10]	HCL Source m_i [a]	Continuum source m_i LPDA (25 μm)	LPDA (50 μm)	LPDA (50 μm)	CCD (25 μm)
As	193.7	17	9.2	9.7	9.9	–	10.5
Se	196.0	30	10	11	10	–	12
Zn	213.9	0.45	0.23	0.50	0.50	1.4	0.17
Pb	217.0	5.0	1.9	1.9	1.8	2.1	3.0
Sb	217.6	38	15	–	–	–	10
Sn	224.6	20	10	9.2	9.4	–	–
Cd	228.8	0.35	0.14	0.27	0.19	0.23	0.29
Ni	232.0	13	5.5	8.0	8.4	6.4	–
Co	240.7	6.0	2.4	2.3	2.3	–	–
Fe	248.3	5.0	1.9	1.6	1.8	–	–
Tl	276.8	10	4.0	–	–	–	11
Mn	279.5	2.0	0.61	0.83	0.88	–	1.4
Pb	283.3	11	4.2	3.9	3.8	–	6.7
Cu	324.7	4.0	1.0	1.0	1.1	–	–

[a] Computed from m_0 as described in Reference 16 using the optimum temperature listed in Reference 10.

where σ_{read} is the read noise, n_{pix} the number of analytical pixels necessary to cover the absorption profile (also equal to the number of reference pixels), and I_b is the measured intensity at each pixel with no analyte present. Equation 4.75 predicts that the noise will decrease if the slit width is increased and fewer pixels are used, as shown in Figure 4.28. The experimental detection limits for CS-LPDA-AAS were comparable to, or better than, those for conventional LSAAS as shown in Table 4.18. These values were obtained with an entrance slit width and height of 500 μm. The resulting spectral bandwidth for a 500 μm entrance slit width ranges from 0.032 nm at 200 nm to 0.060 nm at 350 nm, still considerably smaller than the bandwidth used in most commercial instruments. As the spectral bandwidth with a 500 μm slit width is much larger than the width of the absorption profile, the absorption profile observed by the array (the convolution of the slit function and the absorption profile) can be expected to resemble the slit function (Figure 4.24(b)). The measurement of reference intensities at pixels further to either side will expand the viewed region of the spectrum by a factor of two. Thus, any spectral structure falling within the range of 0.062 nm at 200 nm to 0.12 nm at 350 nm can be expected to cause an interference (see Table 4.3). The 500 μm entrance slit width of the echelle provides improved detection limits for the read noise-limited case, but increases the possibility of spectral interferences.

The calibration curves and relative concentration error plots obtained for CS-LPDA-AAS are shown in Figure 4.29. The model data presented in

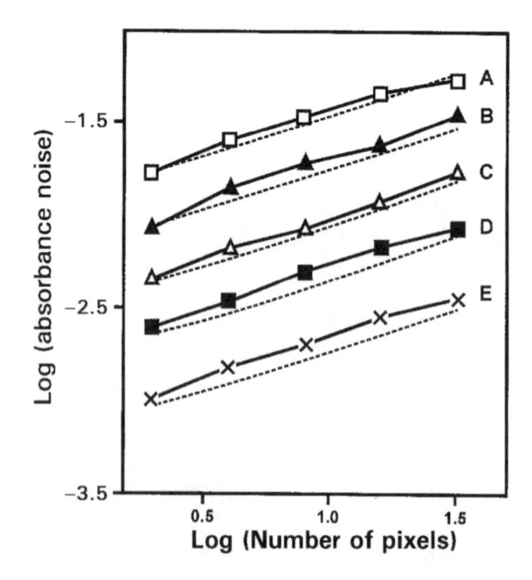

Figure 4.28 Wavelength integral absorbance noise σ_{A_λ} as a function of the number of pixels and entrance slit widths: curve A 25 μm, B 50 μm, C 100 μm, D 200 μm, and E 500 μm [94].

Table 4.18 Continuum source detection limits (pg)

Elements	*Wavelength (nm)*	*LS* [23]	*SIMAAC* [95]	*CSLPDA* [16]	*CSCCD* [96]
As	193.7	20	–	28	24
Se	196.0	30	–	50	49
Zn	213.9	0.1	18	2	–
Pb	217.0	5	–	6	9
Sb	217.7	15	–	–	12
Sn	224.6	20	–	26	–
Cd	228.8	0.4	–	0.4	1
Ni	232.0	10	40	11	–
Co	240.7	2	40	4	–
Fe	248.3	2	20	2	–
Tl	276.8	10	–	–	2
Mn	279.5	1	8	0.5	0.3
Pb	283.3	5	–	0.9	0.7
Al	309.3	4	16	–	–
Mo	313.3	4	40	–	–
V	318.4	20	60	–	–
Cu	324.7	1	8	0.6	–
Cr	357.9	1	8	–	–

this figure have been verified experimentally for eight elements. It can be seen that, as the intensity at the profile center reaches a minimum due to stray light (10 per cent), the absorption in the wings becomes important and results in a slope of 0.5 as expected [95]. It can also be seen that the use of intensities far in the wings of the absorption profile permits calibration over a much greater concentration range than for ZAAS or SRSAAS (Figure 4.16 and 4.23). If taken to a very high concentration, the curve in Figure 4.29(a) would also reverse, because the reference intensity is taken at a finite distance from the line center. With the LPDA, however, the reference intensities can be chosen a considerable distance from the line center. There are two limitations to the location of the reference pixels: they must be close enough to the line center to accurately reflect the non-specific background attenuation, and they should not be selected to overlap a spectral interference. The relative concentration error plot in Figure 4.29(b) is based on shot noise and 1 per cent analyte fluctuation noise. It can be seen that when the calibration curve changes slope from 1.0 to 0.5, the minimum concentration error increases from about 1.6 per cent to about 3 per cent. The relative concentration error increases rapidly as the calibration curve approaches a horizontal line.

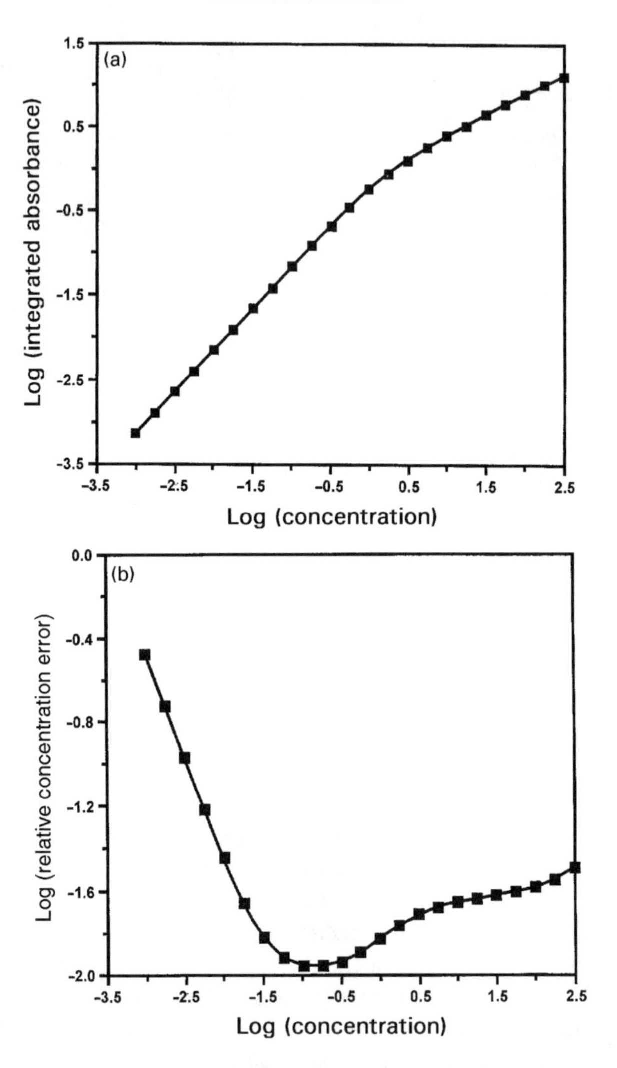

Figure 4.29 (a) Modeled calibration curves and (b) relative concentration error plots for CS-LPDA [96].

A prototype instrument using a larger LPDA has also been reported [103]. This instrument used a 300 W xenon arc lamp (Cermax LX300UV, ILC Technology, Sunnyvale, CA, USA), a 1.33 m monochromator (Model 209, McPherson), and a 2048 pixel LPDA (Model 2048, Princeton Instruments). The graphite atomizer was an HGA 2200 (Perkin-Elmer Corp., Norwalk, CT, USA). The length of the LPDA (2048 pixels on 25 μm centers) and the dispersion of the spectrometer ($R_d = 0.20$ nm mm^{-1}) provided a 10 nm window. A slit width of 25 μm was used to provide a spectral bandwidth of 0.005 nm. The absorption profiles typically covered two to three pixels. Numerous window positions were available, offering multielement detection, but these inevitably covered only one of the most sensitive analytical lines in combination with many less sensitive lines. Thus, for any position, the detection limits for all but one element were far from the best. A 10 nm window was too small to accommodate the primary wavelengths for two or more elements. Design of a true multielement instrument would require multiple large arrays. To date only a single large array has been used. Detection limits obtained with the large array prototype, in the single-element mode, were comparable to CS-LPDA-AAS with a short array and to conventional LSAAS. Unfortunately, exposure times for the array ranged from 100 to 250 ms (a read frequency of 4–10 Hz). Consequently, the absorbances computed for each exposure were based on integrated intensities. Ideally, the exposure time should not exceed 20 ms. As a result of this long exposure time, the measured peak heights and peak areas were dependent on the peak shape. This provides a potential for systematic variations (interferences) between the samples and standards.

4.7.2 Continuum source – CCD detection

The low read noise of the CCD (5–25 e^-) is ideally suited to spectroscopic applications, providing a shot noise limit for all but the lowest intensities. Compared to the read noise of 2800 e^- for the LPDA, the shot noise limited case would provide improved detection limits for most elements below 250 nm. Unfortunately, large two-dimensional arrays suitable for covering a large spectral range on an echelle spectrometer require a long read, which is prohibitive for graphite furnace atomization. In addition, these large arrays are also very expensive. These concerns for size and read time have also influenced the instrument design for ICP atomic emission spectrometry (ICPAES). The recently developed Optima instrument uses an echelle spectrometer and a segmented array CCD detector (SCD) [73,80]. The SCD consists of a monolithic chip containing 251 short, linear CCD arrays to monitor the wavelengths of primary interest to ICPAES. Thus, wide elemental coverage is achieved with far fewer pixels and a much shorter read time. This SCD detector is potentially useful for GFAAS. In a brief study, the SCD was evaluated as a detector for GFAAS (CS-SCD-AAS) [96]. A 300 W xenon arc lamp and a Perkin-Elmer HGA-500 graphite furnace were mounted in front of the SCD spectrometer in place of the ICP and the first collimating mirror. Collimated radiation from the xenon arc lamp passed through the furnace, and was focused by the focusing mirror onto the entrance slit of the SCD spectrometer. The entrance slit of the SCD spectrometer is 250 μm high and 62.5 μm wide in the normal mode of operation, and 31 μm wide in the high resolution mode. The image of the entrance slit is reduced by a factor of 2.5 in the echelle. Each array of the SCD consists of 32 or 64 pixels with a width of 12.5 μm. In the normal mode, the pixels are paired to provide a detection unit which is 25 μm wide, corresponding to the entrance slit width (i.e. $62.5/2.5 = 25$). In the high-resolution mode, each pixel is read independently. The pixels have an aspect of 7.2 or 13.6 (90 or 170 μm high and 12.5 μm wide). The charges on the pixels are shifted vertically into a parallel array of storage registers for reading. The lack of any overlying control lines for the one-dimensional array results in a much larger quantum efficiency (approximately 60 per cent at 200 nm) at wavelengths below 350 nm, even without a phosphor coating. In addition, the optics of the echelle were optimized for transmission at wavelengths below 250 nm.

A primary obstacle to CS-SCD-AAS was the suitability of the wavelength positions of the 251 arrays, selected for ICPAES, for furnace absorption measurements. It was found, however, that of the 43 most commonly used AA lines [96], 77 per cent were available on the SCD spectrometer. The major elements for which there were no wavelengths suitable for absorption measurements were cobalt, iron, molybdenum, and vanadium. The wavelengths available on the segmented array spectrometer, for these elements, were all ion lines. No data were available on the characteristic masses of any of these wavelengths, although it can be reasonably anticipated that they would be very poor.

Initial studies focused on verifying the performance of the SCD. The read noise for at least eight of the arrays of the SCD was determined to be approximately 15 e^-. At higher intensities, the variance of the intensity signal increased linearly with increasing intensity, verifying that photon shot noise was dominant. At very high intensity levels, the intensity noise levels exceeded that predicted for photon shot noise, as a result of the nonlinear response of the pixels as they approached saturation. Of particular interest was a comparison of the measured intensity levels (without amplification) for the SCD and the short LPDA. The entrance slit of the SCD spectrometer is 25 μm high and 62.5 μm wide. This is an entrance slit area 16 times smaller than that used with the echelle with the CS-LPDA-AAS (500 μm by 500 μm). It can be seen in Table 4.19 that the combination of the greater quantum efficiency of the linear CCDs and the improved optical transmission below 250 nm offsets the difference in aperture size at 190 nm. At higher wavelengths, the improving quantum efficiency of the LPDA and optical efficiency of the echelle produced increasing signal levels.

Wavelength and time-integrated absorbances were computed as before (Equations 4.72–4.74) except smaller numbers of pixels were used at low concentrations. Fewer pixels were required because of the higher resolution. In the optima, the 62.5 μm entrance slit is reduced to a 25 μm wide image at the focal plane. In theory, the baseline width of the absorption

Table 4.19 Comparison of intensities for two continuum source/echelle instruments using a 10 ms integration time and 20 A lamp current [96].

		Electrons (e^-) from 25 μm pixels	
Element	*Wavelength (nm)*	*Spectraspan/LPDA*[a]	*Optima SCD*[b]
As	193.7	192 000	198 000
Se	196.0	270 000	192 000
Pb	217.0	1 150 000	312 000
Cd	228.8	3 110 000	780 000

[a]Entrance slit 500 μm × 500 μm.
[b]Entrance slit 250 μm × 62.5 μm.

should then be 50 μm. The absorption profile (at low concentrations) typically covered two or three pixels. Thus, the wavelength absorbance for the absorption profile in Figure 4.24(c) is computed as

$$A_\lambda = \Delta\lambda_{\mathrm{pix}} \log\left(\frac{I_{-20} + I_{19}}{I_{-1} + I_0}\right) \tag{4.76}$$

Of course, more than two pixels can be used for the reference or analytical intensities but three was the largest value for the analytical intensity in this study (extended calibration was not studied). The absorbance noise is dependent on the intensity of the source and the number of pixels used in the calculation. Using the propagation of errors, it can be shown that the standard deviation of the wavelength and time integrated absorbance σ_{A_i}, with no analyte present, is

$$\sigma_{A_{\lambda,\mathrm{t}}} = \frac{0.43}{\sqrt{I_\mathrm{b}}}\sqrt{\frac{n_\mathrm{a}(n_\mathrm{r}+1)}{n_\mathrm{r}}} \tag{4.77}$$

where n_a is the number of pixels used for the analytical intensity, n_r is the number of pixels used for the reference intensity, and I_b is the measured intensity with no analyte present. Thus, the SNR improves if fewer pixels are used for the analytical intensity and more pixels are used for the reference intensity. As expected, the intrinsic masses for the elements determined by CS-SCD-AAS were comparable to those obtained with CS-LPDA-AAS (Table 4.17). After normalization by the spectral width of the pixel, the mass necessary to give an absorbance of 0.0044 pm s should be dependent only on the efficiency of the graphite furnace atomizer.

Table 4.18 shows the experimentally determined detection limits for CS-SC-DAAS and CS-LPDA-AAS. In general, there is no difference between CS-SCD-AAS, CS-LPDA-AAS, and conventional LSAAS. The experimental detection limits do not show the anticipated improvement for CS-SCD-AAS for two reasons. The data acquisition system for the CS-SCD-

AAS system was optimized for emission measurements, not furnace measurements. A 50 Hz absorbance computation frequency allowed 10 ms of exposure of the arrays and required another 10 ms to read the arrays for eight elements. Thus, at 50 Hz the exposure time for the CCD (10 ms) was exactly half that of the LPDA (20 ms). In addition, the photon shot noise levels were almost a factor of three higher than predicted by the pixel intensity levels. As previously explained, this is partially due to the nonlinear response of the pixels at higher intensity levels.

The high resolution of CS-SCD-AAS provided an extremely useful tool for evaluating spectral interferences. With the continuum source and the CCD arrays, it was possible to examine the wavelength region around each element as a function of time. Several interferences were identified for arsenic and selenium, which could be eliminated by proper selection of pixels and/or by proper selection of the start and stop of the integration interval. An example of this is shown in Figure 4.30, where the most sensitive selenium line (196.026 nm) is separated from a potentially interfering, but very insensitive, palladium line (196.011 nm) by just 15 pm. The palladium line can contribute to a high blank value when palladium is used as a chemical modifier.

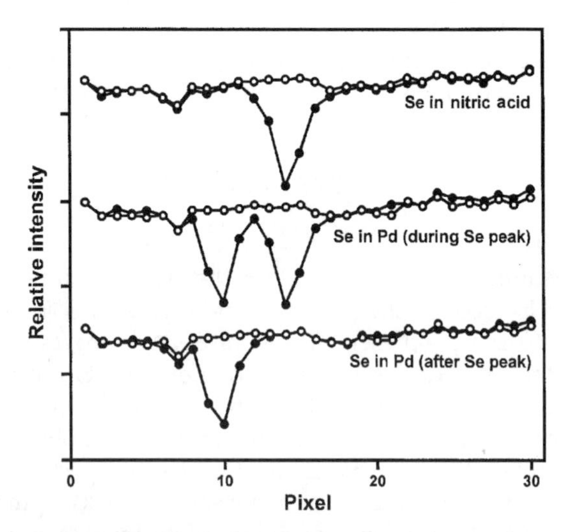

Figure 4.30 Absorption spectra around the Se (196.0 nm) line for the atomization of Se in 1 per cent nitric acid and with Pd as a chemical modifier. Open circles show the intensity prior to the start of the atomization cycle. Filled circles show the intensity during the atomization cycle. From the top, the first two profiles were obtained at the time of the maximum Se signal. The third profile was obtained 2 s after the second profile [96].

4.7.3 Continuum source – CID detection

Although CIDs are potentially useful detectors for GFAAS, no published data are available for this use. However, a reasonable extrapolation can be made based on the specifications of the latest arrays and experience with LPDAs and CCDs. The only commercially available CIDs are found on the Iris ICPAES instrument (Thermo Jarrell Ash, Franklin, MA, USA) [79]. This instrument uses a large CID array in an echelle spectrometer to cover the wavelength region from 200 to 450 nm. The CID array is 512 pixels wide by 512 pixels high and the pixels are 23 μm square. The random access capability of a CID makes it possible to select a series of subarrays (approximately 13 pixels wide and 3 pixels high) for measuring the analytical and off-line reference intensities. The read noise of the CID is approximately 180 e^-, but the reads can be nondestructive. Repetitive reads of the array reduce the effective read noise by the square root of the number of reads. Consequently, a read noise of 18 e^- can be reached with 100 nondestructive reads. It is the random access and repetitive reads that make the CID such a versatile detector.

The intrinsic mass for CS-CID-AAS would be expected to be comparable to that for CS-LPDA and CS-SCD shown in Table 4.17. With wavelength-integrated absorbance, the analytical signal would be solely a function of the efficiency of the graphite furnace atomizer. With 100 nondestructive reads of a series of subarrays, the noise levels of the CID would be comparable to that of the SCD, and the detection limits would be expected to be comparable. Unfortunately, the time required for the necessary 100 reads is currently too long to be useful for GFAAS. As discussed in section 4.3.4.2 the CID is accessed in a pseudo-random manner. Consequently, locating the proper starting address, reading a 39 pixel sub-array, and clearing the address register, can take up to 1 ms. Multiple reads would be too long to be compatible with transient furnace signals. With the development of a true random access array and the use of a fast ADC, however, the CID could potentially achieve the same performance of other array detectors with considerable versatility in the selection wavelengths.

REFERENCES

[1] A. Walsh, *Spectrochim. Acta* **7B**, 103 (1955).

[2] C.T.J. Alkemade and J.M.W. Milatz, *J. Opt. Soc. Am.* **45**, 583 (1955).

[3] B.V. L'vov, *Inzh. Zh.* No. 2, 44 (1959); reprinted in English in *Spectrochim. Acta, Part B* **39**B, 159 (1984).

[4] H. Falk, E. Hoffman and Ch. Ludke, *Spectrochim. Acta, Part B* **39B**, 283 (1984).

[5] J.D. Winefordner, G.A. Petrucci, C.L. Stevenson and B.W. Smith, *J. Anal At. Spectrom.* **9**, 131 (1994).

[6] J.D. Ingle and S.R. Crouch, *Spectrochemical Analysis*, Prentice Hall, Englewood Cliffs, NJ, pp 34–35, 1988.
[7] International Union of Pure and Applied Chemistry, *Pure Appl. Chem.* **64**, 253 (1992).
[8] B. Welz, *Spectrochim. Acta, Part B* **47B**, 1043 (1992).
[9] The Perkin-Elmer Corporation, *Recommended Conditions for Continuum Background Correction*, Part Number 0993-3199 Rev. B, 1984.
[10] The Perkin-Elmer Corporation, *Recommended Conditions for Zeeman Background Correction*, Part Number 0993-8199 Rev. B, 1984.
[11] The Perkin-Elmer Corporation, *Recommended Conditions for THGA Furnaces*, Part Number B050-6158, 1991.
[12] The Perkin-Elmer Corporation, *SIMMA 6000 Atomic Absorption Spectrometer*, Part Number B050-4270/8.94, 1994.
[13] Varian Associates, *Varian Guide to ICP/AAS Analytical Values*, Part Number 86 10100900, 1995.
[14] I.L. Shuttler, G. Schlemmer, G.R. Carnrick and W. Slavin, *Spectrochim. Acta, Part B* **46B**, 583 (1991).
[15] E.G. Su, A.I. Yuzefovsky, R.G. Michel, J.T. McCaffrey and W. Slavin, *Microchem. J.* **48**, 278 (1993).
[16] C.M.M. Smith and J.M. Harnly, *Spectrochim. Acta, Part B* **49B**, 387 (1994).
[17] G. Horlick, *Anal. Chem.* **47**, 352 (1975).
[18] J.D. Ingle and S.R. Crouch, *Spectrochemical Analysis*, Prentice Hall, Englewood Cliffs, NJ, pp. 548–549, 1988.
[19] J.D. Ingle and S.R. Crouch, *Spectrochemical Analysis*, Prentice Hall, Englewood Cliffs, NJ, p. 150, 1988.
[20] International Union of Pure and Applied Chemistry, *Spectrochim. Acta, Part B* **33B**, 242 (1978).
[21] G.L. Long and J.D. Winefordner, *Anal. Chem.* **55**, 721A (1983).
[22] W. Slavin, *Graphite Furnace AAS: A Source Book*, The Perkin-Elmer Corporation, Part Number 0993-8139, 1984.
[23] The Perkin-Elmer Corporation, *Recommended Conditions for THGA Furnaces*, Publication B31 10.06, Uberlingen, Germany, 1991.
[24] Varian Associates, *Varian Guide to ICP/AAS Detection Limits*, Part Number 85 101009 00, Wood Dale, IL, 1995.
[25] Leeman Labs, Inc., *Leeman Letter* **33**, 3 (1994).
[26] R.R. Williams, *Anal Chem.* **63**, 1638 (1991).
[27] C.L. Stevens and J.D. Winefordner, *Appl. Spectrosc.* **45**, 1217 (1991).
[28] F. Voigtman, *Appl. Spectrosc.* **45**, 237 (1991).
[29] J.M. Harnly, *J. Anal At. Spectrom.* **3**, 43 (1988).
[30] B.R. Culver and T. Surles, *Anal. Chem.* **47**, 920 (1975).
[31] W. Slavin and G.R. Carnrick, *Crit. Rev. Chem.* **19**, 95 (1988).
[32] U. Kurfurst and J. Pauwels, *J. Anal At. Spectrom.* **9**, 531 (1994).
[33] R.E. Sturgeon, *Fresenius' Z. Anal Chem.* **337**, 538 (1990).
[34] E. Lundberg and W. Frech, *Anal Chem.* **53**, 1437 (1981).
[35] Z. Grobenski, R. Lehmann, B. Radziuk and U. Voellkopt *At. Spectrosc.* **5**, 87 (1984).
[36] P.R. Liddell, N. Athanasopoulos, R.G. Grey and M.W. Routh, *Am. Lab.*, November, 68 (1986).
[37] D.D. Siemer and J.M. Baldwin, *Anal Chem.* **52**, 295 (1980).
[38] R.A. Newstead, W.J. Price and P.J. Whiteside, *Prog. Anal. At. Spectrosc.* **1**, 267 (1978).

[39] T.C. O'Haver, In *Trace Analysis*, (Ed.) J.D. Winefordner, John Wiley and Sons, New York, pp. 15–62, 1978.
[40] C.G. Enke and T.A. Nieman, *Anal Chem.* **48**, 705A (1976).
[41] B. Chapman, *Glow Discharge Processes* John Wiley and Sons, New York, 1980.
[42] P.L. Larkins, *Spectrochim. Acta, Part B* **40B**, 1585 (1985).
[43] M.L. Parsons, W.J. McCarthy and J.D. Winefordner, *Appl. Spectrosc.* **20**, 223 (1966).
[44] D.O. Cooke, R.M. Dagnall and T.S. West, *Talanta* **19**, 1309 (1972).
[45] B.V. L'vov, N.V. Kocharova, L.K. Polzik, N.P. Romanov and Yu.I. Yarmak, *Spectrochim. Acta, Part B* **47**B, 843 (1992).
[46] S.V. Baranova, B.D. Grachev, I.A. Shikkeeva and I.A. Zemskova, *J. Anal Chem. (USSR)* **37**, 1756 (1982).
[47] S.B. Smith and G.M. Hieftje, *Appl. Spectrosc.* **37**, 419 (1983).
[48] C. Schnurer-Patschan, A. Zybin, H. Groll and K. Niemax, *J. Anal At. Spectrom.* **8**, 1103 (1993).
[49] H. Groll, and K. Niemax, *Spectrochim. Acta, Part B* **48B**, 633 (1993).
[50] R.L. Cochran and G.M. Hieftje, *Anal Chem.* **49**, 2040 (1977).
[51] C.M.M. Smith and J.M. Harnly, *J. Anal At. Spectrom.* **9**, 419 (1994).
[52] T.C. O'Haver, J.M. Harnly and A.T. Zander, *Anal Chem.* **50**, 1218 (1978).
[53] O. Guell and J.A. Holcombe, *Spectrochim. Acta, Part B* **47B**, 1535 (1992).
[54] M. Berglund and D.C. Baxter, *J. Anal At. Spectrom.* **7**, 461 (1992).
[55] H. Falk and A. Glissmann, *Fresenius' Z. Anal. Chem.* **323**, 748 (1986).
[56] S.K. Gin, D. Littlejohn and J.M. Ottaway, *Analyst* **107**, 1095 (1982).
[57] W. Frech, D.C. Baxter and B. Hutsch, *Anal Chem.* **58**, 1973 (1986).
[58] W. Frech, B.V. L'vov and N.P. Romanov, *Spectrochim. Acta, Part B* **47B**, 1471 (1992).
[59] W. Frech and B.V. L'vov, *Spectrochim. Acta, Part B* **48B**, 1371 (1993).
[60] N. Hadgu and W. Frech, *Spectrochim. Acta, Part B* **49B**, 445 (1994).
[61] S. Xiao-quan, B. Radziuk, B. Welz and V. Sychra, *J. Anal At. Spectrom.* **7**, 389 (1992).
[62] M. Suzuki and K. Ohta, *Prog. Anal. At. Spectrosc.* **6**, 49 (1983).
[63] E. Lundberg, W. Frech, D.C. Baxter and A. Cedergren, *Spectrochim. Acta, Part B* **43B**, 451 (1938).
[64] J.J. Sotera, L.C. Cristiano, M.K. Conley and H.L. Kahn, *Anal. Chem.* **55**, 204 (1983).
[65] B. Welz, *Microchem. J.* **45**, 163 (1992).
[66] R.E. Sturgeon, S.N. Wilie, G.I. Sproule, P.T. Robinson and S.S. Berman, *Spectrochim. Acta, Part B* **44B**, 667 (1989).
[67] F.J. Langmyhr and G. Wibetoe, *Prog. Anal. At. Spectrosc.* **8**, 193 (1985).
[68] N.J. Miller-Ihli, *Anal. Chem.* **64**, 964A (1992).
[69] The Perkin-Elmer Corporation, *Instructions: Zeeman/5000 System*, Part Number 0993-8001 Rev. November 1983.
[70] Spectrametrics, Inc., *Spectraspan IIIB Emission Spectrometer Operators Manual*, Part Number 1506029A, 1977.
[71] W. Snelleman, *Spectrochim. Acta, Part B* **23B**, 403 (1968).
[72] A.T. Zander, M.H. Miller, M.S. Hendrick and D. Eastwood, *Appl. Spectrosc.* **39**, 1 (1985).
[73] T.W. Barnard, M.I. Crockett, J.C. Ivaldi and P.L. Lundberg, *Anal. Chem.* **65**, 1225 (1993).
[74] B. Radziuk, G. Rodel, H. Stenz, H. Becker-Ross and S. Florek, *J. Anal At. Spectrom.* **10**, 127 (1995).

[75] B. Radziuk, G. Rodel, M. Zeiher, S. Mizuno and K. Yamamoto, *J. Anal. At. Spectrom.* **10**, 415 (1995).
[76] D.G. Jones, *Anal. Chem.* **57**, 1057A (1985).
[77] G.R. Sims, In *Charge-Transfer Devices in Spectroscopy,* (Eds.) J.V. Sweedler, K.L. Ratzlaff and M.B. Denton, VCN Publishers, Inc., New York, pp. 9–58 (1994).
[78] J.V. Sweedler, R.B. Bilhorn, P.M. Epperson, G.R. Sims and M.B. Denton, *Anal. Chem.* **60**, 282A (1988).
[79] M.J. Pilon, M.B. Denton, R.G. Schleicher, P.M. Moran and S.D. Smith, Jr., *Appl. Spectrosc.* **44**, 1613 (1990).
[80] T.W. Barnard, M.I. Crockett, J.C. Ivaldi, P.L. Lundberg, D.A. Yates, P.A. Levine and D.J. Sauer, *Anal. Chem.* **65**, 1231 (1993).
[81] Hamamatsu Photonics K.K., Solid State Division, *MOS Linear Image Sensors - 1990 Catalog.*
[82] Photometrics Ltd., Product Literature, Tucson, AZ, 1995.
[83] J.M. Harnly and J.A. Holcombe, *Anal. Chem.* **57**, 1983 (1985).
[84] J.A. Holcombe and J.M. Harnly, *Anal. Chem.* **58**, 2606 (1986).
[85] A. Gilmutdinov, A.Kh. Zakharov and A.V. Voloshin, *J. Anal At. Spectrom.* **8**, 387 (1993).
[86] A. Gilmutdinov, K.Yu. Nagulin and A.Kh. Zakharov, *J. Anal. At. Spectrom.* **9**, 643 (1994).
[87] A. Kh. Gilmutdinov, B. Radziuk, M. Sperling, B. Welz and K. Yu. Nagulin, *Appl. Spectrosc.* **49**, 413 (1995).
[88] S.R. Koirtyohann and E.E. Pickett, *Anal Chem.* **37**, 601 (1965).
[89] H.L. Kahn, *At. Absorpt. Newsl.* **7**, 40 (1968).
[90] B. Welz, *Atomic Absorption Spectrometry, 2nd Completely Revised Edition*, VCH Publishers, Weinheim, Germany, p. 141, 1985.
[91] P.L. Larkins, *Spectrochim. Acta, Part B* **43B**, 1175 (1988).
[92] L. de Galan and M.T.C. de Loos-Vollebregt, *Spectrochim. Acta, Part B*, 1984, **39A**, 1011 (1984).
[93] J.M. Harnly, *Anal Chem.* **58**, 933A (1986).
[94] J.M. Harnly, *J. Anal At. Spectrom.* **8**, 317 (1993).
[95] J.M. Harnly, C.M.M. Smith and B. Radziuk, *Spectrochim. Acta, Part B* **51B**, 1055 (1996).
[96] J.M. Harnly, C.M.M. Smith, D.N. Wichems, J.C. Ivaldi, P.L. Lundberg and B. Radziuk, *J. Anal At. Spectrom.* **12**, 617 (1997).
[97] M. Retzick and D. Bass, *Am. Lab.* December, 24 (1989).
[98] G. Dulude, R. Moseley and S. Martin, 1995 Pittsburgh Conference, Paper 239, March 1995.
[99] R. Mavrodineanu and R.C. Hughes, *Appl. Opt.* **7**, 1281 (1968).
[100] V.A. Fassel, V.G. Mossotti, W.E.L. Grossman and R.N. Knisely, *Spectrochim. Acta, Part B* **22B**, 347 (1966).
[101] T.C. O'Haver and J.D. Messman, *Prog. Anal. Spectrosc.* **9**, 483 (1986).
[102] A.T. Zander, T.C. O'Haver and P.N. Keliher **48**, 1166 (1976).
[103] R. Fernando and B.T. Jones, *Spectrochim. Acta, Part B* **49B**, 615 (1994).

5

Modifiers in electrothermal atomic absorption spectrometry

David L. Styris

5.1 INTRODUCTION

Chemical modification is a widely used procedure for inhibiting interferences in electrothermal atomic absorption spectrometry (ETAAS). As first introduced by Ediger [1], the concept was envisaged as deliberately altering the thermal pretreatment/temperature vaporization properties of either the analyte or the matrix (matrix modification) through the procedure of chemical additions. This easily applied technique involves the inclusion, within the sample or standard of interest, of particular substances (chemical modifiers) that induce: (1) a lowering of preatomization (thermal pretreatment) temperatures necessary for depletion of the matrix without loss of analyte; (2) an enhancement of analyte retention (stabilization) at any of the more elevated preatomization temperatures that might otherwise be necessary for matrix depletion; or (3) even in the absence of a sample matrix, retention of analyte to higher temperatures where analyte (M)-containing species (e.g. MO_n, M_mC_n, etc.) will dissociate.

Although this general application of chemical modification for ETAAS involves the addition of the modifier within each sample, a means of permanent modification has been recently introduced by Rademeyer *et al.* [2]. These investigators demonstrated that, when iridium is sputter-deposited onto a pyrolytic graphite-coated graphite tube, the tube stabilizes the volatile metals cadmium, lead, and selenium for up to 700 atomization firings. The controlling mechanism(s) remain a question, but the approach is a worthy addition to the chemical modification arsenal,

Electrothermal Atomization for Analytical Atomic Spectrometry. Edited by K. W. Jackson
© 1999 John Wiley & Sons Ltd

and must necessarily be included in any description of the modification process. Consequently, chemical modification for ETAAS may be described as a chemical addition treatment of the sample, standard or atomizer surface for purposes of promoting either sample matrix depletion at lower preatomization temperatures, analyte retention beyond sample matrix depletion temperatures or, if interfering matrices are absent, thermal dissociation of analyte-containing molecules.

This description emphasizes the effects associated with the use of chemical modifiers and with the simplicity of applying such modifiers. Unfortunately, the description belies the actual, complex causes pertaining to the modification. The mechanisms that control the release of the matrix or the retention of the analyte are, therefore, not even alluded to. In this regard, even designating the modification as 'chemical' is misleading, because it can be interpreted as indicating that the mechanisms responsible for the modification are of a purely chemical nature. Although such an interpretation is justifiable for some modifiers, it is unwarranted in general. The influence of physical interactions must also be considered, as they can play a dominant role in the modification. A more complete definition should at least address the fact that physical, as well as chemical, interactions can be involved in the modification process.

These shortcomings reflect the inadequacy of the knowledge pertaining to the mechanisms that control the chemical modification processes. This deficiency is disturbing, but considering the numerous contributing components and interaction possibilities that are associated with these modifiers, and the present inadequacies in the understanding of basic electrothermal atomization processes themselves, it is not surprising that a more complete understanding has failed to emerge. The mechanisms that control electrothermal atomization processes of even the simplest compounds are only beginning to be understood, and are now known to be considerably more complex than those initially anticipated [3]. Introduction of additional variables associated with modifiers can be expected to interject additional complexities in the chemical and physical processes that prevail in the already complex, and uncharacterized, dynamic environments of ETAAS. Attaining even a rudimentary understanding of such processes is a formidable task.

The various modifiers have evolved primarily through trial and error, and serendipity. Tsalev and Saveykova [4] reported that, by mid-1991, over 150 different qualitative compositions of modifiers had been reported in the literature. These same investigators used multivariate techniques in an effort to classify and organize these existing data. From a pragmatic view, modifier oxide interactions with the graphite atomizer and carbon monoxide were found to provide the most logical classification. Under this scheme, Groups IIA (magnesium, calcium, strontium, and barium) and IIIA (scandium, yttrium, lanthanum, and cerium), and the element aluminum were classified as 'oxides–salt-like carbides'. Groups IVA (titanium, zirconium, and hafnium), VA (vanadium, niobium, and tantalum), VIA (chromium, molybdenum, and tungsten), and the element manganese were given an 'oxides–metal-like carbides' designation. The transition metals (iron, cobalt, nickel, rhodium, palladium, iridium, and platinum), the noble metals (copper, silver, and gold), and the element zinc were classified as metals. Multivariate analysis, employing maximum thermal pretreatment temperatures of the analyte elements in presence of various chemical modifiers, placed the analyte in three groups. The highly volatile elements involving the semimetals (phosphorus, gallium, arsenic, selenium, indium, tin, antimony, and tellurium) are stabilized to higher temperatures by all of the above groups of modifiers. The metal group (zinc, silver, cadmium, gold, thallium, lead, and bismuth) are stabilized to lower temperatures by the oxide-type modifiers, and to higher temperatures by the metal-type modifiers; phosphate modifiers also stabilize these metal elements to higher temperatures.

All of the interaction possibilities that comprise these processes must be considered in the elucidation of the chemical and physical mechanisms that occur in high-temperature atomizer environments. Homogeneous and heterogeneous gas-phase interactions, interactions between the condensed and the adsorbed phases, and those interactions and effects involving the atomizer surface, subsurface and bulk regions, including the associated lattice defects, are among these possibilities. Additions of chemical modifiers have the potential of enhancing or inhibiting any of these interactions and effects, depending on the species employed for the particular modification.

The modifier may, of course, interact directly with the analyte species, or with the matrix. Redox reactions and hydrolysis exemplify possibilities should this interaction be of a chemical nature; the product can then have bulk properties, or it can be adsorbed on the surface of the atomizer substrate. If the interaction between the modifier and the analyte or matrix is a physical one, the candidate mechanism of interaction may involve occlusion of the analyte by the modifier, or it may entail formation of modifier–analyte or modifier–matrix solid solutions. These solid solutions contain dispersed analyte or matrix at the lattice interstices of the modifier crystallite (interstitial solid solution), or the dispersion is at the modifier lattice sites (substitutional solid solution) as illustrated in Figure 5.1. This requires that the analyte or the matrix be soluble in the bulk modifier.

It should be noted that extensive solid solubility occurs in bulk materials only if the Hume–Rothery rules are obeyed [5]. These rules state that, for interstitial solutions, the ratio of solute to solvent atom diameters must be smaller than 0.59 and, for substitutional solutions, the diameters cannot differ by more than 15 per cent. These are not sufficient conditions,

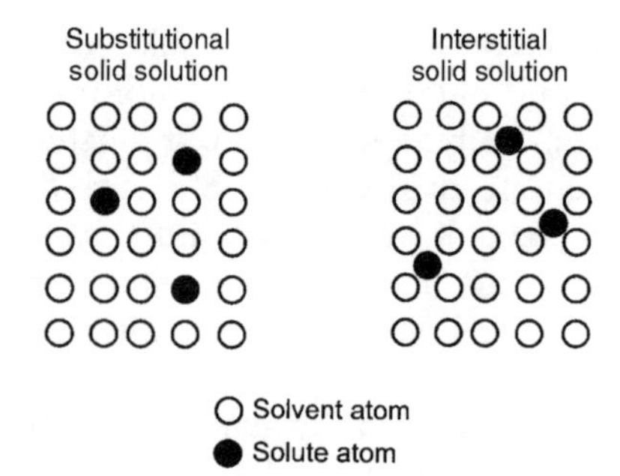

Figure 5.1 The two basic forms of solid solution

and the Hume–Rothery rules stipulate that substitutional solid solution formation requires that the solvent and solute have a similar electronegativities and similar crystal lattice structures. It should also be noted that, for the interstitial solid solutions, the transition metals more readily dissolve atoms to their interstitial sites than do the nontransition metals. This is a result of the directional characteristics of the d orbitals in the transition metal atoms. As there are incomplete d orbital fillings, ligands that bind to these atoms can lower the energy by forcing the accommodation of the orbitals to the unfilled energy levels.

An important point to remember is that the semiempirical requirements established by Hume–Rothery refer to solute placement within an otherwise periodic lattice structure, and that the crystallite lattices associated with modifiers in ETAs are far from being perfectly periodic. The large concentrations of defects associated with the lattice strains in such crystallites can induce significant migration of analyte. For example, a dislocation, which is a lattice defect that displaces large numbers of atoms from the positions they would normally possess in a periodic crystal lattice, can act as sinks for solute atoms. This is readily seen for the edge-type dislocations illustrated in Figure 5.2. By placing an extra half-plane of

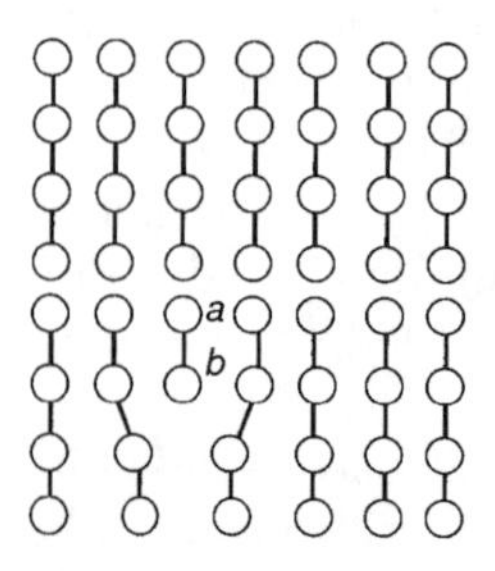

Figure 5.2 Diagram of an edge dislocation: (above) a perfect crystal; (below) the crystal is distorted by the insertion of an extra half plane (*ab*), which creates the dislocation line perpendicular to the page and located at (*b*).

atoms in a normally perfect lattice, an edge-type dislocation (line defect) is formed in a normal direction to the plane of the figure. The stress vectors and stress patterns around this edge dislocation are indicated in Figure 5.3, and illustrate the asymmetry in the stresses induced by this dislocation. The free energy of the crystal will be decreased by a contraction at the (compressed) portion of the defect, and an expansion at the portion that is under biaxial tension. Consequently, this energy is lowered when atoms of smaller diameter are substituted for lattice atoms in the compressed region and when atoms of larger diameter than those of the lattice are substituted in the region that experiences biaxial tension. Interstitial atoms being placed in this tension region of the dislocation also lower the energy of the lattice. This results in solute atoms being attracted toward the dislocation, and an excess of solute atoms forming around the dislocation. A second crystallite state can form if the solute atoms exhibit mutual attraction. Hence, the dislocations in the modifier crystallites can attract the solute species by means of the chemical potential created by the asymmetrical dislocation stress fields. The quantities of solute atoms that can form a steady-state solute atmosphere around the dislocation are temperature dependent, and are depleted with increasing temperature.

The modifier may also alter, physically or chemically, the substrate surface, and this surface alteration can influence the adsorption or desorption properties. For example, dissociative adsorption can be enhanced by increasing the desorption energies associated with particular surface sites, or the modifier can participate in redox reactions with the substrate, and hence modify surface states that affect analyte desorption.

Physical interactions with the substrate can result in the modifier species either replacing point defect sites, or filling surface defects such as kinks, jogs, and grain boundaries (dislocations). Consequently, the modifier can physically influence the participation of these defects in the transport of the analyte within and on the substrate, i.e. the bulk and the surface diffusion properties of the substrate can be modified by the presence of the modifier. Transport-dependent phenomena such as decomposition and oxidation of solids, and nucleation and sintering are, in turn,

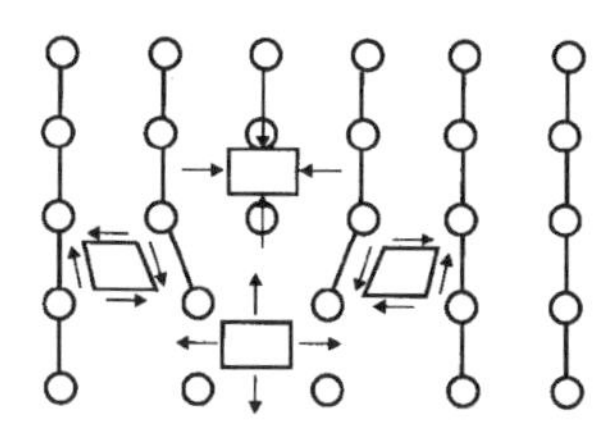

Figure 5.3 Stress patterns surrounding an edge dislocation. Compressive stresses are introduced in the vicinity of the extra half-plane. Biaxial tension is produced below the extra half-plane. Shear stresses appear on either side of the dislocation.

subject to change. Lattice interstitial sites in the substrate can also be substituted by the modifier species. Formation of analyte-related intercalation compounds will then be affected by the depletion of intercalation sites that are otherwise accessible to the analyte. The reader is referred to Section 2.2 for a discussion of intercalation compound formation.

The available data on these interactions in the high-temperature environment of electrothermal atomizers are presently insufficient to theoretically deduce which of the above interactions, reactions or processes, or combination thereof, controls the modification. Fortunately, progress in establishing the responsible mechanisms has been, and continues to be, achieved empirically. The understanding that has developed is, for the most part, qualitative. Two different philosophies have guided the research, so the experiments responsible for the contemporary knowledge of these systems have necessarily taken two approaches. Both involve investigations of the effects of specific combinations of atomizer substrate, analyte, and modifier species. The approaches differ in that the principal data are obtained either after controlled thermal processing and quenching (postliminary investigations), or during the thermal processing and vaporization phases (real-time investigations).

Both approaches provide valuable information. However, for the postliminary experiments, the potential for experimental artifacts during, or after, thermal quenching is considerable. As these artifacts have a significant bearing on the interpretation of these experiments, they cannot be neglected. Such artifacts result from condensation, adsorption and/or nucleation-related alterations in the analyte environment during the thermal quench period. Unfortunately, these alterations are far from being understood and are, by themselves, potential subjects for considerable research. Careful consideration of these quench-induced changes is necessary when extrapolating the interpretations of the postliminary results to the higher temperature, prequench periods of interest. The credibility of these interpretations can be no better than the extent of understanding of the thermal quench-induced alterations themselves. The dilemma is that the understanding of these alterations is inadequate.

Real-time measurements are less likely to introduce these particular difficulties in interpretation. This is because such measurements are made during the heating (temperature) cycle of interest, so extrapolations to temperatures of an earlier cycle are unnecessary. Consequently, the real-time monitoring approaches are best suited for the determination of the mechanisms associated with the inherent and rapid high temperature changes in electrothermal environments of interest. However, even during the period of real-time sampling, creation of experimental artifacts remains possible. Artifacts that are introduced during the observation period, rather than during a quenching period, are more easily identified and, therefore, more reasonably addressed and incorporated into the interpretation of the results. Results from real-time and from postliminary measurements can be mutually supportive, but interpretations of post-liminary measurement data continue to be complicated by the difficult task of developing an understanding of the associated thermal quench-induced interactions.

The most serious problem associated with real-time techniques that have been employed to-date is that they are capable of monitoring only the gas-phase species. Surface species, condensed or adsorbed, are often implicated through this approach, but without direct verification. Corroboration of mechanisms that are proposed from the results of real-time measurements on high-temperature surface-related phenomena awaits the application of a direct method for monitoring such species at these temperatures. Secondary ion mass spectrometry (SIMS) techniques have recently been applied to this problem, but with limited success [6].

It is interesting that the mechanisms that have been proposed to explain the successful applications of various chemical pretreatment approaches generally rely on hypothesizing the appearances of condensed and/or adsorbed species during the atomization heating stage. The fact that these species remain hypothetical entities is a manifestation of the difficulties associated with direct identification of these species. Until such identification can be made the existence of these species will remain conjecture, and questioning of the associated mechanisms must continue. Nonetheless, these proposed mechanisms contribute positively towards the development of the basic principles governing chemical modifier effects in electrothermal atomization, but only as long as the existence of the uncertainties is recognized and the questioning continues. As with any research endeavor, mistaken acceptance of hypothesized mechanisms will seriously jeopardize the advancements necessary for developing the desired understanding.

As the content of this chapter is dictated by the existing level of understanding of modifier mechanisms, it must necessarily focus on the results from the empirical evaluations of various analyte–modifier species. The intent is: (1) to provide a perspective on the chemistry and physics that control modification processes; and (2) to furnish the necessary information for the reader to critically evaluate the available knowledge and determine the areas that require additional basic research on chemical modification.

Recall that real-time monitoring approaches are the most appropriate for mechanistic investigations of transient behavior in ETA. Results from such approaches will, therefore, be used to guide these discussions. For the most part, atomic absorption (AA) and mass spectrometries (MS) have been the principal techniques used for real-time monitoring of ETA phenomena; both kinetics and the evolution of gas-phase species have thus been elucidated. The research results discussed in this chapter on modifier mechanisms are primarily those obtained from modifier–analyte investigations that have incorporated, at the very least, both of these real-time spectrometric techniques. This ensures that results from the most thoroughly investigated modifier–analyte systems are addressed. Therefore, the discussions of empirical results focus on some of the most complete sets of data available for this research, and provide a basis for evaluating the experimental methods as well as the interpretations. The modifiers discussed include magnesium, nickel, oxygen, palladium, phosphorus, and ascorbic acid.

Before addressing the mechanisms involved in the control of atomization by the modifiers, the physics and chemistry of the primary interactions that regulate the associated processes are reviewed. This provides the foundation for the discussions of modifier mechanisms. The chapter concludes with a discussion of two empirical techniques that can provide answers to the condensed phase interactions, and thus establish either the credibility of their involvement as postulated from gas-phase investigations, or provide new or more detailed evidence of the influence of the condensed phase.

5.2 SURFACE INTERACTIONS

Cognizance of the basic physical and chemical surface phenomena associated with high-temperature atomizers is necessary to elucidate the mechanisms responsible for chemical modification. The extremely complex nature of electrothermal atomization is indicated by the strong dependencies of these mechanisms on temperature, lattice strain, and lattice defect concentration. Furthermore, the strains and defect structures are temperature dependent. It should be expected then, at the high temperatures associated with the graphite atomizers, that effects of the atomizer surface, excluding any bulk effects, will involve multiple interactions. Considering the additional coupling with the temperature-dependent gas-phase composition or with the bulk lattice through bulk diffusion-controlled processes, it is doubtful that a realistic, simple mechanism of atomization from high-temperature, high-defect density surfaces exists. In some instances, however, it may be possible to isolate a dominating set of processes at temperatures well below the appearance temperature of the free atoms.

Under high-temperature conditions, the surface characteristics of atomizers vary continuously because of bulk defect migration to and from the surface and defect migration over the surface (reconstruction). These surface defects can be point lattice defects composed of lattice vacancies, lattice species or solute species at the interstices, or solute species at the lattice sites. They may also be line defects such as dislocations, or grain boundaries, which act as sources and sinks for point defects and which can emerge as highly reactive sites on the surface of interest. Surface defects in the forms of extra half-planes of atoms can form steps where two different crystallographic planes are exposed on the surface, as illustrated in Figure 5.4. (Section 2.2 discusses the crystallography of graphite used in electrothermal atomizers.) If the step changes direction so that it becomes irregular, more active kinks or jogs are formed at the position of the direction change. The new corners formed in this manner are preferred adsorption sites that provide multiple bonding opportunities with the host atoms. At a step defect on the surface, adsorbate species can interact with the lattice atoms from each of the two crystallographic planes at the step discontinuity. This interaction is in contrast to that of the adsorbate species interacting with a single, clean, homogeneous defect-free surface. The probability of adsorption is, therefore, enhanced by the presence of the step, and increases with the step density [7]. A similar situation arises for the more reactive kinks and jogs, and sites of dislocation line emergence at the surface–increasing the density of these surface defects further enhances the adsorption characteristics of the substrate. Each of these surface heterogeneities contributes to particular reactivities on the surface.

There is also the potential for a given defect contribution and associated reactivity to be influenced by the presence of a given chemical modifier. In this manner, the modifier can control the surface reactivity. Indeed,

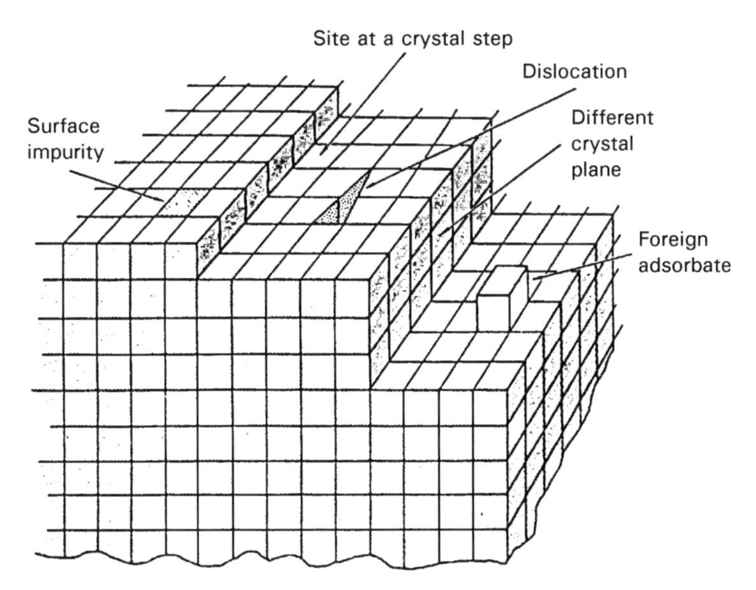

Figure 5.4 Surface defects that can act as preferential adsorption sites. (Reprinted from S.R. Morrison, *The chemical properties of surfaces*, p. 23, with kind permission from Plenum Publishing Corp., New York, NY, USA.)

scanning electron microscopy (SEM) studies on quenched graphite surface show that alteration or corrosion of surfaces of pyrolytic graphite platforms is related to eventual changes in analytical performance [8]. However, similar experiments for pyrolytic graphite tubes failed to produce any general correlation between changing surface morphology and analytical sensitivity [9].

Reactive sites can exist even on clean surfaces that are free of emerging dislocations, as when the lattice atoms at the surface contribute unoccupied bonding orbitals with high electron affinities. These atoms can also introduce occupied bonding orbitals of low ionization energies at the surface, or they might provide dangling bonds, associated with electron orbitals out of the surface. The electrons in these cases are shared with the bulk lattice energy bands and can be considered as providing electronic energy levels that are localized at the surface. These localized energy levels participate in the exchange of electrons with the bulk lattice energy bands, and are commonly referred to as surface states. Such states are intrinsically associated with the discontinuity that the surface establishes in the otherwise periodic lattice. These intrinsic states are composed of: Tamm states, if the electron affinities between the bulk and the surface states differ; and/or Shockley states, if the surface state is associated with dangling bonds.

Surface states need not be intrinsic, however, as adsorbed species or impurities near the surface can participate in electronic exchanges with the nonlocalized, bulk-lattice energy band to produce surface energy levels that are localized; that is, the adsorbed species can themselves induce surface states. It should then be expected that intrinsic surface states will be influenced significantly by the presence of adsorbed species. However, the electronic exchange will also perturb the valence electron energy level that the free species possessed prior to adsorption. This can be of fundamental importance to the understanding of the mechanisms of chemical modification. The bonding between an adsorbate and adsorbent modifies the energy levels of each and has, therefore, the potential of modifying the mechanism controlling analyte release in electrothermal atomization.

The influence that adsorption of one species has on another (co-adsorption) is also pertinent to the mechanisms of modification, although investigations of such effects have not been directed towards this application, but rather towards the field of catalysis. The results are, however, no less interesting; for example, an enhancement in hydrogen sticking on platinum by the presence of an oxide has been attributed partially to compound formation [10]. The dominating effect of sulfur on the behavior of hydrogen on ruthenium is that it interferes with the surface mobility [11]. The increased adsorption energy of carbon monoxide on platinum when potassium is present has been shown to result from the potassium having the ability to migrate to an adsorbed carbon monoxide molecule [12]. Such results indicate that it is not unreasonable to expect the presence of small quantities of some modifiers, such as hydrogen and oxygen, to influence the adsorption of other species that participate in the atomization process. Discussions of hydrogen and oxygen chemisorption on graphite are presented in Section 2.2.

Impurities in the near-surface region can also provide reactive sites. For example, high electron affinity sites are inherent qualities of surface cations that are necessarily lacking the totality of the neighboring anions associated with the lattice configurations in an associated bulk ionic crystal. This type of surface effect may contribute significantly to modifier induced stabilization.

The bonding between the surface and foreign species on the surface controls the surface state energy level. This is illustrated in Figure 5.5 for surface impurities M^{n+} and $M^{(n+1)+}$ on a surface of a semiconductor such as graphite. The redox reaction is

$$M^{n+} = M^{(n+1)+} + ne^- \tag{5.1}$$

where $n \geq 1$, describes the electron exchange between the illustrated substrate and these ions. The occupation of an electron by the surface state indicates the surface state is associated with M^{n+}. However, if the surface state is unoccupied (transfer of an electron to the solid), the associated species is $M^{(n+1)+}$. The energy of this surface state (the state is designated $M^{n+}/M^{(n+1)+}$ in the figure) is directly related to the probability for electron release by the M^{n+}; i.e., both ionic species are represented by the same surface state as seen in the figure. If the couple is located sufficiently above the Fermi level E_F the surface state is unoccupied, and electron transfer can only be made from the adsorbate to the unfilled conduction band of the substrate. Below E_F the energy band is essentially filled up to a few tenths of an electron volt ($2kT$) below the Fermi level, and electron transfer can only occur from the substrate; the surface state is occupied. Thus, the position of the surface state, relative to the Fermi energy, determines the oxidation state of the adsorbate. Chemical modifier type adsorbates can have, in this manner, a significant influence on the chemical properties at the substrate surface.

It might be expected that the surface state is a sensitive function of temperature, as the Fermi–Dirac electron energy-distribution function for the substrate is given by

$$f = [1 + \exp\{(E - E_F)/k_B T\}]^{-1} \tag{5.2}$$

where the subscript B denotes the Boltzmann constant. This function indicates that, at elevated temperatures, the probability f of an electron energy E level (orbital) in the substrate being occupied is unity, for electron energies $k_B T$ below the Fermi energy. The probability diminishes to

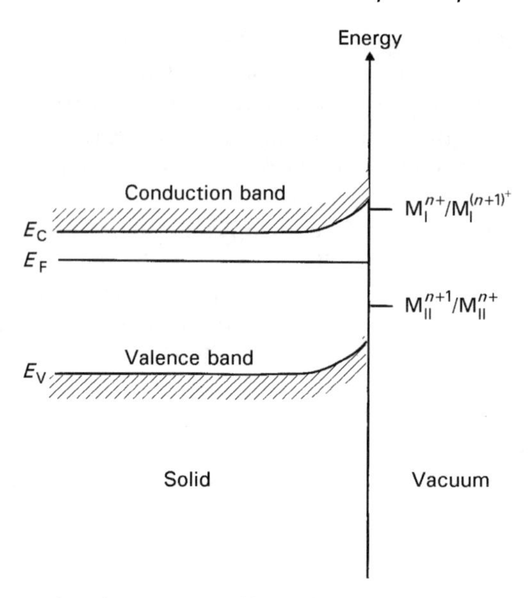

Figure 5.5 Diagram of surface states and band bending in a solid with a band gap. An unoccupied state, containing an acceptor species, and an occupied state, containing a donor species, are shown as examples. Shown are energies of the conduction band (E_C); valence band (E_V); and Fermi level (E_F).

0.5 when $E = E_F$, and further diminishes to zero for electron energies in the neighborhood of k_BT above E_F. Therefore, increasing the substrate temperature provides an increased opportunity for the surface state (adsorbate and localized energy state of the substrate) to become occupied. The result is that substrate electron donation to the adsorbate becomes possible from substrate energy levels that would otherwise be unpopulated, and the adsorbate oxidation state changes.

The curvatures in the conduction and the valence bands in Figure 5.5 are due to the double charge layer that forms when charge is transferred between the substrate and the adsorbate. As an example, assume the surface state energy E is an occupied state below E_F. This state is created by placing an electron-acceptor species on the surface. As $(E - E_F)$ in the above distribution function is negative, the fractional occupancy f is diminished as E approaches E_F. Consequently, the occupancy requirement for equilibrium is lowered as the state E approaches the Fermi level. However, the actual transfer of charge to the adsorbate increases the occupancy of the surface state and forms a double layer (electron plus hole) of surface charge. This charge layer creates a surface potential barrier to the further transfer of charge and lowers the probability of transfer (fractional occupancy) to the surface state. The lower fractional occupancy implies a lower f which, from the above distribution function, suggests that the level E of the surface state moves closer to E_F, in order to satisfy the equilibrium condition. Near the surface, the valence band is also pulled towards the Fermi level. The charge transfer to the surface state reduces the near-surface fractional occupancy of electrons in the band. This lowering of the fractional occupation causes bending of the conduction band but, because E is greater than E_F, the distribution function dictates that the bending is away from the Fermi level.

The electric fields induced by the double charge layer at the surface should be expected to influence adsorption properties, particularly those of ionosorption described in the following paragraph. These fields will be influential in oxygen-induced modification mechanisms, or conceivably in water-induced modification. For the latter, the OH^- can share electron pairs with Lewis acid sites; consequently the remaining weakly bound proton becomes readily available for reaction. The influence that adsorbed water has on the electron transfer is significant, but not understood. As a modifier of the substrate surface properties, the adsorbed water may result in the surface being hydroxolated during the drying step. Depending on the substrate material, further thermal treatment can leave relatively high electrical potential regions, which are neutralized by surface reconstruction as the temperature is elevated; the reactive properties of the surface would change accordingly.

As indicated above, the bonding of the adsorbate with the surface can be related entirely to the electron transfer to and/or from bands in the solid. The bonding can also be purely chemical, where only localized interactions occur between the adsorbate and either solitary or a few, at the most, surface atoms. Combinations of these two are also likely. The first of these types of bonds occurs for ionosorption, where ionized adsorbates and the counterions are separated by numerous lattice spacings, but are not directly involved in sharing of electrons. Thus, an O_2^- ion can be held on the surface through the electrostatic potential established between it and a donor (perhaps the modifier) within the substrate. The other type, the chemical bond, may be covalent, where an electron is shared between the adsorbent and the localized group of surface atoms, or an acid–base chemical bond can form at sites attracting gaseous acids or bases.

Surface diffusion is also related to the adsorbate-to-surface bond strength and bond localization. In moving from adsorption site to adsorption site the adsorbate must overcome the energy of the activated state where the adsorbate is located between two adsorption sites. This energy is expected to be high if the bonding of the adsorbate to the surface is strong and highly localized (diffusion is relatively slow). For a non-localized weak bond, the energies of the activated state and the adsorption site will have similar magnitudes. As the activation energy for diffusion is the difference of these two energies the diffusion can be relatively rapid. Because surface defects provide regions of elevated energies for the activated states, these defects act as diffusion barriers and as sites of nucleation.

Surface diffusion of adsorbed analyte species can control nucleation rates and the sintering of existing clusters of analyte species on the surface. However, such diffusion may also be associated with the rate at which adsorbed species are incorporated into active surface sites. Consequently, during heating of an atomizer, when active sites are being created, there can be competition between the nucleation (cluster growth) rate and filling of the active sites. This site creation can commence through reactions with oxygen and subsequent desorption of carbon monoxide and carbon dioxide, or through surface morphology changes induced by thermal stress in the lattice. When the active site formation rate is greater than the rate of analyte adsorbate inclusion into clusters, the site-filling can dominate cluster growth; subsequently, significant desorption of analyte from the atomizer becomes possible.

Although it is possible to obtain pyrolytic graphite relatively free of surface defects, the surface of pyrolytic graphite-coated graphite common to atomizers used in ETAAS is not even close to this quality. Stresses associated with the bonding to the small radius of the graphite atomizer may well accommodate those needed for slip in the basal planes of the pyrolytic graphite. This, by itself, introduces considerable twinning-type dislocations, even before the atomizer is thermally stressed. The thermal strain and shock associated with these atomizers during the atomization cycle introduces additional defect generation, as well as migration; both effects drastically alter the surface reactivities. In addition, defects formed by condensation and nucleation of carbon species during the atomization step defy characterization, but may have considerable effect on analyte release.

5.3 GAS-PHASE INTERACTIONS

The multiple collisions of gas-phase species with the atomizer surface provide numerous opportunities for these species to interact with the various surface sites. Modifier species may, in some instances, influence the surface states in this manner. However, interactions between the gas and the surface can also influence the migration of gaseous species away from the surface into the central region of observation. Such interactions can induce formation of a dense gas film (layer) that covers the solid surface. If the activation energy is close to the vaporization enthalpy, equilibrium vapor pressures can be established at the solid–gas interrace. This Langmuir film is named after the scientist who established that such a film develops over the filaments of incandescent light bulbs and slows the sublimation of the filament material [13]. Formation of such a film is the result of the balance attained between pressure and gas viscosity. When a Langmuir film forms during electrothermal atomization, species must diffuse through the film to reach the observation volume of the atomizer. This diffusion is driven by gaseous concentration gradients within the film.

The quantity of modifier species in the Langmuir film can be relatively significant when large quantities of modifier are employed. Consequently, there is, for such quantities of modifier, ample opportunity for the gas phase of the modifier to interact with analyte, either through homogeneous or heterogeneous gas-phase interactions. For a given starting composition of condensed phases and a given set of reactions, it is possible to estimate the partial pressures and mass fractions of the gaseous species produced from these condensed phases and contained in the layer [14].

Mechanisms by which modifiers control gas-phase analyte release may be evaluated by the activation energy values obtained from transient electrothermal absorption profiles. However, such thermodynamic approaches should always be questioned, because the mechanisms may be dominated by the associated kinetics. Even though thermal equilibrium is likely during atomization at 1 atm pressure, collision frequencies between analyte and modifier may be insufficient to ensure this. This possibility can be easily investigated for the general homogeneous gas-phase reaction involving analyte M and matrix AB:

$$M_g + (AB)_g = (MA)_g + B_g \tag{5.3}$$

The rate expression for this reaction is given by

$$-\partial[M]/\partial t = k[M][AB] \tag{5.4}$$

and, assuming AB remains constant (i.e. $[AB] \gg [M]$), integration yields the time necessary to deplete the analyte from an initial value $[M_0]$:

$$t = -(1/k[AB])\ln([M]/[M_0] \tag{5.5}$$

Assuming that 99% of the analyte has reacted, and that $k = 10^{-13}$ ml per molecules per second, Holcombe *et al.* [15] showed the magnitude of t to be in the neighborhood of several hundred milliseconds. Consequently, if MA (Eqn (5.3)) is thermodynamically favored over M by 100 : 1, the upper limit on time to reach equilibrium will be of the order of hundreds of milliseconds. These investigators therefore concluded that it is reasonable for the reaction to be limited by kinetics.

Although gas-phase interactions can be evaluated theoretically, the influence that the dynamic defect structures (in uncharacterized pyrolytic graphite coatings) will have on surface reactivities cannot be so evaluated. Neither can the effects of the given chemical modifiers on these reactivities be accurately predicted. Consequently, knowledge of modifier mechanisms must rely heavily on experimental evidence, and on an understanding of the chemical and physical possibilities that can explain the observations.

The following section discusses data from some experimental observations that have been made for the purposes of elucidating the chemical modifier mechanisms. The interpretations of these data are, more often than not, best guesses based on the present knowledge. They do not necessarily represent the last word on the mechanisms. As new data emerge in the future, some of these interpretations will, undoubtedly, be proved inaccurate. Nevertheless, the present interpretations form a nucleus of a knowledge base that is useful for research in this field. However, it is important to remember that these interpretations must be questioned continually.

5.4 MODIFIER MECHANISMS

This section introduces and discusses the relatively few investigations that have been made to determine mechanisms that control chemical modifier processes. It commences with an overview to provide a perspective of the research, and concludes with detailed discussions, under subheadings that denote the modifiers of interest.

5.4.1 Overview

One of the earliest modifier mechanism investigations provided AA data involving phosphate salt modifiers. These data were obtained and interpreted by Czobik and Matousek [16] as being indicative of formation of metal pyrophosphate which, at higher temperatures, decomposes to allow formation of the free metal. Welz *et al.* [17] obtained AA data on the nickel–selenium system. They concluded, from these data, that nickel-induced stabilization of selenium is dependent on the valence state of the selenium. Teague-Nishimura and Tominaga, using electron microscopy and microprobe techniques [18], investigated the palladium-induced stabilization of selenium on pyrolytic graphite. The coincidence of selenium and palladium in these spectra was interpreted as evidence that a 1 : 1 mole ratio palladium–selenium alloy had formed. Interestingly, the palladium–selenium pair fails to obey the Hume–Rothery rules for significant solid solution. Formation of such an alloy is therefore unanticipated, but may be possible in trace quantities. Majidi and Robertson [19] applied Rutherford backscattering spectrometry (RBS) to the palladium–selenium system on pyrolytic graphite. They deduced that a compound involving selenium and palladium had formed, and that both species diffused into the graphite substrate. Data from the same investigative approach on mechanisms of phosphorus-induced modification of cadmium and lead were interpreted as indicating formation of analyte–phosphate compounds [20]. Qiao and Jackson [21] used AAS and SEM to investigate the palladium-induced modification of selenium. Based on the resulting data, they suggested the possible formation of a selenium–palladium compound and bulk diffusion of selenium from molten palladium. From AAS data on the thallium–palladium system, and the fact that additional palladium quantities had little influence on the stabilization, these same investigators deduced that bulk diffusion from molten palladium also controlled the formation of free thallium [22].

The mechanisms by which ascorbic acid removes chloride interferences, in the presence of high concentrations of chloride, was addressed by Regan and Warren [23]. They suggested that the carbon residues, formed from the pyrolysis of ascorbic acid, enhance the reduction of the metal oxides at the atomizer surface. Hydes [24] argued later that this was not the mechanism that is responsible for the removal of the interferences, as his results indicated that chloride interference with copper atomization in seawater was suppressed whether or not oxalic acid was present. Hyde's argument was based on the fact that oxalic acid suppresses the formation of carbon residues. Consequently, contributions from these residues are unlikely to be involved in the modifier processes. Gilchrist *et al.* [25] also used gas chromatography (GC) and ETAAS to investigate the ascorbic acid modification mechanism for the atomization of lead in the presence of low concentrations of chloride. They interpreted their findings as evidence that the hydrogen product of the decomposition of ascorbic acid reacts with the chloride. The resulting hydrogen chloride gas escapes and thus depletes the store of chloride.

These same modifiers (phosphorus [6,26], nickel [27,28], palladium [29,30], ascorbic acid [31], and the modifiers oxygen [28,32–34] and magnesium [35,36], have also been investigated in real-time by MS techniques. With the exception of oxygen, the data for all of these modifiers were interpreted as indicating that modification mechanisms are controlled by formation of modifier–analyte compounds or complexes. The oxygen modifier data indicated that modifier-induced active site formation on the graphite is responsible for enhanced reduction of analyte oxide.

5.4.2 Discussion

5.4.2.1 Nickel modifier

Real-time MS investigations, to elucidate the mechanisms controlling nickel-induced modification, have been made solely on the selenium–nickel system. Alloy formation between these two species is unexpected, because the Hume–Rothery rules are not satisfied for such a system; however, compound formation is possible. There is evidence of the latter in the MS thermal histories of the gas-phase species that evolve from pyrolytic graphite-coated graphite and from tantalum tube atomizers heated *in vacuo*, and from pyrolytic graphite-coated graphite heated in nitrogen at 1 atm [27]. Droessler and Holcombe [28] also used MS thermal histories of species vaporized *in vacuo* from pyrolytic graphite-coated graphite, but the graphite substrates were in the form of flats as opposed to tube atomizers. These latter investigators interpreted their data as indicating that the modifier, in the form of $NiO_{(ad)}$, acted as an adsorption site for selenium dioxide. Their experiments were done using aqueous samples containing 100–200 ng of selenium as selenium dioxide and 100–400 ng of nickel as nickel nitrate. It was observed that the $SeO_{2(g)}$ appearance was shifted to higher temperatures (from 850K to 1300K) when the modifier was present. Carbon monoxide and dioxide were also observed to be released at 1200K when nickel oxide was present without selenium. Thus, it was proposed that carbon reduction of this oxide occurred at this temperature. Furthermore, Droessler and Holcombe surmised that some of the selenium dioxide may be attached to the adsorbed nickel oxide, so that when the reduction of the nickel oxide occurs at 1200K, the selenium dioxide desorbs as $SeO_{2(g)}$, $SeO_{(g)}$, and $Se_{2(g)}$. Formation of this latter species ($Se_{2(g)}$) was not addressed, but it was deduced that the presence of an atmospheric pressure ambient would cause these molecular species to dissociate near the 1575K selenium appearance temperature associated with ETAAS. It is surprising that the apparent activation energies (obtained by AAS), that are associated with this appearance, were found to be independent of the modifier's presence in the atomizer.

The atmospheric pressure vaporization of selenium was done with 50 ng of selenium as dilute selenium(IV) acid, codeposited with 200 ng of nickel as a nitrate solution [27]. It was observed that formation of $Se_{2(g)}$, $SeO_{2(g)}$, and $SeO_{(g)}$ was quenched by the presence of this modifier. Presence of the modifier as a chloride (nickel chloride) was observed to have little influence on the appearance of these precursors during atmospheric pressure vaporization. Under vacuum vaporization conditions in graphite atomizers, an otherwise absent $Se_{(g)}$ appeared when the dichloride modifier was employed; the carbide was the only precursor that was quenched by this modifier.

The suppression of $Se_{2(g)}$ by the nitrate indicates that the nitrate aids dispersion of selenium on the surface. This dispersion interferes with the formation of $Se_{(s,1)}$ species, from which polymeric selenium sublimes. Support for this argument was obtained from Auger electron microscopy investigations of 50–100 ng of selenium/100–200 ng of nickel on pyrolytic graphite-coated graphite substrates [27]. The modifier, in the form of a nitrate, evidently provides adsorption sites for the selenium, or it ties up the selenium in the microcrystals of modifier on the substrate surface. It was also found that the use of a tantalum atomizer resulted in formation of free selenium *in vacuo* whether or not the nickel modifier was present [27]. Carbon must, therefore, be a principal interferent with the formation of free selenium. Styris [27] concluded, based on the observation that free selenium appears only when the $SeC_{2(g)}$ precursor is absent, that nickel blocks the formation of carbide in graphite atomizers, and that this allows free selenium formation. This same investigator observed that a nickel selenide sample, in the presence of excess nickel, produced mass spectra similar to those obtained when the nickel nitrate modifier was used. It was proposed, therefore, that the modifier forms a selenide through the nickel reduction of $SeO_{2(s)}$, and that free selenium is produced by the thermal decomposition of this selenide at higher temperatures. The excess nickel may be necessary to block active sites on the graphite substrate so that ternary alloys, having selenium carbide and nickel components, might be prevented from forming.

Although there is a disagreement regarding the blocking mechanism associated with the nickel modifier in the two sets of experiments discussed above, the hypothesis that nickel acts to interfere with selenium interactions with the substrate surface explains both sets of results. Consequently, the substrate surface plays a significant role in the modification process. Resolving the discrepancy in the mechanisms that have been proposed requires identification of the condensed-phase species that form during the atomization cycle. Unfortunately, the ability to make this identification has yet to be demonstrated, even though the need extends beyond this particular analyte–modifier system. It becomes evident, from these discussions, that the requirement for real-time condensed-phase species identification is generally applicable to the development of valid modifier mechanisms.

It can be argued that differences in experimental parameters in the above selenium–nickel experiments explain the conflicting results; for example, the differences in observations can be attributed to differences in the base vacuum pressures associated with the vacuum vaporization experiments. The two orders of magnitude greater base pressure in the flat substrate experiments could account for the paucity of gaseous carbides in the spectra from the flat substrate experiments. Surface characteristics, such as active site densities in the tube and the flat substrate experiments, may be significantly different; higher oxygen partial pressures associated with the flat substrate experiments can modify the substrate surfaces themselves. Resulting changes in the heats of adsorption on these surfaces may then account for the increased appearance temperatures.

5.4.2.2 Oxygen modifier

The influence of oxygen as a modifier for selenium, and the mechanisms that control the modification, have been investigated mass spectrometrically, in addition to applying ETAAS [28]. The investigations again demonstrate the importance of the substrate surface; desorption and readsorption are hypothesized as the principal mechanisms controlling the modification. It was shown, by AAS, that the temperature of free selenium formation under an argon sheath gas increased by approximately 400K when pyrolytic graphite-coated graphite atomizers were oxygenated. Such an increase parallels that discussed above for the nickel modifier. Droessler and Holcombe [28] attributed the oxygen-induced shift to adsorption and higher temperature redesorption of the oxide. This oxide, which desorbed initially at 450K, could participate in multiple collisions with the surface in the 1 atm environment. The mass spectra indicated that *in vacuo* vaporization from pyrolytic graphite-coated graphite flat substrates results in $SeO_{2(g)}$ spectra that are independent of the presence of modifier; the gaseous oxide appears to be unchanged by the presence of an oxygenated surface. Free selenium was absent during vaporization, *in vacuo*, from either oxygenated or unoxygenated surfaces. It was hypothesized that newly created surface defect sites form as a result of carbon monoxide and dioxide desorption (i.e. $C_{(s)} + O_2 \rightarrow CO_{2(ad)} + C_{(s)} \rightarrow CO_{2(g)} + C_{(s)} + C_{defect}$, where C_{defect} implies an active surface defect). These defect sites can then provide the adsorption sites for the low-temperature $SeO_{2(g)}$ that is released at 450K (i.e. $SeO_{2(s,1)} + C_{defect} + C_{(s)} \rightarrow SeO_{2(ad)} + C_{(s)}$); desorption from these sites could be shifted to higher temperatures if the sites correspond to higher desorption energy sites. The mass spectra from the vacuum vaporization experiments on oxygenated surfaces indicate that $CO_{(g)}$ and $CO_{2(g)}$ were released continuously, starting at 900K; the higher desorption energy sites could, therefore, be manufactured continuously above 900K. The fact that $Se_{(g)}$ and high-temperature (above 450K) $SeO_{2(g)}$ were absent during vaporization *in vacuo* indicates that thermal dissociation of $SeO_{2(g)}$ may be responsible for the free selenium that is observed during atmospheric pressure vaporization. So the suggestion is that, during vaporization at atmospheric pressure, the dioxide is adsorbed at the sites created by $CO_{(g)}$ and $CO_{2(g)}$ desorption from the surface. The free selenium arises from desorption and thermal dissociation of $SeO_{2(g)}$. However, dissociative adsorption of the dioxide at these sites is another possibility that should not be ignored.

The type of dissociation phenomenon provides additional detail on the type of adsorption sites that are created. Dissociative adsorption is likely to occur if the adsorption sites involve surface states, so that electronic exchange can take place between adsorbate and sorbent. Investigations by MS at atmospheric pressure should provide the evidence that is necessary to determine which of these possibilities is responsible. Relatively little high-temperature $SeO_{2(g)}$ should be observed if the adsorption site production creates surface states. It should be noted that Droessler and Holcombe [32] observed the activation energies associated with selenium

AA at 1 atm of argon increased considerably when atomization occurred from oxygenated surfaces. This differs from what these same investigators observed for the nickel modification of selenium; the activation energy was then independent of the presence of nickel. Consequently, the mechanisms that control the nickel- and oxygen-induced modification of selenium differ from one another. The activation energy increase observed for the oxygenated surface supports the contention that desorption from higher desorption energy sites is responsible, at least in part, for the observed increase in the appearance temperature of free selenium when the surface is oxygenated.

Mechanisms controlling the influence of oxygen on the formation of free lead have also been investigated by ETAAS and MS. The results of these investigations indicate that the mechanisms differ from those hypothesized, and discussed above, for oxygen-induced modification of nickel or selenium. However, consensus among the investigators of these mechanisms has yet to be attained. The existing data have been interpreted as indicating either heterogeneous or homogeneous gas-phase reactions are involved [31,37], and that the release of free lead is related to the ratios of carbon monoxide to carbon dioxide [33]. Oxygen adsorption on active sites has been postulated as being responsible for the enhancement of partial pressures of the oxidants carbon dioxide and oxygen [38]. This again indicates the important function of the surface in the control of modification mechanisms.

The interest in mechanisms associated with oxygen effects on free lead formation arose because an associated temperature shift in this formation was reported by Salmon and Holcombe [34]. They observed a substantial shift, toward higher temperatures, in the appearance of the free lead absorption signal whenever 1 per cent oxygen was contained in the argon sheath gas. A smaller shift was also observed when atomization was performed in an inert atmosphere after the graphite atomizer had been thermally pretreated at 1000K in a 1 per cent oxygen sheath gas (surface oxygenation). L'vov and Ryabchuk [37] proposed that heterogeneous reactions involving $O_{2(g)}$ and the condensed metal were responsible for the observed temperature shift. Byrne *et al.* [31] suggested that the temperature shift results from homogeneous gas phase reactions between $O_{2(g)}$ and free lead. Support for involvement of either homogeneous or heterogeneous gas-phase interactions is provided by the observation that identical lead mass spectra evolved from pyrolytic graphite-coated graphite flats, oxygenated and nonoxygenated substrates, heated *in vacuo* [33]. This implies that the shifts observed during atmospheric pressure vaporization must be related to gas-phase interactions, and are not solely dependent on the surface.

Bass and Holcombe [33] did observe, however, that the carbon monoxide and carbon dioxide mass spectra from their graphite flat substrates, heated *in vacuo*, were enhanced when the surface was oxygenated. These same investigators calculated the quantity of carbon dioxide by correcting for instrument response and then proportionally relating the corrected carbon dioxide and lead signals to the number of lead atoms released from the surface. By incorporating the difference in surface areas associated with a hypothetical graphite atomizer and the actual graphite flat, they deduced that the maximum carbon dioxide concentration produced in a tube atomizer should be about 0.84 per cent. Loss of carbon dioxide through diffusion was included in the estimate by employing a finite series evaluation of the quantity retained in the atomizer at given temperatures and times. Interestingly, when these investigators added carbon dioxide to their atomizer, to duplicate the appearance temperature shift that was observed for the oxygenated surface, it was determined that a 30 per cent carbon dioxide concentration was needed for the shift to occur. This is considerably greater than the estimated 0.84 per cent. However, it was also found that addition of carbon monoxide to the sheath gas resulted in the appearance of free lead being shifted in the opposite sense, i.e. towards a lower temperature. It was argued that, instead of a minimum carbon dioxide concentration requirement, the ratio of carbon dioxide to carbon monoxide may be the controlling factor in the mechanism of modification. Assuming formation of carbon monoxide evolves through carbon dioxide reactions with the graphite, an initial 30 per cent carbon dioxide concentration may actually produce the same critical carbon dioxide to carbon monoxide ratio as that associated with the oxygenated surface. Consequently, it may be the increases in this ratio, rather than the carbon dioxide concentration alone, that are interfering with the reduction or dissociative adsorption of lead oxides.

Sturgeon and Falk [38] provided evidence in support of this carbon dioxide to carbon monoxide ratio hypothesis. They found, from molecular absorption measurements in pyrolytic graphite-coated graphite atomizers, that oxygenated surfaces increase the partial pressures of carbon dioxide and oxygen. It was postulated that this increase is produced by oxygen being adsorbed in sufficient quantities to cause blockage of the active surface sites. This blockage would interfere with the reduction or dissociative adsorption of oxidants, such as the lead oxide investigated by Bass and Holcombe [33] as discussed above.

If this is indeed the controlling mechanism of modification, the surface of the substrate necessarily has a significant, but not exclusive, role in the oxygen-induced modification of lead. Both chemical and physical interactions are, evidently, involved in this modification process. The question as to whether the atomization of lead oxide is induced in the condensed, gaseous or adsorbed phases can be addressed by identification of the species on the surface during the atomization cycle. As discussed earlier, the capability for achieving this has yet to be demonstrated. Mass spectra of the gas-phase higher oxides and dimers of lead during this cycle are also needed. Unfortunately, quadrupole MS, presently used for these purposes, inherently discriminates against the higher masses; the sensitivity is sufficiently low that the transient release of analytical quantities of these oxides and dimers of lead, from atomizer surfaces, cannot be detected.

5.4.2.3 Magnesium modifier

Evidence of graphite surface interactions during a specific modification process has also been reported for magnesium-induced modification of beryllium [35]. Mass spectrometric investigations of the vaporization of beryllium (10–20 ng of beryllium as aqueous nitrate), from pyrolytic graphite-coated graphite tube atomizers heated at atmospheric pressure in the presence of 500–1000 ng of magnesium as a nitrate, showed that this modifier induced an approximate 400K downward shift (to about 2000K) in the appearance temperature of the gas-phase oxides and carbides of beryllium. Now, the mass spectra associated with the atomization of nonmodified beryllium indicate that the free beryllium appears near 2000K and is the result of thermal dissociation of its adsorbed monoxide [35]. Consequently, it was suggested that the magnesium oxide blocks the surface sites of greatest heats of adsorption for the beryllium oxides [35]. Support for this argument is given in the observation, made by these investigators, that a similar downward shift in appearance temperature of the molecular beryllium species occurs when aluminum is used as a modifier instead of magnesium.

Although this mechanism explains the downward temperature shift, it fails to explain the magnesium-induced thermal stabilization of the beryllium. This stabilization occurs because the magnesium inhibits the formation of the relatively volatile $Be(OH)_2$ [35]. The mass spectra of species effusing from pyrolytic graphite-coated graphite tube atomizers indicate that this hydroxide is present in the gas phase, during thermal pretreatment (near 1200K), if the magnesium modifier is absent. The gaseous hydroxide of beryllium is not present in the spectra if there is sufficient magnesium (magnesium–beryllium weight ratio of 385) in the sample. Instead, $HMgOH_{(g)}$ and $Mg(OH)_{2(g)}$ were observed to appear at 385K. It was postulated, therefore, that hydration of magnesium oxide depleted the water in the atomizer, and thus inhibited the formation and low-temperature loss of the beryllium as a dihydroxide. Partial support for this argument is provided by the absence of this dihydroxide when the thermal pretreatment was done on 1 μl volume samples of beryllium instead of the 40 μl volume samples used for most of the thermal pretreatment investigations. Such a volume effect on the hydroxide formation was predicted by Frech *et al.* in their high-temperature equilibrium calculations [39].

Stabilization of aluminum by magnesium was found to involve the same mechanism as described above for beryllium [36]. Mass spectra from pyrolytic graphite-coated graphite tube atomizers, containing both 10 ng of aluminum as an aqueous nitrate and the modifier in 1 : 500 (aluminum : magnesium) wt : wt concentration, inhibited the formation of gaseous $Al(OH)_2$; consequently, the low-temperature loss of aluminum, as the hydroxide, was inhibited. This dihydroxide was also absent from the spectra when volumes of water were restricted to only 1 μl. These results are identical to those observed for beryllium. The same arguments that were made for the stabilization of beryllium apply, therefore, to the mechanism controlling the stabilization of aluminum, i.e. hydroxide formation is inhibited, at relatively low temperatures, by depletion of water through the hydration of magnesium oxide.

Evidently, magnesium-induced modifications of beryllium, aluminum, and perhaps other easily hydroxylated elements are primarily controlled by chemical mechanisms, as opposed to the physical mechanisms that were discussed earlier for nickel- and oxygen-induced modifications of selenium and lead, respectively.

The effectiveness of the magnesium-induced modification of aluminum has also been determined. Redfield and Frech [40] quantified the loss of aluminum during thermal pretreatment, by using AA calibration curves that they developed through carefully controlled vaporization of 200–300 pg of aluminum deposited as an aqueous nitrate on pyrolytic graphite-coated graphite platforms in Massman-type tube atomizers and in pyrolytic graphite-coated graphite cups of two-step atomizers. The latter type of atomizer allowed sole low-temperature vaporization of the hydroxide from the cup into the high-temperature atomization tube. In this manner, it was determined that as much as 20 per cent of the sample can be lost as the hydroxide. The MS data of Styris and Redfield indicate that this loss is completely eliminated by the addition of appropriate quantities of magnesium [36].

The mechanism by which magnesium modifies the atomization of oxides of aluminum has been demonstrated to differ from that involving atomization of beryllium oxide (see above discussion). The MS results of the atmospheric pressure atomization (10 ng of aluminum and 500 ng of magnesium as nitrates on pyrolytic graphite-coated graphite tube atomizers) show that $Al_{(g)}$ and $Al(CN)_{(g)}$ are shifted to higher temperatures by the presence of magnesium [36]. During the appearances of these species the $Mg_{(g)}$ and $MgO_{(g)}$ spectra were observed to decay. These observations were explained as being a result of oxidation, by $MgO_{(g)}$, of adsorbed aluminum and of the lower oxides that form from the rapid decomposition of aluminum oxide. Free aluminum is produced once the

$MgO_{(g)}$ is depleted. This oxidation of the aluminum species occurs through the following reactions:

$$MgO_{(g)} + 2Al_{(ad)} = Al_2O_{(ad)} + Mg_{(g)} \quad (5.6)$$

$$MgO_{(g)} + Al_2O_{(ad)} = Al_2O_{2(ad)} + Mg_{(g)} \quad (5.7)$$

$$MgO_{(g)} + Al_2O_{2(ad)} = Al_2O_{3(ad)} + Mg_{(g)} \quad (5.8)$$

These reactions were determined to be thermodynamically feasible under the assumption that the adsorbed species acts as a two-dimensional gas [36].

From the above work on magnesium modifiers, it can be concluded that, although the same mechanisms are associated with the magnesium-induced stabilization of aluminum and beryllium, the proposed modifier mechanisms that control the atomization of these analytes differ. These latter mechanisms involve: (1) occlusion of active surface sites from the condensed beryllium species; and (2) oxidation of the adsorbed aluminum species. The existence of these condensed-phase species, during the atomization cycle, is only implied. Corroboration will require their identification on the graphite surface. However, the importance of this surface on the magnesium-induced modification is strongly indicated.

5.4.2.4 Phosphorus modifier

Chemical mechanisms are implicated in the phosphorus-induced modification of cadmium. Hassel *et al.* [6] applied temperature-programmed, static, secondary ion mass spectrometry (TPS-SIMS) on 260 ng of cadmium (aqueous nitrate), deposited on pyrolytic graphite-coated graphite flat substrates in the presence of 1200 ng of ammonium dihydrogen phosphate. The mass spectrum of $CdPO_x$ was observed *in vacuo* at 298K, even though thermal pretreatment had not been performed. It was also observed that the fragment pattern of the nitrate was present only in the absence of the phosphorus modifier. Evidently, as was suggested, a cadmium–oxyphosphorus compound develops on the surface during the desolvation step. When the substrate was heated, between 340 and 410K *in vacuo*, the mass spectra indicated a rapid increase in $CdPO_2$ and a decrease in free cadmium. This implies that the $CdPO_2$ that binds to the surface, in the region of observation, was converted fully from the CdPO at these temperatures [6]. During thermal vaporization *in vacuo*, the mass spectra show the concomitant presence of $P_3O_{7(g)}$ and $Cd_{(g)}$ near 1200K. It was hypothesized that this results from the decomposition of CdP_xO_y species, postulated as being present on the substrate prior to the atomization [6]. Although the condensed-phase species were not confirmed by the SIMS data, the implication of compound formation suggests that the modification is controlled by the condensed-phase chemistry. Compound formation is expected rather than a solid solution, because extensive solid solubility is prevented by the Hume-Rothery rules. As emphasized earlier, it is imperative to the confirmation of the postulated mechanism that the condensed phase be identified.

Involvement of the graphite substrate and of chemical interactions in the phosphorus-induced modification process has been demonstrated by Bass and Holcombe [26]. The mass spectrometric data obtained on the modification of lead, using 400 ng of lead (lead nitrate with 1 per cent ammonium dihydrogen phosphate) on pyrolytic graphite-coated graphite flat substrates, indicate that free lead appears concurrently *in vacuo* at 1150K, with $PO_{(g)}$ and $PO_{2(g)}$; without the modifier, the free lead appears at 750K. The thermal decomposition of a condensed-phase lead–oxy-phosphorus compound was implicated by the temporally coincident formation of these species. Similarly, coincident carbon monoxide and carbon dioxide were also observed and reported as being indicators for the reduction of graphite-bound lead oxide; they were not present in the spectra if the sample simply involved the modifier, without analyte. The substrate is involved in the reduction or dissociative adsorption of the lead oxide, and compound formation is indicated as being the mechanism responsible for the modification; identification of this compound remains unachieved.

These oxyphosphorus compound hypotheses are supported by results from the results of RBS investigations on phosphorus-induced modification mechanisms (cadmium and lead atomization from pyrolytic graphite-coated graphite substrates) [19]. These experiments employed a 2.0 MeV He^+ incident beam on flat pyrolytic graphite-coated graphite substrates that contained 1 μg of analyte as a nitrate and 10 μg of PO_4^{-3}; the substrates were heated *in situ* and analyzed *in vacuo* at various temperatures between ambient and 1010K. When the modifier was absent, a lead species was observed to exist on the surface and in the bulk of the substrate at 360K. This surface-bound species disappeared at 770K. Lead species were also observed in the bulk to 920K. For cadmium (without modifier), the sole species observed at 370K was contained on the surface. Either desorption or surface migration of the cadmium resulted at temperatures up to 880K. Introduction of the phosphate modifier resulted in cadmium or lead on the surface with phosphorus and oxygen (the oxygen : phosphorus atom ratio was estimated to be about 2.5); this ratio remained constant with varying metal : phosphorus ratios. The surface-bound analytes vanished between 910K and 950K, and the analytes in bulk were present to temperatures of 1010K and 1030K, respectively. When the substrate was heated to temperatures equal to or exceeding 370K, the phosphorus was retained only when the analyte was present. The formation of vitreous analyte–phosphates ($x\mathrm{MO} \cdot P_2O_5$) was postulated as the modification mechanism, based on the constant oxygen : phosphorus and varying metal : phosphorus ratios, and the vast concentrations associated with the lead or cadmium phosphate glasses.

5.4.2.5 Palladium modifier

The mechanisms associated with the palladium modifier have been the most extensively investigated of the modifiers. In addition to real-time spectrometric evaluations, RBS and SEM techniques have provided data needed to establish these mechanisms [18,19]. Yet, the indicated mechanisms remain somewhat tenuous and await corroboration by additional methods or interpretations.

Compound formation has been indicated as being responsible for palladium-induced modification of arsenic and selenium. The MS data of gaseous species that evolve from samples of 115 ng of palladium (aqueous nitrate) that have been codeposited with 20 ng of arsenic (aqueous nitrate) on pyrolytic graphite-coated graphite tube atomizers, and then heated in one atmosphere pressure of nitrogen, indicate that quantities of the gaseous species that precede the evolution of free arsenic are attenuated by the modifier [29]. The $AsO_{(g)}$ mass spectrum was observed to be about a tenth of that observed when the modifier was absent; $As_{2(g)}$ was completely inhibited by the palladium modification. Any appearances of gaseous precursors to free arsenic were eliminated by thermally pretreating (reducing) the palladium nitrate modifier to 1300K, on the atomizer surface, prior to analyte introduction. The appearance temperature of free arsenic was independent of whether or not the modifier was reduced, and was about 150K above the 1450K appearance associated with the nonmodified sample. Because $AsO_{(g)}$, and the higher oxides still evolved when the nonreduced (palladium oxide) was present, it was suggested that the palladium oxide immobilizes release from $As_{(s,1)}$, so that $As_{2(g)}$ cannot form, but decomposition of arsenic oxide is allowed to proceed [29]. The palladium oxide must, therefore, have a stronger association with the $As_{(s,1)}$ than with the $As_2O_{3(s)}$. It was suggested that formation of a solid solution, or a stoichiometric compound that may contain oxygen, was formed by the reactions

$$xPdO_{(s)} + yAs_{(s,1)} \rightarrow [Pd_xAs_yO_z] + (x - z)O_{(g)} \quad (5.9)$$

$$xPdO_{(s)} + yAsO_{(g)} \rightarrow [Pd_xAs_yO_z] + (x + y - z)O \quad (5.10)$$

where the solid solution/compound is indicated by the square brackets. Decomposition of the $[Pd_xAs_yO_z]$ at more elevated temperatures (1600K) then accounts for the appearance of free arsenic. It should be noted that the arsenic–palladium pair fail to comply with the Hume-Rothery rules for extensive substitutional solubility, and are unlikely to form an interstitial solution because the arsenic radius is too large.

The same $[Pd_xAs_yO_z]$ species must form when the modifier is reduced (Pd^0), but this modifier quenched formations of both $As_2(g)$ and the oxides. These same investigators therefore proposed the involvement of the following interaction between palladium and the condensed phase oxide:

$$2xPd_{(s)} + yAs_2O_{3(s)} \rightarrow 2[Pd_xAs_yO_z] + (3y - 2z)O \tag{5.11}$$

It should be noted that mass spectra obtained during vaporization *in vacuo* showed that the modifier quenched the formation of $AsC_{(g)}$ [29]. This carbide necessarily forms from the condensed phase, because the vacuum conditions and the sampling through the sample introduction hole of the atomizer minimizes the probability of involvement of gas-phase interactions. The above interactions between the species of palladium and arsenic must interfere, therefore, with the condensed-phase interactions that are necessary to form this carbide.

Results from the investigations of Rettberg and Beach [30] on the influence of various palladium quantities on atomization of arsenic provide some support for the above solid solution/compound hypothesis. Their atomic absorption results indicate a 210K upward shift in appearance temperature of free arsenic when 500 ng of palladium (aqueous nitrate) are deposited in a pyrolytic graphite-coated graphite tube atomizer prior to the introduction of 1 ng of arsenic (aqueous nitrate). When 1000 ng of palladium were used, the shift increased to 240K. The peak area associated with free arsenic increased monotonically with the quantity of the deposited palladium. These investigators reasoned that, because of the large excess of palladium relative to arsenic, these increases could not result from simple enhancement of the metallic bonding between the two species. It was postulated, therefore, that bulk effects were influencing the atomization. This supports, but does not confirm, the concept of solid solution formation suggested by Styris *et al.* [29]. Identification of the condensed phase is sorely needed to clarify the question of solid solution versus compound formation, and to corroborate the proposed mechanism.

A mechanism similar to the above solid solution/compound formation for arsenic is also indicated from the results of several types of spectrometric investigations of mechanisms responsible for the palladium-induced modification of selenium. There have been considerably more experimental efforts directed toward determining mechanisms of modification of the selenium–palladium system than for any other analyte–modifier system. The large number of different experimental approaches that have been applied to this system have, consequently, provided a relatively large portion of the information base that will be needed to successfully evaluate these mechanisms.

One of the real-time spectrometric approaches involved mass spectrometry of selenium species vaporized *in vacuo* and at atmospheric pressure from pyrolytic graphite-coated graphite flat substrates and tube atomizers; reduced palladium (Pd^0) and nonreduced palladium (palladium oxide, PdO) modifiers were used [41]. For the Pd^0 experiments, the analyte was deposited as selenium(IV) acid, after deposits of palladium nitrate pyrolytic graphite-coated graphite were treated thermally at 1300K. The vacuum vaporization experiments involved 40 ng of selenium and 6000 ng of palladium, whereas the atmospheric pressure vaporization experiments employed samples consisting of 25 ng of selenium and 115 ng of palladium.

The mass spectra from both vacuum and atmospheric pressure experiments show that Pd^0 alters the temperature profiles of the oxides that, otherwise, appear near 400K [41]. This implies that the Pd^0 interacts with some of the oxide at temperatures below 400K. It was also observed that $Se_{2(g)}$ was present only if the atomization step occurred *in vacuo*; this dimer failed to appear in the atmospheric pressure (nitrogen) environment. This was interpreted as an indication that the Pd^0 and the analyte may interact through a heterogeneous reaction. However, the mass spectrum also shows that $SeC_{(g)}$ formation under vacuum conditions was unperturbed by the presence of Pd^0. There cannot be, therefore, any significant interaction between the selenium and the Pd^0, because the carbide formation takes place in the condensed phase; gaseous carbide is, after all, observed under vacuum conditions that prohibit significant gas-phase interactions. It was reasoned, therefore, that the Pd^0 reacts principally with the selenium dioxide [41]. These investigators concluded that, as the releases of free selenium and free palladium *in vacuo* were coincident at 1150K, a stoichiometric compound [Pd,Se,O] was formed, and that the appearances of these free species result from thermal dissociation of this compound. Stoichiometry, rather than an extended solid solution, was proposed because simultaneous appearances of free selenium and palladium from solid solution should not be expected.

The mass spectra from atmospheric pressure vaporization show that, instead of the 1150K appearances that were observed during vaporization *in vacuo*, the appearance temperatures for free selenium and palladium are increased to 1550K and 1950K, respectively [41]. This was attributed to the trapping of these species at higher energy retention sites on the graphite surfaces; release from these sites is initiated at the more elevated temperatures. It was reasoned that these higher-energy-site locations could be in the proximity of the palladium itself, or are at the surface defects (pits or channels) that may be created on the graphite by the palladium. Finally, it was observed that the Pd^0 also inhibits the formation of selenium hydroxides, that are otherwise observed at the atmospheric pressure MS experiments. The mechanisms associated with the Pd^0-induced modification of selenium dioxide were described as follows:

1. inhibition of the hydration of selenium oxides:

$$Pd^0_{(s)} + [C_{(g)} \text{ or } CO_{(g)}] + H_2O \xrightarrow{400K} Pd^0_{(s)} + H_{2(g)} + CO_{x(g)} \quad (5.12)$$

   where $x = 1$ or 2;
2. formation of the [Pd,Se,O] compound:

$$Pd^0_{(s)} + SeO_{2(s,1)} \xrightarrow{400K} [Pd,Se,O] + SeO_{(g)} \quad (5.13)$$

   and

$$Pd^0_{(s)} + SeO_{2(g)} \xrightarrow{>400K} [Pd, Se, O] + SeO_{(g)} \quad (5.14)$$

3. dissociation followed by substrate trapping of selenium and palladium contained in the Langmuir film:

$$[Pd,Se,O] \xrightarrow{1200K} Se_{(g)} + Pd_{(g)} \rightarrow (Se-Pd)_{(ad)} \quad (5.15)$$

   where $(Se–Pd)_{(ad)}$ implies readsorbed selenium and palladium on the graphite at sites associated with the presence of palladium;
4. decomposition and desorption to the free species:

$$(Se-Pd)_{(ad)} \xrightarrow{1500K} Se_{(g)} + Pd_{(ad)} \quad (5.16)$$

   and

$$Pd_{(ad)} \xrightarrow{1900K} Pd_{(g)} \quad (5.17)$$

The MS experiments on palladium oxide-induced modification, as opposed to the above Pd^0 experiments, utilized the same quantities of analyte and modifier as those in the Pd^0 experiments [41]. The principal selenium species observed in the gas phase, for the palladium oxide experiments, were the monoxide and the dioxide; considerable analyte loss in the form of these oxides was indicated by the magnitudes of these spectra. The $Se_{2(g)}$ spectral component vanished when palladium oxide was used. Consequently, the presence of palladium oxide must inhibit formation of, and/or sublimation of, the condensed-phase selenium. Recall the experiments done *in vacuo*, that show the condensed phase is responsible for the free polymer formation. The palladium oxide was also observed to inhibit formation of $Se(OH)_2$ at atmospheric pressure, probably because hydration of palladium oxide competes with the hydration of SeO. Finally, the free palladium and selenium mass spectra, associated with the application of the palladium oxide modifier, compare closely to those obtained when Pd^0 is present. It was reasoned, therefore, that the same [Pd,Se,O] was produced for both modifiers, and that free selenium and palladium originate from the decomposition of this solid solution or compound [41]. Summarizing, the postulated modification mechanisms are as follows:

1. inhibition of $Se(OH)_2$ formation:

$$PdO_{(s)} + H_2O_{(g)} \rightarrow Pd(OH)_{2(s)} \quad (5.18)$$

so that

$$SeO_{2(s)} \rightarrow SeO_{2(s)} + SeO_{g)} + Se_{(s,1)} \quad (5.19)$$

2. compound formation at low temperature:

$$PdO_{(s)} + Se_{(s,1)} \rightarrow [Pd,Se,O] \quad (5.20)$$

and

$$PdO_{(s)} + SeO_{2(s,1)} \rightarrow [Pd,Se,O] \quad (5.21)$$

3. thermal dissociation:

$$[Pd, Se, O] \xrightarrow{1200K} Se_{(g)} + Pd_{(g)} \rightarrow (Se{-}Pd)_{(ad)} \quad (5.22)$$

where $(Se{-}Pd)_{(ad)}$ refers to the recondensed stabilized species;
4. desorption to the free species:

$$(Se{-}Pd)_{(ad)} \xrightarrow{1675K} Se_{(g)} + Pd_{(ad)} \quad (5.23)$$

and

$$Pd_{(ad)} \xrightarrow{1900K} Pd_{(g)} \quad (5.24)$$

The proposed palladium-induced modification mechanisms for selenium are identical to those discussed earlier for arsenic. For both of these analytes, the modification appears to occur by formation of similar analyte–palladium compounds or solid solutions. Because of the failure to meet the Hume-Rothery criteria for extensive substitutional solubility, and because of the relatively large atomic radii of both of these analytes, it might be anticipated that compound formation prevails. To corroborate the existence of either, it will be necessary to identify the condensed-phase that evolves during the atomization heating cycle.

Teague-Nishimura and Tominaga [18] provided the initial data that indicated formation of a selenium–palladium compound during the modification process. These investigators used SEM and X-ray fluorescence (XRF) analysis to show that there was near-spatial coincidence between selenium and palladium on pyrolytic graphite-coated graphite, and that the ratio of palladium to selenium was unity. These same investigators concluded that palladium and selenium formed an alloy on the graphite surface, and that this alloy prevailed at 1000K. The MS results certainly support this argument, but the possibility of artifact formation remains. Recondensation processes associated with the thermal quenching (from 1000K) that were employed in obtaining these results should not be ignored. Once again, real-time identification of the condensed-phase species is needed in order to make the confirmation.

Qiao and Jackson [21] also reported the possibility of compound formation occurring during the palladium-induced modification of selenium. Using AAS and SEM, these investigators showed that 100 ng of palladium are necessary to stabilize 2 ng of selenium, deposited as a nitrate on pyrolytic graphite-coated graphite platforms. Based on the observation that larger quantities of palladium had little additional influence on the stabilization, it was concluded that stabilization resulted from physical effects. The selenium, it was posited, dissolves in molten droplets of palladium or its oxide, and possibly forms a compound with the palladium. Atomization would then be rate limited by the bulk diffusion of analyte through these droplets. Again, it must be remembered that the thermal quenching associated with these experiments is an additional perturbation on the system, and should not be ignored in the interpretation of the results.

Compound formation has also been implicated through the RBS results reported by Majidi and Robertson [19]. Backscattering of 2 MeV He^{++} from 750 ng of selenium (aqueous SeO_2) and 1000 ng of palladium on pyrolytic graphite flat substrates indicated that these species diffuse readily into the graphite. Backscattering was monitored after heating the substrate (with deposits) for 60 s *in vacuo* at a constant temperature between 310 and 1770K, and then cooling the substrate in an argon atmosphere. When Pd^0 was used, the selenium was observed to diffuse instantaneously into the graphite. Diffusion of selenium was stopped when palladium oxide was used as a modifier. As the palladium diffusion was observed to commence at 372K and to increase with increasing temperature, it was hypothesized that selenium diffusion must be occurring along channels created by the diffusion of palladium. When both species were present the selenium and palladium migrated toward the surface whenever temperatures exceeded 1100K. Palladium diffusion back into the bulk was observed to occur at 1500K. It was suggested that $[Pd_xSe_yO_z]$ compound formation occurs within the channels produced by the palladium diffusion. The proposed compound included oxygen because it was observed to be present in the depth profiles of the other diffusing species. It was proposed that migration of the compound to the surface occurs near 1100K. The compound decomposes and releases free selenium at more elevated temperatures; palladium then diffuses back into the bulk graphite. The chemical potential that drives such compound migration has yet to be addressed. It should be noted that even though thermal quenching procedures were used in these backscattering experiments, there is little likelihood that quench-induced artifacts pose a problem at the low temperatures where diffusion was observed.

Mechanisms controlling palladium-induced retardation of chloride interferences have recently been investigated by Qiao and Jackson [21]. These investigators made AA spectrometric evaluations of the vaporization of 1 ng of thallium, in the presence of 20 μg of palladium, deposited as aqueous nitrates on pyrolytic graphite-coated graphite platforms; 100 μg of sodium chloride were included to introduce the interferences. Recovery of thallium was significant only when the palladium was thermally pretreated above 900K. It was hypothesized that thallium forms deposits by a redox reaction (underpotential deposition) on the palladium metal; the energy for the desorption of these thallium deposits is, evidently, sufficient to retain the thallium at temperatures extending to 1000K. It was reported that, at this temperature, the sodium chloride interferent volatilized, and that there was co-volatilization of the adsorbed thallium that failed to bulk-diffuse into the palladium metal. Free thallium is produced as the thallium diffuses through, and escapes, the molten palladium; this type of release mechanism was proposed earlier [21]. Such a modification mechanism illustrates, again, the importance of materials properties in the modification process. In this case these properties involve the surface and bulk properties of the palladium modifier itself. The diffusion of palladium that was reported by Majidi and Robertson [19] does not appear to be a factor in the mechanism controlling the chloride interferences. The palladium diffusion occurs at temperatures near 370K, and the recovery of thallium was not influenced until temperatures exceeded 900K. However, the question of a compound being formed in the bulk of the graphite and then migrating to the surface, as suggested by Majidi and Robertson, is pertinent; it has not been addressed for the thallium system investigated by Qiao and Jackson. Rutherford backscattering experiments on this system are necessary in order to address this diffusion question.

5.4.2.6 Ascorbic acid modifiers

Another modifier mechanism that has been investigated mass spectrometrically is the mechanism that controls ascorbic acid-induced retardation of chloride interferences. Byrne *et al.* [31] used inductively coupled mass spectrometry (ICPMS) to monitor species that vaporized from flat pyrolytic graphite-coated graphite substrates that supported the modifier and manganese analytes. In these experiments, the sample was generally composed of 0.2–0.4 ng of manganese analyte, as an aqueous nitrate with 100–200 μg of magnesium chloride in 1 per cent hydrochloric acid; 100–200 μg of aqueous ascorbic acid composed the modifier. Results from these experiments show that a thermal pretreatment above 970K is necessary before the ascorbic acid has any influence on the atomization of the manganese. Above this threshold, the chloride interference is eliminated completely. These investigators proposed that ascorbic acid retards the otherwise rapid hydrolysis of the $MgCl_{2(s)}$ matrix that was included with the analyte [31]. To confirm this, the magnesium to chloride ratios that remained in pyrolytic graphite-coated graphite tube atomizers, after thermal pretreatment at various temperatures, were determined using flame AAS and ion chromatography on the atomizer extract (from dilute nitric acid washing). This revealed a modifier-induced enhancement of residual chloride. The formation of a complex (between ascorbic acid and manganese ions in the melt) which facilitates the removal of the water of hydration during thermal pretreatment was postulated. The retardation of hydrolysis, and consequently of the associated formation of $HCl_{(g)}$ in the atomizer, reduces the probability of interaction between manganese and $HCL_{(ad)}$. Formation of manganese chloride is impeded, and loss of manganese as a gaseous dichloride is restricted [31].

Gas chromatography and ETAAS were also used by Gilchrist *et al.* [25] to investigate the mechanisms that control ascorbic acid-induced elimination of chloride interferences. These investigators were concerned, however, with the modification of the analyte chloride, as opposed to the manganese-in-presence-of-chloride modification discussed above. The chromatography data indicated that, during heating of the atomizer, whenever the quantity of residual chloride matrix is small (e.g. 1 per cent hydrochloric acid), significant quantities of hydrochloric acid appeared in the carrier gas. This indicated to the investigators that the chloride is depleted by interacting with the modifier to form $HCl_{(g)}$. This $HCl_{(g)}$ is produced through hydrogen (from ascorbic acid decomposition) reactions with gaseous chloride.

Both of these ascorbic acid modifier mechanisms are driven by chemical, rather than by physical, properties. For the first mechanism, formation of hydrogen chloride is suppressed by the modifier through formation of a condensed-phase complex. The second mechanism involves enhanced hydrogen chloride formation through gas-phase reactions. Evidently, the chemical pathways that distinguish these mechanisms are analyte species dependent and/or matrix dependent. Indeed, Lamoureux [42] showed that when sodium chloride was used, the mechanism responsible for modification of manganese differed from that described for the $MgCl_{2(s)}$ matrix. It was proposed that the sodium chloride reacts with the acidic carbon char remains of pyrolyzed ascorbic acid, the product of this reaction being $HCl_{(g)}$. The resulting reduction in gaseous chloride formation, is then responsible for reducing the quantity of manganese chloride formation. To confirm this mechanism it is necessary to experimentally identify the condensed-phase species that are involved.

5.4.3 Summary

Table 5.1 summarizes the above modifier mechanisms. They are sufficient for our purposes and are not meant to be all-inclusive. It should be recognized that there exist additional contributions to the elucidation of modifier mechanisms. Information regarding these additional investigations is given in the excellent review by Tsalev *et al.* [43]. It is apparent from the above results that the present knowledge concerning modifiers is insufficient to associate a given modifier with a general mechanism. Adsorption sites are dependent on the concentrations of oxygen, carbon monoxide, and carbon dioxide in the Langmuir film. Furthermore, formation of these sites may be promoted or inhibited by presence of some modifier species.

Formation of solid solutions, or of compounds, is now associated with palladium-induced modification of selenium and arsenic. It remains to be shown whether such formations can be extended to other semimetals. At thermal pretreatment temperatures, both magnesium and palladium modifiers deplete water of hydration through hydration of magnesium oxide or palladium oxide; this inhibits the formation of gaseous analyte hydroxides. Palladium may retard chloride interferences for some species through a redox reaction on the palladium metal surface. The proposed mechanisms by which ascorbic acid controls chloride interferences are matrix as well as analyte-species dependent; for lead as a chloride, the quantity of chloride is decreased through formation of hydrogen chloride. For a chloride matrix, the ascorbic acid can slow the hydrolysis of the chloride so that $HCl_{(g)}$ is not formed in quantities sufficient to produce a gaseous chloride of the analyte. The mechanisms by which adsorbed oxygen acts as a stabilizer remain obscure, and apparently differ for the two analytes (selenium and lead) that have been investigated by MS. Proposed formation of oxyphosphorus compounds may control the phosphorus-induced modifications, but identification of the condensed-phase species have yet to be made to confirm this.

The research to understand modifier mechanisms is only beginning to a provide the basic results needed to develop and to guide further efforts; the database remains insufficient to draw general conclusions. The more recent approaches (RBS, SIMS, ICPMS) that have been applied to atomization mechanism elucidation can provide considerable insight. Yet, these techniques remain to be fully utilized in addressing the general modifier question. Considering the complexities involved, the present level of knowledge concerning modifier mechanisms is comparable to the level of understanding associated with nonmodified atomization mechanisms in the early 1980s. Extending the understanding of modifiers will require more concentrated efforts involving real-time and spatial distribution analyses. It is also no less true for modifiers than it is for elucidation of

Table 5.1 Proposed modifier mechanisms.

Modifier/analyte and experimental approach	*Mechanism*	*Reference*
Nickel–selenium		
MS and AAS; pyrolytic graphite-coated graphite and Ta tubes in vacuum.	Selenide formation; decomposition of selenide at more elevated temperatures.	27
MS; pyrolytic graphite-coated graphite flats in vacuum.	SeO_2 adsorption on $NiO_{(s)}$; SeO_2 desorption and dissociation as the $NiO_{(s)}$ is reduced by the carbon.	28
Oxygen–selenium		
MS and AAS; pyrolytic graphite-coated graphite, MS done in vacuum on flats.	$SeO_{2(g)}$ readsorption at sites created by desorption of CO and CO_2; desorption and thermal dissociation of the oxide at more elevated temperatures.	28
AAS; pyrolytic graphite-coated graphite tubes.	Higher energy desorption sites are responsible for observed increases in appearance temperatures when surface is oxygenated.	32
Oxygen–lead		
MS; pyrolytic graphite-coated graphite flats in vacuum.	Gas-phase interactions controlled by the CO to CO_2 ratios.	33
MS; pyrolytic graphite-coated graphite tubes at atmospheric pressure.	Homogeneous gas-phase reactions between O_2 and free lead.	31
Magnesium–beryllium		
MS; pyrolytic graphite-coated graphite tubes in vacuum.	Occlusion of sites of greatest heats of adsorption for oxides. Thermal pretreatment inhibits hydroxide formation by depleting water through hydration of MgO.	35
Magnesium–aluminum		
MS; pyrolytic graphite-coated graphite tubes in a vacuum.	Oxidation of adsorbed aluminum and adsorbed lower oxides by $MgO_{(g)}$. During thermal pretreatment, aluminum hydroxide formation is inhibited by water depletion through hydration of MgO.	36

Table 5.1 (*continued*)

Modifier/analyte and experimental approach	*Mechanism*	*Reference*
Phosphorus–cadmium		
Secondary ion MS on pyrolytic graphite-coated graphite flats in vacuum.	Cadmium oxyphosphorus compound formation suggested, but not found in the MS.	6
Phosphorus–lead		
MS on pyrolytic graphite-coated graphite flats in vacuum.	Formation of a condensed-phase lead oxyphosphorus compound and thermal decomposition of this compound at higher temperatures.	26, 33
Phosphorus–cadmium, lead		
RBS on pyrolytic graphite-coated graphite substrates in vacuum.	Formation of vitreous analyte–phosphates.	20
Palladium–arsenic		
MS; pyrolytic graphite-coated graphite tubes at atmospheric pressure.	Stoichiometric compound formation through $PdO_{(s)}$ reactions with condensed phases of As and AsO and $Pd_{(s)}$ reactions with condensed-phase arsenic trioxide.	27
Palladium–selenium		
MS; pyrolytic graphite-coated graphite tubes at atmospheric pressure and in vacuum, and flats of this same material in vacuum.	Selenium hydroxide formation inhibited by water depletion through hydration of $PdO_{(s)}$. Stoichiometric compound formation through $PdO_{(s)}$ interactions with condensed phases of selenium and SeO_2. $Se_{(g)}$ produced from decomposition of this compound readsorbs at sites associated with the presence of adsorbed palladium.	41
AAS; pyrolytic graphite-coated graphite tubes at atmospheric pressure–relatively large masses of palladium.	Selenium dissolves in molten palladium or its oxide; possible compound formation. Atomization is controlled by bulk diffusion of selenium in the palladium matrix.	21

(*contined*)

Table 5.1 (*continued*)

Modifier/analyte and experimental approach	*Mechanism*	*Reference*
SEM and XRF; pyrolytic graphite-coated graphite; postliminary experiments.	Surface alloy formation between selenium and palladium (1 : 1 ratio).	18
RBS; pyrolytic graphite, flat substrates in vacuum.	Compound of selenium, palladium, and oxygen forms within channels that evolve in the graphite as a result of palladium bulk diffusion. This compound migrates to the surface and decomposes at more elevated temperatures.	19
Ascorbic acid–manganese in presence of chloride		
ICPMS; pyrolytic graphite-coated graphite flats at atmospheric pressure.	Formation of an ascorbic acid–manganese ion complex in the melt facilitates removal of water of hydration during thermal pretreatment. This retards formation of hydrogen chloride, thus reducing probability of manganese–hydrogen chloride interactions.	31
Ascorbic acid–lead as a chloride		
GC and AAS; pyrolytic graphite-coated graphite at atmospheric pressure.	Hydrogen, from decomposition of the ascorbic acid, reacts with chloride. The resulting hydrogen chloride escapes, thus depleting the chloride.	25

nonmodified atomization, that condensed-phase identification is necessary to complete the investigations. Unfortunately, the existence of particular condensed-phase species in the atomizers are, presently, only implicated in the gas-phase data. Identification of the condensed-phase species involved is imperative to the verification of the associated mechanisms, and to help direct further research into mechanism determination. Although it is clear that the investigations that have been made indeed elucidate the modifier mechanisms, without the above identification, the interpretations will necessarily remain open to question.

5.5 FINAL COMMENTS

The discussions have, until now, refrained from detailing the experimental methods used to analyze the modifier mechanisms of interest. Although these details are pragmatically important, little can be gained in understanding the mechanisms by including such detail. Approaches that will provide condensed-phase identification are, however, another matter. Relatively little work has been devoted to this identification. However, the identification of the condensed-phase species is often crucial to the full elucidation of these mechanisms. It is not generally clear how this identification might be accomplished. To confront this difficulty, one must be aware of the techniques that are available and what might be achieved by using them. This section addresses the problem by providing some perspective on the techniques that might be successfully exploited for these purposes. The discussion is not intended to be all inclusive, but provides a representative identification and understanding of the existing techniques, and details the most promising.

As discussed above, composition measurements must generally be made at the temperature of interest in order to minimize the possibility of introducing artifacts from the otherwise necessary thermal quenching process. Because of species migration into the bulk of the substrate, composition measurements will generally have to be made beyond the surface region. The depth of the measurement will depend on the probing particle used in a given technique and on the energy of this particle. Table 5.2 identifies the techniques that are available for composition analysis, and the depth generally sampled by the probing (incoming) or the analyzed (exiting) particles. The techniques that monitor composition can be separated into strictly surface-type probes (those capable of sampling only a few nanometers below the surface), and those that have capabilities of sampling more than a few thousand nanometers below the surface. The first category comprizes ESCA or XPS (electron spectroscopy for surface analysis or X-ray photoelectron spectroscopy) [44], LEIS (low energy ion scattering) [45], and REELS (reflected electron energy loss spectroscopy) [46]. These powerful surface analysis techniques can be used to evaluate surface compositions on the substrates of interest here, and graphite in particular. The ESCA is particularly valuable in that it provides information on valence states from the chemical shifts indicated in the spectra. However, these three approaches require ultrahigh vacuum operating conditions which, because of out-gassing of the substrate at high temperatures, is very difficult to maintain.

The remaining approaches sample well below the surface and are, therefore more appropriate for substantiating the species in the graphite substrate. The equation of fast atom bombardment mass spectrometry (FABMS) [47] has yet to be used in the modifier mechanism inquiries.

Table 5.2 Summary of composition probe techniques.

Technique	*Physics*	*Capability*	*Reference*
ESCA, XPS	Incident high-energy electrons excite atoms; characteristic X-rays are monitored.	Depth from which photoelectron can be removed from solid is about 10 nm. Composition and valence state can be determined. Ultrahigh vacuum is required.	42
EMP	Scattering of low-energy noble gas ions; angle of scattering depends on mass of scattering atom.	Composition monitored to depths of 10^3 nm.	49
LEIS	Incident ion-beam sputters surface atoms; the fraction that are ionized are mass analyzed.	Composition monitored to about 0.5 nm below sputtered surface. Ultrahigh vacuum required.	45
SIMS	Incident ion-beam sputters atoms in the near-surface region; those that are ionized are mass analyzed.	Composition sampled as function of depth as sputtering progresses. Ultrahigh vacuum required.	48
FABMS	Incident atom beam sputters atoms in the near-surface region; those that are ionized are mass analyzed.	Composition depth profiles obtained as sputtering progresses. Ultrahigh vacuum required.	47
RBS, HEIS	High-energy ion scattering. Energy of scattered ion is dependent on mass of scattering species.	Composition sampled to depths of several μ. Ultrahigh vacuum required.	45
REELS	Energy loss from exciting the core electron by electron bombardment is characteristic of the excited atom species.	A low incident-angle technique. Composition sampled to a depth of 0.5 nm. Ultrahigh vacuum required.	46
EXAFS, NEXAFS	X-ray induced ejection of core electron. Fine structure in the absorption edge spectrum depends on nearest neighbor interactions with the ejected electron.	Species composition averaged over depths of hundreds of μm. Spectra obtainable under nonvacuum conditions.	50, 51

However, the other three of these techniques have already been applied to such investigations. Hassel *et al.* [6] employed SIMS [48] in their investigations of phosphorus-induced modifications. Teague-Nishimura *et al.* [18] utilized EMP (electron microprobe) [49] for their palladium modifier studies, and Eloi *et al.* [20] used the RBS or high energy ion scattering (HEIS) [45] approach in their modifier studies. These latter (RBS) efforts have provided the best available data on composition of the condensed phases formed in ETAs, although there have been relatively few applications of the method to the elucidation of modifier mechanisms. Although nonavailability of appropriate particle accelerators (Van de Graaf generator) undoubtedly contributes to this, it is also likely that the RBS technique is insufficiently recognized as suitable for these purposes. The principle that guides RBS is relatively simple. It can be shown, from momentum and energy conservation, that the energy E of a scattered incident particle of mass M_i depends on the mass of the particle that is being bombarded and on the scattering angle. For the scattering angle Θ_s shown in Figure 5.6, the relationship is given by

$$(E/E_i)^{1/2} = k = M\cos\Theta_s/(M+M_i) + \{M^2\cos^2\Theta_s/(M+M_i)^2 + (M-M_i)/(M_i+M)\}^{1/2} \tag{5.25}$$

where E_i represents the energy (the order of a few MeV) of the incoming species. Consequently, peaks appear in the scattered energy (E) spectrum when scattering occurs from different masses M. Corrections are necessary for scattering by any species not located on the surface, as energy losses are involved in penetrating and leaving the bulk material. A

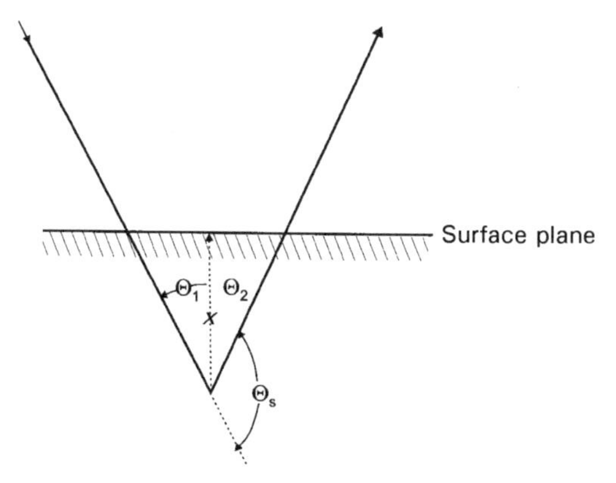

Figure 5.6 Geometrical relationships for incoming and scattered particles associated with Rutherford backscattering analysis: x =the particle depth below the surface; Θ_s=scattering angle; Θ_1 =the angle between the normal to the surface and incoming particles; Θ_2 =that for outgoing particles.

particle that penetrates incremental depths dl and which is finally scattered at a depth x below the surface will emerge with an energy

$$E' = \left[E_i - \int_0^{x \sec \Theta_1} S(E)\mathrm{d}l\right]k^2 - \int_{x \sec \Theta_2}^0 S(E)\mathrm{d}l \qquad (5.26)$$

where $S(E)$ is the stopping power (rate of energy loss) for a given substrate material, and Θ_1 and Θ_2 represent the angles between the normal to the surface and the incoming and outgoing (scattered) particles. Thus, the depth x of the impurity can be determined from measured angles and energy E if the identity M of impurity and the stopping power of the substrate are known. The number of impurity atoms N involved in the scattering can then be obtained from the measured intensities (C and C_s) associated with impurity and substrate scattering. This is achieved through the expression

$$C/C_s = (Z^2NE_i)/Z_s^2N_s\left(E_i - \int_0^{x \sec \Theta_1} S(E)\mathrm{d}l\right)^2 \qquad (5.27)$$

where Z and Z_s are the atomic numbers of the impurity and substrate, and N_s is the number of substrate atoms involved in the scattering. Note that the impurity atoms are not identifiable from the scattering data because the energy of the backscattered particle depends on both x and the mass of the scattering particle. Figure 5.7 illustrates how an RBS spectrum might appear for a given scattering angle. The higher energy peak (1) corresponds to scattering from a surface impurity having greater atomic mass than the substrate. The discontinuity at E_s is due to scattering from the substrate atoms at the surface ($x = 0$). The lower energy continuum is due to scattering at progressively greater depths; increasing x results in the decreasing of E. As indicated by the energy, Eqn (5.25), the peak labeled 2 is due to surface atoms that are of lower mass than those of the substrate; a peak of this type is usually not resolvable. The discontinuity due to scattering from atoms that compose an interface at $x > 0$ may be resolvable, however, and is shown to appear in the figure at the energy E_{impurity}.

Another technique that offers promise is X-ray absorption spectroscopy (XAS), which includes the spectroscopies of X-ray absorption fine structure (EXAFS) [50] and near-edge X-ray absorption fine structure (NEXAFS) [51]. This technique has become an element-specific probe, with capabilities that give local structure such as bond distance, and coordination number. The principle of XAS is based on core electron production, by electronic absorption of photons, when the incident photons have energies in the range from 0.5 keV to 100 keV. For this process, changes in the absorption coefficient manifest themselves in several ways. Of course, the transmissions of the X-rays vary, but the photoelectron intensities and

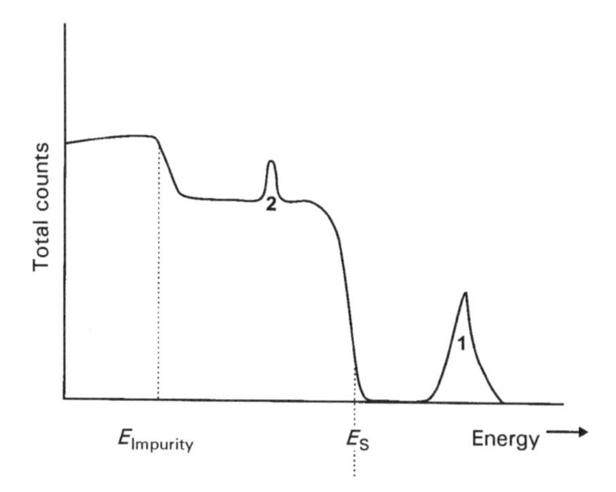

Figure 5.7 Illustration of a Rutherford backscattering spectrum for a given scattering angle. Peaks 1 and 2 represent respective scattering from impurities having greater and smaller masses than the atoms composing the substrate. These peaks (1 and 2) represent atoms located below and on the surface, respectively. E_s represents scattering from substrate surface atoms. The low energy continuum results from scattering at increasing depths. $E_{impurity}$ represents scattering from atoms that form an interface below the surface.

the X-ray fluorescence, induced by the decay of the excited atom to the ground state, follow these changes in the absorption coefficient; the probabilities of photoelectron and of excited state formation are, after all, directly related to the absorption cross-section. However, the cross-section is modulated by the presence of the electric fields of the neighboring species, and this modulation appears in the above absorption-related phenomena (transmission, fluorescence, and photoelectron intensities). These phenomena provide, therefore, probes of the perturbations induced by the nearest neighbor species; the determinations of bond lengths, coordination numbers and types of neighbors becomes possible.

The interaction of an outgoing photoelectron wavefunction is shown schematically in Figure 5.8. The photon is absorbed by a core-level electron and, if neighboring species are absent, the photoelectron outgoing radial wave is as depicted by the solid circles in Figure 5.8(a). When a neighboring atom is present, as shown in Figure 5.8(b), the outgoing portion of the radial wave is reflected by the neighbor and creates the incoming radial wave (one for each nearest neighbor atom) shown as dashed circles. The interaction of the incoming and the outgoing waves produce constructive or destructive interferences, depending on the relative phases. This influence on the wavefunction modulates the density of states at the absorbing center and, therefore, modulates the probability of absorption.

An illustration of the type of effect neighboring species have on the results of a transmission experiment is shown in Figure 5.9, where the magnitudes of the logarithms of the ratio of incident I_0 to transmitted I intensities are plotted against the incident photon energies. The discontinuity represents the K-shell absorption edge in this figure. In the absence of neighboring atoms the postedge region will appear smooth, as depicted in Figure 5.9(a). When neighbors are present, the effect of the density of states modulation (described in the preceding paragraph) will appear as a sinusoidal type variation, similar to that shown in Figure 5.9(b). It is this variation that has been labeled as the EXAFS. It is the frequency and the amplitude of the EXAFS that provide the quantitative information. The frequency is dependent on the distance between the absorbing atom and its neighbor, i.e. the photoelectron wave travels from the absorber to the neighbor (the scatterer) and back to the absorber. Note that two phase shifts are experienced by the absorber (one for the outgoing trip and one for the return). If the phase shifts are known, the interatomic distances can be determined by calculation or by using model compounds. It is important to note that, if the Fourier transformation technique is applied to EXAFS, the photoelectron scattering profile will appear as a function of the radial distance from the absorber to scatterer. The amplitudes of the EXAFS are dependent on the number and type of neighboring atoms; therefore, number, spacing, and types of neighbors are contained in the EXAFS. Only the immediate environment (about 0.6 nm) surrounding each absorbing species is involved in EXAFS and, because

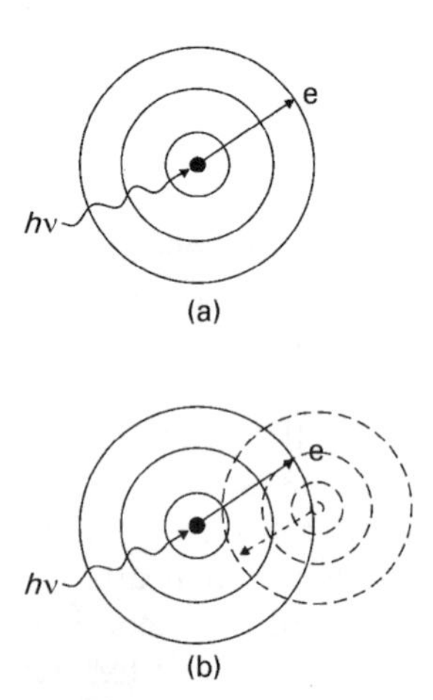

Figure 5.8 Diagram of core-level photoelectron wavefunctions when (a) the target atom is free, and (b) a neighboring atom scatters the photoelectron back towards the target species. Shown are the photon ($h\nu$) and electron (e).

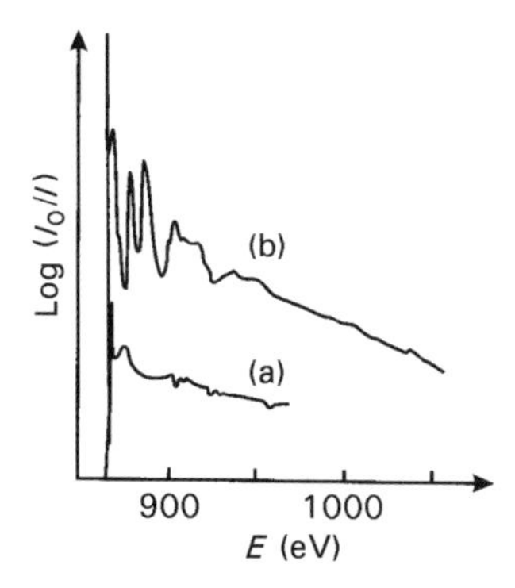

Figure 5.9 X-ray absorption spectra for (a) a free target-atom, and (b) a target atom in the vicinity of neighboring atoms. (Reprinted from P. Lagarde, EXAFS studies of amorphous solids and liquids, in *Amorphous solids and the liquid state*, (Eds.) N.H. March and M. Torsi, 1985, pp. 365–393, with kind permission from Plenum Publishing Corp., New York, NY, USA.)

the absorption is tuned to a particular atom species of interest, EXAFS is highly specific.

In practice, EXAFS $\chi(E)$ is normalized against the background so that it can be represented by the relationship

$$\chi(E) = \{\Phi(E) - \Phi_0(E)\}/\Phi_0(E) \tag{5.28}$$

where Φ and Φ_0 represent the measured and the background signals. To obtain structural information, $\chi(E)$ must then be transformed from energy E space to reciprocal wavelength k space using

$$k = \sqrt{(2M/\hbar^2)(E - E_0)} \tag{5.29}$$

and the EXAFS expression (from short-range single-electron single-scattering theory)

$$\chi(E) = \sum_j N_j S_0(k) F_j(k) \exp(-2\sigma_j^2 k^2) \exp(-2r_j/\lambda(k) \times [\sin(2k_j + \Phi_j(k))]/kr_j^2 \tag{5.30}$$

In Eqn (5.29), E_0 is the threshold energy for a given adsorption edge. In the EXAFS expression (Eqn (5.30)), N_j and the r_j represent the number of type-j neighboring atoms and their distance from the absorbing atom. The expression $F_j(k)$ is the backscattering amplitude of each neighbor, and the exponent containing $\sigma_j(k)$ is the Debye–Waller factor that accounts for static disorder and thermal vibrations. The term $\sigma_j(k)$ represents the root mean-square fluctuation in interatomic spacing. The exponent containing $r_j/\lambda(k)$ accounts for inelastic scattering losses, $\lambda(k)$ being the mean free path of the photoelectron. Losses by many body effects (excitation and ionization induced by the photoelectron) are included in $S_0(k)$. The expression $\Phi(k)$ represents the total phase shift in the photoelectron wave.

It is quite evident that EXAFS involves the backscattering amplitude, which is given by N_jF_j, and the modulation of this amplitude by the Debye–Waller factor, the inelastic scattering, many body effects, and the sinusoidal function of the interatomic spacing and phase shifts. It is assumed, in the extraction of chemical and structural information from EXAFS, that the phase and the scattering amplitude functions are relatively insensitive to the chemical bonding (at least from 50 eV to 1000 eV above the absorption edge). Thus, the determination of these functions for a known system is sufficient for the functions to be applied to unknown systems. Then, assuming a k-dependent relationship for λ, and suitable values for σ and S_0, it is possible to determine N_j and r_j. This is usually accomplished by Fourier transformation (to radial distance space) of the k-weighted EXAFS. The resulting peak in the radial distribution function provides r_j, and the amplitude of the peak, relative to that of a model compound provides N_j. By choosing a window about this peak (i.e. by filtering) and then making a reverse Fourier transform (back to k space) within this window, a filtered EXAFS spectrum is obtained that fits more easily with model compounds.

Unfortunately, application of the EXAFS technique for the present purposes suffers because the high energy (GeV) electron storage ring facilities, necessary to produce the high X-ray photon fluxes, have not generally been readily accessible. Modifier mechanism-type experiments require such facilities, with appropriate X-ray beam lines, because relatively high X-ray intensities (10^4 greater than those characteristic of rotating anode sources) are needed for EXAFS spectroscopy of analytical quantities of analyte in modifier matrices. Certainly, the ability of EXAFS spectroscopy to address condensed-phase modifier mechanisms, in more detail than by any other technique, is sufficient motivation to utilize these facilities for such purposes. It should be noted that allocation of beam time for a given experiment is obtainable simply by contacting the facility of interest and then submitting a description of the proposed experiment.

Lamoureux *et al.* [52] used the X-ray beam facilities of the Stanford Synchrotron Radiation Laboratory to examine the condensed-phase species associated with the palladium-induced modification of the oxide of selenium. Their EXAFS experiments were done on selenium–palladium-doped, powdered graphite samples that were heated *in situ*. They provided the first observations of condensed-phase molecular species related to modifier mechanisms. It was observed that formation of metallic selenium ($Se_{n(ad,s)}$) was inhibited by the presence of the modifier. An Se–Pd compound was also observed. Interestingly, oxygen was not an identifiable component in the compound. Although interatomic spacings of the compound were determined, known palladium selenides do not fit the resulting EXAFS; compound identification was, therefore, not possible. These results corroborate some of the postulates that have been made to describe data from gas-phase investigations of these mechanisms (Section 5.4.2.5), and illustrate the value of EXAFS spectroscopy to the study of modifier mechanisms.

There can be little doubt that RBS, SIMS, and EXAFS-type experiments have a contribution to make to the understanding of the complex mechanisms that control chemical modifiers. Condensed-phase composition information can be provided by any of these three approaches, whereas speciation is possible only by application of the latter.

REFERENCES

[1] R.D. Ediger, *At. Absorpt. Newsl.* **14**, 127 (1975).
[2] C.J. Rademeyer, B. Radziuk, N. Romanova, N.P. Skaugset, A. Skogstad and Y. Thomassen, *J. Anal. At. Spectrom.* **10**, 739 (1995).
[3] D.L. Styris and D.A. Redfield, *Spectrochim. Acta Rev.* **15**, 71 (1993).
[4] D.L. Tsalev and V.I. Slaveykova, *J. Anal. At. Spectrom.* **7**, 147 (1992).
[5] W. Hume-Rothery and B.R. Coles, *Advances in Phys.* **3**, 149 (1954).
[6] D.C. Hassell, V. Majidi and J.A. Holcombe, *J. Anal. At. Spectrom.* **6**, 105 (1991).
[7] B. Fain and S.H. Lin, *Surf. Sci.* **147**, 497 (1984).
[8] B. Welz, G. Schlemmer and H.M. Ortner, *Spectrochim. Acta, Part B* **41B**, 567 (1986).
[9] B. Welz, G. Schlemmer, H.M. Ortner and W. Birzer, *Spectrochim. Acta, Part B* **44B**, 1125 (1989).
[10] R.W. McCabe and L.D. Schmidt, *Surf. Sci.* **60**, 85 (1976).
[11] J.L. Brand, A.A. Deckert and S.M. George, *Surf. Sci.* **194**, 457 (1988).
[12] G. Pirug and H.P. Bonzel, *Surf. Sci.* **199**, 371 (1988).
[13] I. Langmuir, *Phys. Rev.* **34**, 401 (1912).
[14] W. Frech, E. Lundburg and A. Cedergren, *Prog. Anal. At. Spectrosc.* **8**, 257 (1985).
[15] J.A. Holcombe, R.H. Eklund and J.E. Smith, *Anal. Chem.* **51**, 1205 (1979).
[16] E.J. Czobik and J.P. Matousek, *Talanta* **24**, 573 (1977).
[17] B. Welz, G. Schlemmer and U. Voellkopf, *Spectrochim. Acta, Part B* **39B**, 501 (1984).
[18] J.E. Teague-Nishimura and T. Tominaga, *Anal. Chem.* **59**, 1647 (1987).
[19] V. Majidi and J.D. Robertson, *Spectrochim. Acta, Part B* **46B**, 1723 (1991).
[20] C. Eloi, J.D. Robertson and V. Majidi, *J. Anal. At. Spectrom.* **8**, 217 (1993).
[21] H. Qiao and K.W. Jackson, *Spectrochim. Acta, Part B* **46B**, 1841 (1991).
[22] H. Qiao, T.M. Mahmood and K.W. Jackson, *Spectrochim. Acta, Part B* **48B**, 1495 (1993).
[23] J.G.T. Regan and J. Warren, *Analyst* **103**, 447 (1978).
[24] D.J. Hydes, *Anal. Chem.* **52**, 959 (1980).
[25] G.F.R. Gilchrist, C.L. Chakrabarti and J.P. Byrne, *J. Anal. At. Spectrom.* **4**, 533 (1989).
[26] D.A. Bass and J.A. Holcombe, *Anal. Chem.* **59**, 974 (1987).
[27] D.L. Styris, *Fresenius' Z. Anal. Chem.* **323**, 710 (1986).
[28] M.S. Droessler and J.A. Holcombe, *Spectrochim. Acta, Part B* **42B**, 981 (1987).
[29] D.L. Styris, L.J. Prell and D.A. Redfield, *Anal. Chem.* **63**, 503 (1991).
[30] T.N. Rettberg and L.M. Beach, *J. Anal. At. Spectrom.* **4**, 427 (1989).

[31] J.P. Byrne, C.L. Chakrabarti, D.C. Gregoire, M. Lamoureux and J. Ly, *J. Anal. At. Spectrom.* **7**, 371 (1992).
[32] M.S. Droessler and J.A. Holcombe, *J. Anal. At. Spectrom.* **2**, 785 (1987).
[33] D.A. Bass and J.A. Holcombe, *Anal. Chem.* **60**, 578 (1988).
[34] S.G. Salmon and J.A. Holcombe, *Anal. Chem.* **51**, 648 (1979).
[35] D.L. Styris and D.A. Redfield, *Anal. Chem.* **59**, 2897 (1987).
[36] D.L. Styris and D.A. Redfield, *Anal. Chem.* **59**, 2891 (1987).
[37] B.V. L'vov and G.N. Ryabchuk, *Spectrochim. Acta, Part B* **37B**, 673 (1982).
[38] R.E. Sturgeon and H. Falk, *J. Anal. At. Spectrom.* **3**, 27 (1988).
[39] W. Frech, F. Lundburg and A. Cedergren, *Prog. Anal. At. Spectrosc.* **8**, 257 (1985).
[40] D.A. Redfield and W. Frech, *J. Anal. At. Spectrom.* **4**, 685 (1989).
[41] D.L. Styris, L.J. Prell, D.A. Redfield, J.A. Holcombe, D.A. Bass and V. Majidi, *Anal. Chem.* **63**, 508 (1991).
[42] M.M. Lamoureux, *A Multi-Technique Strategy for Study of Some Mechanisms of Atom Formation and Dissipation in Electrothermal Atomic Absorption Spectrometry*, Ph.D. Thesis, Department of Chemistry, Carleton University, Ottawa, Canada (1993).
[43] D.L. Tsalev, V.I. Slaveykova and P.B. Mandjukov, *Spectrochim. Acta Rev.* **13**, 225 (1990).
[44] C.R. Brundle, *Surf. Sci.* **48**, 99 (1975).
[45] D.P. Smith, *J. Appl. Phys.* **38**, 340 (1967).
[46] A.W. Czanderna, *Solar Energy Mat.* **5**, 349 (1981).
[47] R.C. Barber, R.S. Bordoli, G.J. Elliot, R.D. Sedgwick and A.N. Tyler, *Anal. Chem.* **54**, 645A (1982).
[48] H.W. Werner, *Surf. Sci.* **47**, 301 (1975).
[49] H. Yakowitz, *J. Vac. Sci. Technol.* **11**, 1100 (1974).
[50] J. Jaklevic, J.A. Kirby, M.P. Klein, A.S. Robertson, G.S. Brown and P. Eisenberger, *Solid State Comm.* **23**, 679 (1977).
[51] F. Zaera, D.A. Fischer, S. Shen and J.L. Gland, *Surf. Sci.* **194**, 205 (1988).
[52] M.M. Lamoureux, J.C. Hutton and D.L. Styris, *Spectrochim. Acta*, Part B **53B**, 993 (1998).

6

Atomization from solids and slurries

Kenneth W. Jackson

6.1 INTRODUCTION

Some of the earliest work described by L'vov on graphite furnace ETAAS involved the direct analysis of solid samples, because the design of L'vov's original graphite furnace atomizer made solid sample introduction quite convenient (see Section 1.4). However, the simplified furnace developed by Massmann did not allow solids to be analyzed easily and, as most commercial atomizers are based on the Massmann furnace, the technique of ETAAS is used predominantly for the analysis of liquid samples. Unfortunately, most samples submitted for analysis are solids, and thus conventional approaches require chemical decomposition of the sample prior to analysis. Decomposition methods such as dry ashing and chemical digestion are time-consuming, and can lead to systematic and random errors that will degrade the overall accuracy and precision of the analysis. The analyst desiring a technique that can determine trace concentrations of metals directly in solid samples is given little choice.

Traditional techniques include spark source atomic emission or mass spectrometry, which have relatively poor precision, X-ray fluorescence spectrometry, which often lacks the required sensitivity for trace determinations, and neutron activation analysis, which requires complex and expensive instrumentation.

It is not surprising, therefore, that there have been many attempts to adapt atomic spectrometric techniques such as flame atomic absorption spectrometry (FAAS), inductively coupled plasma atomic emission spectrometry (ICPAES) and ETAAS to the direct analysis of solid samples.

Electrothermal Atomization for Analytical Atomic Spectrometry. Edited by K. W. Jackson
© 1999 John Wiley & Sons Ltd

Solids, in the form of slurries or suspensions, have been nebulized into flames for AA spectrometry. However, nebulization provides very low sample transport efficiency, and this is even poorer when solid particles are introduced compared with solutions. Those particles that reach the flame are then more difficult to dissociate, leading to low atomization efficiency. A few successful applications of FAAS to the analysis of slurries have been reported, but the low transport and atomization efficiencies mean that systematic errors are introduced unless carefully matched solid calibration standards are used.

The nebulization of slurries into ICPs has been more successful. High-solids nebulizers improve sample transport efficiency, and the high energy of an ICP, compared with a flame, improves the atomization efficiency. However, there have been reports of low recoveries from slurries compared with calibration standards in solution, and this has been attributed to lower atomization efficiency from the solid particles due to their short residence time in a plasma. Particle diameters must be no more than a few micrometers to assure complete volatilization. Also, with slurry nebulization into flames or plasmas, the effect on transport efficiency of nonuniform distribution of analytes throughout the particle size range should be considered. This can arise through grinding of the material; for example, in finely ground soils and sediments, the smaller sized particles are mostly softer clay material, and the larger particles are mostly harder silica or carbonate materials. Many metals are associated predominantly with the clay particles, so their transport to the plasma may be quite efficient. However, any metal associated more with the harder (larger) particles might be transported with much lower efficiency. The only way of successfully overcoming the above problems of sample transport and atomization efficiency is to analyze the samples against solid calibration standards that are in a similar matrix and can be nebulized in an identical manner to the samples. Although this approach is used, it is not usually feasible for routine analysis, as several calibration standards over an appropriate concentration range are required, and the availability of solid reference materials is severely limited. Successful approaches to the direct analysis of solids by ICPAES have included introduction of the sample on a probe by injecting it coaxially through the plasma torch, or by interfacing an electrothermal vaporization chamber to the plasma torch. Unlike nebulization, both these approaches can provide 100 per cent sample transport efficiency and hence enable calibration standards in solution to be used.

Compared with FAAS and ICPAES, ETAAS is inherently more suitable for the direct analysis of solids. Solids may be weighed directly into the electrothermal atomizer, with the amount weighed depending on the concentration of the analyte in the solid sample (in order to ensure the signal is within the linear dynamic range of the instrument). Solid chips can be weighed into the atomizer, provided the analyte is distributed relatively homogeneously through the sample. This applies to the determination of trace metals in many (not all) metal alloys, glasses, etc. Typically, sample masses of 0.1–0.5 mg are used, but larger masses (up to 40 mg) have been used in larger induction furnaces that were designed specifically for solid samples. Samples in which analytes may not be distributed homogeneously, such as soils, sediments, rocks etc., are usually ground to a fine powder (less than 20 μm particle diameter) in order to make them more nearly homogeneous, and sample sizes typically in the range 0.1–2 mg are used. Obviously, careful weighing of such small amounts is required, and various means have been devised to facilitate the introduction of solid samples into furnaces. These are discussed in Section 6.4.

The other method of introducing solid samples involves the preparation of slurries. This is only applicable to samples that are conveniently ground to a powder. However, most samples can be dealt with in this way, a notable exception being metal alloys, which are generally too soft to grind easily unless they are cooled to liquid nitrogen temperatures. Typically, a known volume of a liquid (usually water) is added to a known weight of the finely ground sample. A uniform suspension is then obtained by mechanical agitation, and an aliquot (e.g. 20–50 μl) is pipetted into the furnace in the same way that solutions are introduced. The mass of solid introduced in this way is typically similar to that introduced by direct weighing. In either case of sample introduction, transport efficiency is 100 per cent, the same as liquid calibration standards.

As discussed in Chapter 2, ETAAS has high atomization efficiency and, under optimized operating conditions using stabilized temperature platform furnace (STPF) conditions (see Section 1.5.1), atomization efficiencies between different instruments are very similar for many elements. If the conditions in the atomizer can be optimized so that samples introduced as solids or slurries have the same high atomization efficiency, then ETAAS will be very well suited to this analysis, and it will be possible to use calibration standards in aqueous solution without the introduction of systematic errors.

Figure 6.1 shows the number of papers published, on a yearly basis since 1974, on the direct analysis of solid samples by the direct introduction of preweighed solid sample aliquots (direct-weighing ETAAS), and the pipetting of solids in the form of a slurry (slurry ETAAS). This figure shows that both techniques increased in popularity through the early 1990s. It is interesting that, in recent years, direct-weighing ETAAS has declined in popularity, whereas the publication rate of slurry ETAAS has decreased only a little. This decrease in slurry ETAAS publications may simply reflect a maturing of the technique. Much of the development of these techniques has been phenomenological in nature, but also

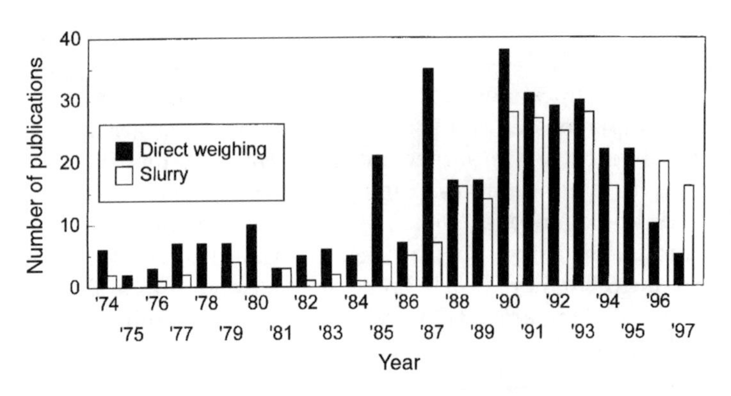

Figure 6.1 Solid-sampling ETAAS publications, 1974–1997.

described in this chapter is some fundamental research that has helped to explain the different atomization processes occurring when solids are atomized, as compared with solutions. This has shown how atomizer conditions can be optimized to remove any systematic errors that might occur due to these differences if aqueous calibration standards are used.

6.2 ATOMIZATION PROCESS

Shown in Figure 6.2 are absorbance signals obtained from lead introduced in aqueous solution as the nitrate, and in a river sediment slurry. Atomization from a platform under STPF conditions was used (see Section 1.5.1), and the two samples have produced distinctly different absorbance profiles. The slurry sample has produced a lead absorbance signal that is delayed in time and broadened compared to the signal from the sample that was introduced as a solution. Such behavior is very common, and the implications of this in the analysis of solids and slurries is discussed later. First, however, it is important to study the reasons for this behavior. There can be many reasons for differences in peak characteristics between samples introduced as solutions and as solid particles, including heat transfer effects, analyte evaporation rates being governed by slow release from molten droplets obtained from solid particles, and physical or chemical forces exerted between the analyte and the solid matrix.

When a sample is introduced in solution form, the drying stage of the ETA heating program results in a fairly uniform deposit of salt microcrystals. Most of these particles are in close contact with the graphite surface and, when their small size is considered relative to the large mass of the ETA, heat transfer effects are negligible. If a powdered sample is introduced as a slurry, there is also a fairly even distribution of particles

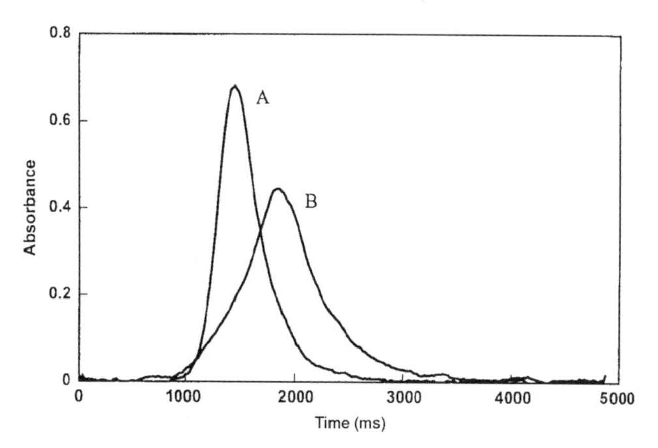

Figure 6.2 Absorbance signals from Pb (1 ng) introduced to the graphite furnace as (curve A) an aqueous solution, and (B) a slurry of a river sediment.

on the ETA surface. Although the particles (typically up to 20–50 μm in diameter) may be substantially larger than the microcrystals resulting from solution deposition, it is still unlikely that slower heating will be significant for such particles. Heat transfer effects may, however, be significant if large chips of solid material are weighed directly into an ETA, as much of the sample will not be in direct contact with the graphite surface. Also, if powders are weighed directly, the particles may not be spread over the surface as efficiently as they are when introduced in slurry form, and atomization may be delayed because some of the particles have not heated as quickly as others. In some cases, this may be a disadvantage of directly weighing solid samples into the ETA compared with slurry introduction.

Bendicho and de Loos-Vollebregt [1] introduced powdered glass samples as slurries into an ETA, and found delayed absorbance signals compared to solutions. Electron microscope studies revealed small particles after the ETA drying stage, but they melted during the pyrolysis stage to produce larger droplets; that is, particle agglomeration occurred, and the slower appearance of analyte from these samples compared with solutions was attributed to slower release from the larger particle agglomerates. Even in this case, slower heat transfer may not be significant, because the atomization rate may be limited by physical processes of analyte migration out of the molten glass droplets. Metal samples are difficult to grind, and are generally weighed directly into the ETA as solid chips. Prior to analyte atomization, the chips melt, and it has been reported [2] that the atomization time will depend on the diffusion rate of the analyte or its compounds through the sample. More recent work by Hinds *et al.* [3] indicates that migration of analyte from a molten metal droplet occurs through processes more complex than diffusion, involving convection currents in the droplet as the temperature increases. In the case of high-melting point materials, a further signal delay may occur, as the migration process will not be significant until the metal chip has reached a high enough temperature to melt. The size of the chip may then be an important factor. Takada and Hirokawa [4] weighed solid steel samples into a cup-in-tube furnace and found that the evaporation rate of the analyte from the chip during the atomization stage was dependent on sample mass and surface area. The migration time will depend on the mean diameter of the molten droplets produced–even if the droplets are small, the temporal width of the absorbance peak is likely to be greater for the sample introduced as a solid particle compared with a solution. Analyte metals in alloys may be distributed homogeneously if there is a solid solution, or heterogeneously if the analyte exists as small solid particles at the grain boundaries or in the grains. In either case, the analyte must migrate from the alloy, but an additional complication may be the appearance of multiple peaks from the solid sample. This has been attributed [5] to cases where there is no solid solution, and analyte at the grain boundaries and near the particle surface is released more quickly than analyte embedded as small particles within the grains.

An important class of sample is geological and environmental materials, such as rocks, soils, sediments, and absorbance signals are generally delayed and broadened compared with the signals from aqueous solutions. Trace metals may be associated with these samples in two ways: metals that have been introduced as environmental contaminants are adsorbed on the smaller active particles (such as the clay particles in soils and sediments); and metals that are naturally present are distributed throughout the solid material.

Karwowska and Jackson, in a two-part study [6,7], simulated both types of sample by preparing particles of alumina with traces of lead either adsorbed on or embedded in the particles. They prepared two types of alumina; (1) γ-alumina, which they made by heating an aluminum salt to about 500 °C, and which consists of a mixture of incompletely deactivated hydroxides; and (2) α-alumina, which is nearly-anhydrous aluminum oxide, made by heating an aluminum salt to 1000 °C. Slurries were prepared from the alumina samples and, on atomization in an ETA operated under STPF conditions, the samples produced distinctly different atomization behavior.

Adsorbed lead was released from the γ-alumina particles later than from the α-alumina particles. For an aqueous solution of a lead salt, the peak was earlier than either slurry peak. Using wall atomization and a slow rate of heating, they used Smets' method [8] (described in Section 2.5.8.3) to measure experimental activation energies for the two samples with adsorbed lead, and concluded from the magnitude of the activation energy that the delayed absorbance signal resulted from a rate-limited release of lead from the particle surface. In the case of α-alumina, they concluded that the rate limiting step in the atomization process was the release of physically adsorbed lead from the slurry particle, whereas in the case of γ-alumina lead was chemisorbed on the active surface sites, and therefore atomization occurred via the breaking of chemical bonds.

Embedded lead was released later than adsorbed lead from both types of alumina. It was not released from γ-alumina until the particle reached its decomposition temperature, but lead was released from the α-alumina particle at a temperature below the decomposition temperature, indicating that weaker forces allowed lead to migrate out of the heated particle.

This work explained why the atomization of trace metals from soil particles is often delayed compared with aqueous solutions, namely that adsorptive forces between the analyte and the matrix particles must be overcome. However, other factors in soil could affect the absorbance peak characteristics, and these were studied systematically by modeling a typical soil matrix [9]. Montmorillonite, a well-characterized clay, is known to have traces of lead adsorbed on its particles. Platform atomization of lead from a slurry of this clay showed a delay in the absorbance peak similar to the effect when lead was adsorbed on alumina. The organic fraction of soil was then simulated by adding increasing amounts of humic acid to the montmorillonite slurry. An earlier shift in the absorbance peak was seen, and the magnitude of this shift was proportional to the humic acid concentration. This effect was confirmed in real soil samples by removing the organic carbon in a muffle furnace, when the absorbance signals obtained on the samples with carbon removed occurred later than the signals from the original soil, as shown in Figure 6.3. Hence, it was shown that the absorbance signal characteristics for environmental samples are highly dependent on the matrix when solid samples are analyzed, compared with solutions. Effects such as these have been reported by several researchers; for example, delayed signals occurred when lead was determined in slurries of iron oxide pigment [10], and coal and fly ash [11].

L'vov [12] defined the overall length of time during which atomization occurs τ_1, the mean length of time spent by an atom in the analysis volume, or residence time τ_2, and the total time for which a signal is recorded τ_3. It is readily apparent from Figures 6.2 and 6.3 that solutions and solids are highly likely to have different values of τ_1. However, τ_2 will be the same provided the temperature remains constant during the total atomization period (τ_3), and that diffusion is the only significant process leading to removal of atoms. This is clear from the equation [12]

$$\tau_2 = \frac{-l^2}{8D_0}\left(\frac{T_0}{T}\right)^{1.6}\frac{P}{P_0} \tag{6.1}$$

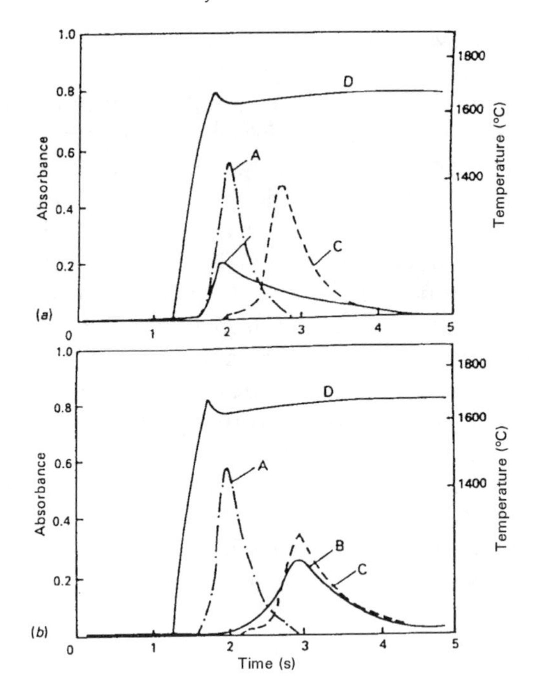

Figure 6.3 Absorbance signals for two soil slurries: (a) untreated soils; (b) soils with organic carbon removed by heating in a muffle furnace. Curve A = aqueous Pb solution (1 ng Pb); B = soil containing 4.7 per cent organic carbon; C = soil containing 0.3 per cent organic carbon; D = the graphite tube wall temperature. (Reprinted by permission of the Royal Society of Chemistry from M.W. Hinds and K.W. Jackson, Lead atomization from soil by slurry introduction electrothermal atomization atomic absorption spectrometry, *J. Anal. At. Spectrom.*, **2**, 441–445 (1987).)

where l is the tube length, D_0 is the gaseous diffusion coefficient, and T and P are temperature and pressure, respectively, with the zero subscript signifying STP values ($P = 1\,\mathrm{atm}$; $T = 273\mathrm{K}$). Both peak height and peak area have been used in analysis by ETAAS. It was shown by L'vov [12] that the number of atoms entering the analysis volume at the peak time N_{peak} is given by

$$N_{\mathrm{peak}} = N_0 \frac{\tau_2}{\tau_1}(1 - \mathrm{e}^{-\tau_1/\tau_2}) \tag{6.2}$$

where N_0 is the total number of atoms in the sample. However, the peak area Q_N is dependent only on τ:

$$Q_N = N_0 \tau_2 \tag{6.3}$$

Clearly, from Eqn (6.2), for peak height measurements to be valid, both τ_1 and τ_2 must be constant unless $\tau_1/\tau_2 \ll 1$, when

$$1 - e^{-\tau_1/\tau_2} \approx \frac{\tau_1}{\tau_2} \quad (6.4)$$

and therefore,

$$N_{\text{peak}} = N_0 \quad (6.5)$$

The peak profiles in Figures 6.2 and 6.3 show that atomization from solids is slow, so the condition $\tau_1/\tau_2 \ll 1$ will not occur. Hence, from Eqn (6.2), it is seen that peak height measurements will only be valid if calibration standards are used that closely match the sample, i.e. solid standards that give identical peak profiles, and hence the same τ_1 values. However, for peak area measurements to be valid, only τ_2 needs to be constant (Eqn (6.3)), and this condition will be realized if the temperature remains constant during the total atomization period τ_3. Hence, it should be possible to use aqueous calibration standards provided that isothermal atomization conditions are used.

6.3 PRACTICAL REQUIREMENTS

6.3.1 Absorbance characteristics

Most modern publications on applications of ETAAS describe operating conditions that attempt to approach isothermality. Commonly, this involves the use of a L'vov platform to delay analyte vaporization until the graphite tube has reached a high and almost steady-state temperature (STPF conditions), but other means include the use of contoured tubes (see Section 3.1.4), probe atomization, etc. Also, there is general agreement that absorbance peak areas should be measured rather than peak heights in ETAAS. Indeed, as every modern commercial AA spectrometer provides electronic peak integration, there seems little point in measuring peak heights. However, much of the earlier work on the analysis of solids used Massmann furnaces with wall atomization and peak height measurements. Under those conditions, from Section 6.2 (above), both τ_1 and τ_2 must be the same for all calibration standards and samples if systematic errors are to be avoided.

Figures 6.2 and 6.3 show clearly that aqueous solutions and solids are likely to have different values of τ_1 and, under nonisothermal operating conditions, the values of τ_2 will also be different. Therefore, the benefit of hindsight tells us it is hardly surprising that most methods using the Massmann furnace with wall atomization required carefully matched solid calibration standards to give comparable values of τ_1 and τ_2. A few early successful applications using wall atomization and aqueous calibration standards for the analysis of solid samples were reported, but this had to be a fortuitous circumstance, even if peak areas were measured, since τ_2 was still required to be constant.

Modern furnace technology (platform atomization) was first applied to the analysis of slurries in 1985 [3], when aqueous calibration standards were used successfully for the determination of lead and cadmium in soil. Since then, there have been many publications of successful applications of the technique with platform atomization. As discussed in Chapter 1, true isothermality is not achieved in a platform furnace. However, if the absorbance peak obtained from the solid sample is not shifted (temporally) too much compared to the aqueous peak, then τ_2 is nearly the same for both samples, and aqueous calibration standards can be used.

It was shown by Hinds and Jackson [14] that addition of certain mixtures containing palladium can cause absorbance signals from solids introduced as slurries to have similar characteristics to those obtained from aqueous standards. Palladium has become extremely popular as a modifier in recent years (see Chapter 5). In the conventional analysis of solutions by ETAAS, it is used to stabilize analytes to higher pyrolysis temperatures, thus enabling volatile matrix components to be removed during the pyrolysis stage of the ETA heating cycle, while the stabilized analyte remains in the graphite furnace.

A particular use of palladium in the case of slurry analysis, however, is its stabilizing effect during the atomization stage. This is seen in Figure 6.4, where the absorbance signal for lead in a sediment slurry is delayed compared to the aqueous standard. However, palladium has delayed both signals so that they have similar characteristic times.

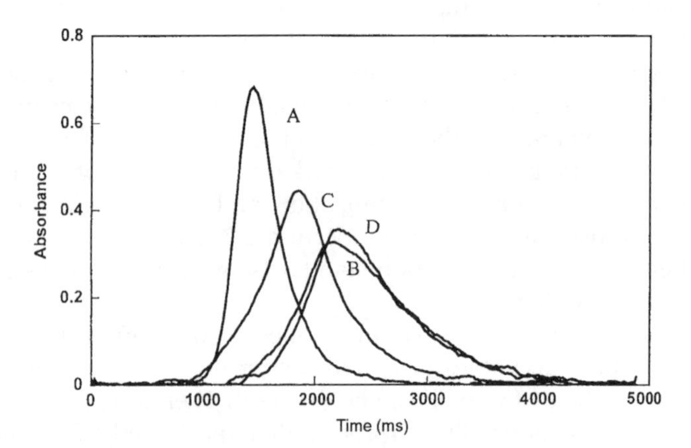

Figure 6.4 Absorbance signals for 1 ng of Pb: curve A = aqueous solution without modifier; B = aqueous solution with a Pd–Mg modifier; C = sediment slurry without modifier; D = sediment slurry with a Pd–Mg modifier.

An explanation for this important effect was suggested by Qiao and Jackson, based on a two-part study of the mechanism of palladium modification. In the first part [15], palladium was considered as a modifier in the conventional analysis of solutions, and a physical mechanism was proposed. During the pyrolysis stage of the ETA heating cycle, the palladium salt (typically the nitrate) is reduced to palladium metal. The analyte deposits on the palladium surface, and then becomes embedded in palladium after it melts, either during the pyrolysis stage if the temperature is high enough, or early in the atomization stage. The rate-limiting step in the kinetically controlled atomization process is then desorption of the analyte from the surface of solid palladium or its migration out of the molten palladium droplet (depending on the temperature). This migration process is probably similar to that described by Hinds *et al.* [3] and, as this process is slow, it accounts for the delay seen in the absorbance signal.

In the second paper [16], this theory was extended to a consideration of the delaying effect of palladium when used with slurries. Electron microscopic studies showed that sediment slurry particles acquire a coating of palladium prior to the atomization stage. This leads to the situation depicted in Figure 6.5, in which the matrix particles have lead adsorbed on their surface, and palladium either covers the particle or is present adjacent to the particle on the graphite surface. During the atomization stage, as the temperature continues to increase, the palladium

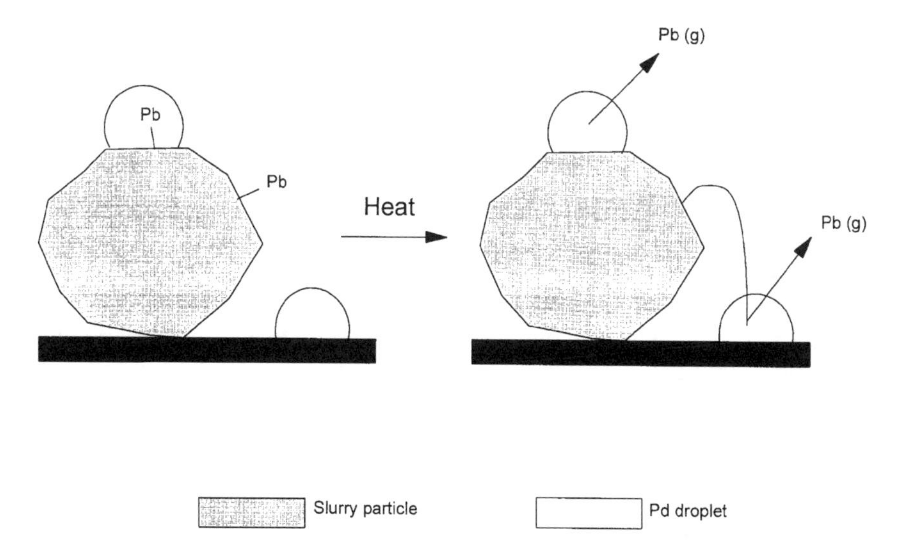

Figure 6.5 Two ways in which Pb may become embedded in a droplet of Pd modifier during furnace heating. In both cases, migration of Pb from the molten Pd droplet delays the absorbance signal.

melts and the analyte either migrates through the layer of palladium covering the matrix particle or, after undergoing multiple collisions with the graphite furnace walls, collides with a palladium droplet on the surface, and becomes embedded in it.

It was shown more recently [17] that volatile elements such as lead are released from slurry particles and then migrate through the gas phase to a palladium modifier, even if the modifier is physically separated from the analyte. Release of the analyte then involves migration from the palladium droplet, which is the same rate-limiting step that occurs when palladium is added to the analyte in solution. Therefore, the rate-limiting step leading to atomization is the same, whether the analyte is introduced in solution form or as a slurry. The resulting temporal characteristics of the aqueous standard and slurry sample peaks are similar (Figure 6.4), so even though atomization does not occur under isothermal conditions, it occurs under nearly identical temperature conditions. Therefore, the mean atom residence time (τ_2) will be the same for standards and samples. Other researchers [18–20] reported this delaying effect of palladium. Phosphate [21–23] has been shown to similarly delay absorbance signals from solid samples and aqueous standards. Friese and Krivan [23] obtained double absorbance peaks for several elements in silicon nitride powders, the first peak being attributed to analyte near the surface of the particle, and the second peak from analyte occluded in the particle. The use of a phosphate–magnesium modifier resulted in a single peak, and this was said to result from the sample reacting with the modifier during thermal pretreatment.

Obviously a better approach would be to use an ETA that is truly isothermal both temporally and spatially. This would then obviate the need for a modifier to control the appearance time, as τ_2 would be the same even if the absorbance signals for aqueous standards and solid samples were temporally separated. Early constant temperature furnaces, discussed in Chapter 1, fulfilled this requirement, but they suffered practical disadvantages, including being too slow to operate.

An isothermal ETA described by Frech and Jonsson [24], and described in Section 1.4, allowed separate control of the vaporization and atomization processes. It was a two-step furnace consisting of a graphite cup fitted beneath a graphite tube to form a T-shape. Samples were dispensed into the cup, vaporized, and the vapors passed into the tube for absorbance measurements. Two separate power supplies heated the cup and tube. Hence, sample vaporization (cup) and atomization (tube) were controlled independently. For normal operation, the tube was first heated to a predetermined optimum atomization temperature, and then the cup was heated to the appropriate temperature for sample vaporization. Thus, the sample vapors entered an environment that was temporally isothermal.

An improved version of this furnace [25] used side-heating of the tube to produce a more uniform temperature along its length, i.e. it did not suffer from the temperature gradient that exists in commercial graphite tube atomizers that are heated from the ends (see Section 3.1.4). Therefore, this furnace also approached spatial isothermality, and so τ_2 should be essentially independent of the sample, provided there is sufficient energy in the ETA for complete dissociation and atomization during the measurement period. It was compared with a commercial system for the direct analysis of solid samples [26], and the two-step furnace was shown to be superior by allowing interference-free analysis using aqueous calibration standards.

6.3.2 Spectral interferences

The dissolution process, required for solid samples that are to be analyzed by conventional ETAAS, most commonly involves chemical digestion of the sample. This converts much of the matrix to a volatile form that is removed during the digestion; for example, the organic carbon in biological samples is removed as oxides of carbon, and halogens may be removed as their volatile halides. Conversely, in the direct analysis of solids, the matrix usually remains intact until the sample is in the ETA. The pretreatment stages of the ETA heating cycle may remove some of the matrix, but it is probable that more matrix is present during the atomization step compared with samples introduced to the ETA as solutions. This is likely to place a greater demand on the background correction system, in order to avoid spectral interference from broad-band molecular absorption and light scattering from matrix particulates.

Shown in Figure 6.6 is an absorbance signal, at 232.0 nm, for nickel in a soil slurry. This was obtained with atomization from a platform, and the nickel atomic absorbance signal is severely overlapped by a strong background signal resulting from vaporization of the silicate matrix at a temperature in the range 2000–2500 °C. The deuterium-lamp background correction system used in this determination [27] was just able to compensate for this high background signal, but clearly there is the danger of spectral interference.

Dobrowolski [28] showed the difficulty of determining nickel at 232.0 nm, as this wavelength coincides with the highest background absorbance signal, as shown in Figure 6.7. Therefore, it was suggested that alternative higher wavelengths, where the background signal is smaller, should be used provided the nickel concentration is high enough to permit its determination at these less-sensitive wavelengths. In many applications of solid sampling ETAAS, the analyte and matrix background signals are temporally resolved. For example, elements such as lead, thallium, and cadmium, with appearance temperatures well below the

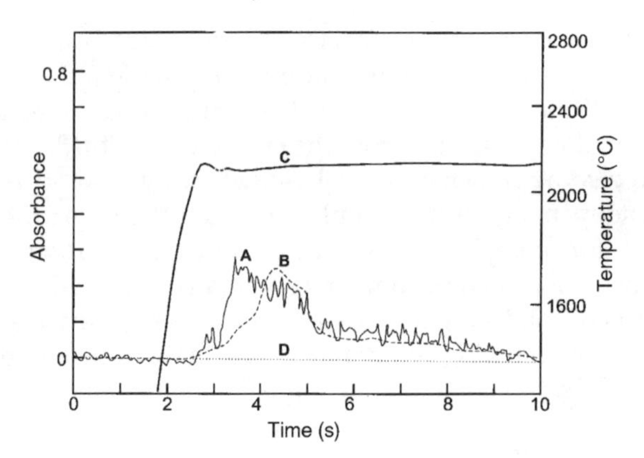

Figure 6.6 Absorbance signals: curve A = 0.8 ng of Ni in a soil slurry; B = background; C = tube wall temperature; D = baseline. (Reprinted by permission of the University of Saskatchewan, Canada, from M.W. Hinds, *Fundamental and applied studies of trace metal determination in soil by direct slurry electrothermal atomic absorption spectrometry*, Ph.D. thesis, 1988, Fig. 39.)

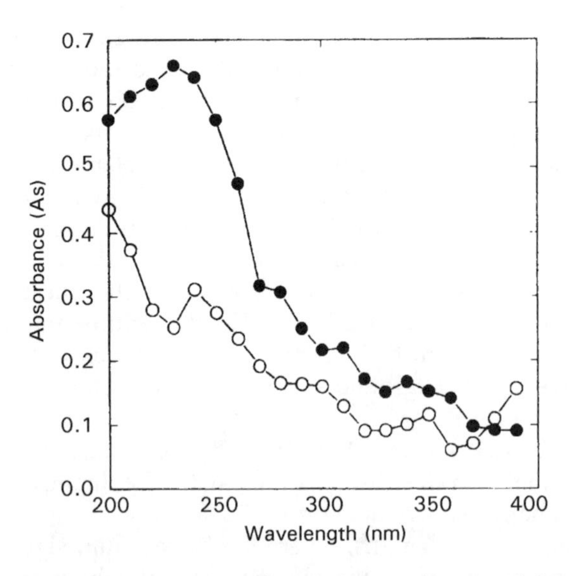

Figure 6.7 Background spectra for two soil slurries. (Reprinted from *Spectrochim. Acta*, Vol. **51B**, R. Dobrowolski, Determination of Ni and Cr in soils by slurry graphite furnace atomic absorption spectrometry, pp. 221–227, 1996, with kind permission of Elsevier Science-NL, Sara Burgerhartstraat 25, 1055 KV Amsterdam, The Netherlands.)

matrix vaporization temperature, are determined in soil with a very small background signal that is easily corrected with a deuterium lamp system.

Slavin *et al.* [29] stated that an important component of the STPF concept is the use of Zeeman-effect background correction, and this may be particularly so if solids are analyzed due to higher background absorbances. Tittarelli and Biffi [30] obtained molecular absorption spectra over the range 200–350 nm from slurry samples using a diode-array detector. Soil samples showed a series of band spectra between 200 and 250 nm (Figure 6.8), and these were attributed to SiO molecular absorption. Zeeman-effect background correction is much more effective than deuterium lamp background correction for structured backgrounds such as this.

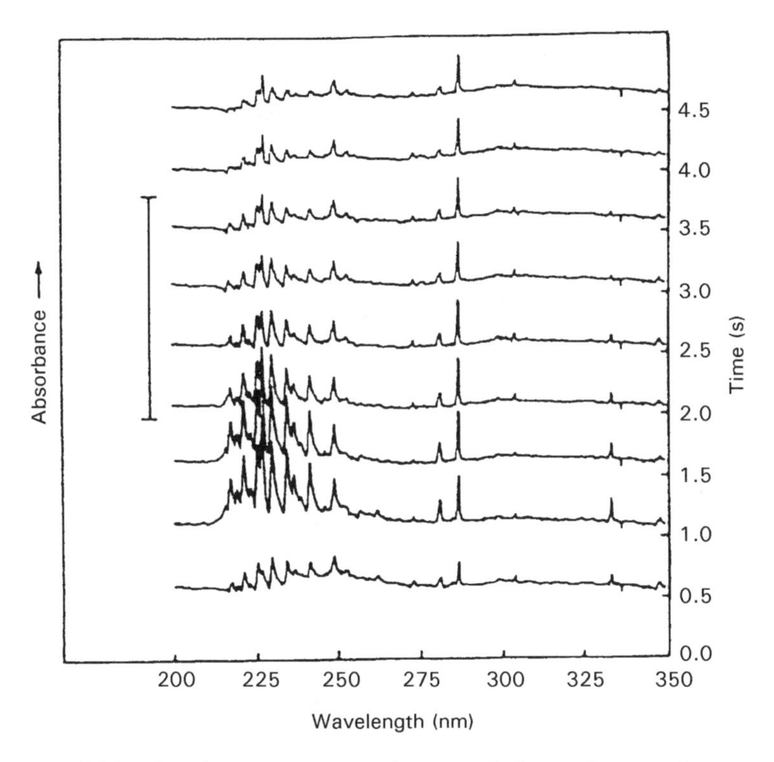

Figure 6.8 Molecular absorption spectra from a soil slurry, showing the presence of band spectra between 200 and 250 nm. (Reprinted by permission of the Royal Society of Chemistry from P. Tittarelli and C. Biffi, Vapor-phase behavior of slurries in electrothermal atomic absorption spectrometry, *J. Anal. At. Spectrom.*, **7**, 409–416 (1992).)

6.3.3 Dynamic range

The dynamic range is the analyte concentration range over which samples may be analyzed. Its lower limit is the practical quantitation limit (PQL), which can be derived by extrapolation from the method detection limit. In order to make quantitative determinations with acceptable precision, an appropriate value for the PQL is five times the detection limit. The upper limit is normally defined as the point at which the calibration curve deviates significantly from linearity. However, it is possible to work at higher concentrations by using a nonlinear graph, provided that care is taken to compute a curve that accurately fits the calibration points.

Shown in Table 6.1 are dynamic ranges for several elements determined by Zeeman-effect ETAAS. In conformance with convention in ETAAS, analyte amounts (picograms) are used rather than concentrations. The highest possible upper limit has been chosen, namely the point at which rollover of the calibration curve occurs. The dynamic range may also be expressed as the upper limit divided by the lower limit, and it can be seen from the table that the dynamic range thus defined extends from a value of 43 (for cadmium) to 254 (for copper). This is a limited dynamic range compared to emission techniques, and it is often cited as a disadvantage of AAS. During the conventional analysis of solutions, it is merely inconvenient when a sample has an analyte concentration above the upper limit of the dynamic range, i.e. the sample can be diluted and the analysis repeated. However, in the direct analysis of solids this flexibility is lost. This is an obvious limitation of the technique, which is best illustrated by comparing the complete analytical procedure when a solid material is analyzed by conventional means (such as dissolution or digestion prior to analysis) and by direct analysis of the solid.

Most solid samples, with the notable exceptions of glasses and some alloys that are solid solutions, are heterogeneous. Many metal alloys contain undissolved trace components that may be distributed fairly uniformly through the grains of the material, but which may accumulate more at the grain boundaries. Biological samples often exhibit a con-

Table 6.1 Typical dynamic ranges for Zeeman-effect ETAAS.

Element	*λ (nm)*	*Dynamic range (pg)*
As	193.7	100–5000
Cd	228.8	1.5–6.4
Cr	357.9	5–1200
Cu	324.7	5–1270
Pb	283.3	25–4090
Zn	213.9	0.1–23

centration gradient. Perhaps the most heterogeneity is seen in geological and environmental samples, such as rocks, soils, sediments etc. Trace metals in soil are frequently associated more with the clay particles than the other particles. It is usually possible to collect a large enough sample so that it is representative of the whole, in spite of its heterogeneity. Then a subsample is required for analysis, and the heterogeneity may be reduced to an acceptable level by thorough mixing and perhaps grinding to a small particle size.

Wilson [31] calculated subsampling errors for silicate rocks. An element may be present in several mineral species, so its variance will depend on the variances and covariances of the distribution of these mineral species. Therefore, an expression for subsampling error was developed that was derived from a multinomial distribution of the mineral particles. Not surprisingly, the variance was inversely proportional to the number of particles sampled (actually a weighted reciprocal mean of the number of particles was used for the real situation of a nonuniform particle size). A reasonable precision of 10 per cent relative standard deviation would be achieved if between 4×10^5 and 8×10^5 particles were sampled. For analysis involving acid digestion of a subsample, the case where the sample is ground to pass a 72-mesh sieve was considered. If the particles were of a uniform size, and they just passed through this sieve, there would be 0.77×10^5 particles of about 200 μm diameter in 1 g of sample. In reality, because a particle size range is present, the number of particles should be replaced by the weighted reciprocal mean, and this was estimated to be in the range 3×10^5 to 6×10^5. Wilson's calculations show that if only 10 mg of material is to be used for analysis, it should be ground to pass a 325-mesh sieve (maximum particle diameter of 44 μm). In reality, the amount of sample used for digestion is usually just a matter of convenience, because it is a simple matter to dilute the digest until the instrument response for the sample falls within the dynamic range of measurement. Therefore, it is generally quite feasible to digest a 1 g subsample, and control the heterogeneity errors. However, for extremely low analyte concentrations, there may be a limit on the maximum amount of subsample that can be dissolved, in order to obtain sufficient analytical sensitivity.

When solids are analyzed directly, this flexibility is lost. The subsample becomes the actual amount of material that is atomized, and it must fall within a narrow weight range to allow an instrument response that is within the dynamic range. In order to meet Wilson's requirement of about 10^5 particles, an aliquot size of at least 0.1 mg would be needed if the particles were 5 μm in diameter. In practical situations, it may not be feasible to grind a sample more finely, so this is probably the smallest aliquot that should be used. The maximum aliquot size is governed by a number of factors. If it is too large, it will block the light path, and this may occur progressively as the sample expands during the ETA heating cycle. Too much matrix material can also produce a nonspecific absorbance signal during atomization that exceeds the capabilities of the background-correction system of the instrument. In practice, aliquot sizes up to about 2 mg are used, although larger furnaces, such as the induction furnace of Headridge [32] were able to accommodate somewhat larger aliquots.

The practical limitations of such a small aliquot size range (0.1–2.0 mg) are seen in Table 6.2, in which three NIST certified reference materials are considered, on the assumption that they are typical of the types of sample that have a heterogeneous distribution of trace element components, and that are introduced to the ETA as a powder if solid sampling is used. From the certified concentration of each analyte element, the range of aliquot mass that can be analyzed to produce an absorbance that is within the dynamic range (from Table 6.1), has been calculated. An aliquot mass range of 0.1–2 mg would allow direct analysis in only nine of the 17 cases presented in the table. In the other eight cases, the sensitivity would be too high, so that a smaller aliquot mass would be required. This is where a solution could simply be diluted, but in the case of direct solid analysis, some means of reducing the sensitivity must be found. The most common way of doing this is probably to perform the analysis at a less sensitive wavelength. This involves the use of nonresonance lines. As they originate from less-populated electronic energy levels above the ground state, their sensitivity is lower.

A problem with the use of nonresonance lines is the strong dependence of the lower energy level population on temperature. As the Boltzmann equation shows, this is an exponential dependence,

$$\frac{N_i}{N_j} = \frac{g_i}{g_j}\exp\left(\frac{-E_{ij}}{kT}\right) \tag{6.6}$$

where N_i and N_j are the populations of the upper and lower energy levels, respectively, g_i and g_j are the statistical weights, E_{ij} is the excitation energy, k is the Boltzmann constant, and T is the absolute temperature. Hence, any energy level i above the ground state will show an exponential dependence of its population on temperature. This means that the sensitivity at the nonresonance wavelength will vary from run-to-run as small changes in ETA temperature occur (the temperature may change slightly as the resistance changes due to aging of the graphite). A greater error will occur is the absorbance signal shifts in time between standards and solid samples, because temporal isothermality is not achieved in Massmann-type graphite furnaces and, as shown in Figures 6.2 and 6.3, there may be a large difference in peak characteristic times for samples and standards when aqueous standards are used with solid samples.

Table 6.2 Aliquot masses that will produce an absorbance signal within the dynamic range.

		NIST 1646 (estuarine sediment)		*NIST 1566a (oyster tissue)*		*NIST 1577b (bovine liver)*	
Element	λ *(nm)*	*Concentration* ($mg\,kg^{-1}$)	*Acceptable aliquot range (mg)*	*Concentration* ($mg\,kg^{-1}$)	*Acceptable aliquot range (mg)*	*Concentration* ($mg\,kg^{-1}$)	*Acceptable aliquot range (mg)*
As	193.7	11.6	0.009–0.43	14.0	0.007–0.36	0.047	2.13–106
Cd	228.8	0.36	0.004–0.18	4.15	0.0004–0.015	0.44	0.0034–0.14
Cr	357.9	76	0.00007–0.016	1.43	0.003–0.84		
Cu	324.7	18	0.0003–0.071	66.3	0.00008–0.019	158	0.00003–0.008
Pb	283.3	28.2	0.0009–0.14	0.37	0.068–11.1	0.135	0.19–30.3
Zn	213.9	138	7×10^{-7}–0.00017	830	1.2×10^{-7}–0.00003	123	8×10^{-7}–0.0002

Baxter and Frech [26] showed that a temperature difference of only 100K would change the sensitivity of the lead 261.4 nm nonresonance line sufficiently to provide a 22 per cent systematic error. These authors also discussed other means of reducing the sensitivity. A normal requisite in modern ETA analysis is to have no flow of protective gas during the atomization stage. This helps ensure that the predominant atom loss mechanism is gaseous diffusion, and it also helps to ensure maximum residence time and hence maximum sensitivity. It follows, therefore, that the use of protective gas flow during atomization will reduce sensitivity by allowing gaseous analyte to be carried from the ETA by the flowing gas, thus decreasing analyte residence time. Baxter and Frech [26] showed that the characteristic mass of 23 pg under stopped flow conditions increased to 76 pg when the argon protective gas was flowing at 50 ml min^{-1}, i.e. more than a threefold reduction in sensitivity was realized. However, the authors cautioned that the use of gas flow conditions also causes reduced vapor temperatures, with increased chance of interference. They also suggested the use of curved calibration graphs to extend the dynamic range (as was done for the values listed in Table 6.1), because deviations from linearity occur at higher amounts. Provided the AA spectrometer has the required software to fit a curve to the data, this is a feasible approach. However, there are practical limitations that must be considered carefully.

Shown in Figure 6.9 is a calibration curve for aqueous lead standards at the resonance line of 283.3 nm. As the graph becomes curved at higher analyte amounts, the sensitivity (slope of the curve) decreases. If a solid or slurry is to be analyzed using this calibration curve, a consequence of making analytical measurements in this curved region of lower sensitivity is a degradation in instrumental precision. A random uncertainty in the

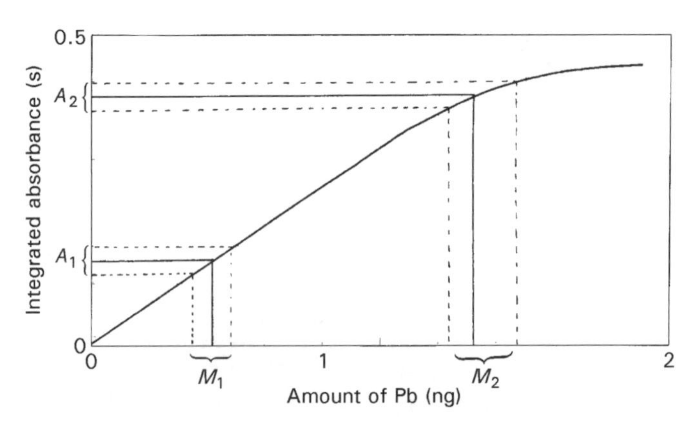

Figure 6.9 Calibration curve for lead at 283.3 nm: A_1 and A_2 show the uncertainty in absorbance measurements; with M_1 and M_2 the resulting uncertainty in the calculated amounts of lead.

absorbance measurement in the linear region of the graph A_1 translates into an uncertainty in the calculated analyte amount M_1. However, the same uncertainty in the absorbance where the graph has become curved A_2 translates into a larger uncertainty in the analyte amount M_2. In the analysis of solids and slurries, instrumental imprecision (noise) is usually insignificant compared to the imprecision arising from subsampling uncertainties, but this may not be so in the upper curved portion of the calibration range. In using this approach, therefore, care must be taken to ensure that significant degradation in the overall precision of the analysis has not occurred. Another approach for extending the calibration range involved operating a graphite furnace under reduced pressure, when sensitivity was also reduced [33].

6.3.4 Sampling errors

As discussed above, the solid sample aliquot must be within a specified narrow mass range if excessive sampling errors are to be avoided, and of course this depends on the size distribution of the sample particles. The influence of particle size on sampling errors must also be considered. In a situation where the analyte is embedded in the particle (rather than being adsorbed on its surface), efficient atomization may not occur until the particle has melted or decomposed. The analyte may migrate slowly from larger particles, and results could be systematically low if quantitative removal of analyte from the particles has not occurred during the atomization stage of the furnace-heating cycle. Fuller *et al.* [34], using slurry sample introduction into a Massmann-type furnace with atomization from the wall, reported low recoveries of trace metals from titanium dioxide powders with large particle sizes. The extent of such errors will depend on the heating characteristics of the atomizer – sufficient energy must be supplied to release the analyte completely, and the temperature must remain constant during this time period so that the mean residence time τ_2 also remains constant for both solid samples and aqueous standards. Modern furnace technology may allow such conditions, and then these systematic errors will not occur (see Section 6.4 below).

A more likely source of systematic error, producing low results, is the existence of rare particles of high analyte content in a powdered sample. This situation was described by Kurfuerst [35], who showed that these nuggets cause skewed distributions of analytical results, as shown in Figure 6.10, and which can be described by the Poisson probability function. As most aliquots will not contain these nuggets, the mean result will be systematically low, unless a very large number of replicate aliquots is taken. These nuggets were even shown to be present in biological reference materials [36], so care should be exercised in their use as calibration standards for the direct-weighing method.

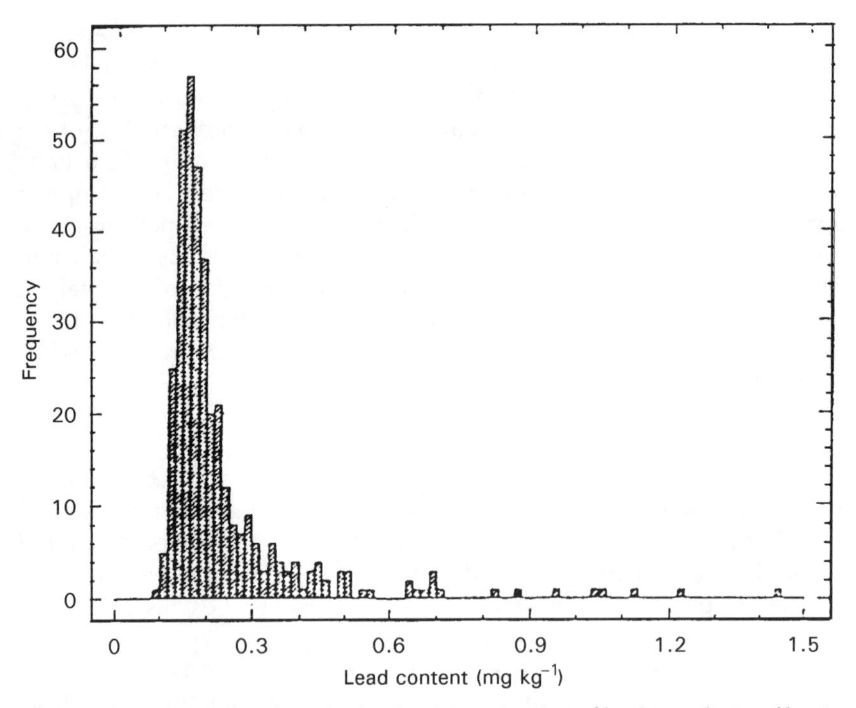

Figure 6.10 Histogram of results for the determination of lead in a slurry of bovine muscle. (Reprinted by permission of the International Union of Pure and Applied Chemistry, from U. Kurfuerst, Statistical treatment of ETA-AAS solid sampling data of heterogeneous samples, *Pure Appl. Chem.*, **63**, 1205–1211 (1991).)

Another situation that may not be so readily correctable was described by Nikolaev [37]. In metal alloys, some analytes are soluble in the matrix. It is then impossible to get complete analyte extraction in the finite distillation time that occurs during the ETA atomization stage. He indicated that a degree of extraction of only 0.9 might be attainable. Miller-Ihli showed that particle size is less important in the analysis of biological and botanical samples, because the analytes are often weakly retained and partially leached from the matrix by a nitric acid diluent [38,39].

Large particle size is more likely to influence the precision due to poor sample homogeneity. Fuller *et al.* [34] also showed that the particle size range of titanium dioxide powders must be below 25 μm in diameter in order to obtain acceptable precision. Several authors have demonstrated the improvement in precision obtained as the particle size is reduced. Nakamura *et al.* [40] showed that, for rock samples, the analytical variance decreased as the grinding time was increased up to 20 min, when the particle size range was 0.3–25 μm; further grinding led to no further improvement in precision (Figure 6.11). The maximum particle size per-

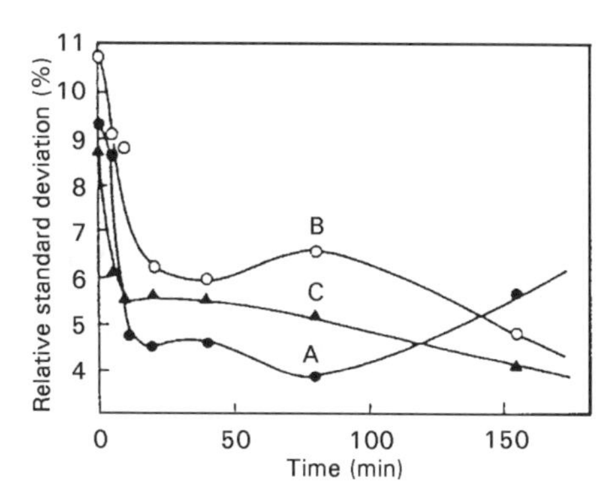

Figure 6.11 Effect of grinding time on the precision of the determination of trace metals in a slurry of silicate rock: curve A = Cu; B = Rb; C = Pb. (Reprinted by permission of the Royal Society of Chemistry from T. Nakamura, H. Oka, H. Morikawa and J. Sato, Determination of lithium, beryllium, cobalt, nickel, copper, rubidium, cesium, lead and bismuth in silicate rocks by direct atomization atomic absorption spectrometry, *Analyst*, **117**, 131–135 (1992).)

missible for adequate precision is highly dependent on the sample, as shown by Hinds *et al.* [13]. They separated an unground soil into several particle size fractions, and determined the lead concentration in each fraction. Their results showed that most of the lead was associated with particles less than 20 μm in diameter. This is because lead (and other metals introduced into the environment as pollutants) is predominantly adsorbed on the small clay particles, rather than the coarser silica particles. Hence, it is possible to have a soil sample with a relatively large mean particle diameter, and still obtain adequate sampling precision.

Holcombe and Majidi published two papers that considered sampling errors that are peculiar to slurry samples. In the first paper [41], the overall sampling variance was shown to include errors associated with uncertainties in the aliquot volume, number of sample particles in the aliquot, and the variation in the mass of the individual particles. By substituting typical values in their equation for the overall precision, it should be possible to design a protocol including particle size considerations and sampling volume to obtain acceptable precision. The authors cautioned, however, that their treatment did not include considerations of nonuniform analyte distribution throughout the particles, or partial dissolution of the analyte in the solvent. The second paper [42] considered sedimentation errors that could occur with slurry sampling (if the sample does not remain agitated during the sampling process). Such errors can be avoided either by using ultrasonic or mechanical mixing immediately before or even during the removal of aliquots for analysis.

The remainder of this chapter is devoted to applications of the direct analysis of solid samples, either by the direct-weighing approach or by slurry sampling. It is useful to bear in mind the above ideal requirements for successful quantitative analysis against aqueous calibration standards. First, unless the ETA is both spatially and temporally isothermal, the absorbance peak characteristics from the solid sample and aqueous standard should be similar. Second, conditions must be arranged so that amounts of analyte that are sufficiently representative of the sample can be analyzed directly. Finally, the sample should be in a suitable form (particle size and homogeneity) to reduce sampling errors to an acceptable level.

6.4 APPLICATIONS OF THE DIRECT-WEIGHING METHOD

Compared with solutions, which are easily pipetted by autosampler into a graphite furnace, solid sample introduction is difficult because a convenient means of introducing a preweighed sample into the furnace must be devised. Interestingly, L'vov's original furnace allowed solid samples to be introduced relatively easily. During his pioneering work in ETAAS, L'vov quickly realized the advantages of the direct analysis of solid samples [43]. He weighed powdered samples directly onto an electrode, which was then inserted into his graphite furnace (see Section 1.4). The Woodriff furnace (described in Section 1.4) was used in a similar way for the analysis of solids, i.e. the solid sample was placed on a graphite cup attached to a rod, and this was inserted into the preheated furnace through a side-arm [44]. Headridge [45] designed a furnace of larger diameter than L'vov's (15 mm compared with 2.5–5 mm), which was heated in an induction coil, and also introduced solid samples on a graphite probe through a side-arm. A later design [46] used a larger furnace, which was preheated prior to the solid sample being dropped in from above. The larger dimensions of the Headridge furnace prevented the solid sample from partially blocking the light path from the lamp, and this furnace was used extensively for the direct analysis of metal alloys [32]. An induction furnace was also used by Langmyhr and co-workers [47] for the analysis of metal samples.

The earlier furnaces described above enabled solid samples to be analyzed quite easily. As discussed earlier, the rate of analyte atomization from solid metal chips may be slower than from samples introduced as solutions, often introducing systematic errors if aqueous calibration standards are used. However, the isothermal operating characteristics of L'vov's furnace permitted the determination of aluminum and zinc in metal chips using aqueous standards [43]. Headridge's induction furnace, however, was typically calibrated with solid standards when trace elements, such as arsenic, antimony, selenium, and tellurium were determined in samples such as nickel-base alloys [48]. Several metals were determined in orchard leaves in the Woodriff furnace using aqueous standards [44]. Compared to solutions, the only difficulty in the use of these furnaces for the analysis of solids involved the tedious (and perhaps error-prone) weighing of each sample aliquot. It is interesting to speculate that solid sample analysis would have gained widespread acceptance many years ago if L'vov's furnace had remained the popular design and had been produced commercially. However, as discussed in Section 1.4, the Massmann furnace quickly became the standard, due to its more compact design and easier operation for the analysis of solutions. Unfortunately, early versions of the Massmann furnace were less suited to the analysis of solids. Besides the inconvenience of introducing the solid sample, vaporization from the tube wall was likely to occur at a different rate for solids and solutions due to the temperature varying with time as the tube was heated. Nevertheless, Massmann analyzed solids by inserting them through the open end of the tube [49]. Others have used this method, but sample introduction is difficult, and background absorption is a problem.

In recent years, using modern furnaces, there have been many publications describing the direct analysis of solids (Figure 6.1), and only a few representative ones are cited here. The modern version of the Massmann furnace generally uses a graphite platform (L'vov platform) inserted in the tube (see Section 1.5), in order to delay analyte vaporization until the tube has reached a higher and more stable temperature. This design is still inconvenient for the analysis of solid samples, but several researchers have used it successfully. The weighed sample is either introduced through the end of the tube on a scoop, weighed onto the platform before inserting it into the furnace [50], or the powdered sample is introduced through the sample-injection hole via a syringe-type injector [51]. In many cases, aqueous calibration standards have been satisfactory, particularly for the analysis of powdered biological and botanical reference materials [51,52]. It was also demonstrated that phosphate [53] and palladium [54] could be beneficial as modifiers in the analysis of solids. (The usefulness of modifiers in solid sampling ETAAS is discussed above in Section 6.3.1.)

An automated dedicated instrument of this kind was developed by Kurfuerst *et al.* [55], and manufactured by Grün Analysengeräte (Figure 6.12(d)). A sample was weighed onto a boat-type platform using an integrated microbalance, and then the platform was inserted automatically into the furnace. For the determination of silver in copper, this instrument produced a delayed absorbance signal compared with the signal from an aqueous solution, so matrix-matched solid standards were required [56].

A method known as extrapolation to zero matrix [57], which does not require the use of standards, was applied to the determination of

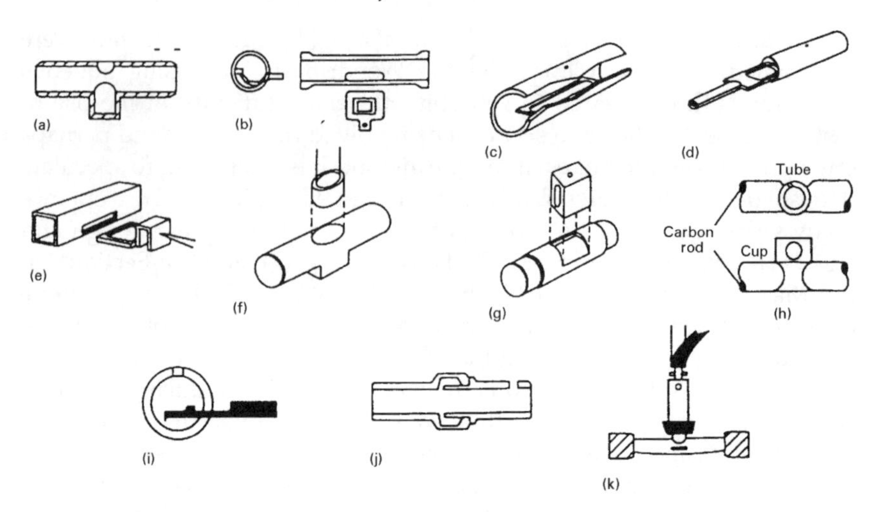

Figure 6.12 Graphite furnace atomizers that have been used for the analysis of solids by direct-weighing methods: (a), Hitachi cup cuvette; (b), platform-tube; (c), L'vov platform, (d), Grün Analysengerate boat; (e) Thermo Jarrell Ash microboat; (f), miniature cup; (g), Perkin-Elmer cup-in-tube; (h), Varian cup; (i) probe; (j), ring chamber; (k), second surface atomizer. (Reprinted by permission of the Royal Society of Chemistry from C. Bendicho and M.T.C. De Loos-Vollebregt, Solid sampling in electrothermal atomic absorption spectrometry using commercial atomizers, *J. Anal. At. Spectrom.*, **6**, 353–374 (1991).)

cadmium, lead, and zinc in various environmental samples using this furnace. A modified Grün furnace was used [58] for the determination of several elements in high-purity tantalum powders, with calibration by aqueous standards added to the residue on the furnace platform after a firing.

Over the years, several innovative designs, based on the Massmann furnace, have been introduced for the direct analysis of preweighed solids, and Bendicho and De Loos-Vollebregt reviewed these in 1991 [59]. An illustration from their review (Figure 6.12) shows some of these designs. The introduction of sample *via* a probe or platform inserted through a slot in the side of the tube is one approach that is more convenient for solids. The microboat system of Thermo Jarrell Ash (originally produced under the name Instrumentation Laboratory) was an early system (Figure 6.12(e)). It was used, with calibration by aqueous standards, for the determination of lead in soil [60] and copper in biological reference materials [61]. A better design is the graphite probe (Figure 6.12(i)) that was originally introduced by L'vov, and developed by Ottaway and co-workers [62]. It does not make close contact with the graphite tube, unlike the Thermo Jarrell Ash microboat, and hence behaves more like a L'vov platform in being heated radiationally, and so delaying analyte vaporization. Indeed, the graphite furnace can be preheated prior to insertion of the probe. A disadvantage compared to the L'vov platform design is the large slot that is needed for introduction of the probe, as increased diffusional loss may occur compared with the loss through the smaller sample introduction hole in the conventional Massmann-type furnace.

Hitachi manufactured a graphite-cup cuvette atomizer (Figure 6.12(a)), that was essentially a Massmann furnace with a cup built into its underside. This allowed solid samples to be atomized without blocking of the light beam by the sample. Numerous applications of this atomizer have appeared in the literature, and some of these have described the use of aqueous calibration standards, such as for the determination of trace metals in bovine liver [63]. However, solid standards were considered necessary for some applications, such as the determination of metals in biological and botanical reference materials [64], steels [4], and powdered rock samples [65], which were diluted with graphite to avoid the double peaks that were otherwise obtained. A phosphate–magnesium modifier was used in this application. Graphite powder was also successful in eliminating matrix suppressive effects when determining copper in rocks, and allowed the use of aqueous standards [66].

Similar in concept is the cup-in-tube atomizer of Perkin-Elmer, shown in Figure 6.12(g), and first described by Voellkopf *et al.* [67]. This was shown to work in a similar way to the L'vov platform by heating up more slowly than the tube wall and hence delaying analyte atomization (in fact the delaying effect was greater than with a platform). Interestingly, there have been more reports of the use of aqueous standards with this design compared with the Hitachi furnace; examples are the determination of trace metals in hay and soil [67], plastic [68,69], and biological samples [69–71]. Several modifiers were used with solid samples, including nickel for the determination of arsenic in soil [67], phosphate for the determination of cadmium in plastic [68], magnesium for the determination of aluminum in biological samples [70], and a copper–palladium mixture for the determination of selenium in biological samples [71]. Solid samples have also been analyzed in this device by using laser-excited atomic fluorescence spectrometry with front-surface illumination [72].

An earlier cup-type atomizer was the Varian design that used a cup clamped between two graphite electrodes (Figure 6.12(h)). Lundberg and Frech [2] showed that this device offered convenience by allowing solid samples to be dropped into preheated cups, similar to the approach used by Headridge with his induction furnace. This allowed atomization to occur under more nearly isothermal conditions, so that the mean residence time of the analyte atoms should be less dependent on the matrix. However, solid standards were still used for the analysis of metallurgical samples [73]. Aqueous standards were used with this device, however, for the determination of several metals in high-purity molybdenum and molybdenum silicide powder [74]. Dolinsek *et al.* [75] used an atomizer of this type for the determination of lead and cadmium in fly ash, sediment, and biological samples using aqueous calibration standards.

A specially designed graphite tube, shown in Figure 6.12(j), was developed by Schmidt and Falk [76]. This ring chamber tube had a sample chamber for the solid material that was separated from the absorption volume. Hence, sample material and residues did not obstruct the light path. It was said to have the advantage over L'vov platform systems that a low ramp rate was possible, allowing the atomic cloud to enter the absorption volume under isothermal conditions. This design of atomizer was also used by Falk *et al.* [77] for the direct analysis of solid samples by furnace atomization nonthermal excitation spectrometry (FANES; see Chapter 7).

A two-step atomizer, developed by Frech and co-workers, consisted of a graphite tube furnace heated transversely to provide an almost uniform temperature along the tube length, and a cup with a separate power supply, mounted under the tube (see Section 1.4). A later version of this furnace was applied to the analysis of solid samples [26], and provided several advantages over the cup-in-tube instruments described above. The sample was weighed into the cup, the tube was preheated to the desired atomization temperature, and then the cup was heated to vaporize the sample. Advantages of this atomizer included higher vapor-phase temperatures, and hence fewer interferences. It was applied to the determination of several elements in biological materials and steels by atomic emission spectrometry and, in most cases, aqueous standards could be used [78].

A semiautomated instrument incorporating a transversely heated integrated contact furnace (described in Section 1.5.1) was used recently for the direct analysis of solids [79]. Rettberg and Holcombe [80] inserted a cooled tantalum plug into the side of a Massmann-type graphite furnace. The analyte was then distilled from the solid material onto the plug. Besides producing a uniform layer of analyte on the surface, it allowed the analyte to be removed from the potentially interfering solid matrix prior to atomization. They used aqueous calibration standards for the determination of trace metals in botanical samples, fly ash, and sediment. Gornuskin *et al.* [81] developed a diffusive atomizer for separating the analyte from the solid matrix. The sample was placed in a graphite boat in a graphite tube, an electrode heated the boat, and analyte vapors diffused through the walls of the tube for analysis by laser-excited AFS. They determined silver in soil with aqueous calibration standards.

In the direct-weighing methods described above, some applications have involved the use of aqueous standards. However, many workers have concluded that the best way to be sure of avoiding systematic errors is to calibrate with solid reference materials having a similar matrix to the samples. Unfortunately, this approach may not always be feasible due to the limited availability of reference materials, and then it is usually only possible to have a single standard concentration, compared with the conventional use of at least 3–5 standard concentrations when aqueous calibration curves are used. Additionally, calibration errors may occur through the inhomogeneity of reference materials. As they are natural materials, it is inevitable that they will contain some particles with a higher analyte content than others and, in order to render this inhomogeneity insignificant, they are designed to be used in larger amounts than the submilligram quantities typically used in ETAAS.

These disadvantages of the use of solid standards have led to other calibration methods being devised. Berglund and Baxter [82] used the method of standard additions of solutions to the solid sample, and avoided the difficulty of weighing an identical mass of solid in each case by developing a method that allowed variable sample masses to be used. The method of standard additions may produce a double absorbance peak, because analyte may atomize from the standard and the solid sample at a different rate. This problem was overcome in the case of biological samples by adding the standard to the solid sample and then lyophilizing so that it behaved as a normal solid sample [83]. A method known as extrapolation to zero matrix [57] does not require the use of standards. It involves plotting the ratio of absorbance peak height to sample mass against sample mass. It is perhaps surprising that such elaborate calibration methods have been found necessary in direct-weighing ETAAS, as aqueous standards are widely used with great success in slurry ETAAS, and the same matrix is present irrespective of the way the sample is introduced to the furnace. This is discussed in more detail in the conclusion of this chapter where direct-weighing and slurry techniques are compared.

6.5 APPLICATIONS OF THE SLURRY METHOD

Slurry analysis involves pipetting a slurry or suspension of the powdered sample into the ETA, and this may be hardly more complicated than pipetting a liquid sample in the conventional analysis of solutions. Therefore, in most cases, conventional ETAs have been used without the need for the modifications in atomizer design described in the preceding section, and with only minor modifications in the sample introduction procedure. The first published application of slurry ETAAS was by Brady *et al.* in 1974 [84]. They determined zinc in ocean sediments by preparing a slurry in water and pipetting aliquots into a conventional Massmann furnace. The only instrument modification compared to solution analysis involved the use of a vortex mixer to ensure that a representative aliquot was transferred to the ETA. Aqueous calibration standards could introduce errors, as described above, in a nonisothermal furnace. However, they used the method of standard additions with aqueous standards and obtained accurate results. Precision was degraded only slightly compared to the analysis of solutions. Brady and Montalvo in a later paper [85] described an even simpler procedure, when aqueous calibration standards, without the need for standard additions, were used for the determination of lead in plant leaves. Following the two papers of Brady and co-workers in 1974, slurry ETAAS was slow to gain acceptance, as shown in Figure 6.1. The more established direct weighing approach remained slightly more popular for many years. However, as with the direct-weighing approach, slurry ETAAS publications increased dramatically around 1990. In recent years, slurry ETAAS has become much more popular than direct-weighing ETAAS. As with the discussion on direct-weighing ETAAS above, only a few representative papers on slurry ETAAS will be cited here.

A particular difficulty in handling slurries, compared with solutions, is in maintaining a uniform suspension during the time that aliquots are removed for injection into the ETA. The extent of this problem depends on the physical properties of the solid particles (radius, density) and the liquid in which they are dispersed (viscosity, density), as discussed by Majidi and Holcombe [42]. If manual pipetting is used, the problem is easily solved either by stirring magnetically or vortex-mixing while aliquots are drawn into the micropipette. This is certainly the case where the density of the solid particles is much higher than the medium (as with soils, sediments, etc. in water), but when low-density powders such as plant tissue are slurried it may be necessary to add a small amount of detergent to aid the dispersion process and prevent the solid floating on the surface. Magnetic stirring can cause problems with samples such as soil and sediment that have high concentrations of iron, as they may become magnetized and adhere to the stirring bar, resulting in low results [86]. Vortex mixing is then preferable. In the routine analysis of real samples, manual pipetting is rarely a practical proposition and an automatic sampler is generally used. A means of stabilizing the slurry that is compatible with commercial autosamplers must then be found. For high-density solids, the use of an additive to increase the viscosity of the slurry and hence slow down the sedimentation process has been a common approach, but this in turn makes pipetting difficult. The most successful application of this approach was probably by Fuller and Thompson [87], who used a thixotropic thickening agent to produce a suspension that they described as a 'highly viscous gel', and which allowed titanium dioxide slurries to remain stable for several days. Pipetting was not a problem, because the thixotropic properties caused an immediate reduction in viscosity due to the shear forces occurring when the suspension was forced through the narrow sampling pipette.

Miller-Ihli [88] developed an automated ultrasonic mixing device, and this has become popular for slurry analysis. The ultrasonic probe is automatically inserted into the sampling cup containing the slurry immediately prior to removing a slurry aliquot for injection into the ETA. This is available as an accessory from Perkin-Elmer. A small magnetic stirring device has also been incorporated into a commercial autosampler [89]. However, it has been shown [90] that ultrasonic mixing can provide better precision and accuracy than other methods involving mechanical mixing. The simplest approach may be to bubble a gas, such as argon, through the slurry immediately prior to removal of the sample aliquot [1,91]. This was applied successfully by Bendicho and de Loos-Vollebregt to the analysis of glass slurries [1], but it was only successful when the particle size range was below 15 μm. Also, it may be less suitable for more dense materials such as soils and geological samples.

Earlier reports on the analysis of slurries generally involved wall atomization from a Massmann furnace, as originally used by Brady and co-workers [84,85]. Some examples are the determination of several elements in alumina using spiked alumina standards [92], lead in spinach using aqueous standards [93], cobalt, chromium, manganese, and lead in biological and botanical reference materials using aqueous standards [94], and more recently chromium and nickel in soil, where aqueous standards were satisfactory for chromium, but calibration by standard additions was required for nickel [28]. The Thermo Jarrell Ash microboat system (described above) was used [60] for the determination of lead in soil using aqueous calibration standards. Most modern applications of slurry ETAAS have been carried out under stabilized temperature platform furnace (STPF) methodology (see Section 1.5), and with few exceptions aqueous calibration standards have been satisfactory. This methodology was first applied to slurries in 1985 by Hinds *et al.* [13], who determined cadmium and lead in soil.

Since that time, many applications have appeared in the literature, demonstrating the high accuracy of slurry ETAAS for the analysis of a wide range of sample types. Reported precision is generally in the range of 2–10 per cent relative standard deviation (RSD), which is a little poorer than the precision attainable with the analysis of solutions, but still quite acceptable for most applications. Included were the determination of lead in alumina [6], and a wide range of elements in biological and botanical reference materials [38,95], and sediments [18]. In some cases, samples have been partially digested and the residue has been made into a slurry. For example, gold, palladium, and platinum were determined in silver after partial nitric acid digestion to produce a slurry [96], and antimony, nickel, and vanadium were determined in slurries of partially digested air filters [97]. More recently, the transversely heated furnace (discussed in Chapter 1) has been successfully applied to the determination of trace metals in boron nitride [98], titanium dioxide [99], silicon nitride [23], graphite powder [100], and high-purity quartz [101]. The slow sample throughput time of STPF-ETAAS has prompted the development of fast-furnace methods, in which some aspect of the conventional atomization cycle, such as sample pretreatment, has been eliminated. This methodology has been found successful for the slurry ETAAS analysis of fly ash and coal [102], biological materials [103], soils and sediments [104,105], and diatomaceous earth [106,107]. The usefulness of slurry ETAAS was also demonstrated in an interlaboratory comparative study, where generally good correlation between laboratories was seen [108,109], and a review on the analysis of foods [110].

During the preparation of slurries, analyte may be partitioned between the solid and liquid phases. Experience shows that this does not occur if soil and sediment samples are slurried with water, but the addition of acid or certain modifier solutions results in partial dissolution of analyte. This phenomenon was studied by Miller-Ihli [38,39], who examined the partition of analyte between the solid and aqueous phases when a 5 per cent nitric acid diluent was used for the preparation of slurries. The improved sample homogeneity resulting when a large percentage of analyte was extracted into the liquid phase led to improved precision compared with the case where the analyte remained in the solid phase.

The use of modifiers has been more common in slurry ETAAS compared with direct-weighing methods. Many successful applications have been published where modifiers were not used but, as discussed earlier in this chapter, modifiers may delay analyte atomization so that slurry and aqueous standard peaks have similar profiles, thus removing the risk of systematic error through sample and standard being analyzed under different furnace temperature conditions. This effect has been noted mostly with palladium as modifier [14,18–20,111]. It was also shown [112] that palladium was more effective when added as the nitrate compared with the chloride. Phosphate also shows this effect as a modifier [21–23]. Other modifiers used in slurry ETAAS include: nickel for the determination of arsenic [113], selenium [114] and gallium [11] in coal; magnesium for the determination of silicon in boron nitride [98]; and calcium for the determination of silicon in titanium dioxide [99]. In order to remove the large amounts of carbon resulting from the analysis of biological samples, oxygen ashing is commonly incorporated into the sample pretreatment stage of the graphite furnace heating cycle [39,92,115]. Vinas *et al.* [116] used hydrogen peroxide as a modifier to alleviate this problem. Hydrofluoric acid was recommended as a modifier to remove background interference by silicon in the analysis of soils and sediments [105].

6.6 CONCLUSIONS

Numerous publications have demonstrated that solid samples may be analyzed accurately by direct-weighing and slurry ETAAS methods. Compared with conventional analysis by ETAAS, requiring sample dissolution, these solid sampling methods have two important advantages. First, the analysis is simplified, especially when samples that are difficult to dissolve or digest are analyzed. Dissolution can take several hours, compared with the relatively short procedures of grinding the sample to a sufficiently small particle size and either weighing aliquots directly into the furnace or preparing a slurry and then pipetting aliquots into the furnace. Second, the reagents used in sample digestion procedures include corrosive mineral acids and powerful oxidizing agents that may be hazardous and may be difficult to obtain in sufficient purity to avoid high blanks. Conversely, pure water or aqueous solutions containing modifiers are the only requirement for the preparation of slurries, and blank concentrations are much more easily controlled.

A review of the literature shows that slurry ETAAS is now more popular than direct-weighing ETAAS. An advantage of the direct-weighing approach is the complete lack of reagents, but it has the complication of devising a convenient way of introducing the solid sample to the furnace, and this has usually involved modifications in furnace design. Also, direct weighing involves the tedious replicate weighing of very small sample aliquots, because it is desirable to analyze in triplicate in ETAAS. In Section 6.5, the shorter discussion of slurry ETAAS applications, compared with direct-weighing ETAAS applications in Section 6.4, is by no means reflective of the relative importance of the two techniques. Rather, it reflects the comparative simplicity of the slurry technique, which does not require changes in atomizer design, and requires only a single weighing of a larger sample amount. This amount may be similar to the amount that would have to be weighed if the sample was to be digested prior to conventional analysis as a solution. Replicate aliquots are then easily introduced by pipetting with an autosampler, as with the analysis of solutions.

Both solid sampling methods are victims of the inherently small dynamic range of AAS. At least in the analysis of solutions, it is possible to dilute a sample until the analyte concentration falls within the dynamic range of the technique. In direct weighing, the obvious recourse if samples have analyte concentrations higher than the upper limit of the dynamic range is to weigh smaller amounts into the furnace. However, this approach is inconvenient and severely limited. As the aliquot amount becomes smaller, it becomes more difficult to obtain a representative sample, because of the inhomogeneity of solid materials. Also, there is a limit to how small an amount can be weighed accurately into the furnace.

Slurries are a little easier in this respect, as they can be diluted the same as solutions, but only to a limited extent, as sample inhomogeneity will also introduce errors. The use of alternative, less sensitive analytical wavelengths has been a common way of increasing the dynamic range of the analysis in both direct-weighing and slurry techniques but, as described earlier in this chapter, the temperature dependence on the lower electronic energy level population can introduce errors. Samples should be aliquoted at least in triplicate when using solid sampling methods, in order to minimize the effects of sample inhomogeneity, and in case of an aliquot containing a solid particle that is particularly rich in the analyte (this is the nugget effect discussed earlier in the chapter). Analysis in this way by direct weighing or slurry ETAAS routinely provides precision that is somewhat poorer than the precision of the conventional analysis of solutions, but still perfectly acceptable for most applications.

If a solid sampling method is to be a viable alternative to sample digestion and subsequent analysis as a solution, it is important that calibration standards in solution can be used. Most applications of slurry ETAAS, at least in recent years, describe the use of calibration standards in aqueous solution. Although this is the case with many applications of direct-weighing ETAAS, there have also been many others where solid matrix-matched standards or standard additions have ben required. This is a surprising situation, because the same solid material is present in the furnace whether it is introduced by direct-weighing or as a slurry aliquot. A key to accurate analysis using aqueous standards is often the use of modifiers, and they have been used much more with slurries than with direct-weighing methods. Modifiers, especially those containing palladium, frequently provide accurate analysis by delaying analyte atomization of more volatile analytes until both solid samples and aqueous standards produce temporally similar absorbance profiles. This modifier is generally effective for the more volatile elements, which are also the most frequently determined. It has been shown [17] that palladium does not need to be in direct contact with the analyte when introduced into the furnace, as more volatile analytes migrate onto palladium during the pretreatment stages. Hence, it follows that a palladium solution should be an effective modifier when pipetted onto samples that have been introduced to the furnace by direct weighing, and therefore aqueous standards should be feasible in many cases. Modifiers are not always required for accurate analysis, but their more widespread use in solid sampling methods should make direct calibration with aqueous standards possible in most cases.

In considering the use of solid sampling methods, the nature of the sample is of paramount importance. If the sample can be ground to a sufficiently small particle size to reduce its homogeneity to an acceptable level, slurry ETAAS is the better approach. For all except the least-volatile analytes, it is highly recommended to use conditions having as many components of STPF technology as possible (see Section 1.5). An important group of samples that cannot easily be ground to a powder are metals, and direct-weighing ETAAS may then be the method to use. However, if aqueous standards cannot be used, or if the sample can be dissolved or digested easily, conventional dissolution and subsequent analysis as a solution may be preferable to either slurry ETAAS or direct-weighing ETAAS.

The future of slurry ETAAS may also include speciation methods, as the feasibility of analyte speciation from the absorbance characteristics of solid samples has been suggested. It was shown by Karwowska and Jackson [6] that a trace metal adsorbed on a particle of alumina atomized with different peak characteristics compared with the metal embedded in the alumina particle. More recently, Wang and Holcombe [117] obtained three distinct absorbance peaks for lead when a solid copper alloy was vaporized at reduced pressure. They considered that the first peak resulted from the release of lead from the surface of the copper, the second from grains, and the third from lead embedded in the bulk of the sample.

REFERENCES

[1] C. Bendicho and M.T.C. De Loos-Vollebregt, *Spectrochim. Acta, Part B* **45B**, 679 (1990).
[2] E. Lundberg and W. Frech, *Anal. Chim. Acta* **104**, 75 (1979).
[3] M.W. Hinds, G.N. Brown and D.L. Styris, *J. Anal. At. Spectrom.* **9**, 1411 (1994).
[4] K. Takada and K. Hirokawa, *Fresenius' Z. Anal. Chem.* **312**, 109 (1982).
[5] K. Takada and K. Hirokawa, *Talanta* **29**, 849 (1982).
[6] R. Karwowska and K.W. Jackson, *Spectrochim. Acta, Part B* **41B**, 947 (1986).
[7] R. Karwowska and K.W. Jackson, *J. Anal. At. Spectrom.* 2, **125** (1987).
[8] B. Smets, *Spectrochim. Acta, Part B* **35B**, 33 (1980).
[9] M.W. Hinds and K.W. Jackson, *J. Anal. At. Spectrom.* **2**, 441 (1987).
[10] I.L. Garcia and M.H. Cordoba, *J. Anal. At. Spectrom.* **4**, 701 (1989).
[11] X. Shan, W. Wang and B. Wen, *J. Anal. At. Spectrom.* **7**, 761 (1992).
[12] B.V. L'vov, *Atomic Absorption Spectrochemical Analysis*, Adam Hilger, London, 1970.
[13] M.W. Hinds, K.W. Jackson and A.P. Newman, *Analyst*, 110, **947** (1985).
[14] M.W. Hinds and K.W. Jackson, *J. Anal. At. Spectrom.* **3**, 997 (1988).
[15] H. Qiao and K.W. Jackson, *Spectrochim. Acta, Part B* **46B**, 1841 (1991).
[16] H. Qiao and K.W. Jackson, *Spectrochim. Acta, Part B* **47B**, 1267 (1992).
[17] G. Chen and K.W. Jackson, *Spectrochim. Acta, Part B* **51B**, 1505 (1996).
[18] M. Hoenig, P. Regnier and R. Wollast, *J. Anal. At. Spectrom.* **4**, 631 (1989).
[19] M. Hoenig, P. Regnier and L. Chou, *J. Anal. At. Spectrom.* **6**, 273 (1991).
[20] P. Bermejo-Barrera, C. Barciel-Alonso, M. Aboal-Somoza and A. Bermejo-Barrera, *J. Anal. At. Spectrom.* **9**, 469 (1994).
[21] M. Hoenig and P. Van Hoeyweghen, *Anal. Chem.* **58**, 2614 (1986).
[22] Z. Yu, C. Vandecasteele, B. Desmet and R. Dams, *Mikrochim. Acta* **I**, 41 (1990).
[23] K.-C. Friese and V. Krivan, *Anal. Chem.* **67**, 354 (1995).

[24] W. Frech and S. Jonsson, *Spectrochim. Acta, Part B* **37B**, 1021 (1982).
[25] W. Frech, A. Cedergren, B. Lundberg and D.D. Siemer, *Spectrochim. Acta, Part B* **38B**, 1435 (1983).
[26] D.C. Baxter and W. Frech, *Fresenius' J. Anal. Chem.* **337**, 253 (1990).
[27] M.W. Hinds, *Fundamental and applied studies of trace metal determination in soil by direct slurry electrothermal atomic absorption spectrometry*; Ph.D. Thesis, University of Saskatchewan, Canada, 1988.
[28] R. Dobrowolski, *Spectrochim. Acta, Part B* **51B**, 221 (1996).
[29] W. Slavin, D.C. Manning and G.R. Carnrick, *At. Spectrosc.* **2**, 137 (1981).
[30] P. Tittarelli and C. Biffi, *J. Anal. At. Spectrom.* **7**, 409 (1992).
[31] A.D. Wilson, *Analyst* **89**, 18 (1964).
[32] J.B. Headridge, *Spectrochim. Acta, Part B* **35B**, 785 (1980).
[33] J.A. Holcombe and P. Wang, *Fresenius' J. Anal. Chem.* **346**, 1047 (1993).
[34] C.W. Fuller, R.C. Hutton and B. Preston, *Analyst* 106, **913** (1981).
[35] U. Kurfuerst, *Pure Appl. Chem.* **63**, 1205 (1991).
[36] U. Kurfuerst, J. Pauwels, K.H. Grobecker, M. Stoeppler and H. Muntau, *Fresenius' J. Anal. Chem.* **345**, 112 (1993).
[37] G.I. Nikolaev, *Zh. Anal. Khim.* **28**, 454 (1973).
[38] N.J. Miller-Ihli, *Spectrochim. Acta, Part B* **44B**, 1221 (1989).
[39] N.J. Miller-Ihli, *J. Anal. At. Spectrom.* **9**, 1129 (1994).
[40] T. Nakamura, H. Oka, H. Morikawa and J. Sato, *Analyst* **117**, 131 (1992).
[41] J.A. Holcombe and V. Majidi, *J. Anal. At. Spectrom.* **4**, 423 (1989).
[42] V. Majidi and J.A. Holcombe, *Spectrochim. Acta, Part B* **45B**, 753 (1990).
[43] B.V. L'vov, *Spectrochim. Acta, Part B* **24B**, 53 (1969).
[44] J.A. Nichols, R.D. Jones and R. Woodriff, *Anal. Chem.* **50**, 2071 (1978).
[45] J.B. Headridge and D.R. Smith, *Talanta* **18**, 247 (1971).
[46] J.B. Headridge and D.R. Smith, *Talanta* **19**, 833 (1972).
[47] F.J. Langmyhr and Y. Thomassen, *Fresenius' Z. Anal. Chem.* **264**, 122 (1973).
[48] J.B. Headridge and R.A. Nicholson, *Analyst* **107**, 1200 (1982).
[49] H. Massmann, *Spectrochim. Acta, Part B* **23B**, 215 (1968).
[50] C.L. Chakrabarti, C.C. Wan and W.C. Li, *Spectrochim. Acta, Part B* **35B**, 547 (1980).
[51] Z. Grobenski, R. Lehmann, R. Tamm and B. Welz, *Mikrochim. Acta* **I**, 115 (1982).
[52] C.L. Chakrabarti, R. Karwowska, B.R. Hollebone and P.M. Johnson, *Spectrochim. Acta, Part B* **42B**, 1217 (1987).
[53] A.A. Brown, M. Lee, G. Kullemer and A. Rosopulo, *Fresenius' Z. Anal. Chem.* **328**, 354 (1987).
[54] A. Isozaki, Y. Morita, T. Okutani and T. Matsumura, *Anal. Sci.* **12**, 755 (1997).
[55] U. Kurfuerst, M. Kempeneer, M. Stoeppler and O. Schuierer, *Fresenius' J. Anal. Chem.* **337**, 248 (1990).
[56] J. Pauwels, L. De Angelis, F. Peetermans and C. Ingelbrecht, *Fresenius' J. Anal. Chem.* **337**, 290 (1990).
[57] J. Pauwels, C. Hofmann and C. Vandecasteele, *Fresenius' J. Anal. Chem.* **348**, 411 (1994).
[58] K.-C. Friese, V. Krivan and O. Schuirer, *Spectrochim. Acta, Part B* **51B**, 1223 (1996).
[59] C. Bendicho and M.T.C. De Loos-Vollebregt, *J. Anal. At. Spectrom.* **6**, 353 (1991).
[60] K.W. Jackson and A.P. Newman, *Analyst* **108**, 261 (1983).
[61] L. Ebdon and E.H. Evans, *J. Anal. At. Spectrom.* **2**, 317 (1987).

[62] J.M. Ottaway, J. Carroll, S. Cook, S.P. Corr, D. Littlejohn and J. Marshall, *Fresenius' Z. Anal. Chem.* **323**, 742 (1983).
[63] I. Atsuya and K. Itoh, *Spectrochim. Acta, Part B* **38B**, 1259 (1983).
[64] I. Atsuya, K. Itoh and K. Akatsuka, *Fresenius' Z. Anal. Chem.* **328**, 338 (1987).
[65] A.M. De Kersabiec and M.F. Benedetti, *Fresenius' Z. Anal. Chem.* **328**, 342 (1987).
[66] T. Nakamura, K. Okubo and J. Sato, *Anal. Chim. Acta* **209**, 287 (1988).
[67] U. Voellkopf, Z. Grobenski, R. Tamm and B. Welz, *Analyst* **110**, 573 (1985).
[68] U. Voellkopf, R. Lehmann and D. Weber, *J. Anal. At. Spectrom.* **2** 455 (1987).
[69] G.R. Carnrick, B.K. Lumas and W.B. Barnett, *J. Anal. At. Spectrom.* **1**, 443 (1986).
[70] K. Nordahl, B. Radziuk, Y. Thomassen and R. Weberg, *Fresenius' J. Anal. Chem.* **337**, 310 (1990).
[71] R. Oilunkaniemi, P. Peramaki and L.H.J. Lajunen, *At. Spectrosc.* **15**, 126 (1994).
[72] R.L. Irwin, D.J. Butcher, J. Takahashi, G.T. Wei and R.G. Michel, *J. Anal. At. Spectrom.* **5**, 603 (1990).
[73] W. Frech, E. Lundberg and M.M. Barbooti, *Anal. Chim. Acta* **131**, 45 (1981).
[74] B. Docekal and V. Krivan, *Spectrochim. Acta, Part B* **50B**, 517 (1995).
[75] F. Dolinsek, J. Stupar and V. Vrscaj, *J. Anal. At. Spectrom.* **6**, 653 (1991).
[76] K.P. Schmidt and H. Falk, *Spectrochim. Acta, Part B* **42B**, 431 (1987).
[77] H. Falk, E. Hoffmann, C. Ludke and K.P. Schmidt, *Spectrochim. Acta, Part B* **41B**, 853 (1986).
[78] D.C. Baxter and W. Frech, *Fresenius' Z. Anal. Chem.* **328**, 324 (1987).
[79] R. Nowka and H. Muller, *Fresenius' J. Anal. Chem.* **359**, 132 (1997).
[80] T.M. Rettberg and J.A. Holcombe, *Anal. Chem.* **58**, 1462 (1986).
[81] I.B. Gornushkin, B.W. Smith and J.D. Winefordner, *Spectrochim. Acta, Part B* **51B**, 1355 (1996).
[82] M. Berglund and D.C. Baxter, *Spectrochim. Acta, Part B* **47B**, E1567 (1992).
[83] C. Hofmann, C. Vandecasteele and J. Pauwels, *Fresenius' J. Anal. Chem.* **342**, 936 (1992).
[84] D.V. Brady, J.G. Montalvo, G. Glowacki and A. Pisciotta, *Anal. Chim. Acta* **70**, 448 (1974).
[85] D.V. Brady and J.G. Montalvo, *At. Absorpt. Newsl.* **13**, 118 (1974).
[86] M.W. Hinds and K.W. Jackson, *At. Spectrosc.* **12**, 109 (1991).
[87] C.W. Fuller and I. Thompson, *Analyst*, **102**, 141 (1977).
[88] N.J. Miller-Ihli, *J. Anal. At. Spectrom.* **4**, 295 (1989).
[89] B. Docekal, *J. Anal. At. Spectrom.* **8**, 763 (1993).
[90] L. Sandoval, J.C. Herraez, G. Steadman and K.I. Mahan, *Mikrochim. Acta* **108**, 19 (1992).
[91] I. Lopez-Garcia, M. Sanchez-Merlos and M. Hernandez-Cordoba, *J. Anal. At. Spectrom.* **12**, 777 (1997).
[92] Z. Slovak and B. Docekal, *Anal. Chim. Acta* **129**, 263 (1981).
[93] S.C. Stephen, D. Littlejohn and J.M. Ottaway, *Analyst* **110**, 1147 (1985).
[94] L. Ebdon, A.S. Fisher, H.G.M. Parry and A.A. Brown, *J. Anal. At. Spectrom.* **5**, 321 (1990).
[95] N.J. Miller-Ihli, *J. Anal. At. Spectrom.* **3**, 73 (1988).
[96] M.W. Hinds, *Spectrochim. Acta, Part B* **48B**, 435 (1993).
[97] M.C. Carneiro, R.C. Campos and A.J. Curtius, *Talanta* **40**, 1815 (1993).
[98] S. Hauptkorn and V. Krivan, *Spectrochim. Acta, Part B* **49B**, 221 (1994).
[99] S. Hauptkorn, G. Schneider and V. Krivan, *J. Anal. At. Spectrom.* **9**, 463 (1994).
[100] U. Schaffer and V. Krivan, *Spectrochim. Acta, Part B* **51B**, 1211 (1996).

[101] S. Hauptkorn and V. Krivan, *Spectrochim. Acta, Part B* **51B**, 1197 (1996).
[102] D. Bradshaw and W. Slavin, *Spectrochim. Acta, Part B* **44B**, 1245 (1989).
[103] P. Jordan, J.M. Ives, G.R. Carnrick and W. Slavin, *At. Spectrosc.* **10**, 165 (1989).
[104] M.W. Hinds, K.E. Latimer and K.W. Jackson, *J. Anal. At. Spectrom.* **6**, 473 (1991).
[105] I. Lopez-Garcia, M. Sanchez-Merlos and M. Hernandez-Cordoba, *Spectrochim. Acta, Part B* **52B**, 437 (1997).
[106] I. Lopez-Garcia, J. Arroyo-Cortez and M. Hernandez-Cordoba, *J. Anal. At. Spectrom.* **8**, 103 (1993).
[107] I. Lopez-Garcia, J. Arroyo-Cortez and M. Hernandez-Cordoba, *Anal. Chim. Acta* **283**, 167 (1993).
[108] N.J. Miller-Ihli, *Spectrochim. Acta, Part B* **50B**, 477 (1995).
[109] N.J. Miller-Ihli, *J. Anal. At. Spectrom.* **12**, 205 (1997).
[110] M.A.Z. Arruda, M. Gallego and M. Valcarcel, *Quim. Anal. (Barcelona)* **14**, 17 (1995).
[111] M.W. Hinds, M. Katyal and K.W. Jackson, *J. Anal. At. Spectrom.* **3**, 83 (1988).
[112] M.W. Hinds and K.W. Jackson, *J. Anal. At. Spectrom.* **5**, 199 (1990).
[113] L. Ebdon and H.G.M. Parry, *J. Anal. At. Spectrom.* **2**, 131 (1987).
[114] L. Ebdon and H.G.M. Parry, *J. Anal. At. Spectrom.* **3**, 131 (1988).
[115] K.O. Olayinka, S.J. Haswell and R. Grzeskowiak, *J. Anal. At. Spectrom.* **1**, 297 (1986).
[116] P. Vinas, N. Campillo, I. Lopez-Garcia and M. Hernandez-Cordoba, *Talanta* **42**, 527 (1995).
[117] P. Wang and J.A. Holcombe, *Appl. Spectrosc.* **48**, 713 (1994).

7

Specialized techniques using electrothermal atomizers

Ralph E. Sturgeon

7.1 INTRODUCTION

It is clear from the preceding chapters that the electrothermal atomizer (ETA) enjoys widespread acceptance as a versatile source for the production of atomic vapor. The inherent advantages of high atomization efficiency, compact design, long residence time of analyte vapor, and controlled environment make it attractive for use with other nonabsorption-based techniques. It has also been widely employed in both integrated and spatially separated tandem source formats. The application of ETAs as atom sources for the detection of elements by atomic emission, fluorescence and ionization are considered here. The focus is on local thermal equilibrium (LTE) and nonthermally excited atomic emission, coherent forward scattering, conventional and laser-source excited atomic fluorescence, and laser-enhanced ionization. Molecular emission/absorption techniques are not addressed, nor are the more traditional uses of ETAs in hyphenated configurations. These include common sample introduction applications such as ETA–microwave-induced plasma atomic emission spectrometry, ETA–inductively coupled plasma atomic emission spectrometry, and ETA–inductively coupled plasma mass spectrometry, as well as their performance as detectors for gas chromatography, high performance liquid chromatography, flow injection analysis, and hydride generation systems.

Electrothermal Atomization for Analytical Atomic Spectrometry. Edited by K. W. Jackson
© 1999 John Wiley & Sons Ltd

7.2 ATOMIC EMISSION

7.2.1 Thermal emission

Emission techniques are inherently multielement in nature and the use of ETAs as atomization/excitation sources for thermally excited atomic emission spectrometry (AES) has been widely explored for this purpose. On a practical level, however, the technique has been little used despite the availability of multichannel spectrometers. The excitation temperature of the source is limited to that of the atomizer wall. As such, effective excitation can only be achieved for transitions above 300 nm, restricting application to those elements having excitation energies less than the high-energy tail of the Maxwell–Boltzmann energy distribution of collision partners or the radiation field density at the wall temperature, as defined by Planck's equation.

Early studies by Ottaway and colleagues at the University of Strathclyde drew attention to the potential and viability of AES using ETAs and highlighted the problems associated with the need for fundamental studies, background correction and atomizer design to improve excitation conditions. The excited state can be described approximately by the Boltzmann distribution, because the furnace is in LTE. Temperatures calculated from vibrational excitation (from CN band emission), atom excitation (from nickel and iron atom emission), and ionization (from atom and ion emission intensities of barium, calcium, europium, strontium, and ytterbium) are sufficiently close to one another, and to the radiation temperature derived from the spectral intensity distribution of the tube wall radiation, that LTE can be considered to exist under gas-stop conditions. As such, the wavelength-integrated intensity of a spectral line I_t ($W\,cm^{-2}\,sr^{-1}$) of wavelength λ_{ij} (cm) emanating from the furnace and viewed along the atomizer axis (not accounting for detection system efficiencies) is given by

$$I_t = [hcg_iA_{ij}N(t)/4\pi\lambda_{ij}Z(T)S]\exp(-E_i/kT) \qquad (7.1)$$

where h (J s) is Planck's constant, c ($cm\,s^{-1}$) is the velocity of light, g_i is the statistical weight of the upper level of the transition, A_{ij} (s^{-1}) is the Einstein transition probability for spontaneous emission between levels i and j, $N(t)$ is the total number of analyte atoms in the furnace at time t, $Z(T)$ is the partition function at the absolute temperature T(K), S (cm^2) is the cross-sectional area of the tube, E_i(J) is the energy of the excited state and k ($J\,K^{-1}$) is Boltzmann's constant. Assuming that atomization is complete and that excitation occurs at constant temperature, the integrated emission intensity Q_I ($W\,cm^{-2}\,sr^{-1}\,s$) from an analyte species is given by

$$Q_I = [hcg_iA_{ij}N_0/4\pi\lambda_{ij}Z(T)S]\exp(-E_i/kT)l^2/8D \qquad (7.2)$$

where simple diffusional loss of vapor is assumed over the length of the atomizer l (cm), governed by a diffusion coefficient D (cm^2 s^{-1}) and N_0 is the total number of analyte atoms introduced as the sample. The above equations were used to compare theoretical and experimental emission intensities for a number of elements in an effort to characterize the efficiency of the graphite furnace as an emission source [1,2]. As can be quickly appreciated, efficiencies are much greater when integrated intensities are considered. Whereas peak intensity efficiency ranged from 7 to 25 per cent, integrated values were found to lie in the range 34–100 per cent for the same suite of elements which covered a range of volatilities (silver to chromium) and excitation energies (2.98–4.38 eV). Although several constant-temperature atomizers were utilized in the study, it was concluded that optimal performance could only be expected with atomizers providing both temporal and spatial isothermality. Massmann-type devices, which exhibit severe temperature gradients (see Section 3.1.4), probably suffer from self-absorption and reduced effective length.

The above calculations characterize ETAs operating in emission to the same extent that characteristic mass calculations do in absorption, i.e. they provide little diagnostic information about the system and are constrained by the numerous assumptions implicit in the approach. Schwab and Lovett [3] attempted a more detailed simulation of the time-dependent emission from several analytes in a graphite furnace, accounting for activation energies and preexponential factors governing release of analyte atoms as well as diffusion coefficients for removal. The time-dependent number of atoms in the furnace was described by the convolution integral of supply $S(t)$ and removal $R(t)$ functions, as discussed in Section 2.5.3:

$$N(t) = \int_0^t S(t')R(t-t')\mathrm{d}t' \tag{7.3}$$

A generalized description of the excited state population was introduced, which accounted for both collisional and radiative excitation and decay processes:

$$\mathrm{M} + \mathrm{e}^- \rightarrow \mathrm{M}^* + \mathrm{e}^- \tag{7.4}$$

$$\mathrm{M} + \mathrm{X} \rightarrow \mathrm{M}^* + \mathrm{X} \tag{7.5}$$

and

$$\mathrm{M} + h\nu \rightarrow \mathrm{M}^* \tag{7.6}$$

where X denotes a molecule and the asterisk represents an excited electronic state of the analyte atom M. Thus,

$$n_i/n_j = (R_{ji} + B_{ji}\rho_{ji})/(R_{ij} + A_{ij} + B_{ij}\rho_{ji}) \tag{7.7}$$

where the partial collisional excitation rate is $R_{ji} = k_{ji}n_c$, the rate constant is $k(s^{-1})$ and n_c (cm^{-3}) is the collision gas density calculated from the ideal gas law, $k = n_c\sigma v$ where v ($cm\,s^{-1}$) is the mean speed of the collider–radiator pair, and σ is their collision cross-section. The expressions B_{ji} and B_{ij} are absorption and stimulated emission transition probabilities (s^{-1}), and ρ_{ji} is the radiation density at the frequency corresponding to the transition considered ($erg\,cm^{-3}\,Hz^{-1}$). The excitation and de-excitation rate constants are related by the Boltzmann equation and the radiation density ($\rho_{ij} = \rho_{ji}$) is given by the Planck equation

$$\rho = [8\pi h v^3/c^3]/[\exp(hv/kT) - 1] \tag{7.8}$$

Simulations based on the above relationships show clearly that furnaces purged with noble gases are continuum-driven fluorescence systems, not collisionally controlled systems. At atmospheric pressure the emission process in molecular gases is due to the combined action of collisional excitation and radiative absorption. Regardless of the mechanism of excitation, the population of the excited state depends on the temperature and its time dependence as determined by the supply and removal functions. Figure 7.1 illustrates the typical time-dependent relationship between the absorbance and emission profile. The emission maximum occurs on the decay portion of the absorbance peak. Although the atom density is decreasing, the emission intensity is increasing, because the rate of excited state population increase (due to increased temperature) is larger than the rate of atom removal. The emission peak occurs when the increase in excited state density due to temperature increases is offset by the atom loss due to diffusion. Absorbance and emission peak maxima

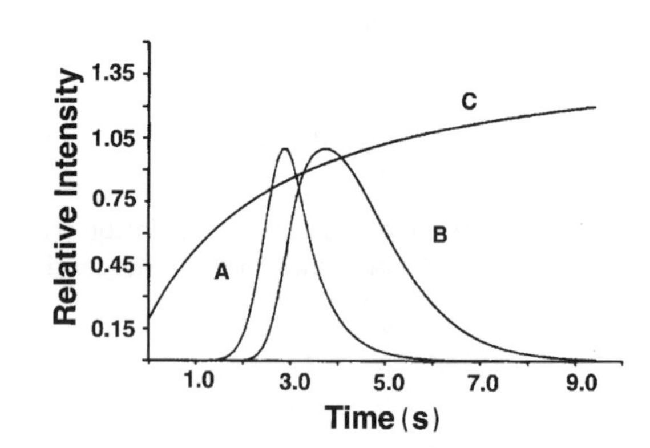

Figure 7.1 Curve (A) simulated absorbance and (B) emission profiles for the 324.7 nm line of copper. Temperature ramp (curve C) = 1100 K s^{-1} [3]. (Reprinted from *Spectrochim. Acta*, Vol. **45B**, J.J. Schwab and R.J. Lovett, Graphite furnace atomic emission spectrometry; computer simulations, pp. 281–300, 1990, with kind permission of Elsevier Science – NL, Sara Burgerhartstraat 25, 1055 KV Amsterdam, The Netherlands.)

would only coincide under the conditions of instantaneous atomization at constant temperature. Elements that are more difficult to atomize have more nearly coincident absorbance and emission peaks. Increased temperature ramp rates serve to coalesce the signals from the more volatile elements, but atom loss processes still occur primarily over the changing portion of the temperature profile. Measurement of the time between the absorbance and emission peak maxima thus provides a measure of the degree of isothermality of the furnace during atomization.

It is clear that significant advantage can accrue with use of the spatially and temporally isothermal furnace based on the constant-temperature two-step furnace of Baxter and co-workers [2,4] (see Section 1.4). These devices are capable of providing rapid atomization into a constant high-temperature environment. Hence, they provide superior analytical performance over the conventional Massmann-based designs, as well as the alternative approaches to achieving constant temperature conditions, as approximated by use of specially contoured tubes, the L'vov platform [5], or probe techniques [6]. Optimum excitation temperatures also lie within a narrow range for all elements, with the result that this system more readily lends itself to multielement determinations [2]. Table 7.1 compares detection limits obtained for several elements using the two-step furnace,

Table 7.1 Detection limits for ETAES in Massmann and two-step furnaces with thermal excitation (data from Reference 6, data rounded to nearest figure).

		Detection limit (pg)			
		Peak height		*Peak area*	
Element	*Wavelength (nm)*	*Massmann*	*Two-step*	*Massmann*	*Two-step*
Ag	328.07	3	2		2
Al	396.15	2	1		1
Co	345.35	80	74		40
Cu	324.75	2	5	7	22
Cr	425.43	1	2	3	1
Fe	371.99	9	18		10
Ga	403.29	2	4		2
In	410.17	0.3	5		2
K	404.41	100	80		65
Mg	285.21	15	11		6
Mn	403.08	0.4	2	3	1
Na	330.23	180	37		50
Ni	341.48	105	41	36	18
Pb	405.78	23	120	1200	230
	283.33		80		110
Tl	377.57	50	30		25
V	437.92	75	25		30

with best values generated using platform, probe or tube wall atomization [7]. In addition to the superior detection limit performance with the two-step furnace approach, spectral and non-spectral interferences should be less severe in this source, as should the degree of self-absorption.

The nature and magnitude of the background signal is often the factor that limits detection limits in emission techniques. Although the ETA presents a comparatively low-temperature source, emission spectra are more complex than those encountered in atomic absorption spectrometry (AAS), particularly when real samples are considered. Spectral interferences from line and molecular band (usually weak) features may occur, in addition to the intense continuum arising from Mie and Rayleigh scattering of the black-body radiation from the high-temperature tube wall (particularly severe in the visible region). For this reason, detection systems capable of greater resolving power are commonly utilized; the échelle polychromator has, based on its performance in continuum source AAS (CSAAS), been particularly attractive. Wavelength modulation techniques have proved to be a versatile choice for background correction–both the conventional oscillating refractor plate configuration, discussed in Section 4.5.5, and a rotating quartz chopper have been shown to be effective for this purpose. Three-step square-wave modulation offers superior signal-to-noise ratios (SNR; theoretically 1.8-fold) over those obtained with sine-wave-driven techniques. As with its continuum source AA counterpart (Chapter 4), the frequency of modulation must be sufficient to permit accurate recording of the transient signal. Refractor plate modulation frequencies ranging from 20 to 165 Hz have been utilized, with the majority of studies being conducted at 50–60 Hz. Detection is achieved in the 2-*f* mode in order to more effectively discriminate between the atomic line intensity and any sloping background continuum intensity.

Wavelength modulation techniques improve electrothermal atomic emission spectrometry (ETAES) detection limits by approximately two orders of magnitude. Recently, attempts have been made to assess the performance of low-resolution monochromators for use in emission. They were based on the premise that, for a wavelength-modulated system, optimum detection limits are achieved when the detection system has a large slit height and effective grating area, a short focal length and a large angular dispersion. A low-resolution monochromator operating with a large slit height may offer a trade-off of poorer angular dispersion for shorter focal length, as compared to the échelle polychromator. Experimentally, detection limits with such a system were found to be about 10-fold poorer than those achieved with the échelle system [8] but highlight the point that an inexpensive alternative to the high-resolution approach may be taken which provides sufficient sensitivity for many applications.

Owing to the finite response time of the torque motor used to drive the wavelength modulation system (particularly with square-wave drive), a very narrow modulation interval must be selected in order to operate at a reasonable frequency. Although this may appear to be a desirable feature, a narrow wavelength interval is limiting in terms of the analyte concentration range that can be accommodated before line broadening causes the signal profile to develop beyond the edges of the modulation interval, and hence introduce background correction errors. Furnace emission line widths increase with analyte concentration, reaching values of 0.1 nm at 1000 mg l^{-1} and self-reversal of atomic emission line profiles has been observed for a number of elements in the concentration range 1–1000 mg l^{-1} [9]. To avoid correction errors, the modulation interval must be increased which, in turn, degrades the SNR at the lower concentrations and detracts from the advantage of achieving background correction in close proximity to the analytical line. The modulation interval should therefore be selected to be in conformity with the concentration range of interest.

As a consequence of the above, and in addition to the non-isothermal nature of conventional ETAs giving rise to self-absorption at the cooler ends of the furnace at lower analyte concentrations, the linear working range for AES is typically 3–4 orders of magnitude. As with CSAAS (Chapter 4), this can be extended to 4–5 orders of magnitude by measurement of emission intensities off the center of the line profile.

Analytical applications of ETAES have been few, primarily because of the limited number of researchers involved as well as the lack of commercial instrumentation and support hardware. Nevertheless, sufficient work with real samples has been accomplished to demonstrate the utility of this potential multielement technique to provide an additional, alternative methodology for trace-element determinations. A range of elements (chromium, copper, manganese, and lead) has been determined in clinical samples, and a number of reference materials has been analyzed successfully for copper, chromium, sodium, calcium, potassium, iron, silver and nickel, including NIST bovine liver, oyster tissue, wheat flour, and rice flour, as well as orange and pineapple juices [7,10]. As with ETAAS, quantitation with integrated emission intensities provides better precision and fewer erroneous results than peak height emission. The latter is a function of the rate of atomization and is strongly matrix dependent.

7.2.2 Nonthermal emission

Nonthermal excitation is superior to thermal excitation when low absolute detection limits are sought, because of the low background associated with the sources [11]. Consequently, to utilize the full capability of AES, the high density and residence time of analyte atoms in the graphite furnace should be combined with a nonthermal excitation medium. Such discharges operating in graphite furnaces have been reviewed by Harnly *et al.* [12]. The concept of using glow discharges in the graphite furnace as excitation sources for emission spectroscopy is now established with both direct current (d.c.) and radiofrequency (rf) powered discharges being evaluated and operated at both reduced and ambient pressures. In the latter case, the descriptive term 'nonthermal' implies that the line radiance associated with the discharge cannot be described by any LTE models. Although this may not be true with the higher-pressure systems, in the interests of conformity these discharges are included for consideration in this category and are discussed separately below. Falk *et al.* [13] were the first to present theoretical considerations concerning the feasibility and potential advantages of such devices. A tandem source, consisting of a low-pressure tubular furnace fitted with a concentric grid tube and axial wire electrode to provide electron impact excitation, was constructed for proof-of-principle.

7.2.2.1 Hollow cathode furnace atomic nonthermal excitation spectrometry

In an effort to improve the practicality of the approach, a second-generation device was designed, in which the conductive furnace tube served as the cathode (held at ground potential) of an electrode pair, with the anode being a metal ring positioned just outside the furnace [14]. This arrangement is illustrated in Figure 7.2. With argon or helium fill gas at a pressure of 1–30 torr and d.c. power applied to the electrodes (typically 300–600 V, 10–100 mA), a hollow cathode discharge is spontaneously ignited which serves as the excitation source. In practice, drying and pyrolysis of the sample occurs as is usual with ETAAS. The pressure in the chamber is then reduced to about 0.01 torr, the device is backfilled with the selected support gas, the discharge is ignited, the sample is vaporized/atomized into the discharge volume, and the emission transient is recorded. Because of the additional nonthermal excitation process (from hot electrons) simultaneously applied with the electrothermal atomization, this new technique was characterized as furnace atomic nonthermal excitation spectrometry (FANES) and this particular electrode arrangement is designated as hollow-cathode FANES (HC-FANES). As the electron energy in the FANES plasma using helium as discharge gas is high enough to populate atomic and ionic levels up to 24.5 eV, all elements in the Periodic Table can, in principle, be determined in a simultaneous multielement mode. The HC-FANES concept thus extends the excitation energy range of the graphite furnace not only to elements with transitions below 300 nm, but provides access to elements having resonance lines in the vacuum ultraviolet region, as these can be efficiently excited in a hollow cathode discharge. As evident from the data in Table 7.2, detection limits achieved with this approach are comparable to or better than those of ETAAS. Because of the independence of the line intensity from the

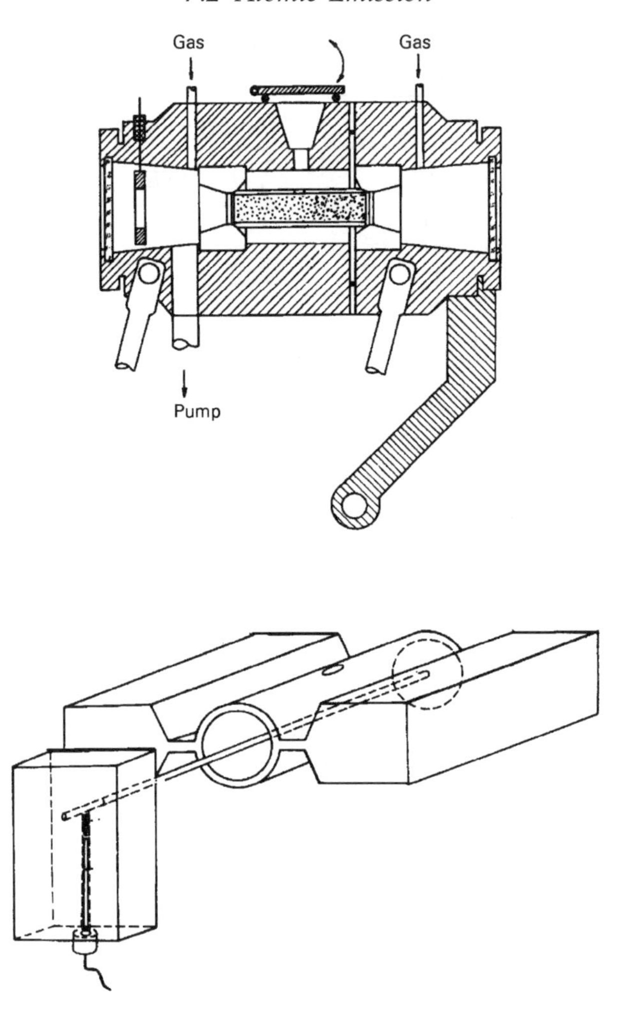

Figure 7.2 Schematic cross-sections of the HC-FANES and HA-FANES emission sources [15,20]. (Reprinted from *Prog. Anal. Spectrosc.*, Vol. **11**, H. Falk, E. Hoffmann and C. Ludke, Experimental and theoretical investigations related to FANES, pp. 417–480, 1988; and *Spectrochim. Acta*, Vol. **46B**, P.G. Riby, J.M. Harnly, D.L. Styris and N.E. Ballou, Emission characteristics of chromium in hollow anode-furnace atomization nonthermal excitation spectrometry, pp. 203–215, 1991, with kind permission of Elsevier Science – NL, Sara Burgerhartstraat 25, 1055 KV Amsterdam, The Netherlands.)

Boltzmann factor found in LTE sources, no systematic dependence of the detection limits on the excitation potentials has been observed.

Falk *et al.* [15] have extensively characterized the physical operating parameters, plasma characteristics, excitation mechanisms and analytical performance of the HC-FANES system. In principle, the HC-FANES

Table 7.2 Detection limits in ETA-based nonthermal excitation sources (best values summarized from the available literature – 3σ-based and rounded off).

	Detection limit (pg)		
Element	*HC-FANES*	*HA-FANES*	*RF-FAPES*
Ag	0.6		0.3
Al	20		
Au	4		
B	2	2	280
Be			5
Bi	50		30
Ca	0.08		
Cd	0.6	0.8	0.2
Co	5		
Cr	0.5	0.8	
Cu	1	0.5	4
Dy	200		
Er	220		
Eu	100		
Fe	0.7		10
Ga	0.6		
Hg	4		
Ho	440		
K	0.04		
Li	0.03		
Lu	11 000		
Mg	3		
Mn			1
Mo	30		
Na	0.06		
Ni	1		15
Pb	4		2
Pt			50
Rb	2		
Sc	30		
Sm	70		
Sn			10
Tb	3100		
Tc	90		
Tm	300		
Ti	1200		
Tl	0.3		3
V	8		
Y	2600		
Yb	20		
Zn	3		2
As			7
Br	6000		
Cl	120		
F	250		

(continued)

Table 7.2 *(continued)*

	Detection limit (pg)		
Element	*HC-FANES*	*HA-FANES*	*RF-FAPES*
I	4800		
P	90		30
S	4000		1100
Sb	14		50
Se			90
Te			60

source is identical to Perkin-Elmer Massmann-type furnace atomizers (see Section 1.5.1), and a commercially available system supports the conventional 6 mm internal diameter by 28 mm long graphite tube in a Massmann configuration [16]. Vacuum seals have been added at the appropriate locations, but are not of sufficient quality to maintain constant reduced pressure, with the result that continuous pumping is required in conjunction with a flow of support gas to maintain the discharge. It has been suggested that detection limits could be improved by at least an order of magnitude if static conditions could be achieved within the atomizer workhead during the initial seconds of vaporization/atomization.

Compared to other spectroscopic emission sources, HC-FANES presents a unique environment of low gas temperature (essentially determined by the ramped cathode temperature) and high ionization temperature (12 000 K for helium). Although the plasma is an acknowledged non-LTE system, the LTE derived temperatures of several thermometric species were used by Falk *et al.* [14] to characterize the discharge. Helium and argon excitation temperatures were found to be essentially independent of the discharge current (20–60 mA) and are approximately 10 000 and 8500K, respectively (at 1 torr). Increased pressure decreases the excitation temperature. As the cathode temperature increases from 500–2500K, the discharge gas excitation temperature increases by approximately 1000K. Electron densities of 3×10^{14} and $4 \times 10^{12}\ cm^{-3}$ were estimated for argon and helium, respectively.

Discharge formation is essentially the same as that for other hollow cathode devices. For typical operating pressures and currents, the current–voltage characteristics of the room temperature (cold) HC-FANES are essentially those of an abnormal discharge, in that the voltage and current density increase with the applied current. In such situations, the discharge covers the entire cathode surface. Only at the lowest pressures and comparatively low currents does a normal cathode fall occur. With rising fill gas pressure the discharge voltage decreases. The primary regions of interest in the discharge are those of the cathode fall adjacent to the cathode and the neighboring negative glow, a region of nearly constant voltage. Because of the higher mobility of electrons, a positive space charge creates a strong electric field in front of the cathode where most of the discharge voltage drop occurs. The discharge is sustained by electron impact ionization of the support gas, as well as by secondary electron emission from the cathode surface due to the photoelectric effect. As with conventional d.c. discharges, the radial extent of both regions diminishes with increasing pressure. For the negative glow, this is a consequence of the smaller mean free path of electrons; collapse of the cathode sheath thickness is a result of increased depth of ion penetration into the positive space charge region, owing to the corresponding decrease in electron temperature.

As the cathode is resistively heated during an atomization temperature ramp, carrier gas density declines (at constant pressure) and emission of thermionic electrons becomes appreciable above 1700K, in accordance with the Richardson equation expressing the current density j_e (A cm^{-2}),

$$j_e = AT^2 \exp(-e\Phi/kT) \tag{7.9}$$

where A is an emission constant of theoretical value about 120 A cm^{-2} K^{-2}, T is the surface temperature, and Φ is the work function of graphite. At 2500K, the thermionic emission current can approach several hundred milliamps. Injection of thermionic electrons into the discharge induces a rapid decline in the discharge voltage (from 350 to 50 V), which achieves an asymptotic value independent of the cathode upper temperature, presumably brought about by space charge considerations limiting thermoelectron current. Discharge characteristics and, consequently, excitation conditions are thus changed. This is reflected in the decreased emission intensities of carrier gas lines, helium being affected to a greater degree than argon because of its higher excitation potential. The high-energy tail of the electron energy distribution function evidently disappears when thermionic emission is dominant. The density of electrons at the low-energy end remains almost constant, despite a substantial decrease in the electrical power consumption of the discharge when the transition from the low- to high-temperature region takes place. This potential loss of excitation capability may have significant consequences for the analytical performance of the system, as discussed later.

These effects are more easily put into perspective if consideration is given to the mechanisms of excitation. Excitation of atoms or molecules in a glow discharge occurs predominantly by direct electron impact, as outlined in Eqn (7.4). The excitation frequency per unit volume k_{ex} (s^{-1} cm^{-3}), is dependant on the density of electrons n_e (cm^{-3}), their velocity v (cm s^{-1}), energy E_e (eV), energy distribution $f_e(E_e)$ (eV^{-1}), excitation cross-section $A_{ex}(E_e)$ (cm^{-2}), and collider density n_a (cm^{-3}), according to [16]

$$k_{ex} = n_a n_e \int_0^\infty v f_e(E_e) A_{ex}(E_e) \mathrm{d}E_e \tag{7.10}$$

As the electron density is proportional to discharge current, the measured intensity of a spectral transition will, to a first approximation (for de-excitation by spontaneous emission only), be proportional to the applied discharge current. With the low gas pressures and low electron densities encountered in the HC-FANES source, the equilibrium indicated by Eqn (7.4) does not play a role. Atoms, ions, and molecules excited by electron impact undergo energy loss almost exclusively by radiative processes and, consequently, single-headed arrows have been used to depict reactions in the ensuing equations. If excitation occurs *via* a step-wise process, such as radiative recombination,

$$M^+ + e^- \rightarrow M^* + h\nu \tag{7.11}$$

then the intensity becomes a quadratic function of the discharge current (assuming plasma quasineutrality). Similarly, stepwise excitation by electron impact of a species in a metastable state yields a radiative intensity, which is either a quadratic function of the discharge current (if the metastable state was initially populated by electron impact), or a cubic function of the current (for a metastable state initially populated by a two-step process). Additionally, selective excitation of analyte species may proceed *via* resonance energy exchange;

$$A^* + M \rightarrow A \rightarrow M^* \tag{7.12}$$

or charge exchange,

$$A^+ + M \rightarrow A + (M^+)^* \tag{7.13}$$

where A denotes discharge-gas species. Penning ionization may also occur in the discharge by interaction of analyte with metastable states of the discharge gas. Thus, the analytical signal is proportional to i^n where i is the discharge current and n represents the number of collisions involved in the excitation process.

It is thus possible to delineate excitation mechanisms from the dependence of the plasma gas emission intensity on discharge current and ambient pressure. In this regard, argon and helium behave differently in the HC-FANES source, as illustrated by Figure 7.3. In general, discharge gas line intensities initially increase as the cathode temperature rises, possibly due to an increase in the mean free path for electrons at reduced gas density (constant pressure), followed by a precipitous decline with the onset of thermoelectron emission at temperatures beyond 1700K. Argon

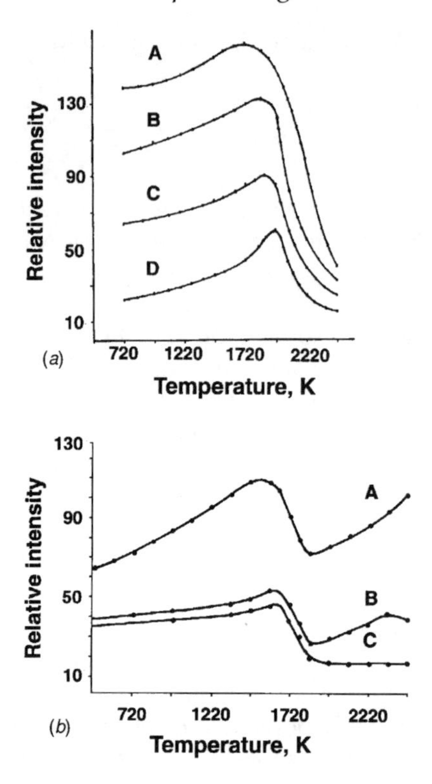

Figure 7.3 (a) Intensity of the He 318.8 nm line as a function of cathode temperature in the HC-FANES source with a discharge current of 40 mA at pressures of (A) 9, (B) 13, (C) 27, and (D) 40 hPa. (b) Intensity of the Ar 451.1 nm line as a function of the cathode temperature in the HC-FANES source with a discharge current of 40 mA at pressures of (A) 10, (B) 20, and (C) 30 hPa [15] (Reprinted from *Prog. Anal. Spectrosc.*, Vol. **11**, H. Falk, E. Hoffmann and C. Ludke, Experimental and theoretical investigations related to FANES, pp. 417–480, 1988, with kind permission of Elsevier Science – NL, Sara Burgerhartstraat 25, 1055 KV Amsterdam, The Netherlands.)

lines are much less affected than those of helium, because of the lower excitation energy requirements.

Although the HC-FANES is based on the hollow cathode effect, analyte is introduced into the discharge almost completely by thermal vaporization during heating of the furnace. Typical rates of removal of the sample by cathode sputtering are 0.1–1 μg s^{-1} at discharge currents of 20 mA. For comparison, thermal vaporization amounts to about 1 mg s^{-1}. A study of the change in analyte emission intensity with discharge current can provide information about the vaporization processes involved, as outlined above. In this manner it has been possible to ascertain whether or not the analyte was vaporized as a molecule with subsequent dissociation to form atoms. As an example, a quadratic dependence for aluminum and cadmium was observed in nitrate and chloride media, respectively, suggesting vaporization of the analyte as a molecular species, with plasma-induced dissociation in the discharge followed by excitation in a second step. Moreover, plasma-induced dissociation of volatile molecular species, followed by excitation of the released atoms, yields detectable analyte emission signals at temperatures well below those at which AAS signals can be detected.

In general, analytical emission signals are dependent on the support gas type, discharge pressure and current as well as the atomization temperature. With respect to the support gas, both argon and helium have been used interchangeably in the FANES source. The higher excitation energy accessible with helium makes it attractive for the excitation of nonmetals. The low operating pressure (1–30 torr) routinely used in FANES arises as a compromise between optimum discharge conditions and analyte residence time considerations. As the diffusion rate is inversely proportional to pressure, to a first approximation the integrated analytical signal should increase with ambient pressure. However, as continuous pumping and a constant support-gas flow are required to maintain the desired pressure, it is not clear whether the analyte residence time is limited by convection or diffusion. Additionally, HC discharges are highly pressure dependent. Transitions requiring high-energy electrons for excitation appear to be the most sensitive to increased pressure. As the pressure increases, the mean free path of electrons decreases and the energy from the electrons, accelerated across the dark space, is dissipated in inelastic collisions closer to the cathode wall. Thus, the region of maximum emission shifts from the axis of the HC toward the cathode wall as the pressure increases. Most conveniently, HC-FANES is viewed along the central axis (for minimum blackbody emission from the furnace walls), and no detailed data characterizing either the spatial dependence of the emission signal or the overall system response as a function of pressure are available.

Commensurate with higher atomization temperatures are increased rates of diffusional loss and decreased integrated emission intensities. Residence times for many elements in the low-pressure source are only 5–10 ms. Atomization temperatures are therefore kept as low as possible. Cathodic temperatures above 1800K can have a significant influence on the discharge due to thermoelectron emission. Fortunately, low-pressure operation supports low-temperature vaporization [15], and no deleterious effects of thermionic emission on the detection of a wide range of elements has been reported for the HC-FANES system. It is assumed that all elements were volatilized at temperatures less than 1800K.

Both spectral and nonspectral interferences have been reported for HC-FANES. In contrast to thermal emission in ETAES systems, limiting noise sources in HC-FANES are not determined by continua, but by atomic line and structured molecular features arising from discharge gas impurities and ambient gases that have entered the furnace through imperfect seals. The most likely of such bands are those belonging to the systems of OH, CO, CO^+, CN, N_2, NO, and NH. Optimum SNRs have been achieved at a resolution of $1\text{–}2\times10^4$ and, although wavelength modulation techniques are useful, the possibility of spectral interferences arising from background structure within the modulation interval has been confirmed. The short duration of the analyte emission transients requires faster modulation frequencies than those associated with ETAES techniques, and square-wave modulation at 130 Hz or sine-wave modulation at 200 Hz has been utilized. Additionally, in certain wavelength regions, the temperature of the cathode (tube) was found to have an influence on the characteristics of the background spectrum. As examples of the above, spectral interference at the chromium 357.9 nm and nickel 352.5 nm lines occurs because CN bands fell within the modulation window of the background correction system. Such bands were located on one side of the spectral region scanned by the refractor plate. Thus, the background measured at these points was greater than that measured at the line center, resulting in overcorrection. Blank correction of the data, although proving satisfactory, increases the limiting noise by a factor of 1.4, assuming a quadratic addition of independent noise sources. At 2000°C the background feature near the 357.9 nm chromium line was quite prominent, but at 2700°C the peak disappeared almost totally. Characterization of the HC-FANES background spectrum, both with and without samples, is urgently required and will facilitate both line selection and interpretation of results.

Although not widely studied, nonspectral or chemical interferences also occur, and may prove more severe than with conventional ETAAAS because their influence also extends to the properties of the discharge. Matrix components may alter the discharge conditions, thereby affecting the energetics of analyte excitation. In particular, easily ionized elements will alter the discharge potential much as thermionic electrons do, leading to a quenching of analyte emission. In connection with Eqn (7.10), discussion centered around the assumption that de-excitation of analyte occurred solely by spontaneous emission. Although this is the primary process in low-pressure discharges, in reality introduction of sample matrix species into the plasma will alter this. The excitation rate must be balanced by the collisional de-excitation rate. For a simple two-level system,

$$n^*/n_0 = \mathrm{CE}/(\mathrm{CD} + \mathrm{RD}) \qquad (7.14)$$

where CE, CD and RD are the rates for collisional excitation, de-excitation and radiative de-excitation, respectively. The maximum concentration of matrix species relative to discharge gas in the graphite furnace ($c_{\mathrm{rel,m}}$) is given by the expression

$$c_{\mathrm{rel,m}} = \varepsilon c_{\mathrm{m}} V_{\mathrm{s}} N_{\mathrm{A}} / n_{\mathrm{c}} M V_{\mathrm{a}} \tag{7.15}$$

where ε is the atomizer efficiency, c_{m} is the sample matrix concentration ($\mathrm{g\,cm^{-3}}$), V_{s} is the sample volume, N_{A} is Avogadro's number, n_{c} is the carrier gas density, M is the molar mass and V_{a} is the atomizer volume. In low-pressure HC-FANES with a typical efficiency of 1 per cent, $c_{\mathrm{rel,m}}$ is about 8 per cent for a 10 μl sample of 1 per cent concentration ($M = 50$) and average working conditions. Assuming the photons emitted from the excited analyte leave the plasma, the power loss P_{ex} by this process becomes

$$P_{\mathrm{ex}} = k_{\mathrm{ex}} r_{\mathrm{dp}} E_{\mathrm{ph}} V_{\mathrm{so}} \tag{7.16}$$

where $r_{\mathrm{dp}} = \mathrm{RD}/(\mathrm{CD} + \mathrm{RD})$, E_{ph} is the photon energy and V_{so} is the source volume. Expressions (7.10) and (7.14)–(7.16) can be used to calculate the matrix concentration of any species in the plasma that would dissipate a given fraction of the input power by excitation losses. Falk [16] estimated this for sodium, where the tolerable excitation loss was assumed to be 10 per cent. The upper limit for sodium that would cause a noticeable influence on the plasma was calculated to be only $5\text{–}10^{-4}$ per cent. The severity of these effects is, as with ETAAS, dependant on the degree of temporal overlap between the analyte and the interferent in the gas phase. Figure 7.4 illustrates the sodium matrix effect for several

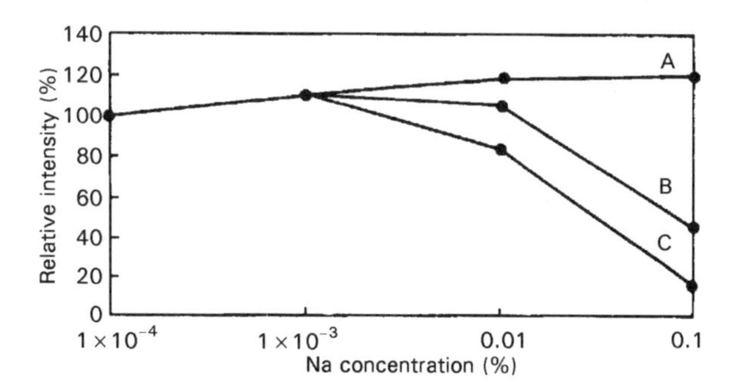

Figure 7.4 Emission intensities from the HC-FANES source as a function of the Na concentration for analytes with different volatilities: (A) Al; (B) Fe, Ni, Co and Cr; and (C) Cu. Volatilization temperatures: Na 800; Cu 800; Fe, Ni, Co and Cr 850–1000; Al 1100°C [16]. (Reprinted by permission of the Royal Society of Chemistry from H. Falk, Tandem sources using electrothermal atomizers: analytical capabilities and limitations, *J. Anal. At. Spectrom.*, **6**, 631–635 (1991).

elements. It is clear that interferences occur at relatively low matrix concentrations, and that their severity is influenced by the relative volatilities of the analyte–matrix pair. Fortunately, the low-pressure environment of the HC-FANES source minimizes this effect. Additionally, there is evidence to support the notion that plasma-induced dissociation of molecular analyte species occurs, thereby raising the tolerable level of matrix interferent. Use of platform techniques to improve detection limits and minimize chemical interferences in this source has only recently been explored [17].

A further source of interference that is unique to HC-FANES is the potential for preatomization losses of analyte because of sputtering processes. Early losses of standards and samples, the severity of which is dependent on the gas pressure and the time interval between the start of the discharge and the high-temperature ramp, have been noted for cadmium and sulfur. Chemical modifiers have been used to alleviate the problem.

There have been a number of applications of HC-FANES to analytical problems, and a commercial version of the source (Fanoquant 100, Jenoptic JENA GmbH, Germany) has been available for several years. However, even with atomization from a platform, matrix interferences arise and essentially all calibrations to date have been obtained using the method of additions to achieve accurate determinations in known materials. Advantage can be taken of the 5-decade linear range for some analytes (silver, sodium, and lead), but other elements such as cadmium and the rare earths exhibit dynamic ranges as short as 2 and 3 orders of magnitude, respectively. The problem stems from the source being neither optically thin nor homogeneous, in that the majority of analyte atoms are in the ground state and capable of absorbing resonance emission. Further, the furnace is not spatially isothermal and this gives rise to self-absorption by the vapor at the cooler extremities.

Samples analyzed include: deproteinized whole blood for cadmium; the determination of copper, iron, and nickel in grass, corn and NIST orchard leaves (using solid sampling and a specially designed furnace [15]); the determination of sodium in aluminum, and silver in gold; the multielement analysis of deionized water and nitric acid for aluminum, cobalt, copper, iron, and nickel; and the multielement analysis of digests of NIST reference materials (citrus leaves, pine needles and river sediment) for aluminum, cadmium, chromium, iron, nickel and lead. These samples were also analyzed for mercury using cold vapor generation and *in situ* collection of the analyte in the furnace. Separation of the analyte from the matrix permitted calibration against simple aqueous standards. Generally, multielement determinations degrade detection capability (by factors of 2–30-fold) because of the compromise nature of both the discharge and furnace thermal parameters.

Nonmetal determinations have primarily been accomplished with the HC-FANES source through the measurement of emission from analyte containing molecules [18] in a technique known as MONES (molecular nonthermal excitation spectrometry). Thus, phosphorus (as HPO and PO), sulfur (as CS), fluorine (as MgF), and chlorine (as MgCl) have been investigated and figures of merit obtained. This technique is not considered further.

7.2.2.2 Hollow-anode FANES

Ballou *et al.* [19] described a source similar to HC-FANES, in that a low-pressure d.c. discharge was established in a graphite furnace, but the polarities were reversed. The furnace comprises the hollow anode held at ground potential (HA-FANES) while the cathode is a coaxial graphite rod traversing the length of the furnace, as illustrated in Figure 7.2. The furnace itself is a 19 mm long integrated contact cuvette (ICC) which is enclosed within a 100 mm diameter six-way vacuum cross. The glow discharge in this design is a bright corona constricted around the cathode. This geometry minimizes electric fields between the furnace and vacuum chamber, simplifies electrical shielding, and promotes ignition of the glow. Use of proper vacuum design permits static operation at low pressures and reduces the complexity of the structured background due to the absence of molecular emission from CO and OH, for example. However, CN and CH emission have been observed and may be unavoidable due to exposure of the high surface area graphite structure to the ambient atmosphere during sample dosing operations. Emission intensities increase linearly with pressure (decreased diffusional loss rate of analyte) and optimum SNRs are found within 1 or 2 mm of the cathode surface. Maximum response from the high-intensity region concentric to the cathode rod is obtained for most elements near 200 torr for argon and 400–600 torr for helium. Under such conditions, analyte residence times are similar to those for ETAAS. Detection limits are comparable to those of HC-FANES (see Table 7.2), as is reproducibility. Linear dynamic ranges of 3–4 decades at discharge pressures of 70–200 torr of argon appear reasonable.

Analyte sputtering does not occur in HA-FANES unless the sample is deliberately loaded onto the cathode rod. As with HC-FANES, the sample is delivered into the discharge primarily by thermal vaporization as the tube wall heats. Use of the ICC with HA-FANES permits use of lower atomization temperatures than with conventional furnaces. Because the ICC has a uniform temperature along its length, there is no need to exceed the optimum atomization temperature to compensate for the cooler ends of the furnace.

As with the HC-FANES, HA-FANES operates as an abnormal discharge, in that the glow completely covers the cathode surface and any increase in current results in an increase in current density and discharge voltage. Because of its smaller surface area, current densities at the cathode surface are, for the same current, much larger in the HA- than the HC-FANES configuration. Integrated analytical signals for HA-FANES exhibit completely different dependence on the discharge current than the low-pressure HC-FANES [12]. Signal intensities initially increase in proportion to i^n, as with HC-FANES, but ultimately reach a plateau at higher discharge currents. The range of discharge currents is determined by the pressure. At a given pressure, the lower current limit is determined by that necessary to cover the entire cathode surface with the glow to establish the abnormal region. The current necessary to reach the abnormal discharge increases with pressure. Higher currents then result in increased current densities and discharge voltages. The upper current is limited by arcing between the anode and the cathode. For a specified pressure, optimum SNRs are found at the intermediate currents, as background emission levels become noisier towards the upper and lower limits, while analyte signals show no variation with current.

Successful operation of the HA-FANES source at pressures of several hundred torr is probably due to the relatively small influence of thermionic electron emission from its cathode. This central rod not only presents a smaller cathodic surface area for electron emission (Eqn (7.9)) but, for comparable furnace wall temperatures, it is much cooler than the cathode in the HC-FANES device. This is because it is completely removed from the wall, is attached to a large, cool radiating mass outside one end of the furnace, and is heated only radiatively and convectively. Thermionic emission from the respective cathodes is thus much smaller with the HA configuration. Although higher-pressure operation of the HA-FANES requires higher atomization temperatures than with HC-FANES, complete volatilization of analytes can be achieved before the temperature of the cathode rises above 1800K. For example, the analytical signal for chromium, atomized at 2500K, has completely returned to baseline before a decrease in the discharge potential is observed [20]. Plasma gas line intensities for Ar I increased with temperature at both 1500°C and 2000°C, but Ar II intensities decreased dramatically at 2000°C, as did both He I and He II intensities at 1500°C just prior to the end of the atomization cycle. This reflects the influence of thermionic electrons in reducing the discharge potential of the plasma.

The temperature lag of the central electrode is responsible for the appearance of double peaks in the emission signals for a number of analytes. Material vaporized from the hot wall condenses on the cooler electrode, which, with continued heating, is ultimately redesorbed (second surface vaporization). As expected, double peaks are particularly severe with the less-volatile elements. As the temperature of the cathode at the start of the atomization step is a function of the discharge current as a result of ohmic heating, its temperature can exceed 1300°C at moderate currents (60 mA). Double peaks can thus be minimized for the more involatile elements by operation at higher discharge currents. In this way, at the start of the atomization step, the cathode is not only much hotter than the furnace wall, thereby minimizing condensation, but the temperature lag between the cathode and the furnace wall is decreased and the double peaks merge [20].

To date there has been only a single application of HA-FANES to analysis, that being the determination of chromium in NIST reference material 1643b, acidified water [20]. The simple matrix permitted direct calibration against aqueous standards with acceptable accuracy and precision. Although the effect of more complex matrices requires extensive investigation, it is expected that the problems highlighted earlier for HC-FANES will be present with this source as well. It is equally possible that these may be exacerbated, because increased residence times at higher pressure increase the temporal overlap between matrix and analyte vapor. Spectral interferences, however, are likely to be less severe considering the quality of the vacuum enclosure.

7.2.2.3 Furnace atomization plasma emission spectrometry (FAPES)

In FANES systems, d.c. glow discharges are used for excitation, and there is a tendency towards arc formation at high pressures. High-intensity high-current-density arc discharges are not useful in ETA, due to the difficulty of reproducible introduction of thermally desorbed analyte into the excitation zone, as well as problems associated with imaging and rapid deterioration of the graphite surface. Radiofrequency discharges, however, can be used to sustain uniform, high-current-density discharges at atmospheric pressure. The geometry of the HA-FANES device is ideal for the formation of capacitively coupled plasmas, wherein the radio frequency (rf) power is supplied to the center electrode. Such a system has recently been described by Liang and Blades [21] and by Sturgeon *et al.* [22]. At operating frequencies of 13.6 or 27.12 MHz and 20–100 W (apparent forward power), plasmas can be self-ignited at atmospheric pressure in helium or argon, respectively, to produce a plasma that appears to uniformly fill the furnace volume. Impedance-matching networks, identical to those used for inductively coupled plasma (ICP) applications, are necessary to permit efficient power coupling into the source. By analogy to FANES, and in the interests of consistency in nomenclature, this technique could be entitled RF-FANES. However, fundamental differences in the operation of this device compared to its FANES precursors (in particular, is nonthermal excitation dominant at atmospheric pressure?) have resulted in this source being christened FAPES (furnace atomization plasma emission (or excitation) spectrometry). Although the original device described by Liang and Blades [21] was based on a modified Thermo Jarrell-Ash (Franklin, MA, USA) furnace workhead (Model IL655), subsequent study has concentrated on the use of a laboratory-built ICC-equipped furnace (identical to the HA-FAPES source) enclosed in a vacuum six-way cube. The FAPES source described by Sturgeon *et al.* [22] uses a modified Perkin-Elmer (Norwalk, CT, USA) furnace (Model HGA-2200). Recently, a commercial version of FAPES has become available (Aurora Instruments, Vancouver, BC, Canada), which is also based on ICC technology and provides a convenient and compact workhead for interfacing to a variety of spectrometers. A diagram of each source is presented in Figure 7.5.

The FAPES technique offers several advantages over FANES, including the ability to operate at atmospheric pressure, increased analyte residence time in the discharge, greater analyte number density (higher source efficiency), potentially improved efficiency of atomization, greater ease of use and comparable detection limits (see Table 7.2). However, the advantageous features of low-pressure glow-discharge devices, which included reduced background complexity and intensity, higher excitation temperature and potentially lower noise, may have been sacrificed. Nevertheless, FAPES also holds promise as a means of providing simultaneous multielement analysis on small liquid and solid samples with detection limits comparable to those of ETAAS.

The FAPES discharge appears similar to that of HA-FANES, in that an intense region surrounds the central electrode and a less intense plasma fills the remainder of the furnace volume. Two-dimensional imaging of the intensities of several support gas lines, using a charge coupled device (CCD) camera detector, have clearly revealed the presence of a second, less intense, ring discharge around the wall of the graphite tube [23]. This is associated with the reversal of polarity of the applied rf power during which time the wall behaves as a cathode, tempered by the electric field strength being much weaker over this larger surface area (compared to that of the center electrode). A cathode dark space between the graphite tube wall and the ring plasma is also present, and for a helium discharge at atmospheric pressure, is about 90 μm in depth. These structures are clearly evident in the two-dimensional image of the He I 667.82 nm distribution shown in Figure 7.6. The discharge is primarily capacitively coupled to the generator, with a blocking capacitor connected in series so that no net current flows through the circuit. As a consequence of the difference in the mobility of positive ions and electrons in the plasma, high frequency rf discharges result in the production of a negative d.c. self-bias potential on the powered electrode, which then responds similarly to the cathode in a d.c. system; hence the visual similarity to the HA-FANES

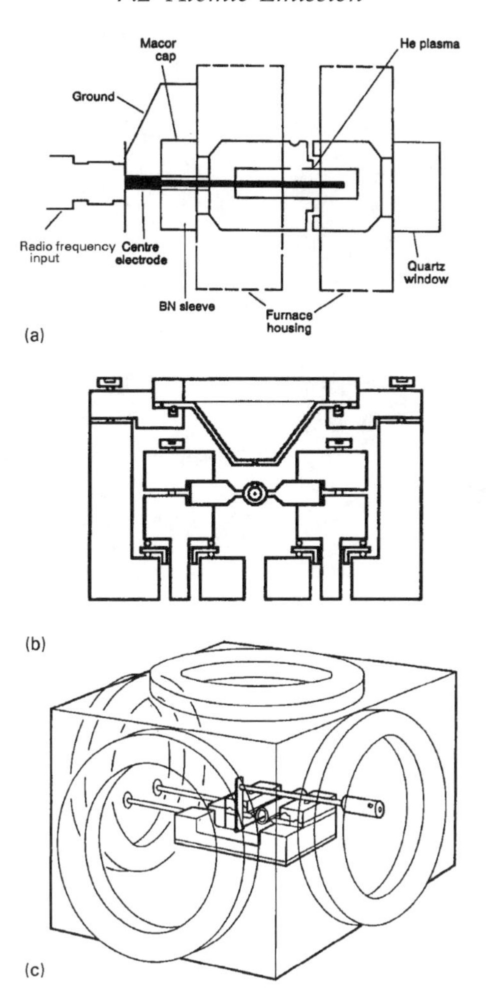

Figure 7.5 Schematic cross-sections of various RF-FAPES sources [22,28]. ((a) Reprinted by permission of the Royal Society of Chemistry from R.E. Sturgeon, S.N. Willie, V. Luong, S.S. Berman and J.G. Dunn, Furnace atomization plasma emission spectrometry (FAPES), J. Anal. At. Spectrom., **4**, 669–672 (1989); and (c) from *Spectrochim. Acta*, Vol. **47B**, T.D. Hettipathirana and M.W. Blades, Furnace atomization plasma emission spectroscopy: spectra, spatial, and temporal characteristics, pp. 493–503, 1992, with kind permission of Elsevier Science – NL, Sara Burgerhartstraat 25, 1055 KV Amsterdam, The Netherlands; (b) reprinted by permission of Aurora instruments Ltd.)

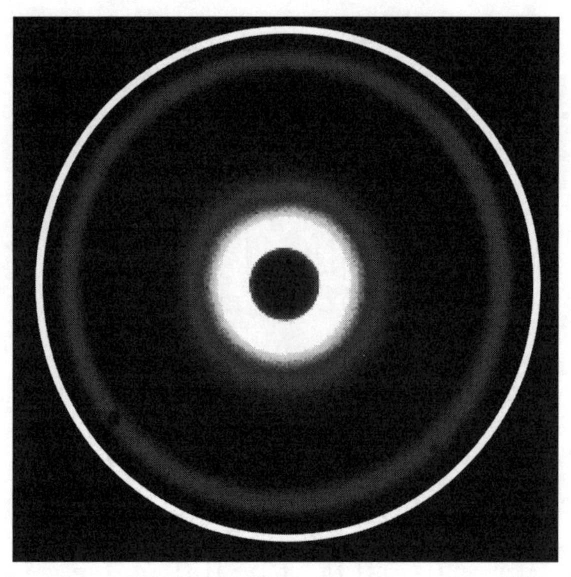

Figure 7.6 False colored CCD image illustrating cross-sectional distribution of the He 667.8 nm line intensity in the FAPES source obtained with the graphite tube at room temperature, a forward rf power of 50 W and a 0 V bias imposed on the center electrode [23]. (Reprinted by permission of the Royal Society of Chemistry from V. Pavski, C.L. Chakrabarti and R.E. Sturgeon, Spatial imaging of the furnace atomization plasma emission spectrometry source, J. Anal. At. Spectrom., **9**, 1399–1409 (1994).)

discharge. With a cold cathode (unheated furnace) the self-bias typically amounts to a few volts negative for a 50 W forward power plasma, increasing to −18 V at 100 W power. Commensurate with this is an increase in the intensity and size of the bright region surrounding the center electrode. During the rapid heating period of an atomization cycle, the bias voltage briefly floats negative to about −120 V at 2600K and then recovers as the system approaches a steady-state temperature [24]. Significant changes in the self-bias potential begin to occur as the temperature of the tube wall approaches 1600K, and are coincident with the onset of a substantial flux of thermionic electrons from the heated tube wall. Because of the conductivity of the plasma, an electron current of approximately 220 mA can be measured between the wall and the center electrode at the moment of the maximum dip in the bias potential. The self-bias on the center electrode responds by going more negative in an effort to curb the flow of electrons to its surface. As the system continues to heat, the temperature of the center electrode rises to the point where significant thermionic emission from its surface begins, creating a space charge of electrons, which repels the flux from the walls. At this point the

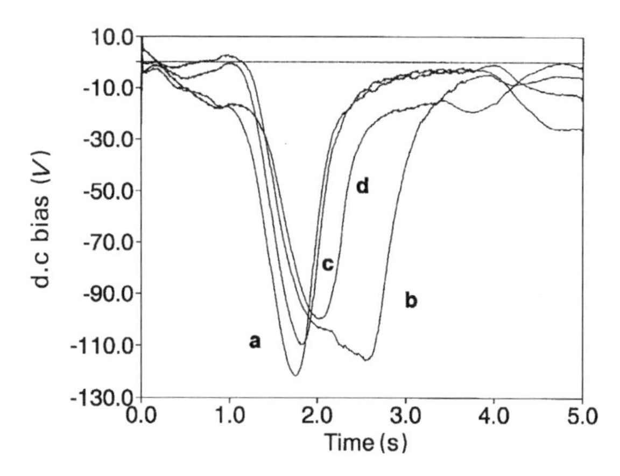

Figure 7.7 Variation in the self-bias voltage during an atomization transient in the FAPES source. He plasma at 50 W forward power: curve (a) 1 mm diameter solid graphite center electrode; (b) 2 mm solid graphite electrode; (c) 2 mm hollow graphite electrode; and (d) 1 mm Ta electrode [24]. (Reprinted from *Spectrochim. Acta*, Vol. **48B**, R.E. Sturgeon, V.T. Luong, S.N. Willie and R.K. Markus, *Impact of bias voltage on furnace atomization plasma emission spectrometry performance*, pp. 893–908, 1993, with kind permission of Elsevier Science – NL, Sara Burgerhartstraat 25, 1055 KV Amsterdam, The Netherlands.)

self-bias begins to reverse and, when the electrode eventually reaches the same temperature as the tube wall, the self-bias returns to its initial value of a few volts. The heating characteristics of the center electrode (size, mass, heat capacity, conductivity, emissivity), as well as its electrical properties (work function), influence the shape of the self-bias transient, as illustrated in Figure 7.7. The bias voltage has a significant effect on the analytical performance of FAPES, and will be addressed later.

In addition to influencing the bias potential, thermionic emission also affects the reflected power characteristics of the FAPES source, in that it alters the impedance characteristics of the plasma. Reflected power transients that develop during the heating of the furnace are shown in Figure 7.8. With the tube wall at room temperature, reflected power can usually be tuned to 1 W for forward powers up to 200 W. As the temperature of the system increases, reflected power rises to a plateau and abruptly returns to 0 W with the cessation of tube wall heating. The autotuning capabilities of impedance matching networks are effective in minimizing this problem, but do not eliminate it because the impedance changes occur too rapidly for these mechanical systems to follow (a free running oscillator may alleviate this problem). The onset of impedance mismatch occurs at a tube wall temperature of about 2000K, and may be the result of minor perturbations to the impedance of the plasma caused by the

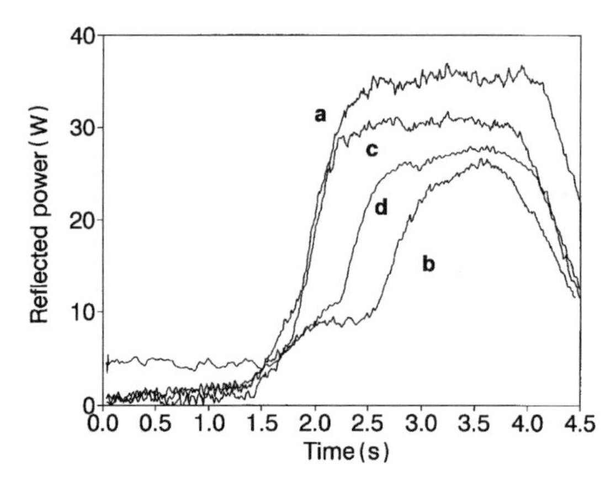

Figure 7.8 Reflected power transients in a self-biased 50 W He FAPES Plasma: curve (a) 1 mm diameter solid graphite center electrode; (b) 2 mm solid graphite electrode; (c) 2 mm hollow graphite electrode; (d) 1 mm Ta electrode [24]. (Reprinted from *Spectrochim. Acta*, Vol. **48B**, R.E. Sturgeon, V.T. Luong, S.N. Willie and R.K. Markus, Impact of bias voltage on furnace atomization plasma emission spectrometry performance, pp. 893–908, 1993, with kind permission of Elsevier Science – NL, Sara Burgerhartstraat 25, 1055 KV Amsterdam, The Netherlands.)

emission of thermoelectrons from the heated tube wall. However, significant rise in the reflected power levels does not begin until the temperature of the center electrode approaches 1700K, at which point significant thermionic emission from this surface commences. This is coincident with the maximum dip in the self-bias potential. As the center electrode continues to heat, thermionic electron emission increases exponentially and the reflected power follows suit. The annulus of plasma closest to the center electrode is the most significant interaction region (highest electric field density), where impedance changes have their most profound effect on the reflected power, and hence, tuning of the plasma. When the electrode reaches a steady-state temperature the reflected power level stabilizes. Increasing the forward power to the plasma results in an increase in the temperature of the center electrode (heated by ion bombardment) and a growth in the volume of the plasma within the tube. Both factors serve to alter the reflected power transient, in that the onset of impedance mismatch occurs earlier and, although the rate of rise of reflected power is greater, the extent of mismatch is similar in that the reflected power levels amount to about 60 per cent of the forward power in all eases. Although high reflected power levels in rf systems suggest degraded performance, no apparent analytical problems related to this have arisen with the FAPES source.

Figure 7.9 illustrates the change in He I line intensities for several transitions as a function of the tube wall temperature. Line intensities increase with the wall temperature; following cut-off of the power to the tube and its rapid cooling, line intensities continue either to increase or decrease, depending on the spectroscopic transition. Thus, despite a decrease in the helium gas density by 3–4-fold due to gas expansion during heating, as well as an increase in the reflected power, line intensities increase. A longer mean free path for electron excitation at reduced density, an increase in the electron density in the plasma due to thermionic emission from heated surfaces, and increased population of the upper levels of all transitions due to more favorable Boltzmann factors may contribute to this. The most significant rise in intensity begins when the center electrode reaches a temperature sufficient for significant thermionic emission. As the observation area is confined to close proximity to the surface of the center electrode, thermoelectrons may foster increased rates of neutralization of He^+ ions in the electrode sheath, and result in increased emission due to the increased production of an intermediate excited-state helium species. Detailed interpretation of these data requires information on rate processes for the excitation and de-excitation of all population levels.

At no time does the plasma collapse due to the presence of thermionic electrons, probably because they are efficiently excited in the rf field and

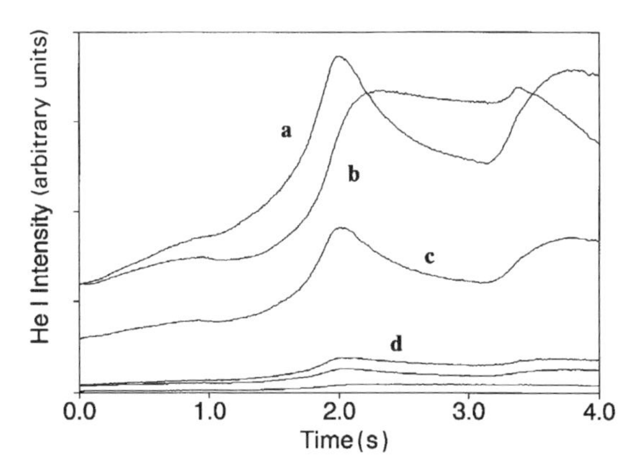

Figure 7.9 Temporal variation of He I line intensities during an atomization transient in a self-biased 50 W forward power He FAPES plasma: curve (a) 388.87 nm; (b) 667.82 nm; (c) 501.57 nm; (d) 492.19 nm [24]. (Reprinted from *Spectrochim. Acta*, Vol. **48B**, R.E. Sturgeon, V.T. Luong, S.N. Willie and R.K. Markus, Impact of bias voltage on furnace atomization plasma emission spectrometry performance, pp. 893–908, 1993, with kind permission of Elsevier Science – NL, Sara Burgerhartstraat 25, 1055 KV Amsterdam, The Netherlands.)

contribute to the overall excitation of spectra. Imposing a negative bias voltage on the system increases the intensity of all species in proportion to the voltage. The cathode glow is associated solely with the electrode which acts as the cathode during a given half-cycle of the rf power. Therefore, imposition of a negative voltage bias on the center electrode serves to make it behave as the cathode for a greater period of time during a given half-cycle [23]. Visual observation suggests that the plasma is more robust in a negative bias mode, in that the intensity is more uniform and plasmoids are absent. Although initially applied in an effort to enhance the excitation conditions within the plasma by augmenting the energy of electrons, thermalization probably occurs because of the high pressure as a significant number of collisions occurs in the electrode sheaths.

Excitation temperatures in the FAPES source are, as expected, much lower than those measured in low pressure FANES. With the graphite furnace tube wall at room temperature, He I excitation temperatures of approximately 3000K have been reported [25]. Deliberate detuning of the system to increase the reflected power level to 40 per cent of incident resulted in no significant change in measured temperature, despite shrinkage of the plasma volume [24]. This is consistent with the relatively small change in the excitation temperature in response to variation in the forward power, suggesting that in this high-pressure source, only plasma volume and electron density respond to power level rather than excitation temperature. Additionally, the excitation temperature is relatively little influenced by the tube wall temperature. As discussed above, thermalization of thermionic electrons injected into the plasma and excited in the rf field occurs quickly at atmospheric pressure. During the course of an atomization cycle, the He I excitation temperature in a free-running self-biased source increased from 2950K to 3600K as the wall ramped through a 600–2600K cycle [24]. A helium ionization temperature of 11 000K has been estimated based on a measurement of the electron density. Ionization temperatures derived for magnesium, iron, cadmium and zinc varied between 4800 and 5900K [25]. A gas kinetic temperature, estimated from the Doppler profile of the Be I 234.861 nm emission line, was calculated to be 7000 K, i.e. far too high to be realistic [25]. This overestimation was probably the result of imaging a high-intensity current arc that formed between the tube wall and the relatively massive deposit of metallic beryllium placed in the source in an attempt to produce a steady-state signal as the line profile was scanned using an oscillating refractor plate. More recently, Le Blanc and Blades have reported on the spatially resolved rotational temperatures of OH and N_2^+ in the FAPES system [26]. A significant thermal gradient exists in the source, as concluded from the range of OH (650–1050 K) and N_2^+ (580–1920 K) rotational temperatures measured for a 60 W He plasma. Highest temperature zones were located near the center electrode, as expected.

Analyte excitation temperatures have also been measured, most commonly employing iron as the thermometric species. Figure 7.10 presents two-line excitation temperature–time curves for iron as the thermometric species, introduced into the source in a steady-state manner as $Fe(CO)_5$ in the discharge gas. Despite a measured increase in the He I excitation temperature of more than 600K over the time period displayed, as well as an increase in the reflected power level to a maximum of 65 per cent of incident (50 W), the Fe I excitation temperature does not vary by more than 110K in the extreme as the wall temperature ramps to 2600K [24]. Temperatures derived from the discrete vaporization of iron, which was dosed onto the center electrode as a solution and then vaporized into the plasma, were about 300K higher than their wall atomization counterparts, and the corresponding emission intensities were eight-fold larger. These data confirm that radial gradients in the temperature exist in this source with more energetic excitation conditions occurring near the center electrode, as noted above for the molecular species. It is evident that not only are the mechanisms leading to excitation of Fe I and He I different, but that stable analyte excitation conditions can be established in the FAPES source, placing the technique on a firm basis for analytical usage.

The FAPES source has also been operated in argon, although this discharge is difficult to ignite and sustain at lower frequencies, despite the

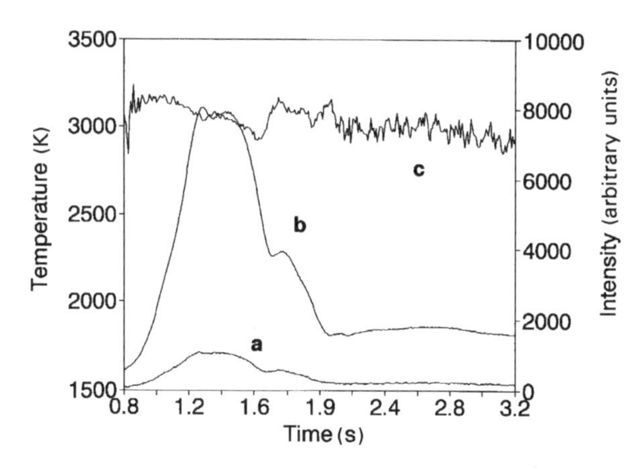

Figure 7.10 Emission transients and two-line Fe excitation temperatures during an atomization transient obtained during the steady-state introduction of $Fe(CO)_5$ into a self-biased 50 W He FAPES plasma. Curve (a) 298.4 nm Fe line; (b) 386.0 nm Fe line; (c) calculated two-line excitation temperature [24]. (Reprinted from *Spectrochim. Acta*, Vol. **48B**, R.E. Sturgeon, V.T. Luong, S.N. Willie and R.K. Markus, Impact of bias voltage on furnace atomization plasma emission spectrometry performance, pp. 893–908, 1993, with kind permission of Elsevier Science – NL, Sara Burgerhartstraat 25, 1055 KV Amsterdam, The Netherlands.)

lower ionization potential of this gas [27] compared with helium. Hettipathirana and Blades [28] rationalized this seeming paradox by noting that the ionization rate in helium is two orders of magnitude greater than in argon. The rate of ion production should increase with frequency, because at higher frequencies, electron loss by collision with the tube walls is lower, and electrons can ionize during more than one half-cycle of the rf field. Thus, easier operation should be expected at higher frequencies and, at 27.12 MHz, argon plasmas are operated with ease [27]. In general, higher operating frequencies have been found to increase the excitation temperature of the plasma and lead to substantial increases in line intensities. However, a corresponding increase in the background, results in only a marginal improvement in the detection limit [27]. Excitation temperatures in argon plasmas operated at the same powers as helium plasmas are higher by several hundred kelvin. This, coupled with the increased residence time of analyte atoms in this gas, results in improved detection power [27].

As with the FANES devices, introduction of analyte vapor into the discharge occurs predominantly *via* thermal vaporization from the hot tube wall. Analyte sputtering in this system is minimal, not only because helium is a weak primary sputtering agent, but also because the small mean free paths at atmospheric pressure promote immediate redeposition of sputtered material. Additionally, as a result of the surface area asymmetry between the electrodes, the impedance sheath potential developed on the larger tube wall (where the sample is deposited) is too small to accelerate helium ions to the necessary energy, with the result that any sputtering that occurs in the system does so primarily from the center electrode. Application of d.c. bias potentials of up to ± 88 V for 150 s, followed by atomization under self-biased conditions, resulted in no measurable sputter loss from samples of silver and copper. However, preatomization loss of phosphorus in the FAPES source has been noted and ascribed to interaction between the plasma and the analyte. Whether sputtering or simple thermal desorption of volatile PO species due to plasma heating is the cause remains unknown. The problem was rectified by use of a chemical modifier (palladium), which was shown to inhibit formation of PO. Additionally, arcing to the site of sample deposition at the initiation of the discharge has been observed, particularly for large amounts of sample. This problem was resolved by placing a graphite platform within the furnace. The edges of the platform probably serve as field concentration sites for the discharge initiation and, once established, appear to add robustness to the plasma.

Residence times for analyte species in FAPES are shorter than for conventional ETAAS, because atomization is conducted primarily in helium as compared to argon. The constant internal flow of helium (75–325 cm^{-3} min^{-1}) in the design of Sturgeon *et al.* (Figure 7.5) has little effect on analyte residence times, suggesting that the emission signals are source function controlled, and that diffusion is the dominant loss mechanism. The vaporization characteristics and temporal response in FAPES is different from that in ETAAS, despite the same furnace being used with both techniques [29]. The presence of the plasma and the center electrode in the FAPES source provides a secondary input of thermal energy, as well as a potential site for condensation and secondary release of analyte species. Double peaks are frequently observed in the analyte emission transients, and can be ascribed to second surface vaporization phenomena in that transfer of atomic and molecular vapor between the furnace wall and electrode occurs.

Analytically, FAPES sources have been operated at between 20 and 100 W forward power. This is comparable to the 20 W and 60 W powers typically dissipated in HC-FANES and HA-FANES sources, respectively. The lower power limit is that which is necessary to sustain the discharge; the upper limit is determined by the formation of current arcs between the electrodes. In general, the analytical signal increases with power, as do most background structural features. It remains for a comprehensive study to determine the optimum SNRs. There is little doubt that they will vary with each element and be highly dependent on the background structure in close proximity to each line. In this respect, relatively little work has been done concerning the selection of optimum lines for each element.

In the FAPES source described by Sturgeon *et al.* [22], the background is highly structured and dominated by the same molecular features discussed in connection with HC-FANES. Figure 7.11 illustrates the primary features in a 200–700 nm window. The lower panel depicts features in an argon plasma operated at 40 MHz, and sustained in the modified Massmann-type furnace. This spectrum is similar to that of a helium plasma in the same furnace workhead. In helium, of course, intense He I lines replace those of Ar I. Additionally, the strong transition noted for C I at 247.9 nm is almost absent in helium, as are the band systems of CN and C_2. Notably, the helium plasma excites the intense emission bands of CO^+ *via* a charge transfer reaction, not attainable in argon [28]. The primary source of these features is the argon-induced sputtering of carbon from the center electrode, followed by its reaction with ambient atmospheric gas that diffuses into the plasma through the sample injection hole in the tube wall. The upper panel illustrates the background features found in a 40 MHz He plasma sustained in an ICC device enclosed within a vacuum six-way cross. With improved exclusion of the ambient atmosphere, the intense molecular bands of NO and N_2, which appear in the Massmann-type furnace, are completely absent. Residual intense bands from OH and CO^+ are probably derived from trace impurities in the helium carrier gas, as well as from water adsorbed on the surface of the graphite tube.

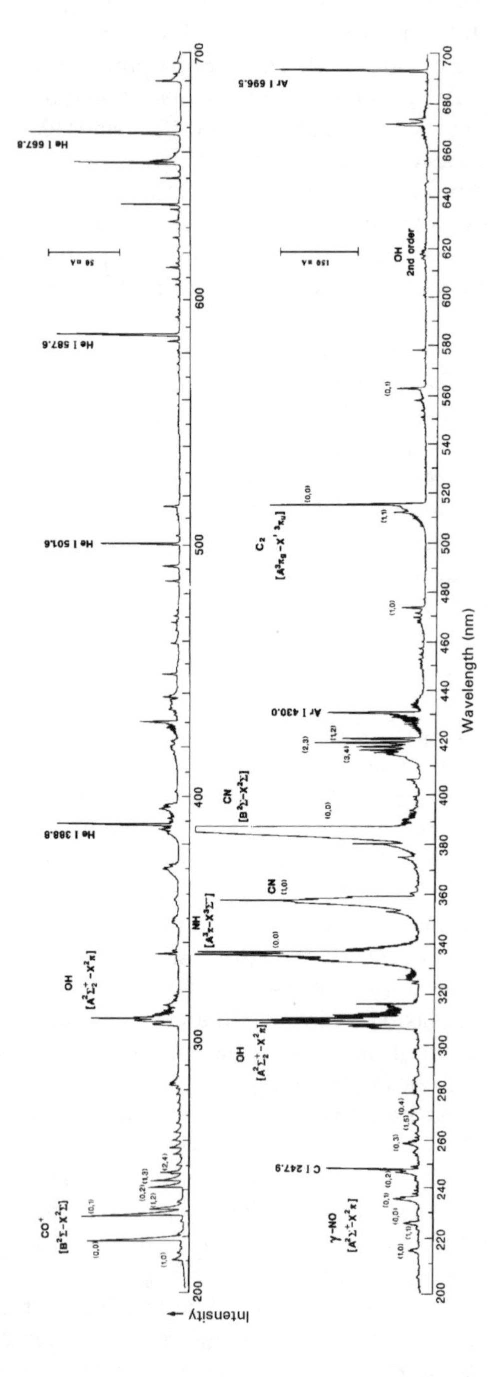

Figure 7.11 Background spectrum in a 50 W FAPES source. Upper spectrum from a 40 MHz He plasma sustained in an enclosed ICC device. Lower spectrum from a 40 MHz Ar plasma in a Massmann furnace [22]. (Reprinted by permission of the Royal Society of Chemistry from R.E. Sturgeon, S.N. Willie, V. Luong, S.S. Berman and J.G. Dunn, Furnace atomization plasma emission spectrometry (FAPES), J. Anal. At. Spectrom., **4**, 669–672 (1989).)

Application of bias voltages to the system alters the intensities of these background features, and increased power serves to intensify all transitions. These molecular features present highly structured background close to the analytical lines of a number of elements [30], which may be spatially nonhomogeneous throughout the plasma volume and exhibit temporal intensity variations as the furnace temperature changes. Wavelength modulation techniques for background correction will thus suffer the same potential problems as outlined for FANES. Fortunately, the demands on the response time of such a system are more relaxed, as the residence time for analytes is significantly longer (hundreds of milliseconds) than in the low pressure HC-FANES source.

FAPES offers detection limits in the low picogram range, comparable to FANES and ETAAS techniques, as evidenced from the data in Table 7.2. Relative standard deviations of 3–10 per cent for manually injected sample replicates compares with the 2–3 per cent precisions for HC-FANES and 2–6 per cent for HA-FANES systems, also operated with manual sample introduction. Linear working ranges cover 2–4 orders of magnitude of concentration, far short of the 5–6 decades reported for HC-FANES. This plasma is obviously not optically thin and suffers from possible self-absorption of atomic vapor at the cooler extremities. Whereas this effect can be more severe in the Massmann-type furnace, recent experience shows that use of the ICC furnace does not significantly improve the situation. This suggests that part of the problem may simply reside with the source function, in that vaporization of the entire sample becomes more problematic as the mass increases. Integrated response always attains a greater linear dynamic range than peak intensity response.

Platform technology and chemical modification can be used with the FAPES source, but not with impunity. The masses of any modifiers must be minimized; otherwise, there is a tendency for arcing to occur between the center electrode and the site of sample deposition on the furnace wall. This situation can be circumvented with the use of a platform but, in the Massmann-type furnace, the limitations of this device for use with involatile analytes are well known. The ICC furnace should be more suitable for operation with a platform, such as the newer forked design or the totally integrated structure now used in the Perkin-Elmer transversely heated graphite atomizer (THGA) furnace. Additionally, being an emission technique, the probability of line overlap from the modifier or a major impurity element in the modifier is of concern, and may prohibit use of specific modifier–element combinations.

Nonspectral interferences are currently problematic with the helium-based FAPES source. The presence of easily ionized elements (EIEs) alters the discharge characteristics of the plasma, as evidenced by the fluctuations noted in the self-bias during introduction of sodium into the plasma

[24]. Sodium appears to have no effect on the reflected power characteristics [24], but power dissipation in the plasma may occur as a result of the excitation of this element, as indicated by Eqn (7.16). This may explain the decreased emission intensity observed for lead in the presence of sodium chloride, and why platform techniques could only partially alleviate the problem. These speculations are supported by recent experimental studies. Classical analyte-molecule formation, gas-phase expulsion, and detuning of the plasma in the presence of large masses of EIEs are not significant factors in determining the extent of interference. Radiative power losses from the plasma due to excitation of the EIE matrix species, and possible alteration of the EEDF (due to EIE ionization and collisional dissociation of molecules), appear to be the major sources of interference [31]. Additionally, analyte ionization, which has been observed in the FAPES source [32,33], can be altered by the presence of EIEs, thus changing neutral atom response.

As with HC-FANES, FAPES has been used for the analysis of real samples, but quantitation by the method of additions was required in order to compensate for chemical interference effects. It is anticipated that use of argon as the discharge gas at high frequency, in combination with ICC technology and high plasma forward powers, will substantially reduce chemical matrix effects. An argon plasma has an intrinsically higher excitation temperature, substantially higher electron density (which may serve as an electron buffer for the EIE interference problem), and provides longer analyte residence times.

A realistic assessment of the capabilities and limitations of nonthermal excitation sources based on ETA devices will require considerable investment of time and additional research. In particular, the spatial distribution of noise and analyte emission over the cross-section of the source requires characterization for optimum SNR and line selection considerations. This is perhaps most easily addressed with nondispersed charge-coupled device (CCD) camera systems [23,26]. The possibility of obtaining multielement detection of both metals and nonmetals in a single, compact and inexpensive source will catalyze these developments, as will the present commercial availability of both HC-FANES and FAPES.

7.3 COHERENT FORWARD SCATTERING

Coherent forward scattering (CFS), sometimes known as atomic magneto-optic rotation, is a relatively well-developed field of atomic spectrometry, at least with respect to theoretical understanding; however, no apparatus based on this principle is commercially available. The technique relies on the rotation of the plane of polarized light, which occurs on passing the radiation of atomic resonance lines through atoms subjected to a magnetic field. Both the Faraday and Voigt configurations have been employed, in which the magnetic field is placed axially or transversely to the direction of light propagation, respectively. The latter arrangement has been used almost exclusively in experimental analytical applications, because it is physically easier to implement. The technique has recently been reviewed by Hermann [34] with an excellent discussion presented of the advantages and shortcomings of its application to trace multielement analysis.

The CFS technique is based on the use of two crossed polarizers (the polarizer and the analyzer) located on either side of the atomic vapor cell, which prohibit transmission through the system when there is no resonance interaction. When atoms are located in the magnetic field, splitting of the atomic energy levels occurs, and in a normal splitting pattern for a transverse magnetic field, for example, the atomic line splits into three components. A single π component remains unshifted at the resonance wavelength, and absorbs light whose plane of polarization is parallel to the direction of the magnetic field; two σ components, shifted to either side of the resonance line, absorb light whose plane of polarization is perpendicular to the direction of the magnetic field. Associated with these absorption lines are the anomalous dispersion functions describing the increase and decrease in the refractive index, which occurs at the wings of the absorption lines. Polarization of the incident light is changed in the spectral range of atomic resonance lines, and the resonance spectrum of the atomic vapor is transmitted. Magnetooptic rotation of the incident light is produced by both linear and circular dichroism (the difference between the absorption of the π and σ components), as well as birefringence induced by the difference between the refractive indices of these components. At low atomic density, dichroism contributes Gaussian profiles centered at the Zeeman components of the line, and birefringence contributes additional intensity at the wings of the Zeeman lines. If the source line is broad compared to the CFS line structure (continuum sources), dichroism and birefringence contribute equally to the output. Expressions for the resulting transmitted intensity have been derived by numerous researchers for a variety of configurations. In general, for the case of the polarizers set orthogonal to each other, an adequate magnetic field strength (0.7–1.5 T) and in the limit of low atomic density, the CFS intensity I_{CFS} transmitted from an initial incident intensity I_0 shows quadratic dependence on the atom concentration:

$$I_{CFS} \propto I_0(n_a)^2 l^2 f_{ij}^2 \tag{7.17}$$

where n_a is the analyte number density, l is the path length of the sample vapor, and f_{ji} is the absorption oscillator strength. In the low density limit, calibration curves exhibit the expected quadratic slope; at higher atomic densities, the dichroic intensity is first saturated, while the birefringent intensity continues to follow the quadratic characteristic. Ultimately, complete saturation occurs and rollover may set in. The CFS intensity can be expressed in terms of the zero field absorbance A as [34]

$$I_{CFS} = \beta(2.3A^2)I_0 = \beta(m/100m_0)^2 I_0 \quad (7.18)$$

where β is a factor (near unity), the value of which depends on the polarizer transmission and the convolution of the source line with the complex birefringence of the Zeeman absorption profile, m is the analyte mass, and m_0 is the characteristic mass for ETAAS (see Section 1.3). At low atom density, the intensity that is transmitted through a magnetooptic cell is proportional to the square of the zero field absorbance in the vapor. As the atom density is reduced to the point where its absorbance is small, the CFS intensity decreases proportionally to $A^2 \propto n_a^2$. In the analytical range, the optical CFS signal is lower, by a factor $\beta 2.3A$, than that of AAS. However, the signal is detected on a much lower level of background and thus on a much lower noise level. A high relative signal response is thus obtained, and determinations near the detection limit do not require the measurement of small variations in the incident light intensity as encountered in AAS. The result is that comparable detection limits can be achieved, as illustrated by the data in Table 7.3.

Hermann [35] has discussed the instrumental requirements for CFS in some detail. Figure 7.12 illustrates the experimental arrangement for the Voigt effect CFS with a continuum source (typically 75–1000 W Xenon lamp). The angle between the first polarizer and the magnetic field is

Table 7.3 Detection limits for ETA-CFS (best values summarized from available literature (3σ based)).

		Detection limit (pg)	
Element	*Wavelength*	*Continuum source*	*Line source*
Ag	328.1	5	–
As	197.2	5000	50
Au	242.8	–	14
Ca	422.7	2	–
Cd	228.8	5	0.2
Cr	357.9	70	15
Cu	324.7	70	5
Fe	248.8	130	50
K	766.5	0.2	–
Mg	285.2	3	–
Mn	279.5	20	2
Na	589.0	3	–
Pb	283.3	3	–
Ti	365.4	1400	2000
V	318.4	500	200
Zn	213.9	10	0.5

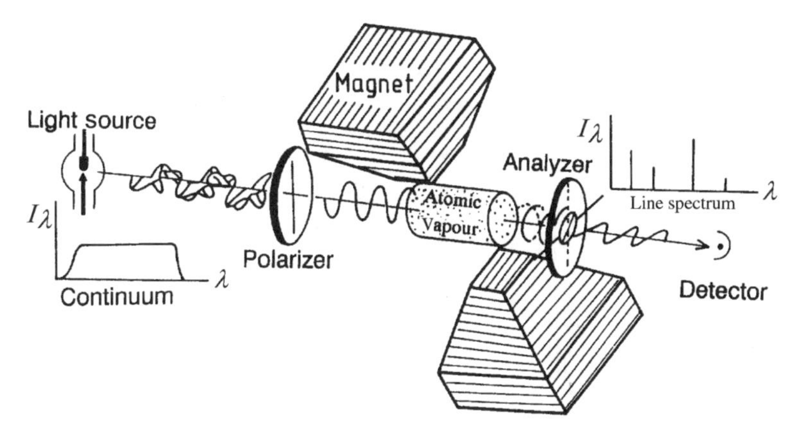

Figure 7.12 Schematic illustration of the Voigt-effect coherent forward scattering spectrometer based on a continuum source [35]. (Reprinted with permission G.M. Hermann, Coherent forward scattering atomic spectrometry (CFS): present status and future perspectives, *CRC Crit. Rev. Anal. Chem.*, **19**, 323–377; copyright 1988.)

usually set at 45°, so that equal amounts of light are available for interaction with the σ and π components. Because the magnetic field strength is the same order of magnitude as that in Zeeman AAS, this portion of the instrumentation is interchangeable for the two techniques, so that off-the-shelf components for the atomizer, magnet and autosampler system are readily available for use. The CFS spectrum contains ground-state resonance lines only, i.e. no lines from ions or excited state species are present. This greatly simplifies the dispersive requirements of the detection systems, which can be remarkably simple, permitting use of acoustooptic tunable filters and low-resolution monochromators, as well as imaging detectors. High intensity continuum sources provide distinct advantages over line sources in that they emit lines that, in hollow cathode sources, are too weak to be of use. The spectral intensity of a continuum source generally exceeds that of a hollow cathode source for such weak transitions, because it does not depend on the oscillator strength of the related element line. Thus, not only is there a multielement capability associated with use of continuum sources, but also a multiline capability. As Hermann has pointed out [34], 'a method with the capability for simultaneous determination and with the dynamic range that depends on lines that differ in strength by many orders of magnitude is advantageous'. Major matrix elements can then be determined simultaneously with trace elements. Thus ETA-CFS with continuum sources can combine the high detection power usually associated with GF techniques with the capability to detect masses up to the higher microgram range simultaneously in the same sample, by use of lines of extremely different strengths, i.e. 0.1 $\mu g\ l^{-1}$

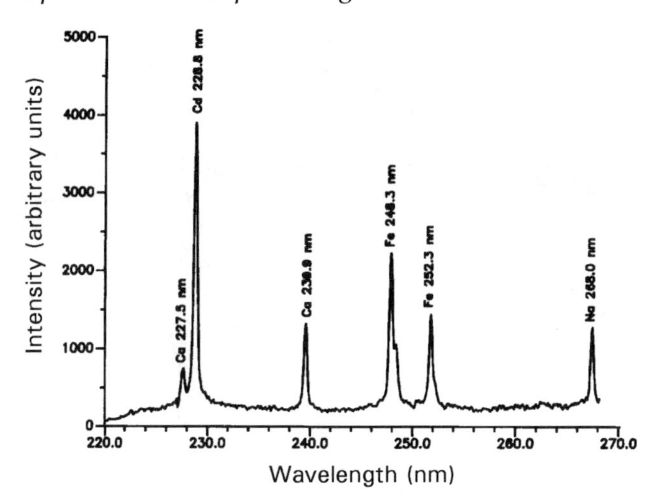

Figure 7.13 CFS spectrum obtained with the use of a continuum source for the simultaneous detection of several elements at widely different concentrations: Cd 10 ng ml^{-1}; Ca 5 mg l^{-1}; Fe 5 ng ml^{-1}; Na 75 mg l^{-1} [34]. (Reprinted with permission from G.M. Hermann, Coherent forward scattering atomic spectrometry, Anal. Chem. **64**, 571A–579A. Copyright 1992 American Chemical Society.)

up to 100 mg l^{-1}. Figure 7.13 illustrates a CFS spectrum obtained with use of a continuum source for the simultaneous detection of several elements at widely different concentrations.

As the atom density is reduced to the point where its absorbance is small, the CFS drops very rapidly towards the detection limit, due to its quadratic dependence on n_a. By offsetting the orthogonal polarizers over a small angle, the transmitted intensity at low concentration can be enhanced over the noise and linearized [36]. The detection limit can be improved by increasing the signal over the background noise level at lower concentrations, as long as the total SNR at the detection limit is not degraded by noise resulting from residual source light. Improvements offered by polarizer offsetting are restricted to use of line sources for excitation [34].

There is a direct relationship between CFS signals and the source intensity (Eqn (7.17)). The spectral intensity of a continuum source is low compared with bright atomic line sources, and it is natural to consider the benefits that may arise from use of intense line sources such as hollow cathode lamps and lasers. In the Voigt-effect configuration, where the line source is centered on the Zeeman π component, the contribution to the CFS signal by birefringence is negligible. In practice, the SNR improvement depends on the extinction ratio of the polarizers used. Thus, whereas Davis *et al.* [37] earlier concluded that CW lasers were restricted due to wavelength range, pulsed lasers were too noisy to be analytically useful for CFS, because the primary source of noise was the amount of light which passed through the polarizers due to their imperfect extinction coefficient. Detection limits obtained with their system were poorer than those fitted with hollow cathode lamps and xenon arc lamps. Improvement in SNR by polarizer offsetting has been demonstrated with lasers and hollow cathode line sources [34]. Multielement detection approaches with line sources have made use of time gated multiplexed sources in a nondispersive format, as well as multiple frequency modulated line sources with Fourier transform deconvolution of the transmitted intensity.

The CFS technique is characterized as being a zero- or low-background method, because without light interaction with resonant atoms, the transmitted intensity is zero under ideal conditions (infinite polarizer extinction ratio) and the signal is directly given by the detected line intensity. Additionally, zero zero-intensity I_0 means zero zero-intensity noise, such that CFS provides inherent background suppression with respect to noise from the source. Experimentally, the flicker noise of the source is reduced to the fraction related to the transmitted signal intensity, and its main contribution is blocked together with the basic intensity by the polarizer pair. Further, source modulation techniques then permit detection and correction of the background due to nonspecific intensity, such as that arising from incomplete polarizer extinction, emission from the atomizer, and background from nonspeciflc absorption. Modulation of the magnetic field permits discrimination of all types of nonspecific intensity, because with the field off, the atomic vapor is isotropic and the intensity transmitted by CFS is zero. Nonspeciflc background, a common problem in AAS, does not cause any optical rotation and hence no response in CFS. Interference by molecular CFS is generally insignificant, because only paramagnetic molecules exhibit activity (e.g. OH and CH), and usually only at high concentration because their oscillator strengths are distributed over the width of a broad vibration–rotation band. Interferences by atomic CFS are usually unimportant, even with low-dispersion detection systems, because of the relatively few lines of concern compared to classical emission spectroscopy. Structured interferences in CFS are expected to be as rare as those encountered in ZAAS, and arise in the Voigt effect as birefringence caused by the overlap of the σ component of the interferent on the π component of the analyte, or *vice versa*. Chemical interference effects in CFS are identical to those encountered in ETAAS, and hence the same measures should be taken to minimize them in order to assure accuracy.

Analytically, ETA-CFS has been used in both single element and multielement application, as well as with line and continuum source excitation [35,38–41], with direct and favorable comparison to Zeeman-based ETAAS. Due to the great similarity of CFS techniques to Zeeman AAS, the two instrumental systems can be used almost interchangeably. Additionally, when high intensity continuum excitation sources are used, the possibility of CS-ETAAS is also readily available with computer-controlled instrumentation. Unfortunately, few researchers are actively involved with this field of study, and it is unclear as to what degree this approach to trace element analysis may evolve.

7.4 ATOMIC FLUORESCENCE

Analytical atomic fluorescence spectrometry (AFS) involves the emission of photons from an atomic vapor that has been excited by photon absorption from a source lamp. Several different excitation-detection schemes have been utilized for analytical purposes. In resonance fluorescence, the same upper and lower levels are involved in the excitation–de-excitation process, so that the absorption and emission wavelengths are the same. This situation is usually avoided, where possible, due to high levels of background noise from scattered source light. Nonresonant Stokes schemes rely on use of excitation wavelengths that are less than the emission wavelength, and may involve collisionally assisted processes to populate a close lying higher state above the radiatively populated level from which de-excitation then occurs. With nonresonance anti-Stokes, the excitation wavelength is greater than the emission wavelength, and the lower state is usually a low-lying metastable level that is thermally populated. It is possible that, at moderate temperatures, the population of such a level will exceed that of the ground state because of a larger degeneracy. A great advantage of anti-Stokes schemes is the lower level of scattered source light at the shorter detection wavelengths. Additionally, double excitation or multiphoton fluorescence schemes arise when two or more photons promote an atom to an excited state, through intermediate levels, which may be virtual or real.

Expressions for the intensity of the fluorescence signal as a function of known fundamental and experimental parameters have been used to interpret the slopes of calibration curves (growth curves) and optimize experimental conditions. A comprehensive treatment of AFS expressions has been presented elsewhere [42]. For AF, the absorption rate generally exceeds the collisional excitation rate and, for optically thin conditions ($n_a l \rightarrow 0$), which are the most analytically significant, the radiant power of an AF line I_F (erg s^{-1} sr^{-1}) is directly proportional to the absorbed radiant power from either a continuum or line source as

$$(I_F)_C = [C(\lambda_0)^2(E_{\lambda 0})Yf_{ji}V]n_a \tag{7.19}$$

or

$$(I_F)_L = [C(\lambda_0)^2(E_\lambda/\Delta\lambda_{eff})Yf_{ji}V]n_a \tag{7.20}$$

where C is a collection of constants (7.05×10^{-14} cm), λ_0 is the center of the absorption line (cm), V is the illuminated volume (cm^3), $E_{\lambda 0}$ is the spectral radiance of the continuum source at the absorption line center (in units of erg s^{-1} cm^{-2} cm^{-1}, and assumed constant over the absorption profile and equal to the value at the wavelength (cm) of maximum absorption), E_λ is the radiance (integrated over the narrow emission line profile, with units of erg s^{-1} cm^{-2}), $\Delta\lambda_{\text{eff}}$ is the effective width of the Voigt absorption profile (cm), and Y is the fluorescence quantum yield or efficiency. The latter is further defined by

$$Y = A_{ij} \bigg/ \left(\sum_j A_{ij} + \sum_I k_i + \sum_m C_{im} \right) \qquad (7.21)$$

The first term in the denominator represents spontaneous emission processes from the upper i state to all lower j states; the second term accounts for collisional de-excitation from the upper i state to all other states (except i); and the final term accounts for losses by molecule formation. The quantum yield is thus simply the ratio of the optical transition probability per second to the total probability of de-excitation per second and per excited atom. Under optically thin conditions, the AFS signal is proportional to the source intensity and the analyte number density.

The same signal can be obtained from a narrow-line source as from a continuum source, if $E_\lambda = E_{\lambda 0}\Delta\lambda_{\text{eff}}$. A full description of the fluorescence curves of growth, even under the conditions of low analyte density, is complex and must account for pre- and postfilter effects which alter the theoretical linear relationship between I_F and the (spectral) radiance of the source. Prefilter effects account for the fraction of exciting radiation that is absorbed before reaching the viewed volume element, whereas the postfilter effect accounts for the self-absorption of fluorescence radiation from the illuminated volume by analyte closest to the detector. Both filter effects are dependent on the optical configuration, cell geometry and analyte concentration. These treatments can be found within the additional reading suggested at the end of this chapter.

Early applications of electrothermal atomization AFS (ETAFS) utilized conventional excitation sources, such as high-intensity continuum lamps, hollow cathode lamps, glow discharge lamps and vapor discharge lamps. Such sources are characterized by relatively low radiances (10^{-6}–10^{-2} W cm^{-2}). It is not surprising that the performance of early experimental systems was relatively disappointing, in that significant gains in sensitivity and detection limit over those offered by state-of-the-art ETAAS were not realized, as illustrated by the data summarized in Table 7.4. The maximum AFS signal was thus not attained, because the rate of excitation by the source is much less than the rate of de-excitation and thus the majority of the atoms remain in the ground state.

Table 7.4 Detection limits for conventional source excited ETAFS. (Data [43] rounded to nearest figure.)

		Detection limit (pg)	
Element	*Wavelength, nm*	*Line source*	*Continuum source*
Ag	328.1	0.5	0.3
Au	242.8	4	
Cd	228.8	0.03	3
Cr	357.9		3
Cu	324.8	1	3
Fe	248.3		5
Mg	285.2	0.0001	0.2
Mn	279.5		2
Ni	232.0		50
Pb	283.3	3	35
Zn	213.9	0.1	1

Lasers are considerably more intense sources, providing radiances up to 10^6 W cm^{-2} (pulsed dye lasers), and the current thrust in AFS research is the utilization of these devices because they are capable of attaining optical saturation of the atomic vapor. The concept of saturation is most easily illustrated using rate equations and a two-level system; in contrast to conventional source excitation, it is necessary to consider stimulated emission when using lasers [44]:

$$dn_1/dt = -B_{12}E_\lambda n_1 + (B_{21}E_\lambda + k_{21} + A_{21})n_2 \tag{7.22}$$

$$dn_2/dt = -dn_1/dt = B_{12}E_\lambda n_1 - (B_{21}E_\lambda + k_{21} + A_{21})n_2 \tag{7.23}$$

$$n_1 + n_2 = n_T \tag{7.24}$$

where n_1 and n_2 are the populations of the lower and upper levels, respectively and B_{21} is the transition probability for stimulated emission. The thermal collisional excitation rate has been neglected as insignificant. The number of fluorescence photons emitted per atom and laser pulse, η_{fl} is given by the time integration of the excited state population during the laser pulse duration τ times the spontaneous emission factor:

$$\eta_{fl} = A_{21}\int_0^\tau n_2(t)dt \tag{7.25}$$

Assuming steady-state conditions for Eqns (7.22) and (7.23),

$$\eta_{fl} = \frac{g_2}{g_1+g_2}\frac{B_{12}E_\lambda}{B_{12}E_\lambda + \left[\dfrac{g_2(k_{21}+A_{21})}{(g_1+g_2)}\right]}A_{21}\tau n_T \tag{7.26}$$

where g_1 and g_2 are the statistical weights of the lower and upper levels of the transitions involved. Two limiting conditions arise with respect to laser radiance. If $B_{12}E_\lambda \ll (k_{21} + A_{21})$, the fluorescence signal is linearly proportional to the laser radiance:

$$\eta_{fl} = [A_{21}/(k_{21} + A_{21})]B_{12}E_\lambda \tau n_T = YB_{12}E_\lambda \tau n_T \quad (7.27)$$

If $B_{12}E_\lambda \gg (k_{21} + A_{21})$, the fluorescence signal is saturated, in that it is independent of the quantum efficiency as well as the laser radiance, i.e.

$$\eta_{fl} = [g_2/(g_1 + g_2)]A_{21}\tau n_T \quad (7.28)$$

Several advantages are evident if saturation is achieved: maximum fluorescence is attained, because the rate of excitation is equal to the rate of de-excitation; the signal is insensitive to variations in the laser power and quenching species (no prefilter effect); and self-absorption is reduced and the linear range is extended. It is to be noted that these remarks strictly apply to a two-level system. In a three-level system, the fluorescence signal is still independent of the source radiance, but is now dependent on the collisional constants for the coupling of levels. It is possible that the signal in such a system is approximately independent of quenching if only the initial portion of the time-gated signal is measured [44].

Saturation can be attained for many elements using laser radiances of about 1 kW cm^{-2} at atmospheric pressure [45,46], which is easily achieved with pulsed dye lasers. By comparison with the output of conventional sources, signal enhancements of up to nine orders of magnitude are to be expected. It must be realized, however, that the expressions for AFS signal intensities given by Eqns (7.19), (7.20), and (7.27) are for the ideal case and must be modified to account for the illumination and collection optics. Whereas the useful radiating area of pulsed lasers and conventional excitation sources is similar, the solid angle illuminated with a laser is only 10^{-6} sr as compared to 0.5 for conventional sources. Assuming the same optical collection efficiencies and atomizer efficiencies, AFS intensities with laser excitation should then be only 10^3-fold greater under optimum conditions. This order of magnitude improvement is borne out experimentally, as revealed by a comparison of the detection limit data presented in Tables 7.4 and 7.5. Although the AFS signal is independent of source intensity under saturation conditions, nonspecific scatter is directly proportional to the source intensity. Thus, the minimum laser intensity required to saturate the transition should be employed, in order to minimize the scatter that will degrade detection limits. Use of nonresonant fluorescence schemes is particularly attractive under such conditions. Use of laser excitation sources for AFS has increased the complexity of the instrumentation, but this is outweighed by the increase

Table 7.5 Detection limits for laser excited ETAFS

Element	λ_1 *(nm)*	λ_2 *(nm)*	λ_{fl} *(nm)*	*Detection limits*[a] *(pg)*	*Reference*
Ag	328.1		338.3	0.008	
Al	308.2		394	0.1	
Au	267.6	406.5	200	0.003	47
As	193.7		245.7	0.05	48
Bi	223.1		299.3	0.003	49
Cd	228.8	643.8	361	0.02	
Co	308.3		345.4	0.004	
Cu	325		325	0.4	
F(MgF)	268.9		358.8	0.3	
Fe	297		373	0.07	
Ga	403.3	641.4	250.0	0.001	
Hg	253.7	435.8	546.1	0.09	50
In	410.1	571.0	271.0	0.002	
Ir	284.9		357.4	475	
Li	670.8		670.8	0.01	
Mn	279.5		279.5	0.09	
Mo	313.3		317.0	100	
Ni	224.5		231.4	0.01	51
Pb	283		405	0.0002	
Pt	265.9		270.2	1	
Rb	794		780	2	
Se	196		204	0.02	48
Sb	212.7		259.8	10	
Sn	286		318	0.03	
Te	214.3		238.3	20	
Tl	276.8		352	0.0001	
V	458.0	578.6	256	10	52
Yb	398.8	666.7	246.4	0.2	

[a]Data from Reference 44 for tube furnaces, except where noted.

in the versatility of such a system, as it is continuously tunable from 217 nm to 950 nm using various dyes and frequency-doubling crystals.

Figure 7.14 depicts a typical experimental set-up for laser excited atomic fluorescence (LEAFS) in a graphite furnace. It should be noted that Enger *et al.* [53] have recently utilized a multichannel intensified CCD device system for simultaneous wavelength coverage in their ETA-LEAFS experiments. Excimer, neodymium:YAG, nitrogen, and copper vapor-pumped dye lasers have all been used as sources, because they offer the wide wavelength coverage into the ultraviolet necessary for practical application to a range of elements. Butcher *et al.* [46] discussed factors to be considered in the choice of laser for ETA-LEAFS. Apart from ultraviolet accessibility, other points to consider include the pulse repetition rate and output power. The transient vapor populations produced in ETAs require that they be sampled more often than in steady-state systems, such as flames and ICPs. Additionally, if alternating laser pulses are to be used to

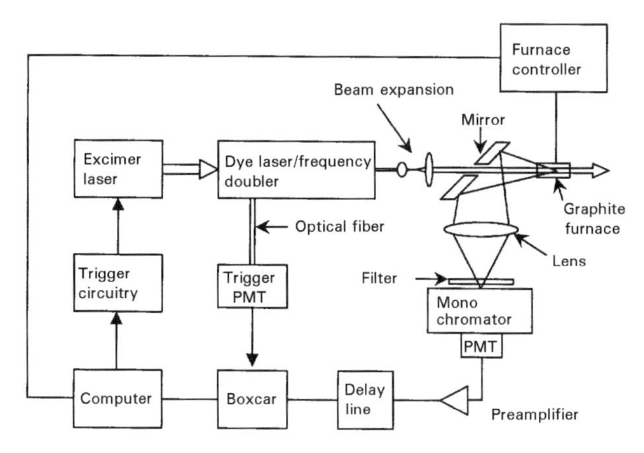

Figure 7.14 Typical experimental set-up for ETA-LEAFS [82]. (Reprinted from *Spectrochim. Acta,* Vol. **48B**, Z. Liang, R.F. Lonardo and R.G. Michel, Determination of tellurium and antimony in nickel alloys by laser excited atomic fluorescence spectrometry in a graphite furnace, pp. 7–23, 1993, with kind permission of Elsevier Science – NL Sara Burgerhartstraat 25, 1055 KV Amsterdam, The Netherlands.)

measure background, such as with Zeeman-effect correction or wavelength-modulation techniques, this rate needs to be further increased. Adequate sampling for ETAAS has been achieved at 30–50 Hz but, as the SNR improves approximately as the square root of the repetition rate, it is advantageous to use as high a rate as possible. Use of the neodymium: YAG laser is thus questionable. Excimer lasers and copper vapor laser pumped systems permit high-frequency operation, but the latter is somewhat restricted in its wavelength coverage (above 280 nm). Laser spectral bandwidths between 0.01 and 0.002 nm are commonly used, as most of the spectral power density then falls within the spectral line width of the atomic absorption line profile.

Semiconductor continuous wave (CW) diode lasers, recently applied to ETAAS, have also been evaluated for application in ETA-LEAFS [54,55]. Their compact size and temperature tunability permit construction of inexpensive and robust LEAFS instrumentation. Simple multiplexing with a chopper gives rise to potential simultaneous multielement detection in a nondispersive system. Preliminary results with this approach appear very promising, and further studies can be expected to be undertaken using laser diodes with higher CW power and improved detection systems.

Graphite rod, tube and cup ETA configurations have all been examined, with tubular devices offering the optimum performance, as expected from their superior overall efficiency [56]. The various excitation–detection geometries examined are summarized in Figure 7.15. Front-surface illumination is the preferred arrangement for tubular furnaces, in which the

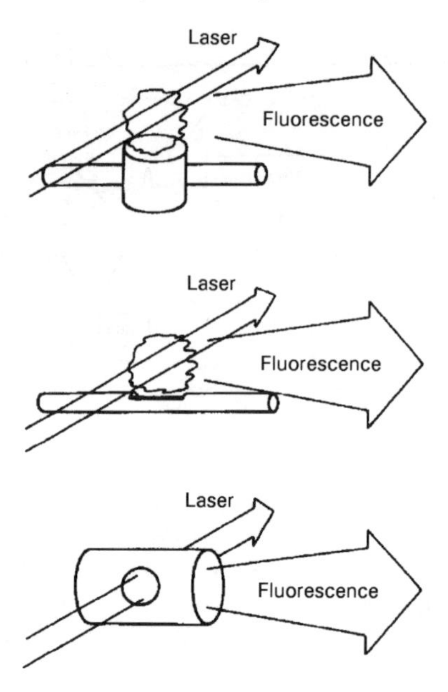

Figure 7.15 Excitation–detection geometries for ETA-LEAFS [46]. (Reprinted by permission of the Royal Society of Chemistry from D.J. Butcher, J.P. Dougherty, F.R. Preli, A.P. Walton, G.-T. Wei, R.L. Irwin and R.G. Michel, Laser-excited atomic fluorescence spectrometry in flames, plasmas and electrothermal atomizers, *J. Anal. At. Spectrom.*, **3**, 1059–1078 (1988).)

fluorescence is detected at 180° to the incident beam, whereas conventional 90° collection optics are used for cup atomizers. The front-surface approach is practical, because it avoids any significant modification to commercial furnaces; only the windows on the furnace housing have to be mounted at an angle (45°) in order to minimize back-reflected light.

Wavelength selection is most commonly achieved with a high-luminosity monochromator. High dispersion is not necessary, because the technique is inherently selective. For this reason, it is also possible to consider nondispersive detection using optical bandpass filters. These offer the advantage of larger light collection efficiency but poorer stray light rejection.

As the duration of the emitted fluorescence radiation is typically of the same order as the duration of the exciting laser pulse, the duty cycle is low (10^{-4}–10^{-6}), and the fluorescence excited by pulsed laser radiation is traditionally recorded using a gating technique where the response from a fast PMT is fed to a boxcar integrator. A variable gate (in duration) is operated by a reference signal (suitably delayed) relative to the laser pulse, offering the possibility of separating the scattered laser light signal (coincident with the laser pulse) from the analytical emission signal [57].

The lowest detection limits ever achieved for optical spectroscopy have been attained using ETA-LEAFS techniques. Due to the extreme sensitivity, it is not uncommon that the detection limits are determined from measurements made with the laser slightly detuned from the resonance transition, in order to obtain a better measure of the statistical fluctuation of the background rather than a measure of a contaminated blank. The latter can be a problem, due to both reagent impurities and impurities vaporized from the graphite substrate of the ETA [52]. So far, only about 30 elements have been investigated by ETA-LEAFS. As shown in Table 7.5, the best detection limits have been obtained for silver, gold, copper, gallium, indium, lead, and thallium which, except for copper, all possess simple atomic spectra convenient for nonresonance detection. Considerably worse detection limits, well above the picogram range, have been reported for the refractory elements and/or those that have a low fluorescence yield. This is still such a new technique that significantly improved detection limits can yet be anticipated for many elements [52]. Nonmetals, having excitation wavelengths too deep in the ultraviolet to be conveniently accessed, have also been addressed using laser-excited molecular fluorescence from stable diatomic molecules, such as MgF, InCl, and AlBr.

Calibration curves for ETA-LEAFS extend over 4–7 decades of linearity above the detection limit. Dougherty *et al.* [58] discussed the causes for nonlinearity of calibration curves in ETA-LEAFS, concluding that the most significant factors were the atomization efficiency of the furnace at high masses of analyte, and postfilter effects (even with nonresonance detection) caused by secondary absorption of fluorescence radiation that occurs within the detection volume but outside the illumination volume. Double-valued calibration curves have not been reported to date. The useful concentration range for ETA-LEAFS may not be limited by the linearity of the calibration graph but by the ability to completely remove high concentrations of analyte from the ETA (memory effects), in order to analyze low concentrations. In this respect, the transversely heated graphite furnace configuration may offer distinct advantages for real samples.

Analytical precisions of 5 per cent (RSD of replicate determinations) can be obtained with ETA-LEAFS at a laser pulse frequency of 80 Hz [45]. Presumably, this can yet be improved by working with higher pulse repetition rates.

Interferences in ETA-LEAFS may be both spectral and chemical. Sources of spectral interference include black-body radiation from the incandescent atomizer walls, stray light from the laser scattered off the instrument, nonanalyte atomic fluorescence, molecular fluorescence, and concomitant scattering of laser light from sample (matrix) particles or carbon particles ejected from the hot wall. Black-body emission (directly imaged as well as the fraction scattered off particles) is dependent on the type of atomizer, the atomization temperature and the detection wavelength, and must be minimized by efficient use of optical baffles between the furnace and detector. Furnace emission is the limiting noise source when the fluorescence wavelength is in the visible region. The lower temperatures that can be used with the ICC may have significant benefit for use with ETA-LEAFS. Black-body radiation can also be significantly reduced with use of shorter detection wavelengths, which necessitate use of double-resonance schemes or the excitation of atoms to higher levels.

Stray light is caused by scatter off any particles in the ETA, or off the walls of the device or other parts of the instrument. Stray light or scatter is often the limiting noise source for resonance detection schemes. It may also be serious in nonresonance LEAFS because of the wide band-pass and poor stray-light rejection detection systems used with this approach, particularly for elements whose fluorescence wavelengths are in the ultraviolet (where black-body emission is small), and when the detection wavelength is close to that of excitation. The main source of stray light in front-surface illumination is the windows of the furnace, particularly the rear one, and this source can be significantly attenuated by mounting the windows at an angle to the laser beam. Source scatter can also be minimized using two photon excitation techniques with detection at much shorter wavelengths, use of polarization filters to discriminate the unpolarized fluorescent signal from the polarized source radiation, and minimization of the spectral radiance required to saturate the transition by decreasing the laser bandwidth (using an intracavity etalon).

Nonanalyte AF is unlikely, particularly for nonresonance detection schemes, because of the double selectivity. As the extent of spectral overlap in LEAFS increases with increasing laser linewidth, these can be minimized when the latter are less than the atomic absorption line widths. No such interferences have been reported for ETA-LEAFS. As molecule formation can be minimized at high temperature, tubular ETAs, especially the ICC, should be superior to cup devices. As molecular transitions are much weaker than atomic transitions and the probability of overlapping fluorescence is small, analyte molecule fluorescence interference would be insignificant. Molecules formed by the much larger mass of matrix volatilized in real samples may contribute to the measured signal, but this is infrequently reported in the analytical LEAFS literature [49].

Being an inherently selective technique, optical interferences in LEAFS are small. Nevertheless, a number of background-correction techniques have been evaluated for ETA-LEAFS, although none have been put to rigorous scrutiny with many real samples because of the lack of researchers. By analogy to ETAAS, three basic approaches have been taken: wavelength modulation [59], effected by rapid, repetitive piezoelectric movement of the tuning mirror of the dye laser (0–0.2 nm); inverse Zeeman modulation to permit measurement of the background at exactly the analyte wavelength [60]; and simultaneous multichannel detection.

Although shown to be relatively easy to implement and capable of high repetition rate, the primary drawback of wavelength modulation techniques is that the background is not measured at exactly the analyte line, and structured backgrounds may result in errors. However, the system described by Su *et al.* [59] permits high spectral resolution, in that the wavelength modulation interval can be controlled precisely.

Zeeman-based techniques (ZETA-LEAFS) provide correction at exactly the analyte wavelength and are thus more immune to the above problem. In the transverse field case, polarization of the laser excitation beam perpendicular to the magnetic lines of flux is required to exclude π component fluorescence during (field on) measurement of the background. Irwin *et al.* [60] collected data at 120 Hz in each of two gated channels dedicated to magnetic field on and field off conditions. A field strength of 10.5 kG (as opposed to the commercial ZAAS field of 8 kG) was used in order to achieve sufficient splitting to minimize sensitivity losses. No significant degradation in detection limit was noted with use of this background correction approach.

Simultaneous multichannel detection offers the benefits of simultaneous detection of signal and background as well as no limitation on the upper limit of repetition of the laser, but usually suffers from the problem that background measurements are made too far from the analyte line so that structured backgrounds cannot be corrected. Because of the few applications of ETA-LEAFS to real samples, it is not yet known whether Zeeman-based correction is necessary or whether or not wavelength modulated correction will be adequate.

Chemical interferences in ETA-LEAFS are similar to those shared by all ETA-based instrumentation, and the ETAAS literature provides information pertinent to its consideration. ETA-LEAFS can accommodate all of the analytical features of modern furnace technology in that atomization from platforms and chemical modifiers may be used, although the latter may present problems with respect to purity and with regard to increased source scatter due to the formation of condensed species in the high-temperature furnace [61]. However, the unique features and capabilities of LEAFS also present several unique solutions to these problems that are not available in AAS. Because of its high detection power, chemical interferences in real samples can be minimized or avoided by simple dilution. Additionally, atomization at reduced pressure (10^{-2} torr) can be used to minimize the temporal overlap of analyte and matrix vapor clouds in the ETA, thereby resulting in almost complete suppression of chemical interferences and concomitant scattering [62,63]. Despite the 100-fold loss of sensitivity, the high detection power of LEAFS permits such action, akin to the removal of interference by dilution of sample. Although this requires a more complex ETA, it serves to avoid potential contamination that may arise during the dilution or use of chemical modifiers. It should be noted that maximum atomization efficiency occurs near atmospheric pressure [63].

As is clear from the above, analytical applications of ETA-LEAFS are distinctly lacking. However, applications to date have clearly demonstrated the selectivity of the technique, as well as its relative freedom from interferences in that background-correction requirements are far less stringent than in ETAAS. Most ETA-LEAFS measurements have been performed using cup atomizers, which are less than ideal from the viewpoint of chemical (vapor phase) interferences. These are clearly less severe in tube-type furnaces. Iridium, cobalt, iron, copper, lead, thallium, bismuth, tin, manganese, and gold have been determined in a variety of environmental samples and reference materials, including the analysis of solid (alloy) and slurried (biological) samples. Butcher *et al.* [46] and Sjostrom [44] summarized analytical applications to 1990. As with any rapidly growing field of technology, ETA-LEAFS is continually improving its performance characteristics. A particularly revealing example of this is the recent performance of the Zeeman-based background-corrected ETA-LEAFS instrument described by Irwin *et al.* [60], in which stabilized temperature platform (STPF; see Section 1.5.1) conditions were used to accurately determine lead and cobalt in slurried samples of NIST estuarine sediment, coal fly ash and citrus leaves by simple aqueous calibration, with precisions in the 6–14 per cent range (4–6 per cent for aqueous standards). Accurate determination of cobalt in the coal fly ash required dissolution of the sample and use of palladium as a chemical modifier because of matrix interferences. Overall performance is comparable to that of state-of-the-art ETAAS techniques. An example of the advantageous use of the high detection power of ETA-LEAFS is the direct determination of lead in whole blood, following 1:21 dilution with high-purity water to yield a detection limit of 10 fg ml^{-1} [64].

The advantages of ETA-LEAFS over ETAAAS are clear, but the future of the technique is intimately linked with developments in laser technology. Problems are related to the cost and inconvenience of operating dye lasers. The availability of CW diode lasers with broader wavelength coverage and higher power will do much to make this an affordable multielement technique, capable of making inroads into laboratories concerned with routine analyses.

Although not a fluorescence-based technique, mention is made here of an approach to the use of a laser for the creation of a hot plasma inside an ETA, focussed above the desorption surface and designed to provide high excitation temperature and multielement detection capability [65]. Although an inefficient system was used, based on a low repetition rate neodymium:YAG laser with collection of analyte emission through the sample introduction hole at right angles to the incident pulse, relatively low determination limits of 5 and 50 pg were obtained for cobalt and cadmium in dilute acid, respectively. Unfortunately, the laser-induced plasma generates a pressure shock wave, which decreases residence times in the ETA. Although the potential for improvement is obvious, no discussion is given of the possible difficulties likely to be encountered in real samples with regard to spectral and chemical interferences.

7.5 LASER-ENHANCED IONIZATION

Laser-enhanced ionization (LEI) is based on the optogalvanic effect, which has as its origin the perturbation of the electrical properties of a plasma in response to the absorption of light via discrete optical transitions in atomic or molecular constituents in the plasma. In LEI, tunable laser(s) selectively excite analyte atoms in a collisional medium to an energy level near the ionization continuum. An enhancement of the normal collisional rate of ionization then ensues, which is detected as a change in the current collected at a voltage biased electrode. The probability of collisional ionization depends on the ionization energy, the lifetime, collisional ionization cross-section, and the temperature of the gas. Excited atoms in states close to the ionization limit have long lifetimes and their collisional ionization probabilities approach unity in a flame. In ETAs, consideration must be given to collisional ionization, as well as radiative ionization in the blackbody field from the heated walls and photoionization (by the absorption of two photons) by the incident laser [66]. When this is combined with the ability to saturate transitions with pulsed lasers and to detect ionization with unit efficiency, the overall efficiency of the technique is seen to be very high.

In contrast, LEAFS emits fluorescence isotropically from the irradiated volume, and only a small fraction of the emitted fluorescence can be optically collected and detected. However, in comparison to LEAFS, the disadvantage of ionization detection is the lack of selectivity, causing LEI measurements to be reliant solely on the selectivity of the excitation process. Photoexcitation in the wings of strong atomic lines (which may be very broad) can occur, and thus matrix concomitants may contribute to the measured signal. Figure 7.16 illustrates the typical instrumental set-up for ETA-LEI. Pulsed tunable dye lasers have been used exclusively in LEI. Excitation of atoms in LEI can be accomplished in several ways, including: one-step excitation from the ground state using ultraviolet light; two-step excitation relying on use of two simultaneously pumped dye lasers (two-color excitation), each tuned to a suitable transition and sharing a close lying (rapidly coupled) common intermediate level; and two photon transitions and excitations to autoionizing states. Most analytical measurements are performed using stepwise laser excitation (with the upper laser level being about 0.2–0.4 eV from the ionization limit), as this gives a substantial increase in the signal strength (100-fold) over that realized

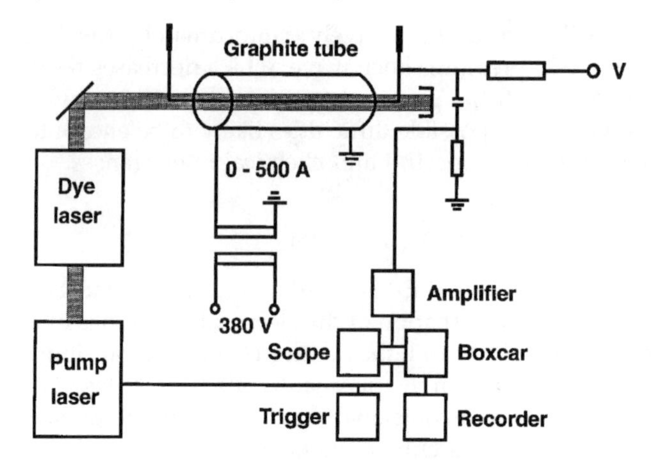

Figure 7.16 Diagram of instrumentation for LEI [72]. (Reprinted from *Spectrochim. Acta*, Vol. **42B**, I. Magnusson, S. Sjostrom, M. Lejon and H. Rubinsztein-Dunlop, Trace element determination by two-color laser enhanced ionization spectroscopy in a graphite furnace, pp. 713–718, 1987, with kind permission of Elsevier Science – NL, Sara Burgerhartstraat 25, 1055 KV Amsterdam, The Netherlands.)

with single-step excitation, as well as increased selectivity. The current is collected, amplified and stored by a boxcar integrator. The signal is the sum of all LEI pulses during atomization. Because both ionization and de-excitation of atoms are governed by collision processes, the largest signal should be obtained for the gas which maximizes the ratio of these competing processes. Experimentally, this is found to be argon. Collection voltages rarely exceed −100 V on the electrode and are governed in their upper limit by electrical breakdown of the gas in the ETA. Positive voltages (up to 100 V) have also been used, but found to suffer from large current background due to the collection of thermionic electrons from the hot wall.

Although LEI has enjoyed rather widespread characterization with flame cells [67], relatively little effort has focussed on use of ETAs, although these offer the advantages of longer analyte residence time for atoms in the interaction region and lower molecular background. Because the ionization yield in the graphite furnace is poorer than in a combustion flame, because of the low collisional rate available in the furnace environment, the ETA has also been attractive as a tandem source, where it is used as a vaporizer with actual LEI detection in a closely coupled flame [68]. Experimentally, several noise sources unique to ETAs severely limit the application of the technique. Through careful triggering of the laser pulse, synchronized to the heating current waveform of the ETA, noise from the furnace heating can be minimized but, at temperatures below

1800°C, this remained a dominant noise source in a system designed to take advantage of atomization of samples from a graphite probe [66]. Above 1900°C it was impossible to obtain any LEI signals because of intense background from thermionic electron emission. This immediately eliminates the prospect of obtaining LEI signals from the moderately volatile and refractory elements. Additionally, the presence of easily ionized elements suppresses the LEI signal to the point where it is difficult to use the graphite furnace for the analysis of a matrix that contains appreciable amounts of EIEs.

A partial solution to the problem connected to thermionic emission has recently been addressed by Chekalin and Vlasov [69], who placed the collection electrode external to the ETA, above the sample introduction hole. The analyte vapor effusing from the hole during heating of the ETA entered the interaction region of the laser pulse (between the hole and the electrode) and, in this manner, background from the ETA was significantly reduced and eliminated from the collector electrode itself. Although a more thorough characterization of this system is needed, the general consensus is that graphite-furnace LEI is not particularly sensitive compared to GFAAS or to flame LEI, because of the high level of noise, and is limited to the atomization of volatile elements. Table 7.6 summarizes the detection limits reported for ETA-LEI. Linear ranges of calibration curves extend 3–4 decades and the precision of replicate measurement varies 4–16 per cent RSD.

Except for the lightest and heaviest elements, the isotope shift of atomic lines is masked by Doppler broadening at atmospheric pressure. Resonant Doppler-free two-photon LEI has been achieved using ETAs with sample atomization into a vacuum or low-pressure noble gas environment [70,71]. Tunable lasers are applied, whose fields are interacting in counter- (or co-)

Table 7.6 Detection limits for one-color and two-color ETA-LEI (data from References 66 and 72).

Element	λ_1 *(nm)*	λ_2 *(nm)*	*Detection limit (pg)*
Fe	302.06		50
In	303.94		2
	271.04		0.7
Li	274.12		1
Mg	285.21		10
Mn	279.48		30
Pb	283.31		60
Co	298.72	480.82	5
Cr	298.60	483.59	5
Mn	280.11	472.97	1
Ni	294.39	497.67	7
Pb	283.31	471.99	0.5

propagating directions in the same volume of analyte vapor, such that one is tuned into the Doppler profile of the first transition to excite those atoms with a velocity component in the direction of the laser beam, and the second excites these further to higher states. Detection of the second transition then reveals a Doppler-free line. Ionization of the analyte is achieved by thermal collisions with the buffer-gas atoms. This approach is very attractive, because it offers the possibility of isotope-selective measurements and the determination of isotope ratios by optical spectroscopy. Additionally, it then becomes possible to calibrate analytical data using the well-known technique of isotope dilution. Detection limits in the femtogram range should be accessible. Obrebski *et al.* [73] have recently discussed the potential and limitations of this approach for sample analysis.

A technique closely related to LEI is resonance ionization spectroscopy (RIS) [74], sometimes referred to as resonantly-enhanced multiphoton ionization spectrometry (REMPI). This technique makes use of the resonant absorption of photons, via allowed electronic transitions, to transfer an electron from some initial state through various excited states to the continuum. Use of the same laser for selective excitation and photoionization is usually discouraged, because the strong field necessary will also photoionize matrix atoms by off-resonant transitions and selectivity will be lost. The most popular scheme is three-laser RIS, where two internal transitions are induced by moderate laser intensities, and a powerful laser for photoionization is then operated at a relatively long wavelength where the probability of multiphoton ionization of matrix atoms is greatly reduced. Few data are available to characterize the analytical potential of the technique, although it is evident that the selectivity is inherently high. The range of elements that can be addressed is large, because RIS schemes can be found for all elements except helium and neon, using presently available lasers. Extreme detection limits can be achieved when the selectively created ion is subsequently analyzed in a mass spectrometer [75].

7.6 OTHER APPLICATIONS

Although the literature on ETAs is dominated by analytical applications, these devices have also emerged as powerful tools for the fundamental study of physical and chemical processes at high temperature. King's initial work [76] with graphite furnaces laid the foundation for subsequent studies relating to thermodynamics, physics and spectroscopy. Many of these applications have already been addressed in earlier Chapters (i.e. studies of thermochemistry of gaseous molecules, measurements of molecular dissociation energies, the kinetics of atom formation and the measurement of adsorbent–adsorbate interaction energies). Although it is not the intent to treat these subjects here, the reader is directed to the following references: measurement of absolute and relative *f* values [77,78]; investigation of the Voigt parameters of spectral lines [77,79]; atomic diffusion coefficients [80] and the measurement of the heats of sublimation and vapor pressures of metals [81].

REFERENCES

[1] D.C. Baxter, *Spectrochim. Acta, Part B* **43B**, 129 (1988).
[2] D.C. Baxter and W. Frech, *Spectrochim. Acta, Part B* **50B**, 655 (1995).
[3] J.J. Schwab and R.J. Lovett, *Spectrochim. Acta, Part B* **45B**, 281 (1990).
[4] D.C. Baxter, W. Frech and E. Lundberg, *Analyst* **110**, 475 (1985).
[5] D.C. Gregoire and C.L. Chakrabarti, *Spectrochim. Acta, Part B* **37B**, 625 (1982).
[6] D.C. Baxter, D. Littlejohn, J.M. Ottaway, G.S. Fell and D.J. Halls, *J. Anal. At. Spectrom.* **1**, 35 (1986).
[7] E. Lundberg, D.C. Baxter and W. Frech, *J. Anal. At. Spectrom.* **1**, 105 (1986).
[8] D.C. Baxter, I.S. Duncan, D. Littlejohn, J. Marshall, J.M. Ottaway, G.S. Fell and O.Y. Ataman, *J. Anal. At. Spectrom.*, **1**, 29 (1986).
[9] J. Marshall, D. Littlejohn, J.M. Ottaway, N.J. Miller-Ihli, T.C. O'Haver and J.M. Harnly, *Spectrochim. Acta, Part B* **39B**, 321 (1984).
[10] J. Marshall, D. Littlejohn, J.M. Ottaway, J.N. Harnly, N.J. Miller-Ihli and T.C. O'Haver, *Analyst* **108**, 178 (1983).
[11] H. Falk, *Spectrochim. Acta, Part B* **32B**, 437 (1977).
[12] J.M. Harnly, D.L. Styris and P.G. Riby, in *Glow Discharge Spectroscopies*, (Ed.) R.K. Marcus, Plenum Press, NY, 1993, Chapter 9.
[13] H. Falk, E. Hoffmann, I. Jaeckel and Ch. Ludke, *Spectrochim. Acta, Part B* **34B**, 333 (1979).
[14] H. Falk, B. Hoffmann and Ch. Ludke, *Spectrochim. Acta, Part B* **36B**, 767 (1981).
[15] H. Falk, E. Hoffmann and Ch. Ludke, *Prog. Anal. Spectrosc.* **11**, 417 (1988).
[16] H. Falk, *J. Anal. At. Spectrom.* **6**, 631 (1991).
[17] D.C. Baxter, R. Nichol, D. Littlejohn, Ch. Ludke, J. Skole and I. Hoffmann, *J. Anal. At. Spectrom.* **7**, 727 (1992).
[18] K. Dittrich and H. Fuchs, *J. Anal. At. Spectrom.* **2**, 533 (1987).
[19] N.E. Ballou, D.L. Styris and J.M. Harnly, *J. Anal. At. Spectrom.* **3**, 1141 (1988).
[20] P.G. Riby, J.M. Harnly, D.L. Styris and N.E. Ballou, *Spectrochim. Acta, Part B* **46B**, 203 (1991).
[21] D.C. Liang and M.W. Blades, *Spectrochim. Acta, Part B* **44B**, 1059 (1989).
[22] R.E. Sturgeon, S.N. Willie, V. Luong, S.S. Berman and J.G. Dunn, *J. Anal. At. Spectrom.* **4**, 669 (1989).
[23] V. Pavski, C.L. Chakrabarti and R.E. Sturgeon, *J. Anal. At. Spectrom.* **9**, 1399 (1994).
[24] R.E. Sturgeon, V.T. Luong, S.N. Willie and R.K. Markus, *Spectrochim. Acta, Part B* **48B**, 893 (1993).
[25] R.E. Sturgeon, S.N. Willie and V.T. Luong, *Spectrochim. Acta, Part B* **46B**, 1021 (1991).
[26] C.W. LeBlanc and M.W. Blades, *Spectrochim. Acta, Part B* **50B**, 1395 (1995).
[27] R.E. Sturgeon, S.N. Willie, V.T. Luong and J.G. Dunn, *Appl. Spectrosc.* **45**, 1413 (1991).
[28] T.D. Hettipathirana and M.W. Blades, *Spectrochim. Acta, Part B* **47B**, 493 (1992).

[29] S. Imai and R.E. Sturgeon, *J. Anal. At. Spectrom.* **9**, 493 (1994).
[30] R.E. Sturgeon, S.N. Willie, V.T. Luong and S.S. Berman, *Anal. Chem.* **62**, 2370 (1990).
[31] S. Imai and R.E. Sturgeon, *J. Anal. At. Spectrom.* **9**, 765 (1994).
[32] R.E. Sturgeon, V. Pavski and C.L. Chakrabarti, *Spectrochim. Acta, Part B* **51B**, 999 (1996).
[33] R.E. Sturgeon and R. Guevremont *Anal. Chem.* **69**, 2129 (1997).
[34] G.M. Hermann, *Anal. Chem.* **64**, 571A (1992).
[35] G.M. Hermann, *CRC Crit. Rev. Anal. Chem.* **19**, 323 (1988).
[36] R. Stephens, *Spectrochim. Acta, Part B*, **38B**, 1077 (1983).
[37] L.A. Davis, R.J. Krupa and J.D. Winefordner, *Spectrochim. Acta, Part B* **41B**, 1167 (1986).
[38] H. Debus, G. Hermann and A. Scharmann, *J. Anal. At. Spectrom.* **4**, 529 (1989).
[39] Z.Tai, L. Jiaxi, Y. Xiaotao, D. Yicheng and G. Tiezheng, *Spectrochim. Acta, Part B* **45B**, 1075 (1990).
[40] G. Hermann, G.F. Lasnitschka, R. Moder and T.W. Szardening, *J. Anal. At. Spectrom.* **7**, 457 (1992).
[41] G. Hermann, B. Kling and B. Koch, *Spectrochim. Acta, Part B* **49B**, 1657 (1994).
[42] J.D. Winefordner and N. Omenetto, *Prog. Anal. At. Spectrosc.* **2**, 1 (1979).
[43] D.J. Butcher, J.P. Dougherty, J.T. McCaffrey, F.R. Preli, Jr., A.P. Walton and R.G. Michel, *Prog. Anal. Spectrosc.* **10**, 359 (1987).
[44] S. Sjostrom, *Spectrochim. Acta Rev.* **13**, 407 (1990).
[45] H. Falk, *Prog. Anal. At. Spectrosc.* **3**, 131 (1980).
[46] D.J. Butcher, J.P. Dougherty, F.R. Preli, A.P. Walton, G.-T. Wei, R.L. Irwin and R.G. Michel, *J. Anal. At. Spectrom.* **3**, 1059 (1988).
[47] G.A. Petrucci, H. Beissler, O. Matveev, P. Cavalli and N. Omenetto, *J. Anal. At. Spectrom.* **10**, 885 (1995).
[48] U. Heitmann, T. Sy, A. Hese and U. Schoknecht, *J. Anal. At. Spectrom.* **9**, 437 (1994).
[49] M.A. Bolshov, S.N. Rudnev, J.-P. Candelone, C.F. Boutron and S. Hong, *Spectrochim. Acta, Part B* **49B**, 1445 (1994).
[50] W. Resto, R.G. Badini, B.W. Smith, C.L. Stevenson and J.D. Winefordner, *Spectrochim. Acta, Part B* **48B**, 627 (1993).
[51] A. Marunkov, N. Chekalin, J. Enger and O. Axner, *Spectrochim. Acta, Part B* **49B**, 1385 (1994).
[52] S. Sjostrom, O. Axner and M. Norberg, *J. Anal. At. Spectrom.* **8**, 375 (1993).
[53] J. Enger, Y. Malmsten, P. Ljungberg and O. Axner, *Analyst* **120**, 635 (1995).
[54] A. Zybin, C. Schnurer-Patschan and K. Niemax, *Spectrochim. Acta, Part B* **47B**, 1519 (1992).
[55] K. Niemax, H. Groll and C. Schnurer-Patschan, *Spectrochim. Acta Rev.* **15**, 349 (1993).
[56] H. Falk and J. Tilch, *J. Anal. At. Spectrom.* **2**, 527 (1987).
[57] N. Omenetto and O.I. Matveev, *Spectrochim. Acta, Part B* **49B**, 1519 (1994).
[58] J.P. Dougherty, F.R. Preli, Jr., G.-T. Wei and R.G. Michel, *Appl. Spectrosc.* **44**, 934 (1990).
[59] E.G. Su, R.L. Irwin, Z. Liang and R.G. Michel, *Anal. Chem.* **64**, 1710 (1992).
[60] R.L. Irwin, G.T. Wei, D.J. Butcher, Z. Liang, E.G. Su, J. Takahashi, A.P. Walton and R.G. Michel, *Spectrochim. Acta, Part B* **47B**, 1497 (1992).
[61] B.V L'vov and W. Frech, *Spectrochim. Acta, Part B* **48B**, 425 (1993).
[62] M.A. Bolshov, A.V. Zybin, V.G. Koloshnikov, I.A. Mayorov and I.I. Smirenkina, *Spectrochim Acta, Part B* **41B**, 487 (1986).

[63] R.F. Lonardo, A.I. Yuzefovsky, R.L. Irwin and R.G. Michel, *Anal. Chem.* **68**, 514 (1996).
[64] E.P. Wagner, B.W. Smith and L.D Winefordner, *Anal. Chem.* **68**, 3199 (1996).
[65] V. Majidi, J.T. Rae and J. Ratliff, *Anal. Chem.* **63**, 1600 (1991).
[66] D.J. Butcher, R.L. Irwin, S. Sjostrom, A.P. Walton and R.G. Michel, *Spectrochim. Acta Part B* **46B**, 9 (1991).
[67] O. Axner, and H. Rubinsztein-Dunlop, *Spectrochim. Acta, Part B* **44B**, 835 (1989).
[68] B.W. Smith, G.A. Petrucci, R.G. Badini and J.W. Winefordner, *Anal. Chem.* **65**, 118 (1993).
[69] N.V. Chekalin and I.I. Vlasov, *J. Anal. At. Spectrom.*, **7**, 225 (1992).
[70] K. Niemax, J. Lawrenz, A. Obrebski and K.-H. Weber, *Anal. Chem.* **58**, 1566 (1986).
[71] J. Lawrenz, A. Obrebski and K. Niemax, *Anal. Chem.* **59**, 1232 (1987).
[72] I. Magnusson, S. Sjostrom, M. Lejon and H. Rubinsztein-Dunlop, *Spectrochim. Acta, Part B* **42B**, 713 (1987).
[73] A. Obrebski, J, Lawrenz and K. Niemax *Spectrochim Acta, Part B* **45B**, 15 (1990).
[74] G.S. Hurst and M.G. Payne, *Principles and Applications of Resonance Ionization Spectroscopy*, Adam Hilger, Bristol, 1988.
[75] G.I. Bekov, V.N. Radaev, D.D. and V.S. Letokhov, *Spectrochim. Acta, Part B* **43B**, 491 (1988).
[76] A.S. King, *Astrophys. J.* **21**, 236 (1905).
[77] B.V. L'vov, *J. Quant. Spectrosc. Radiat. Transfer* **12**, 651 (1972).
[78] P. Hannaford, *Spectrochim. Acta, Part B* **49B**, 1581 (1994).
[79] A.M. Nemets, G.I. Nikolaev and V.G. Flisyuk, *Zh. Prikl. Spektrosk.* **21**, 212 (1974).
[80] J.P. Matousek, *Spectrochim. Acta, Part B* **39B**, 205 (1984).
[81] A.M. Nemets and G.I. Nikolaev, *Zh. Prikl. Spektrosk* **21**, 405 (1974).
[82] Z. Liang, R.F. Lonardo and R.G. Michel, *Spectrochim. Acta, Part B* **48B**, 7 (1993).

ADDITIONAL READING

M.W. Borer and G.M. Hieftje, Tandem Sources for Analytical Atomic Spectroscopy, Spectrochim. Acta Rev. **14**, 463 (1991).
C. Th. J. Alkemade, T. Hollander, W. Snelleman and P.J. Th. Zeegers, *Metal Vapours in Flames*, Pergamon Press, Oxford, 1982, Chapter II.
J.D. Ingle, Jr. and S.R. Crouch, *Spectrochemical Analysis*, Prentice Hall, Englewood Cliffs, New Jersey, 1988.
B. Chapman, *Glow Discharge Processes*, John Wiley and Sons, NY, 1980.
R. Stephens, Magneto-optic Rotation in Atomic Vapours, Prog. Anal. At. Spectrosc. **12**, 277 (1989).
J.D. Winefordner and N. Omenetto, in *Analytical Application of Lasers* (Ed.) E.H. Piepmeier, John Wiley and Sons, New York, 1986, Chapter 2.
D.J. Butcher, J.P. Dougherty, J.T. McCaffrey, F.R. Preli, A.P. Walton and R.G. Michel, Conventional Source Excited Atomic Fluorescence Spectrometry, Prog. Anal. Spectrosc. **10**, 359 (1987).
J.C. Turk, G.C. Travis, J.R. DeVoe and P.K. Schenck, Principles of Laser-Enhanced Ionization Spectrometry in Flames, *Prog. Anal. At. Spectrosc.* **7**, 199 (1984).
B.A. Bushaw, High-Resolution Laser-Induced Ionization Spectroscopy, Prog. Anal. At. Spectrosc. **12**, 247 (1989).

8

Future trends

8.1 INTRODUCTION

The preceding chapters attest to the enormous amount of fundamental and applied research in ETAAS that has been undertaken over a period of almost 40 years. As a consequence, ETAAS is a mature technique. Accurate methods have been developed for a wide range of analytes in an even wider range of sample types, and ETAAS methods are accepted by regulatory agencies that require the use of approved methods. Questions that might now be posed are whether ETAAS has reached its ultimate level in terms of accuracy, precision, detection limits, ease of operation, and applicability.

It could be argued, rightly, that one person's views would be biased. Therefore, in an attempt to provide a more balanced insight into the future of ETAAS, several of the authors of this book were invited to respond to five specific questions. Perhaps surprisingly, their answers are quite similar. Everyone agreed that, even after almost 40 years, there is much to learn about the chemistry and physics of electrothermal atomization through fundamental research. Also, it is agreed that commercial ETAs are less than ideal, and that two-step furnaces with independent control of sample vaporization and atomization can provide enhanced performance with fewer matrix interferences. There is also general agreement that recent developments in light sources and solid-state detectors make continuum source multielement ETAAS a technique that is ready to move from the research laboratory into the real world.

The questions, and the authors' individual responses, are given below.

Electrothermal Atomization for Analytical Atomic Spectrometry. Edited by K. W. Jackson
© 1999 John Wiley & Sons Ltd

8.2 QUESTIONS AND ANSWERS

Question 1. Can further fundamental research offer improved performance in terms of better sensitivity and reduced interferences?

Holcombe: Let us first consider research directed at improving instrumentation. There is little likelihood of increasing sensitivity with current atomizer design (e.g. size and shape), and with current detectors and light sources. New designs in furnace geometries show the most promise for improving sensitivity and, possibly, immunity to interferences. More intense and/or more stable line sources will lower detection limits, as has already been shown for diode laser sources, but sensitivity (i.e. the slope of the calibration curve) will go virtually unchanged. The light source stability and significant improvement in detection limits will, by definition, increase the dynamic range. Additionally, improved narrow band light sources may extend the linear dynamic range, when it is limited by source line widths. However, in cases where stray light reduces the linear range, new sources will help if the source intensity increase is significant relative to the stray light. Further development of new continuum sources and high resolution, high throughput spectrometers with array detectors may turn AA into a true simultaneous multielement technique. Current limitations rest primarily in poorer detection limits at short wavelengths, and this is due, for the most part, to the source irradiance.
Sturgeon: The answer is yes, and this is clearly indicated by recent work of Frech and co-workers [1]. In considering the movement of gases past the exterior of the sample injection hole of the transversely heated furnace, they showed that there are venturi current effects that can be eliminated by redesign of the support cones, thereby enhancing sensitivity. This is just one recent example of how further research into the characterization of the atomizers may yet enhance their performance.
Gilmutdinov: Another example is our recent research on spatially resolved AAS [2], which has resulted in a means of absorbance detection that is more sensitive, provides greater dynamic range, and suffers less from interferences compared with conventional detection. The key idea is to detect intensities transmitted through the atomizer with spatial resolution and compute local absorbances. The improvements are based on better understanding of signal formation in AAS.
Holcombe: Improved fundamental knowledge of the chemistry and physics of atom generation, atom loss, and condensed-state and gas-phase reactions will continue to add to the foundation of how ETAAS operates. Some improvements in the minimization of matrix effects will probably result from this information. Current technologies and a suite of chemical modifiers provide interference-free analysis for most samples; however,

our understanding of some of the key modifiers, such as palladium or metal nitrates, is conspicuously thin.

Jackson: Although ETAAS is relatively free of chemical and physical interferences, some troublesome ones persist, such as interference by chloride. Many workers have chosen a phenomenological approach to reducing interferences and this has been partly successful. However, for complete control of interferences, a fundamental understanding of their mechanisms would be required. There has been considerable fundamental research directed at understanding chloride interference (see Section 1.5.2.1), and this has undoubtedly helped in devising the means for its reduction. I agree with Jim Holcombe that chemical modifiers play an important role in reducing interferences, and they also can be much more effective if their chemical or physical action is understood. I also agree that this remains one of the least researched areas, and further fundamental research on modifiers will enable them to be used optimally to further reduce interferences, and may also lead to new, more effective modifiers being developed.

Styris: As with fundamental research in general, the derived benefits are far from obvious. What is obvious is that our understanding of mechanisms that control atomization is deficient. It is likely that increased understanding will eventually lead to enhanced performance, but this will not occur in the near future. Unfortunately, atomization mechanisms are too complex to expect a significant, near-term improvement in the associated knowledge base. Considerable effort is needed just to understand the associated high-temperature surface effects. Certainly, a monumental effort is required to close the gap between superficial and complete understanding of atomization control mechanisms. Such an effort is unlikely, as bottom-line perception appears to be the guiding influence for funding of fundamental research efforts to elucidate these mechanisms. The lengthy duration expected before pay-off and the uncertainties regarding extent of impact the research will have on performance is not a persuasive argument for funding from either commercial or from federal sources. The result is that funding for and interest in this research is jeopardized. Without such funding and the necessary perseverance we will never comprehend the extent to which fundamental studies can improve performance levels.

Question 2. What are the prospects for absolute analysis?

Styris: Absolute analysis implies that some measured value can be correlated with the analyte mass through fundamental physical constants. The problem with achieving this for ETAAS is two-fold: (1) the necessary constants are generally not known with sufficient accuracy, such as diffusion coefficients and oscillator strengths; and (2) the model that is used in the correlation may not contain all of the controlling influences, such as surface states, adsorption–desorption parameters. It should not be expected that an absolute technique that is applicable to most analytes will be forthcoming without a more thorough understanding of the atomization mechanisms or without sufficient evaluation of the implicated physical constants.

Jackson: Certainly, the accuracy of absolute analysis continues to be limited by uncertainties in many of the fundamental parameters that are used in the theoretical calculations. However, these uncertainties are becoming less. For example, in the last few years Wiese [3] has provided accuracy ratings of oscillator strengths, and a new compilation of statistical weights and oscillator strengths was described for 2249 spectral lines [4]. Undoubtedly, the accuracy of absolute analysis will then improve. Obviously, there will be a limit, because there will always be instrumental factors that cannot be controlled completely, and absolute analysis will not be as accurate as calibration methods. However, analysts and data users may finally be starting to realize that highly accurate (and precise) analysis is not always needed. The United States Environmental Protection Agency is moving towards performance-based measurement systems (PBMS) where the method will be chosen based on data quality objectives. This means that, for many environmental analysis applications, it will be realized that lower accuracy and precision than those provided by the currently used standard methods is acceptable. That's where absolute analysis, with its reduced accuracy, could have a place.

Holcombe: The prospects are excellent and, indeed, a number of papers testify to this point. I also want to emphasize that routine use of absolute analysis must rest primarily on acceptance of somewhat lower expectations on accuracy than are currently available with the use of calibration curves and standard additions techniques. For many questions that are being asked where ultratrace quantities need to be determined, the accuracy available with ETAAS using absolute analysis mode is sufficient. Unfortunately, many regulatory agencies and uninformed individuals requesting quantitative information on a sample continue to have little appreciation for the correlation between their question and the accuracy necessary to answer the question.

Gilmutdinov: The prospects are bright. Current achievements already allow AA standardless determinations within 20–25 percent of accuracy for the majority of samples. It can be further improved in *multidimensional AAS* where spectrally, spatially, and temporally resolved absorbances are used. The instrument that is capable of such measurements includes: a continuum source, a high-resolution echelle spectrometer, a two-dimensional solid-state array detector, and a proper means of data handling. This concept has been presented in a recent paper [5].

Harnly: I think we are now ready to make dramatic improvements in the fundamental nature of the absorbance measurement and calibration, and to finally approach absolute analysis. We have recently shown that, with a continuum source and array detection, we can integrate absorbance with respect to wavelength, furnace height, and time. Integrated absorbances, when normalized for the pixel width and height, and the time interval between absorbances, and when combined with an appropriate optical design, will provide accurate measurements (corrected for stray light and nonspecific background absorption) which are independent of all instrumental parameters except the furnace atomization efficiency.

Question 3. Are the current commercial atomizers (end-heated Massmann furnaces with platforms and side-heated tube furnaces with platforms) the best we can get?

Gilmutdinov: They are not. Much better performance can be achieved with the two-step atomizers where volatilization and atomization are spatially separated and controlled independently. The price for the improvements is higher cost and complexity. Another way of improving atomizer design is through pressure control. Increased pressure results in an increased sensitivity, whereas decreased pressure leads to less interference and the possibility of isotope analysis.
Sturgeon: Clearly, the two-step system originally described by Frech and Jonsson [6] seems to be the last frontier in pursuing this goal. Irrespective of how nearly isothermal an atomizer can be, a simple examination of the problem from the viewpoint of equilibrium (which is the best state the system can achieve) reveals that there is no such thing as the complete elimination of all interferences (that is, the formation of stable molecules with the analyte will always occur), and it is just the concentration of the interferent in the gas phase and its temperature that will determine the extent of the problem.
Styris: I agree that existing commercial atomizers with platforms do not qualify as the best possible performers among the atomizer concepts. The Frech two-step atomizer is superior for absolute analysis, and generally achieves better performance [7,8]. Complexity in operating this atomizer has understandably kept it from the marketplace, but its ability to better control vaporization/atomization is significant. The two-step atomizer concept could be a major player and should not be ignored indefinitely.
Harnly: I agree with my colleagues that the two-step furnace will provide a significant improvement over the Massmann furnaces.
Jackson: Our own experience with the Frech two-step atomizer has shown us its real advantages over current commercial atomizers. The ability to independently control sample vaporization (cup) and atomiza-

tion (tube), together with temporally and spatially more nearly isothermal atomization, leads to significant reductions in interferences. It is unfortunate that the advantages have not proved sufficient to persuade a manufacturer that the increased cost and complexity of a two-step furnace is justified. The biggest potential advantage of two-step furnaces may not yet have been fully exploited. Ultratrace amounts of analytes vaporize at very low temperatures in graphite furnaces, much lower than predicted by their vapor pressures when present in bulk form. For example, thallium vaporizes from the inside wall of a Massmann furnace at less than 300 °C [9], which is lower than the normal melting point of the metal or its volatile suboxide. It then condenses on cooler surfaces in the furnace. Hence, a two-step furnace with independently heated components could provide controlled separation of analyte and matrix prior to the atomization stage, resulting in removal of interferences. Unfortunately, the Frech design does not allow this to happen. The geometry of the cup is such that analytes vaporize and then condense higher up the cup, with the result that they do not leave the cup until they have reached their atomization temperatures, which is 850 °C for thallium. Therefore, an improved design of two-step furnace may be called for, where analytes would not be retained in this way. This would overcome another difficulty we have encountered with the Frech furnace. Samples with large amounts of refractory matrix are difficult to remove from the cup, even with a high-temperature clean-out step. In particular, we encounter build-up problems when analyzing soil and sediment slurries in this furnace.

Holcombe: The answer must be 'no', but it is difficult to state what is better until someone clever presents it to us for our evaluation. As an example, Frech and co-workers have demonstrated encouraging results with endcapped tubes [1], as well as the two-step furnace noted above. As with any improvement, it is not clear whether the magnitude of the improvement will be sufficient to justify adoption of a new design (with all its requisite engineering) by a commercial firm. In theory, one would like a totally enclosed tube of very small diameter and long length. This would obviate the need of a platform and permit steady-state measurements for a user-determined time period. The furnace material should exhibit no adsorptive behavior toward the analyte but should assist in, for example, oxide reduction and matrix decomposition. Similarly, with this perfect high-temperature container (with equally hot, transparent windows of course), many elements could be determined within the same microsample. With this idea in mind, one wonders if it is only a simple matter of engineering.

Question 4. What improvements are likely to be made in order to make ETAAS more competitive for ultratrace metal determinations?

Sturgeon: Clearly the throughput factor, i.e. how many determinations and how many elements per minute. Inductively-coupled plasma mass spectrometry (ICPMS) poses the greatest threat to the continuance of ETAAS on a large scale. As costs come down, this will become more apparent. The performance of ETAAS can be improved by the use of a continuum source with a solid-state detection system, perhaps combined with a temporal, spatial and wavelength-integrated detection system as well as a two-step cup-tube furnace for highest effective temperature environment, and minimum interferences and diffusional loss.
Holcombe: Faster throughput, even lower detection limits, improved dynamic range, and a universal modifier (if one is needed at all).
Jackson: To answer this, you have to consider the competition–mainly ICPMS which is a multielement technique and much faster. Of course, ICPMS has its disadvantages, especially cost, and it may be fair to say it suffers more from matrix effects than modern ETAAS. Harnly and co-workers have demonstrated over the years that continuum-source multi-element ETAAS is a successful technique. Unfortunately, no instrument manufacturer has yet picked up on this idea with a commercial instrument. Some commercial multielement instruments using line sources are now available, but they only allow 4–6 elements to be determined simultaneously. A multielement instrument, used with fast-furnace technology (see section 1.5.2.3) would have many advantages over its competitors, though fast-furnace techniques may be more difficult to apply when compromise conditions are required for the simultaneous determination of several analytes.
Harnly: I think it is now possible to construct a multielement instrument that will compute fundamentally accurate absorbances, will provide detection limits better than existing state-of-the-art line-source instruments, and calibration ranges of 5–6 orders of magnitude. In reality, practical limitations, such as clean-up of the furnace between firings, and spectral interferences, will limit the calibration range. This instrument will allow direct inspection of spectral interferences, and will make absolute analysis possible. Computation of instrument-independent absorbances, as discussed above in the answer to Question 2, will greatly facilitate quality assurance for ETAAS, allowing all instruments and analyses to be compared directly.
Gilmutdinov: The technique of ETAAS suffers from two major limitations: the dynamic range is limited to less than three orders of magnitude; and it is in essence a single-element technique (despite successful employment of multielement AA spectrometers allowing simultaneous

determinations of up to six elements). A truly multielement AA spectrometer with dynamic range comparable to that provided by ICP spectrometry can be constructed on the basis of a continuum source, high-resolution echelle and a solid-state array detector (see the answer to Question 2).

Styris: Matrix interferences are a major problem. Until there is a better understanding of how modifiers work, the community will have to rely on present knowledge and serendipity to direct the efforts to minimizes these interferences. An existing solution to this problem is implementation of the two-step atomizer concept. This technique eliminates or minimizes the interference problem and provides a constant temperature environment for the volume of and the duration of the measurement. By coupling the two-step atomizer with multielement detection approaches, ETAAS could, conceivably, become a top competitor in the area of ultratrace-metal determinations.

Question 5. What are the biggest remaining problems in applying ETAAS to the analysis of real samples? How will these problems be minimized or eliminated?

Styris: One can never ignore the problem of matrix interferences, particularly with real samples. If these samples contain unknown elements, multielement detection becomes essential. Then there is the problem of calibration range. For real samples, it is imperative that this range be extended as far as possible. It is anticipated that matrix interferences and multielement detection will, respectively, be thwarted and improved significantly by incorporating the two-step atomizer approach with multielement detection techniques that incorporate continuous light sources [10] with CCD detection techniques [11]. The solution to extension of the calibration range is not obvious, but nor is the extent of the calibration problem for a combined two-step atomizer/continuous light source, multielement detection scheme. Such a scheme may, in itself provide an extended calibration range.

Sturgeon: Clearly throughput and interferences; and the need to carefully calibrate. Perhaps the linear range problem can be addressed with software flags to identify an out-of-range sample. An intelligent spectrometer can then perform a rerun with a dilution. Similarly, if the sample is below the detection limit it can again be flagged and the instrument can undertake some kind of on-line separation/concentration in a flow injection (FI) manifold with unattended operation. The use of Frech-type two-step furnaces will minimize interferences as well as the carryover from sample to sample or the accidental running of a sample with a large concentration of analyte causing contamination problems.

Holcombe: The perception of users that the technique is difficult to use and very prone to interference effects. Another problem is the inability to readily develop a method or tune the instrument while watching the signal. Most atomic spectroscopists are used to working with a system that produces a continuous signal and are more uncomfortable in working with transient signals and relatively slow turnaround. It is possible that a good fundamental understanding of vaporization (this may also fit with Question 2) could allow a smart instrument to flag the user when a problem is present through analysis of the peak shape, etc. This chemometric approach to a smart instrument would give the user more confidence. However, the algorithm and knowledge of what to monitor would be a major research investment. Picking up on the latter thought, consider not only a highly automated ETAAS system, but a system with smart software (as suggested above) so the instrument could really operate unattended for days, or as long as the sample carousel remains filled.

Gilmutdinov: Limited dynamic range, matrix interferences and limited multielement possibilities. The most radical way of coping with the problem is the way towards absolute AA analysis as described in the answer to Question 2.

Jackson: I don't think matrix interferences are a major problem anymore, when compared to competitive techniques. As Jim Holcombe has indicated above, this may be more of a perception than a reality. The judicious use of modifiers in modem furnaces, with their improved heating characteristics, has shown that most interferences can be eliminated or reduced to an acceptable level (perhaps with the notable exception of interference by large amounts of chloride). Increased automation must be the answer to this question. An intelligent instrument with automated sample pretreatment using FI, that could be left unattended for a long time, would suffer less from the limitations of being a single-element technique. In recent years, there has been a tremendous interest in the use of FI with ETAAS, and so the technology to realize the above instrument is probably here already. Somewhere I feel that I have to tout the advantages of ETAAS for the direct analysis of slurries. I don't want to repeat what is written in Chapter 6, but I believe that this technique has many advantages over its competitors for slurry analysis, and numerous publications have shown that it works.

REFERENCES

[1] N. Hadgu and W. Frech, *Spectrochim. Acta, Part B* **52B**, 1431 (1997).
[2] A.K. Gilmutdinov, B. Radziuk, M. Sperling and B. Welz., *Spectrochim. Acta, Part B* **51B**, 1023 (1996).
[3] W.L. Wiese, *Phys. Scr.* **T65**, 188 (1996).

[4] D.A. Werner, P.D. Barthel and D. Tytler, *Astron. Soc. Pac. Conf. Ser.* **78**, 145 (1995).
[5] A.K. Gilmutdinov and J.M. Harnly, *Spectrochim. Acta, Part B* **53B**, 1003 (1998).
[6] W. Frech and S. Jonsson, *Spectrochim. Acta, Part B* **37B**, 1021 (1982).
[7] E. Lundberg, W. Frech and J.M. Harnly, *J. Anal. Atom. Spectrom.* **3**, 1115 (1988).
[8] W. Frech, B.V. L'vov and N.P. Romanova, *Spectrochim. Acta, Part B* **47B**, 1461 (1992).
[9] G. Chen and K.W. Jackson, *Spectrochim. Acta, Part B* **53B**, 981 (1998).
[10] J.M. Harnly, T.C. O'Haver, B. Golden and W.R. Wolf, *Anal Chem.* **51**, 2007 (1979).
[11] B. Radzuik, G. Rodel, H. Stenz, H. Becker-Ross and S. Florek, *J. Anal At. Spectrom* **10**, 127 (1995).

Coventry University

Index